Chris D. Geddes
Editor

Reviews in Fluorescence 2009

Springer

Editor
Chris D. Geddes
Institute of Fluorescence
University of Maryland
701 E. Pratt St.
Baltimore, MD 21202
Suites 3017-21
USA
geddes@umbc.edu

Joseph R. Lakowicz
Center for Fluorescence Spectroscopy
725 Lombar Street
Baltimore, MD 21201
USA

ISSN 1573-8086
ISBN 978-1-4419-9671-8 e-ISBN 978-1-4419-9672-5
DOI 10.1007/978-1-4419-9672-5
Springer New York Dordrecht Heidelberg London

Printed on acid-free paper

Springer is part of Springer Science+Business Media (www.springer.com)

Reviews in Fluorescence

Editor:
Chris D. Geddes

For further volumes:
http://www.springer.com/series/6946

Preface

This is the sixth volume in the *Reviews in Fluorescence* series. To date, five previous volumes have been both published and well-received by the scientific community. Several book reviews, in the last few years, have also favorably remarked on the series.

We thank the authors for their very timely and exciting contributions again this year. We hope you will find this volume as useful as past volumes, which promises to be just as diverse with regard to fluorescence-based content.

Finally, in closing, I would like to thank Caroleann Aitken, Manager, The Institute of Fluorescence, for help in coordinating content with authors and Michael Weston at Springer for help in publishing this current volume.

Chris D. Geddes

Contents

Contributors

Martin Štefl J. Heyrovský Institute of Physical Chemistry v.v.i., Academy of Sciences of the Czech Republic, Dolejškova 3, Prague 18223, Czech Republic

J.R. Albani Laboratoire de Biophysique Moléculaire, Université des Sciences et Technologies de Lille, Bâtiment C6. 59655, Villeneuve d'Ascq Cédex, France

Wajih Al-Soufi Departamento de Química Física, Facultade de Ciencias, Universidade de Santiago de Compostela, E-27002 Lugo, Spain

Jon P. Anderson LI-COR Biosciences Inc., 4647 Superior Street, Lincoln, NE 68504, USA

Andrea Balogh Department of Immunology, Eötvös Lorand University, Pazmany Peter setany 1/C, 1117 Budapest, Hungary

Buddhudu Srinivasa Department of Physics, Sri Venkateswara University, Tirupati 517502, Andhra Pradesh, India

Peter J. Butler Department of Bioengineering, The Pennsylvania State University, 205 Hallowell Building, University Park, PA 16802, USA

Zhi Cao Department of Chemistry, New Mexico Institute of Mining and Technology, Socorro, NM 87801, USA

Nihav Dhawale School of Physical Sciences, Jawaharlal Nehru University, New Delhi 110 067, India

Tarek A. Fayed Chemistry Department, Faculty of Science, Tanta University, 31527 Tanta, Egypt

Michael C. Ferko Stryker Orthopaedics, 325 Corporate Drive, Mahwah, NJ 07430, USA

Daniel L. Grone LI-COR Biosciences Inc., 4647 Superior Street, Lincoln, NE 68504, USA

Ramachandra R. Gullapalli Department of Bioengineering, The Pennsylvania State University, 205 Hallowell Building, University Park, PA 16802, USA

Michael D. Heagy Department of Chemistry, New Mexico Institute of Mining and Technology, Socorro, NM 87801, USA

Martin Hof J. Heyrovský Institute of Physical Chemistry v.v.i., Academy of Sciences of the Czech Republic, Dolejškova 3, Prague 18223, Czech Republic

G. Krishnamoorthy Department of Chemical Sciences, Tata Institute of Fundamental Research, Mumbai 400 005, India

Sakari Kulmala School of Chemical Technology, Aalto University, Espoo, Finland

Bhaskar Kumar Grandhe Department of Physics, Sri Venkateswara University, Tirupati 517502, Andhra Pradesh, India

Radek Macháň J. Heyrovský Institute of Physical Chemistry v.v.i., Academy of Sciences of the Czech Republic, Dolejškova 3, Prague 18223, Czech Republic

János Matkó Department of Immunology, Eötvös Lorand University, Pazmany Peter setany 1/C, 1117 Budapest, Hungary

Immunology Research Group of the Hungarian Academy of Sciences, Eötvös Lorand University, Budapest, Hungary

Manuel Mosquera Departamento de Química Física, Facultade de Química, Universidade de Santiago de Compostela, E-15782 Santiago de Compostela, Spain

Péter Nagy Department of Biophysics and Cell Biology, University of Debrecen, Debrecen, Hungary

Premchendar Nandhikonda Department of Chemistry, New Mexico Institute of Mining and Technology, Socorro, NM 87801, USA

Punya Navaratnarajah Department of Biochemistry and Biophysics, University of North Carolina at Chapel Hill, Chapel Hill, NC 27599-7260, USA

Michael G. Nichols Physics Department, Creighton University, 2500 California Plaza, Omaha, NE 68178, USA

Mercedes Novo Departamento de Química Física, Facultade de Ciencias, Universidade de Santiago de Compostela, E-27002 Lugo, Spain

Flor Rodríguez-Prieto Departamento de Química Física, Facultade de Química, Universidade de Santiago de Compostela, E-15782 Santiago de Compostela, Spain

Saswata Sankar Sarkar Department of Chemical Sciences, Tata Institute of Fundamental Research, Mumbai 400 005, India

Anoop Saxena Department of Chemical Sciences, Tata Institute of Fundamental Research, Mumbai 400 005, India

Alexander Schiller Institute for Inorganic and Analytical Chemistry, Friedrich-Schiller-University Jena, Humboldtstr. 8, 07743 Jena, Germany

Bakthan Singaram Department of Chemistry and Biochemistry, University of California, 1156 High Street, Santa Cruz, CA 95064, USA

Johanna Suomi School of Chemical Technology, Aalto University, Espoo, Finland

János Szöllősi Cell Biology and Signaling Research Group of the Hungarian Academy of Sciences, University of Debrecen, Debrecen, Hungary

Tristan Tabouillot Department of Bioengineering, The Pennsylvania State University, 205 Hallowell Building, University Park, PA 16802, USA

Nancy L. Thompson Department of Chemistry, University of North Carolina at Chapel Hill, Chapel Hill, NC 27599-3290, USA

Department of Biochemistry and Biophysics, University of North Carolina at Chapel Hill, Chapel Hill, NC, USA

Jayant B. Udgaonkar National Centre for Biological Sciences, Tata Institute of Fundamental Research, Bangalore 560 065, India

Boaz Vilozny Department of Chemistry and Biochemistry, University of California, 1156 High Street, Santa Cruz, CA 95064, USA

Xiang Wang Department of Chemistry, University of North Carolina at Chapel Hill, Chapel Hill, NC 27599-3290, USA

Ritchie A. Wessling Department of Chemistry and Biochemistry, University of California, 1156 High Street, Santa Cruz, CA 95064, USA

John G. Williams LI-COR Biosciences Inc., 4647 Superior Street, Lincoln, NE 68504, USA

Metal Enhancement of Near-IR Fluorescence for Molecular Biotechnology Applications

Jon P. Anderson, John G. Williams, Daniel L. Grone, and Michael G. Nichols

Abstract Metal-enhanced fluorescence (MEF) can increase the overall emissions of a multitude of fluorophores by positioning the fluorophore in close proximity to an appropriate metal-coated surface. Near-infrared (near-IR) fluorophores placed near these metal surfaces combine the increased emissions of MEF with the low background characteristics of near-IR fluorescence. Together, this combination of high emission, low background detection may provide a powerful tool in the analysis of biological samples. In this brief review, we will outline the feasibility of using near-IR MEF in biotechnology research, will cover the types of experiments required to bring this technology from the feasibility stage to a commercial product, usable by molecular biologists, and will investigate the sources of background emissions that may be further reduced in the future.

1 Introduction

Fluorescence is a highly sensitive and convenient method of detection that has altered the landscape of molecular biotechnology over the past few decades. An increasingly wide variety of fluorophores are used in biotechnology, genomics, immunoassays, array technologies, imaging, and drug discovery [11, 32, 60, 61, 69]. Fluorescent molecules can easily be attached to a cast of target molecules, including DNA, RNA, antibodies, peptides, and proteins, and have the distinct advantage of being small compared to several other molecular labels [16, 20, 62, 71, 73]. Fluorescence-based technologies are used in a wide variety of biotechnology applications, including automated DNA sequencing, real-time PCR, microarray analyses, and immunoassays. The small size of fluorescent dyes provides little interference with the properties of the labeled molecule and allows the dye to

J.P. Anderson (✉)
LI-COR Biosciences Inc., 4647 Superior Street, Lincoln, NE 68504, USA
e-mail: jon.anderson@licor.com

C.D. Geddes (ed.), *Reviews in Fluorescence 2009*, Reviews in Fluorescence 2009,
DOI 10.1007/978-1-4419-9672-5_1,

infiltrate cellular regions that cannot be labeled by larger molecules (i.e., quantum dots and metal colloids) [14, 63]. Fluorescent dyes provide a predictable red-wavelength Stokes shift from the excitation to the emission spectra, allowing for very efficient collection of the emission photons away from the excitation beam [14, 52]. In addition to measuring the excitation wavelength, fluorescence can provide information by the way of polarization, lifetime, fluorescence resonance energy transfer (FRET), and quenching. Discovery and development of green fluorescent protein (GFP) has also allowed for the direct labeling and expression of a fluorescent biomolecule by a protein [12]. A major drawback, however, in using fluorescent dyes is the fact that the dyes can easily photodegrade, unlike quantum dots and metal colloids. Fluorescent molecules with small Stokes shifts may also have the disadvantage of self-quenching when located near another fluorophore by means of homo-FRET, reducing the overall signal.

Near-infrared (near-IR)fluorescence has unique advantages over visible fluorescence and is gaining popularity in biological imaging and molecular applications [64]. Fluorescence in the near-IR offers significantly lower background signals from scatter than those generated by visible wavelength excitation [50, 55]. Near-IR wavelengths are not readily absorbed by water and biological compounds, allowing deep penetration through tissues and cells, and reducing the possibility of photodamage to biological samples. Instruments using these longer wavelengths have a distinct advantage, with the near-IR spectral region showing little interference from biological molecules [51, 65].

2 Metal-Enhanced Fluorescence

Though highly successful, fluorescent-based methodologies have a continuing need for increased sensitivity and reduced detection limits [34]. To improve the sensitivity and reliability of fluorescent probes, the signal emissions from target molecules need to increase without increasing the signal from nonspecific molecules. In other words, the sensitivity and detection limit can be improved by increasing the quantum yield of the fluorophore, increasing its absorption cross section, increasing its photostability, and reducing the background signal of the system.

Metal-enhanced fluorescence (MEF) has been highly successful at increasing fluorescence sensitivity through an increase of the quantum yield and photostability of the fluorophore [4, 7, 35, 36]. The act of using a metal-coated surface to increase signal intensities is not a recent endeavor, for metal island films have been used for over two decades in Raman spectroscopy for signal enhancement [1, 17]. However, this field has only recently begun to show a large following in the field of fluorescence detection [35, 37, 40, 44, 47, 48]. Research on MEF has primarily focused on using silver island films (SIFs) [3, 5, 38, 40, 45, 46, 48] for enhancement of fluorescence, although recent studies have utilized metal colloids [6, 43], light-deposited silver [23], silver fractals [24], nanosphere lithography [29], and ordered arrays [15]. SIFs are composed of subwavelength sized patches of silver located on an inert substrate. Randomly seeded metal islands that ranged in size from 20 to 500 nm have been successfully used to enhance a variety of fluorophores [38, 40, 45, 46, 48]. The size and the shape of the metal islands may also play an important role in the ability to enhance the fluorescent signal [26, 59]. Periodic metal island nanoarray structures may further enhance the fluorophore emissions and could result in directional instead of isotropic photonic emissions [37].

SIFs have been largely studied for their ability to increase fluorescence intensities in low quantum yield and visible fluorophores [28, 40, 44, 45]. Much less information, however, exists on the ability of these metal islands to enhance the performance of the longer-wavelength, near-IR fluorophores commonly used in proteomic and genomic applications. The combination of increased emissions from MEF and the low background signal from near-IR fluorescence provides a means for lowering the limit of detection (LOD), providing the increased sensitivity that is required for many biological assays.

3 MEF Design Considerations

In MEF, the distance between the fluorophore and the metal is critical for enhancement of fluorescent emissions. The metal will quench the fluorophore if it is located within 50 Å of the metal surface. At a distance greater than 200 Å from the metal surface, the fluorophore displays free space characteristics. Therefore, the fluorophore must be located roughly between 50 and 200 Å from the metal surface for metal enhancement to take place (Fig. 1). It is this nanometer sized zone of enhancement that provides the greatest effect to a fluorophore's quantum yield and may aid in signal specificity, essentially amplifying only those fluorophores that are specifically located within this enhancement zone.

Metallic surfaces placed near a fluorophore can alter the fluorophore's properties. MEF uses metallic surfaces to modify the radiative decay rate (Γ) in a

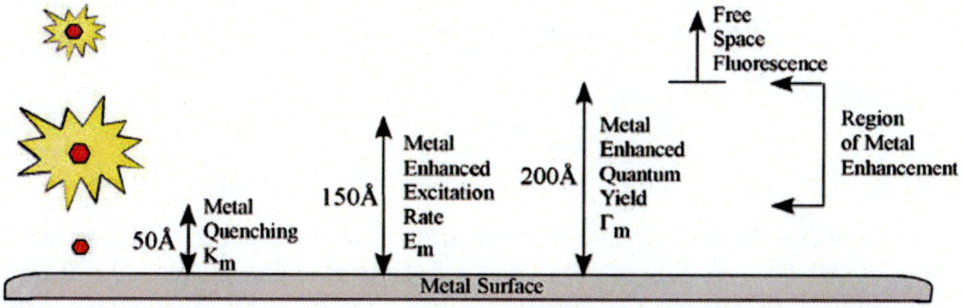

Fig. 1 Schematic representation of fluorescent properties near a metal surface. MEF properties are shown, as the particle moves from near the surface where it is quenched by the metal, through the region of metal enhancement and then into free space, where the metal has no effect on the particle

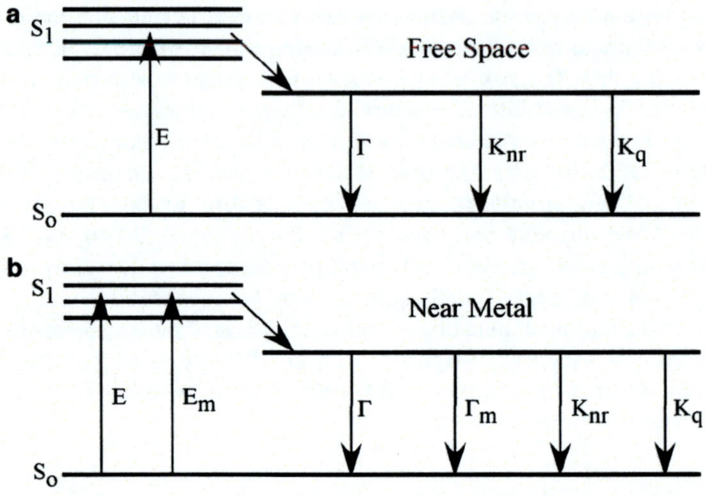

Fig. 2 Simplified Jablonski diagram showing fluorophores in (**a**) free space and (**b**) in the presence of a metallic nanoparticle. The diagram highlights the radiative decay rate (Γ), the radiative decay rate in the presence of metal nanoparticles (Γ_m), the nonradiative decay rate (k_{nr}), and the rate of quenching processes (k_q)

useful manner [35, 40]. An excited fluorophore can emit or return to a ground state by means of some nonradiative decay at a rate (k_{nr}) or by other quenching processes at a rate (k_q) (Fig. 2a). Fluorophores placed near a metal show both a decreased lifetime and an increased quantum yield, indicating an effect on Γ (Fig. 2b). Alterations in Γ are due to interactions of the excited state fluorophores with the oscillations of the free electrons in the metal [18, 33, 40]. The excited fluorophores induce surface plasmons which in turn emit photons [9, 49]. The fluorescence lifetime (τ_m) and the quantum yield (Q_m) for a fluorophore placed near a metal surface are given by the equations:

$$\tau_m = \frac{1}{\Gamma + \Gamma_m + k_{nr} + k_q}, \qquad (1)$$

$$Q_m = \frac{\Gamma + \Gamma_m}{\Gamma + \Gamma_m + k_{nr} + k_q}. \qquad (2)$$

By reducing the lifetime and increasing the quantum yield, metallic surfaces may increase the overall radiative rate of a fluorophore by more than 10,000-fold [33], although this increase is greatest for low quantum yield fluorophores [40]. Naturally, high quantum yield fluorophores may reveal only moderate gains in the overall quantum yield when placed near a metal surface; however, these dyes may still show a decrease in their fluorescent lifetime [40]. Producing a decrease in the fluorescence lifetime near a metal surface can result in a greater number of total photons emitted and an increased photostability. The maximum number of photons emitted by a fluorophore per second is limited by the inverse of the lifetime. The photostability of the fluorophore is affected by the amount of time the fluorophore remains in the excited state, where it may photodegrade. By reducing the lifetime (i.e., the average time spent in the excited state), the fluorophore can effectively pass through more excitation-emission cycles and thus produce an increased number of photons before photodegrading.

Though MEF has been successfully demonstrated by multiple groups [18, 26, 33, 40], there remains little experimental data on the effects of MEF on near-IR fluorescence. Furthermore, even with the successful demonstration of this technology, there still remain several challenges in transferring this technology from the research stage to the commercial stage. The process of demonstrating feasibility using MEF is only the first stage in making this technology useful for molecular biologists. Studies on the reproducibility and linearity of the fluorescence emissions, along with investigations on the longevity of theses metal structures, need to be performed before they can be reliably used to quantitate labeled biological samples. Along with these verification studies, a fundamental understanding of the sources of background will help find ways to reduce noise in the assay and lower the LOD beyond current methodologies.

4 Silver Island Films

Preliminary results with LI-COR IRDye®800CW (ex 774 nm; em 789 nm) showed that near-IR fluorophores can be enhanced by both silver and gold nanostructures. The silver structures were shown to enhance the dye significantly more than the gold, a result that is not unexpected and has been reported for visible fluorophores [49]. From these initial results, we decided to focus on silver structures and proceeded to produce silver coatings on glass slides, including SIFs and colloid-coated surfaces.

Experiments were performed to quantitate the relative enhancement of near-IR fluorophores on the deposited silver-coated glass surface as compared to the uncoated glass. Throughout our experiments, a known quantity of fluorophore-labeled protein or DNA was spotted onto the SIF-coated and uncoated portions of the glass slides and then detected using a LI-COR Odyssey® near-IR fluorescence imager (Fig. 3). Spotting a known amount of fluorophore onto the SIF-coated or uncoated glass surface allows us to remove any affinity biases that may occur when depositing a layer of fluorophores over the entire surface of a slide. Analyzing individual spots on a scanned image also allows the Odyssey® software to account for any increase in background fluorescence signal caused by the silver surface, with the near-IR excitation wavelengths producing almost no increase in the background signal.

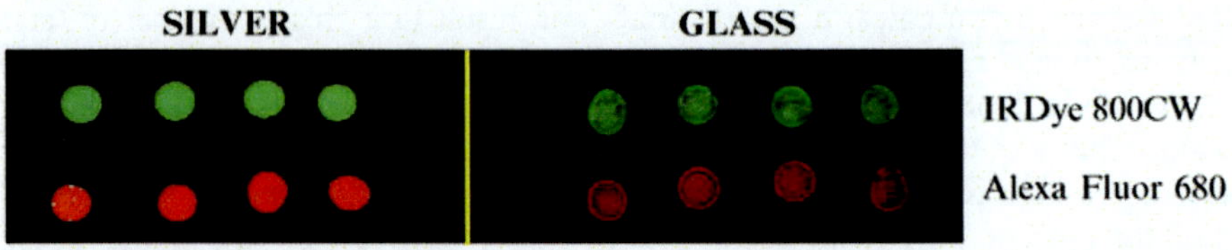

Fig. 3 Typical results of fluorophore enhancement by SIF. Equal amounts of LI-COR IRDye®800 (*green*) and Molecular Probes Alexa Fluor® 680 (*red*) labeled streptavidin are spotted onto either the silver-coated (SILVER) or uncoated (GLASS) regions of a glass slide and allowed to dry. The slide is imaged using a LI-COR Odyssey® imager. The SIF-coated regions show increased fluorescence intensity as shown by the brighter spots

4.1 Fluorophore to SIF Distance Dependence

The distance between the fluorophore and the SIF is critical for metal enhancement to take place. Previous studies have concluded that at distances greater than 30–50 nm from the SIF, the fluorophore and metal surface generate little measurable interaction and the fluorophore displays mainly free space characteristics [46]. When the fluorophore is in close proximity to the metal surface, however, the excited fluorophore induces surface plasmons in the SIF, which then also radiate at the same wavelengths as the excited state fluorophore [13]. Using both LI-COR IRDye®700 (ex 689 nm; em 700 nm) and IRDye®800 near-IR fluorophore-labeled oligos, we observed this distance-dependent relationship by spotting 1 μl (10 fmol) of dye-labeled DNA oligo onto the SIF-coated and uncoated portions of the glass slide, and detecting the fluorescence emissions generated from each droplet as it dried. The drying of the droplet moved the dye-labeled DNA oligos from a free space emission mode, far from the SIF surface, to a metal-enhanced mode, near the SIF surface (Fig. 4). Integrated intensities were measured for each of the spots and the relative amount of enhancement for fluorophores spotted over silver was calculated. The integrated intensity calculations on the Odyssey® system are independent of the spot size and scan resolution, and account for any increased background signal generated by the silver surface [42]. Measured over time as the spot dried, signal from dye-labeled oligos spotted on an uncoated region of the glass slide showed little change, while oligos spotted on the SIF-coated region of the slide showed increased fluorescence emissions.

SIFs are generated by a chemical reduction of silver nitrate in solution [39, 40, 45–47, 59]. The process of producing the SIF creates a heterogeneous population of metal nanostructures. The size, shape, and position of the silver nanostructures are somewhat random along the glass surface (Fig. 5). By specifying the coating time, however, the size and density of the silver structures can be roughly controlled. To determine the optimal SIF density for enhancement of near-IR fluorophores, we first generated SIF gradients on glass slides. The SIF gradients were constructed by slowly retracting a glass slide from the reduced silver nitrate solution as it is heated in a 40°C waterbath. The slow removal of the slide was done by using a programmable

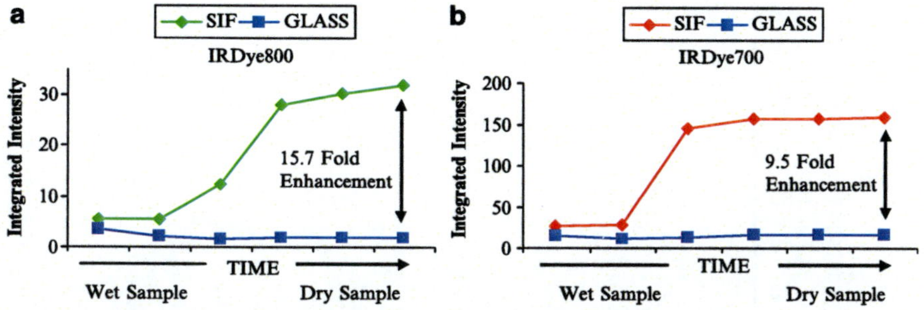

Fig. 4 Fluorescence integrated intensity of an (**a**) IRDye®800- and (**b**) IRDye®700-labeled DNA oligo spotted onto either a plain glass slide (GLASS) or onto a silver island film coated glass slide (SIF). The integrated intensity is monitored over time as the liquid spot dries onto the slide, bringing the fluorophore in close proximity to the silver surface, and producing an increase in the fluorescence intensity for the SIF sample. The data was generated by implementing consecutive scans on the Odyssey® imager, and with the focus plane of detection adjusted to accommodate for scanning through the bottom of the glass slide to allow for the imaging of wet spots

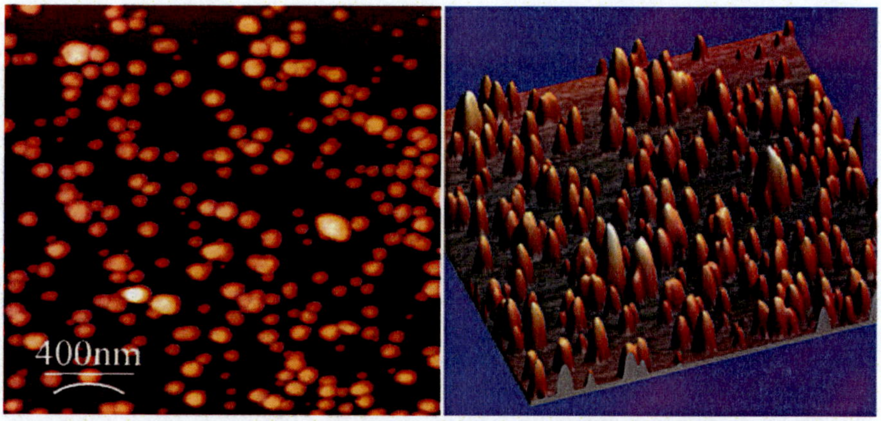

Fig. 5 Atomic force micrograph (AFM) image of a typical silver island film (SIF) produced by LI-COR and imaged at the University of Nebraska-Lincoln. Image on the *right* is a 3D reconstruction of the AFM image. Note variations in size and random distribution of silver nanostructures

stepper motor that would retract the slide from the silver solution at a rate of approximately 5 mm/min, creating a reproducible SIF gradient. Absorption spectra were taken at various regions along the SIF gradient, with the peak absorption correlating to the silver density on the slide surface. Near-IR fluorophore-labeled DNA oligos were then spotted along the SIF gradient and the emissions were recorded and analyzed using an Odyssey® imager. These experiments determined that a SIF containing a peak absorbance of 0.6–0.8 absorbance units (AU) would produce the greatest enhancement of near-IR fluorophores (Fig. 6).

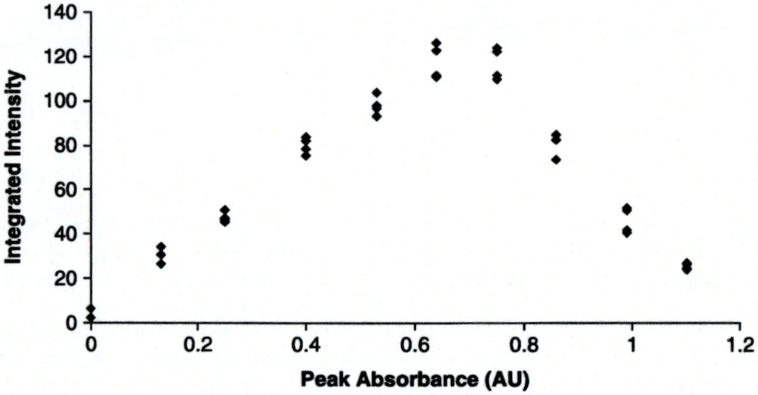

Fig. 6 A plot of the relationship between SIF coating density is shown, as measured by peak absorbance, and the integrated intensity measurement of 10 fmol of an IRDye®700-labeled DNA oligo. A total of four measurements were taken at each silver density along the SIF gradient slide. The results indicate that a SIF coating that produces an absorption intensity peak between 0.6 and 0.8 AU produced the highest integrated intensity

Using SIFs optimized for near-IR enhancement, we were able to obtain an average of 18-fold enhancement of IRDye®700-labeled DNA oligo spotted on plain glass using a manual glass slide microarrayer system (V&P Scientific inc.). IRDye®800-labeled DNA oligo was enhanced 15-fold over dye on plain glass. These enhancement results are slightly better than what has been achieved using SIFs on visible fluorophores [4, 40, 46].

5 Colloid-Coated Surfaces

Producing metal islands of a specific size may be critical for efficient MEF. Although chemical dip coating can successfully create SIFs over a range of sizes, controlling the reaction and halting the growth of the islands at an exact size and density can be difficult. Silver colloids were therefore generated to create colloid-coated surfaces containing islands of a specific size across the entire slide. The process uses metal nanoparticles (colloids) to effectively coat a glass slide, creating a metal island film. A clean glass substrate is surface coated with 3-aminopropyltrimethoxysilane (APS), producing a monolayer of reactive groups that can bind colloidal silver nanoparticles [19, 58]. The silanized glass substrates is then immersed in a colloidal solution, immobilizing the metal nanoparticles on the surface [66].

The particle spacing on the glass surface is random, with the interparticle spacing (i.e., the particle density) being dependent on both the amount of time the substrate is immersed in the colloidal solution and the concentration of the colloidal solution. Because the size of the islands is determined by the colloids, sparse to densely packed colloid-coated surfaces that maintain a unique metal island size by simply varying the amount of time used to attach the colloids to the glass surface can be produced. This finer control of the nanostructure size and density was investigated to determine if these colloid-coated surfaces could provide even greater enhancement for near-IR fluorophores.

Reports have indicated that colloid-coated glass surfaces can enhance the fluorescence of some visible fluorophores more than SIFs, with the best results showing 16.7-fold enhancements [43]. Several methods of producing silver colloids were employed, including sodium citrate reduction [43], the polyol process using ethylene glycol as the reductant [67, 68], small silver seed production followed by repeated rounds of controlled growth [8, 10], silver reduction in the presence of gum arabic [27, 72], as well as the photoinduced production of triangular silver particles [30, 31]. Using these methodologies, we were able to generate colloids from 4 nm to greater than 120 nm in diameter. Sizes of the colloids were determined at Brookhaven Instruments Corporation (Holtsville, NY) using a Brookhaven 90Plus dynamic light scattering particle size analyzer. The colloids were attached to APS-coated glass slides by dip coating the slides for up to 24 h in the colloid solution. These various sized colloid-coated surfaces generated enhancements of up to 11-fold for IRDye®800CW and up to 5-fold for IRDye®700. These results did show enhancement by the colloid-coated surfaces, but the enhancements obtained were less than that obtained from the SIFs. The longer-wavelength, near-IR fluorescence may not be efficiently enhanced with small (4–120 nm) colloids that were tested, whereas the shorter, visible fluorescence may have been better enhanced. Producing larger colloids that are highly monodispersed might improve the enhancement for the near-IR, but we have not yet tested this hypothesis. Based on these results, we moved away from colloid-coated surfaces and began to focus our attention on characterizing the near-IR metal enhancement from SIFs.

6 Validation Tests for Near-IR MEF

Apart from demonstrating feasibility of MEF on near-IR fluorescence, we set out to further test the characteristics of the SIF-coated slides and to validate their properties. These validation tests include assays on longevity, reproducibility, and linearity.

Longevity studies involve storing SIF-coated slides for several months and then assaying them for their ability to enhance near-IR fluorescence. These shelf-life studies may not be routinely done for determining the feasibility of a product, but are crucial for the development of a commercially viable product.

SIF-coated slides are also tested for their ability to produce consistent, reproducible results across a coated slide surface. An array of near-IR fluorophore-labeled protein spots is first generated across a slide using a SpotBot personal microarrayer. Fluorescence emissions from the spots on the SIF-coated and uncoated regions of the slide are then compared, with the goal of maintaining % CV values not statistically different than that of plain glass slides.

Linearity measurements are performed on the slides to validate that the fluorescence intensities recorded by the Odyssey® scanner remain linear over several orders of magnitude based on the amount of fluorophore spotted on the slide. These types of measurements remain important to the utility of the slides as a commercial product.

6.1 Longevity of SIF-Coated Surfaces

Silver surfaces and films are prone to oxidation and are far less stable than gold surfaces [25, 75]. Frequently, SIFs are used immediately after they are produced, but have routinely be stored for up to 1 week in deionized water before being used [54]. In order to protect and stabilize the silver, others have applied a thin coating of silica to the silver surface [2, 53]. This thin layer of silica should help protect the silver surface, but could however adversely affect the ability of the SIF to enhance fluorescence by MEF. To first determine if a protective layer will be needed on the silver-coated glass slides, we performed a longevity study on the SIF-coated slides.

The longevity study was designed to measure the ability of the SIF surfaces to continue to enhance near-IR fluorophores over time. For this investigation, we generated SIF-coated slides and stored them under three different conditions for a period of 9 months. It should be noted that the stored slides were prepared before we determined the best silver density for near-IR fluorophores and contained an average peak absorbance of 0.3 AU, which is suboptimal for enhancement of near-IR fluorophores. All newly prepared SIF-coated slides that were used in the longevity study were also made with similar (0.3 AU) absorbances. The storage conditions for these slides included: wet (in dH_2O), dry under argon, and dry under ambient conditions. After 9 months, the stored slides, along with newly prepared SIF-coated slides, were freshly spotted with dye-labeled oligo and tested for their ability to enhance near-IR fluorophores. The results showed that IRDye®800-labeled DNA oligos were enhanced by an average of 6.6-fold for newly prepared slides, while the slides stored dry under Argon and Ambient conditions enhanced the dye by an average of 10.9-fold and 8.4-fold, respectively. The IRDye®700-labeled DNA oligos were enhanced by an average of 13.8-fold for newly prepared slides, while the slides stored dry under Argon and Ambient conditions enhanced the dye by an average of 4.0-fold and 4.4-fold, respectively. The slides stored for 9 months in H_2O had a visually distinct alteration in their appearance and were unable to enhance either fluorophore.

Our longevity studies demonstrated that the SIFs were able to be stored dry for 9 months and still sustain their ability to enhance the near-IR fluorophores. The reduction in the overall enhancement of these stored and newly prepared SIF-coated slides can be accounted for by the suboptimal coating of SIF, which contained an absorbance of approximately 0.3 AU. The optimal absorbance for SIF using LI-COR dyes was later determined to be between 0.6 and 0.8 AU (Fig. 6). This difference in the amount of silver coating on the slides can generate the slightly lower enhancements that were observed in the longevity studies as compared to our best enhancements using SIFs, which were 15-fold for IRDye®800- and 18-fold for IRDye®700-labeled DNA oligos. The results, however, show that the slides stored dry remain able to effectively enhance near-IR fluorophores even after 9 months of storage, although a reduction in enhancement was observed using IRDye®700.

These results demonstrate that we may not need to overcoat the silver nanostructures with silica or other self-assembled monolayers (SAMs) to preserve their function over time. Silica or SAM coatings may, however, still be useful in providing specific surface coatings over the SIF-coated slide. Experiments on SAMs with the protein BSA have been shown to work well with silver-coated surfaces and allow for enhancement of visible fluorophores [46, 70]. Others have also performed research on overcoating silver surfaces with increasing thicknesses of silica. These results indicate that silica can be successfully used to coat the silver surfaces with a thickness of 10–15 nm silica being optimal for providing fluorescence enhancement [74].

6.2 *Reproducibility of Near-IR MEF*

The ability to reproducibly generate a fluorescence signal across the surface of a silver-coated glass slide is paramount in developing a commercial product. Simply producing a SIF surface that can enhance fluorescence will not be of use unless the enhancement is reproducible across the surface of the slide. Reproducibility of the silver-coated glass surfaces were determined by measuring the percent coefficient of variation (%CV) of the intensity from fluorescently labeled protein or DNA spotted along either uncoated or SIF-coated glass slides. The %CV is calculated as the standard deviation of the sample intensities divided by the mean sample intensity, multiplied by 100. For each slide that was half coated with a SIF, a SpotBot microarrayer (Telechem International Inc.) was used to spot approximately 350–450 individual spots on both the silver-coated and uncoated halves of the slide (Fig. 7). Fluorescence intensities were collected by scanning the spotted slides on an Odyssey® imaging system and analyzing the data using either Odyssey® software or Media Cybernetics Array-Pro Analyzer version 4.5 software. The fluorescence intensities from the silver-coated and uncoated portions of each slide were then used to compute %CVs. A paired T-test was then used to statistically analyze the silver-coated and uncoated regions of a slide. The results indicate that there was no significant difference in the %CVs for the coated and uncoated regions of the slide, with the uncoated glass maintaining an average CVs of 12.84 ± 5.28%, and the silver-coated glass having an average CV of 13.39 ± 5.41%.

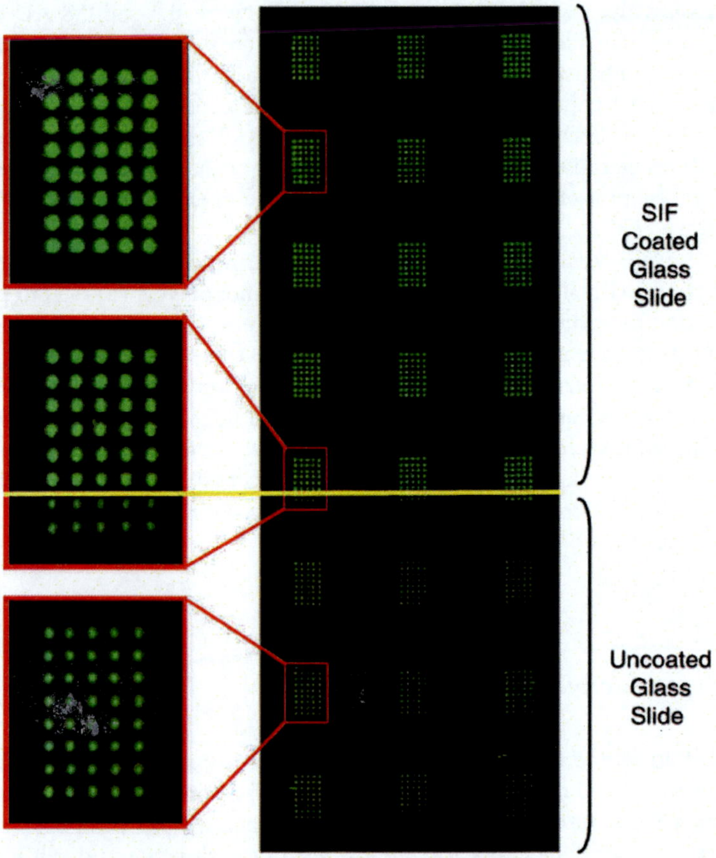

Fig. 7 Fluorescence reproducibility of SIF-coated glass slides. An Odyssey® image of a glass slide spotted with IRDye®800CW-labeled streptavidin is shown. The upper portion of the slide (above the *yellow line*) is coated with a SIF, while the lower portion of the slide is uncoated. A SpotBot personal microarrayer was used to spot grids of 40 individual spots along the glass slide, which were then analyzed using either Odyssey® or Array-Pro software. Enlarged sections of the SIF-coated region (*top*), transition region from SIF coated to uncoated (*middle*), and uncoated region (*bottom*) of the slide are shown, with fluorophores spotted over SIF-coated glass showing enhanced fluorescence signals

6.3 Linearity of Enhanced Fluorescence

The SIF-coated slides have been shown to generate up to an 18-fold increase in fluorescence emissions of near-IR dyes, but these enhancements alone cannot determine the overall utility of using SIFs in molecular biology applications. For this methodology to be a useful quantitative tool, it would be advantageous to show that the system is linear over a wide range of fluorophore concentrations. To show that SIFs can be used as a molecular biology tool for enhancing near-IR

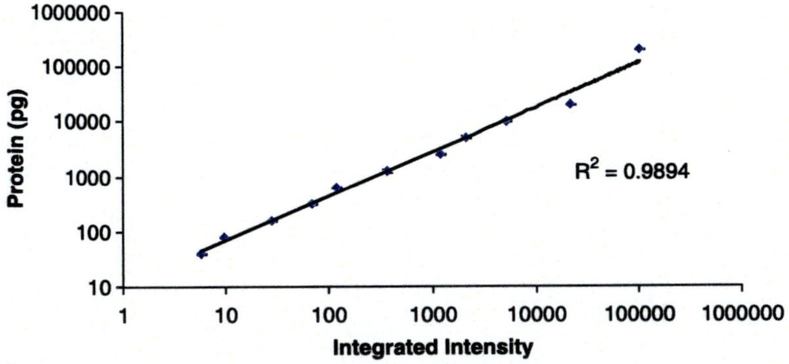

Fig. 8 Linearity of SIF fluorescence enhancement. Twofold serial dilutions of IRDye®800CW-labeled Goat-IgG antibody are plotted. Each plotted point represents the average of two measurements, with one standard deviation shown above and below the average by error bars

fluorescence, we determined the linearity of our assay over four orders of magnitude. A series of twofold serial dilutions were prepared using IRDye®800CW-labeled Goat-IgG antibody (LI-COR). The dilution series was spotted in duplicate onto a SIF-coated glass slide and imaged using an Odyssey® scanner at two intensity settings. A low-intensity setting was used to avoid saturation of fluorescence signal in concentrated sample spots and a higher scan intensity setting was used to detect highly dilute samples. The linearity measurements were performed on the slides to validate that the fluorescence intensities recorded by the Odyssey® scanner remain linear, based on the amount of fluorophore spotted on the slide. The results demonstrate that the signal is linear with dye amount over four orders of magnitude with an R^2 value of 0.9894 (Fig. 8).

7 Background Reduction

The use of MEF for producing increased fluorescence intensity has been successfully demonstrated using a variety of fluorophores [38, 40, 45, 46, 48]. However, along with the dramatic increase in signal, these metal island films may also display an increased amount of background or scatter, hindering detection of a desired signal. One of the main goals of using MEF to enhance near-IR fluorescence is the ability to significantly lower the LOD. The ability to improve the LOD can greatly expand the utility of a technique. Greater sensitivity allows the use of smaller samples, the ability to test for more analytes in a single small sample and, in the case of immunoassays, an improved LOD would expand the repertoire of available antibodies to include those with lower binding affinities.

For our experiments, the LOD is defined as the smallest concentration of analyte that can be reliably detected above the system noise, producing a signal that is three times the standard deviation (3σ) of the system noise level. Since the LOD is directly related to the background noise level of the system, the ability of the Odyssey® to detect in the near-IR is advantageous for silver-coated slides, with the silver surfaces producing little added background signal. Any further reduction in the background, however, will aid in the reduction of the overall LOD of the system. Therefore, we set out to develop a better understanding of the sources of background coming from silver particle coated surfaces, with the aim of reducing such background and lowering the LOD.

7.1 Silver Storms

One source of background noise emanating from the silver-coated surface is a bright, but short-lived luminescent blinking that is produced by the silver nanoparticles, an observation which we designate "silver storms." These silver storms have been previously observed for both gold and silver nanoparticles, and remain an interesting phenomenon [21, 22, 41, 57]. Peyser et al. have indicated that the silver storms strongly occur at wavelengths <520 nm. However, we have observed these storms using evanescent excitation via a total internal reflection (TIR) microscope with near-IR excitation at 680 nm, one of the excitation wavelengths currently used in our Odyssey imagers [57]. To view a movie of these silver storms, see http://www.licor.com/silverstorm. Because the Odyssey scans large areas over long time periods, it is not possible to observe silver storms in movie format. However, we believe storms likely occur and contribute to background noise in Odyssey images, particularly since excitation power is greater in the Odyssey than in our microscopic observations (1,000 W/cm^2 vs. 10 W/cm^2). To further investigate silver storms, we have made observations using two-photon excitation. This setup allowed the emission spectrum to be obtained at visible wavelengths without interference from the near-IR excitation light. It also allowed emission lifetimes to be measured, in order to distinguish between prompt scatter and delayed emissions. The two-photon excitation work was performed at the Nebraska Center for Cell Biology at Creighton University, Omaha, NE.

7.2 Two-Photon Microscopy Imaging of Silver Nanoparticles

Silver nanoparticles were deposited by reduction of silver ion with DMF [56] on ITO-coated coverglass (ZC&R, 140 Ω/sq). Silver storms were observed under two-photon excitation as shown in Fig. 9.

Silver nanoparticles were imaged using intense near-IR illumination at 740 and 950 nm by two-photon laser scanning microscopy. Femtosecond pulses of near-IR illumination from a tunable Chameleon XR laser (Coherent Inc., Santa Clara, CA) were scanned across the sample by an LSM510 META NLO laser scanning microscope with a 63× Plan Apochromat 1.4 NA oil immersion objective

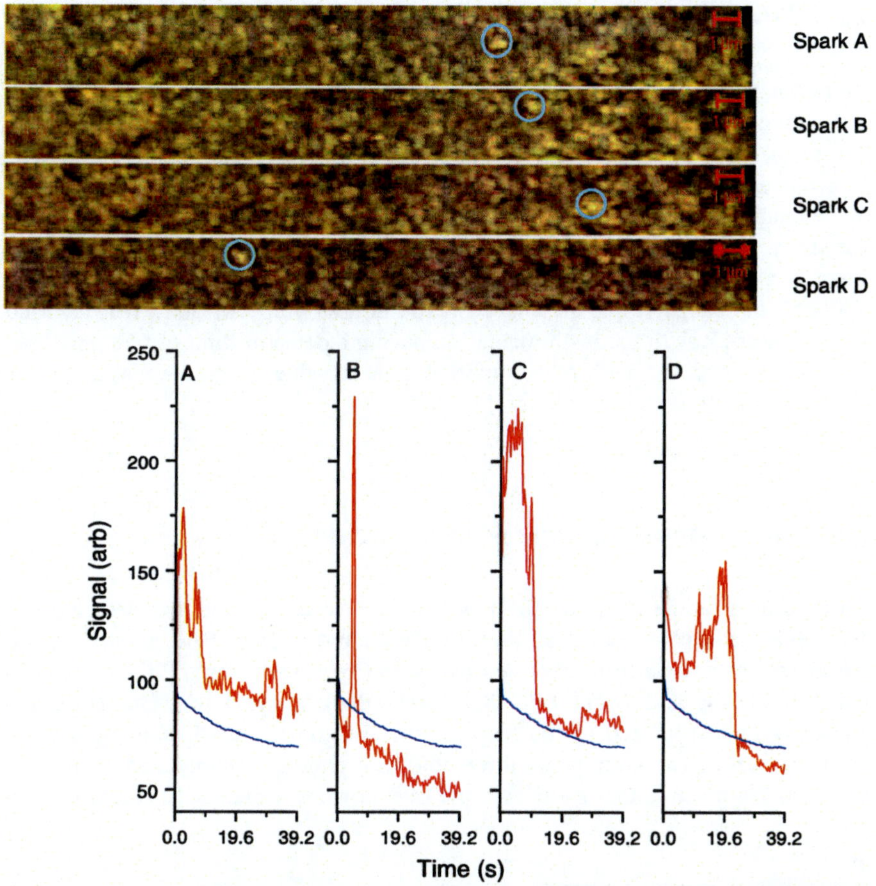

Fig. 9 Silver nanoparticles "sparking" on ITO-coated glass coverslips. A 2.9 × 29 μm region was imaged at a frame rate of 2.55 Hz. The *blue line* indicates the average pixel intensity for the entire image, while the *red lines* show the average pixel intensity in the sparking regions circled in the frames shown above. Scale bar is 1 μm

(Carl Zeiss Inc., Thornwood, NY). The repetition rate of the pulse train was 90 MHz. We estimate the spatial resolution to be approximately 0.3 μm radially (in the image plane) by 0.75 μm axially (depth of the imaging plane). An average laser power of approximately 0.4 mW was directed to the sample for most of the measurements. The light emitted from the sample was initially separated into two detection channels using a 500-nm long pass dichroic (500 DCXR, Chroma Technology, Brattleboro, VT) and detected by photomultiplier tubes without descanning. Blue emission was isolated using a custom-made bandpass filter (HQ460/80, Chroma Technology, Brattleboro, VT) and green emission was isolated using an HQ540/60 bandpass filter (Chroma Technology, Brattleboro, VT).

7.3 Silver Storm Emission Spectra

To obtain emission spectra from individual pixels, the emission was diverted to a diffraction grating and detected in 24 wavelength bands centered from 404 to 650 nm using the META detector of the LSM510 microscope. The spectral resolution was approximately 10.8 nm. No attempt was made to measure or correct for any nonuniform spectral response of the META detector. The emission results from the silver nanoparticles for both 740 and 950 nm two-photon excitation show that there is a broad emission spectra from the silver nanoparticles, throughout the 400–650 nm range (Figs. 10 and 11). The emission spectra may well continue into the near-IR with the instrumentation having a detector limit of 650 nm. These results are consistent with other reported data showing a broad emission spectra from silver nanoparticle excitation [22, 57].

7.4 Silver Storm Lifetime Measurements

To perform lifetime imaging, the blue emission was diverted and detected by a H7422-40 GaAs photomultiplier detector (Hamamatsu Corp., Bridgewater, NJ) and the emission lifetime was measured by Time Correlated Single Photon Counting using a Becker and Hickl SPC 830 card (Becker and Hickl GmbH, Berlin, Germany), synchronized to the laser pulses and the LSM510 scan signals. The field was scanned approximately three times and photons accumulated for 90 s. To avoid double pulse counting artifacts, the laser power was reduced so that much less

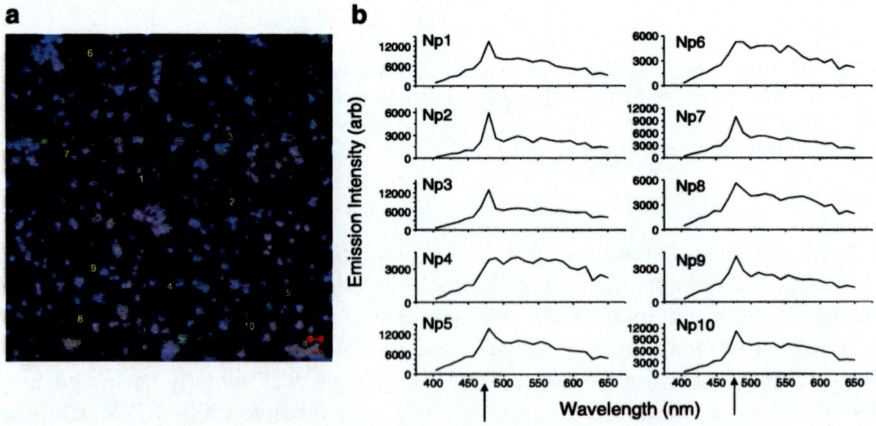

Fig. 10 (**a**) Spectral image showing blue emission from silver nanoparticles using 950 nm two-photon excitation. Scale bar is 1 μm. Spectra from individual nanoparticles are shown in (**b**). SHG and Hyper-Raleigh scattering expected at 475 nm (*arrow*). The continuous spectrum is probably silver-ion luminescence. All spectra are background subtracted

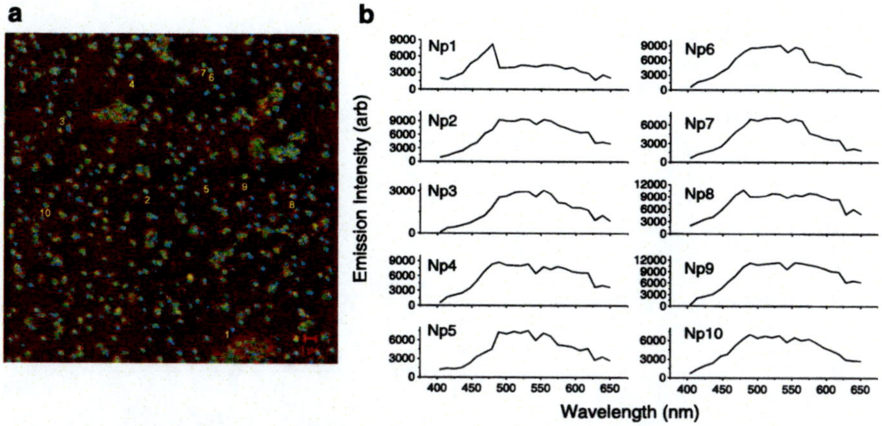

Fig. 11 (a) Spectral image showing emission from silver nanoparticles using 740 nm two-photon excitation. Scale bar is 1 μm. Spectra from individual nanoparticles are shown in (b). SHG and Hyper-Raleigh scattering expected at 370 nm, and not detectable in the spectra. The continuous spectrum is probably silver-ion luminescence. All spectra are background subtracted

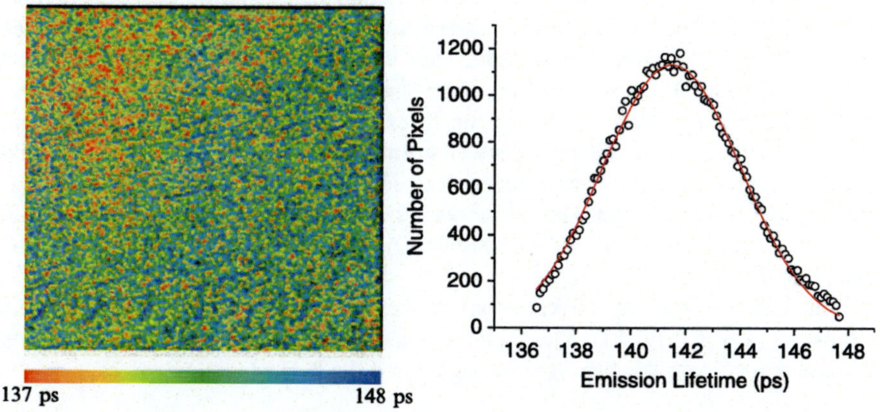

Fig. 12 Lifetime measurements for silver nanoparticle emissions using two-photon excitation. A single Gaussian fit to lifetime histogram gives an average emission lifetime of 141 ± 5 ps

than one photon was detected per pulse. Photon count rates did not exceed 1 MHz. The analysis of the lifetime measurements was performed using the Becker and Hickl SPC Image Software. The lifetime measurement fit is a convolution of the instrument response function (assumed Gaussian) and a single exponential decay. FWHM of the IRF is estimated to be 196 ps from the same fit. The results from the lifetime measurements show that the silver nanoparticles display a very short (140 ps) lifetime (Fig. 12). We believe that this may be the first reporting of the emission lifetimes for these silver storms.

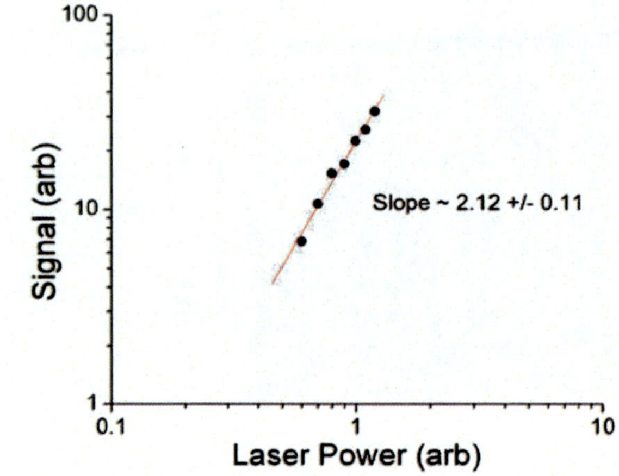

Fig. 13 The emission intensity ("signal") appears to increase quadratically (I^2) with the average laser power as the laser power was increased

7.5 Silver Storm Power Function Analysis

To further investigate the properties of the silver storms, we performed emission measurements on the silver storms as a function of excitation power. Previous studies have indicated that silver nanoparticles will cease to emit fluorescence below a specific excitation threshold, with the threshold partially dependent on the nanoparticle structure [21]. We therefore wanted to determine if the silver storms display a nonlinear relationship to excitation power. Nonlinearity would allow the differential reduction of silver storms relative to fluorescence by simply reducing the excitation power of the system. For two-photon excitation, "linearity" means the emission signal should increase quadratically (I^2) with laser power I, whereas "nonlinearity" would be revealed by a stronger power dependence, I^n with $n > 2$.

The results show that silver emissions increase as I^2 (i.e., "linear") with the excitation power (Fig. 13). Therefore, a reduction in the excitation power would equally reduce both the fluorescence signal and the background noise.

8 Conclusions

Fluorescence measurements at near-IR wavelengths generate significantly lower background noise than fluorescence in the visible spectrum. Near-IR light only weakly interacts with biological material, making it highly advantageous for biological applications. MEF has been utilized to enhance the fluorescence signal of several different dyes but little information exists on its use in enhancing the signal for near-IR fluorescence. Combining low-background, near-IR fluorescence with MEF should provide a powerful tool in the analysis of biological samples.

Using SIFs, we were able to effectively increase the fluorescence emissions of near-IR fluorophores. We produced an 18-fold increase in the signal of IRDye®800 and a 15-fold increase in IRDye®700. The fluorescence intensities produced on SIFs were also shown to be linear over four orders of magnitude. The ability of the SIFs to enhance the near-IR fluorophores yet maintain a linear relationship is crucial for applying this technology to biological assays. Furthermore, we were able to demonstrate that the signal produced across a SIF-coated slide remains reproducible, as determined by the %CV of the detected signals, and that the SIF-coated slides can be stored for 9 months and remain able to enhance near-IR fluorophores. The ability to produce SIF-coated slides with a long shelf life that can enhance fluorescence while maintaining linearity and reproducibility is essential for producing a class of new tools that can benefit molecular biologists by increasing the sensitivity of fluorescence-based assays.

In addition to studies on the enhancement of near-IR fluorescence using MEF, we also investigated the phenomenon of silver storms with the goal of determining techniques to reduce the background signal coming from these emission events. These studies provided us with information on the emission spectrum, the emission lifetime, and the signal to laser power relationship. Broad emission spectra produced by the silver nanoparticles may make it difficult to selectively filter out this source of background.

Acknowledgments This work was supported by the NIH National Center for Research Resources, SBIR Grant number RR021785. We also thank the Center for Fluorescence Spectroscopy which is supported by an NCRR grant number RR08119 for their assistance.

References

1. Albrecht MG, Creighton JA (1977) Anomalously intense Raman spectra of pyridine at a silver electrode. J Am Chem Soc 99:5215–5217
2. Alejandro-Arellano M, Ung T, Blanco A, Mulvaney P, Liz-Marzan LM (2000) Silica-coated metals and semiconductors. Stabilization and nanostructuring. Pure Appl Chem 72:257–267
3. Aslan K, Geddes CD (2006) Microwave-accelerated metal-enhanced fluorescence (MAMEF): application to ultra fast and sensitive clinical assays. J Fluoresc 16:3–8
4. Aslan K, Gryczynski I, Malicka J, Matveeva E, Lakowicz JR, Geddes CD (2005) Metal-enhanced fluorescence: an emerging tool in biotechnology. Curr Opin Biotechnol 16:55–62
5. Aslan K, Huang J, Wilson GM, Geddes CD (2006) Metal-enhanced fluorescence-based RNA sensing. J Am Chem Soc 128:4206–4207
6. Aslan K, Lakowicz JR, Geddes CD (2005) Metal-enhanced fluorescence using anisotropic silver nanostructures: critical progress to date. Anal Bioanal Chem 382:926–933
7. Aslan K, Lakowicz JR, Geddes CD (2005) Plasmon light scattering in biology and medicine: new sensing approaches, visions and perspectives. Curr Opin Chem Biol 9:538–544
8. Aslan K, Lakowicz JR, Geddes CD (2005) Rapid deposition of triangular silver nanoplates on planar surfaces: application to metal-enhanced fluorescence. J Phys Chem B 109:6247–6251
9. Aslan K, Leonenko Z, Lakowicz JR, Geddes CD (2005) Annealed silver-island films for applications in metal-enhanced fluorescence: interpretation in terms of radiating plasmons. J Fluoresc 15:643–654

10. Aslan K, Leonenko Z, Lakowicz JR, Geddes CD (2005) Fast and slow deposition of silver nanorods on planar surfaces: application to metal-enhanced fluorescence. J Phys Chem B 109:3157–3162
11. Brumbaugh JA, Middendorf LR, Grone DL, Ruth JL (1988) Continuous, on-line DNA sequencing using oligodeoxynucleotide primers with multiple fluorophores. Proc Natl Acad Sci USA 85:5610–5614
12. Chalfie M, Tu Y, Euskirchen G, Ward WW, Prasher DC (1994) Green fluorescent protein as a marker for gene expression. Science 263:802–805
13. Chowdhury MH (2006) Metal-enhanced chemiluminescence: radiating plasmons generated from chemically induced electronic excited states. Appl Phys Lett 88:173104
14. Cognet L, Tardin C, Boyer D, Choquet D, Tamarat P, Lounis B (2003) Single metallic nanoparticle imaging for protein detection in cells. Proc Natl Acad Sci USA 100:11350–11355
15. Corrigan TD, Guo S, Phaneud RJ, Szmacinski H (2005) Enhanced fluorescence from periodic arrays of silver nanoparticles. J Fluoresc 15:777–784
16. Dubertret B, Skourides P, Norris DJ, Noireaux V, Brivanlou AH, Libchaber A (2002) In vivo imaging of quantum dots encapsulated in phospholipid micelles. Science 298:1759–1762
17. Fleischmann M, Hendra PJ, McQuillan AJ (1974) Raman spectra of pyridine adsorbed at a silver electrode. Chem Phys Letts 26:163–166
18. Ford GW, Weber WH (1984) Electromagnetic interactions of molecules with metal surfaces. Phys Rep 113:195–287
19. Freeman GR, Grabar KC, Allison KJ, Bright RM, Davis JA, Guthrie AP, Hommer MB, Jackson MA, Smith PC, Walter DG, Natan MJ (1995) Self-assembled metal colloid monolayers: An approach to SERS substrates. Science 267:1629–1632
20. Frey PA, Frey TG (1999) Synthesis of undecagold labeling compounds and their applications in electron microscopic analysis of multiprotein complexes. J Struct Biol 127:94–100
21. Geddes CD, Parfenov A, Gryczynski I, Lakowicz JR (2003) Luminescent blinking from silver nanoparticles. J Phys Chem B 107:9989–9993
22. Geddes CD, Parfenov A, Gryczynski I, Lakowicz JR (2003) Luminescent blinking of gold nanoparticles. Chem Phys Lett 380:269–272
23. Geddes CD, Parfenov A, Lakowicz JR (2003) Photodeposition of silver can result in metal-enhanced fluorescence. Appl Spectrosc 57:526–531
24. Geddes CD, Parfenov A, Roll D, Gryczynski I, Malicka J, Lakowicz JR (2003) Silver fractal-like structures for metal-enhanced fluorescence: enhanced fluorescence intensities and increased probe photostabilities. J Fluoresc 13:267–276
25. Geddes CD, Parfenov A, Roll D, Uddin MJ, Lakowicz JR (2003) Fluorescence spectral properties of indocyanine green on a roughened platinum electrode: metal-enhanced fluorescence. J Fluoresc 13:453–457
26. Gersten JI, Nitzan A (1985) Photophysics and photochemistry near surfaces and small particles. Surf Sci 158:165–189
27. Graf C, Vossen DLJ, Imhof A, van Blaaderen A (2003) A general method to coat colloidal particles with silica. Langmuir 19:6693–6700
28. Gryczynski I, Malicka J, Gryczynski Z, Geddes CD, Lakowicz JR (2002) The CFS engineers the intrinsic radiative decay rate of low quantum yield fluorophores. J Fluoresc 12:11–13
29. Jensen TR, Malinsky MD, Haynes CL, Van Duyne P (2000) Nanosphere lithography: tunable localized surface plasmon resonance spectra of silver nanoparticles. J Phys Chem 104:10549–10556
30. Jin R, Cao Y, Mirkin CA, Kelly KL, Schatz GC, Zheng JG (2001) Photoinduced conversion of silver nanospheres to nanoprisms. Science 294:1901–1903
31. Jin R, Cao YC, Hao E, Metraux GS, Schatz GC, Mirkin CA (2003) Controlling anisotropic nanoparticle growth through plasmon excitation. Nature 425:487–490
32. Kaiser RJ, MacKellar SL, Vinayak RS, Sanders JZ, Saavedra RA, Hood LE (1989) Specific-primer-directed DNA sequencing using automated fluorescence detection. Nucleic Acids Res 17:6087–6102

33. Kummerlen J, Leitner A, Brunner H, Aussenegg FR, Wokaun A (1993) Enhanced dye fluorescence over silver island films: analysis of the distance dependence. Mol Phys 80:1031–1046
34. Lakowicz JR (1999) Principles of fluorescence spectroscopy. Kluwer Academic/Plenum Publishers, New York
35. Lakowicz JR (2001) Radiative decay engineering: biophysical and biomedical applications. Anal Biochem 298:1–24
36. Lakowicz JR, Gryczynski I, Malicka J, Gryczynski Z, Geddes CD (2002) Enhanced and localized multiphoton excited fluorescence near metallic silver islands: metallic islands can increase probe photostability. J Fluoresc 12:299–302
37. Lakowicz JR, Gryczynski I, Shen Y, Malicka J, Gryczynski Z (2001) Intensified fluorescence. Photonics Spectra 35:96–104
38. Lakowicz JR, Malicka J, Gryczynski I (2003) Increased intensities of YOYO-1-labeled DNA oligomers near silver particles. Photochem Photobiol 77:604–607
39. Lakowicz JR, Shen B, Gryczynski Z, D'Auria S, Gryczynski I (2001) Intrinsic fluorescence from DNA can be enhanced by metallic particles. Biochem Biophys Res Commun 286:875–879
40. Lakowicz JR, Shen Y, D'Auria S, Malicka J, Fang J, Gryczynski Z, Gryczynski I (2002) Radiative decay engineering. 2. Effects of Silver Island films on fluorescence intensity, lifetimes, and resonance energy transfer. Anal Biochem 301:261–277
41. Lee TH, Gonzalez JI, Dickson RM (2002) Strongly enhanced field-dependent single-molecule electroluminescence. Proc Natl Acad Sci USA 99:10272–10275
42. LI-COR (2006) Odyssey infrared imaging system: users guide version 1.2. LI-COR Inc.
43. Lukomska J, Malicka J, Gryczynski I, Lakowicz JR (2004) Fluorescence enhancements on silver colloid coated surfaces. J Fluoresc 14:417–423
44. Malicka J, Gryczynski I, Fang J, Lakowicz JR (2003) Fluorescence spectral properties of cyanine dye-labeled DNA oligomers on surfaces coated with silver particles. Anal Biochem 317:136–146
45. Malicka J, Gryczynski I, Geddes CD, Lakowicz JR (2003) Metal-enhanced emission from indocyanine green: a new approach to in vivo imaging. J Biomed Opt 8:472–478
46. Malicka J, Gryczynski I, Gryczynski Z, Lakowicz JR (2003) Effects of fluorophore-to-silver distance on the emission of cyanine-dye-labeled oligonucleotides. Anal Biochem 315:57–66
47. Malicka J, Gryczynski I, Lakowicz JR (2003) Enhanced emission of highly labeled DNA oligomers near silver metallic surfaces. Anal Chem 75:4408–4414
48. Malicka J, Gryczynski I, Maliwal BP, Fang J, Lakowicz JR (2003) Fluorescence spectral properties of cyanine dye labeled DNA near metallic silver particles. Biopolymers 72:96–104
49. Matveeva EG, Gryczynski I, Malicka J, Gryczynski Z, Goldys E, Howe J, Berndt KW, Lakowicz JR (2005) Plastic versus glass support for an immunoassay on metal-coated surfaces in optically dense samples utilizing directional surface plasmon-coupled emission. J Fluoresc 15:865–871
50. Middendorf LR, Amen J, Bruce RC, Draney D, DeGraff D, Gewecke J, Grone DL, Humphrey P, Little G, Lugade A, Narayanan N, Oommen A, Osterman H, Peterson R, Rada J, Raghavachari R, Roemer SC (1998) Near-infrared fluorescence instrumentation for DNA analysis. In: Daehne S (ed) Near-infrared dyes for high technology applications. Kluwer Academic Publishers, The Netherlands, pp 21–54
51. Middendorf LR, Bruce JC, Bruce RC, Eckles RD, Grone DL, Roemer SC, Sloniker GD, Steffens DL, Sutter SL, Brumbaugh JA et al (1992) Continuous, on-line DNA sequencing using a versatile infrared laser scanner/electrophoresis apparatus. Electrophoresis 13:487–494
52. Moerner WE, Orrit M (1999) Illuminating single molecules in condensed matter. Science 283:1670–1676
53. Muniz-Miranda M (2002) SERS effect from silver photoreduced on to silica colloidal nanoparticles. J Ram Spec 33:295–297

54. Ni F, Cotton TM (1986) Chemical procedure for preparing surface-enhanced Raman scattering active silver films. Anal Chem 58:3159–3163
55. Olive DM (2004) Quantitative methods for the analysis of protein phosphorylation in drug development. Expert Rev Proteomics 1:327–341
56. Pastoriza-Santos II, Liz-Marzan LM (2000) Reduction of silver nanoparticles in DMF. Formation of monolayers and stable colloids. Pure Appl Chem 72:83–90
57. Peyser LA, Vinson AE, Bartko AP, Dickson RM (2001) Photoactivated fluorescence from individual silver nanoclusters. Science 291:103–106
58. Prikulis J, Murty KV, Olin H, Kall M (2003) Large-area topography analysis and near-field Raman spectroscopy using bent fibre probes. J Microsc 210:269–273
59. Pugh VJ, Szmacinski H, Moore WE, Geddes CD, Lakowicz JR (2003) Submicrometer spatial resolution of metal-enhanced fluorescence. Appl Spectrosc 57:1592–1598
60. Randolph JB, Waggoner AS (1997) Stability, specificity and fluorescence brightness of multiply-labeled fluorescent DNA probes. Nucleic Acids Res 25:2923–2929
61. Schena M, Shalon D, Davis RW, Brown PO (1995) Quantitative monitoring of gene expression patterns with a complementary DNA microarray. Science 270:467–470
62. Schultz S, Smith DR, Mock JJ, Schultz DA (2000) Single-target molecule detection with nonbleaching multicolor optical immunolabels. Proc Natl Acad Sci USA 97:996–1001
63. Seydel C (2003) Quantum dots get wet. Science 300:80–81
64. Shah K, Weissleder R (2005) Molecular optical imaging: applications leading to the development of present day therapeutics. NeuroRx 2:215–225
65. Shealy DB, Lipowska M, Lipowski J, Narayanan N, Sutter S, Strekowski L, Patonay G (1995) Synthesis, chromatographic separation, and characterization of near-infrared-labeled DNA oligomers for use in DNA sequencing. Anal Chem 67:247–251
66. Shipway AN, Katz E, Willner I (2000) Nanoparticle arrays on surfaces for electronic, optical, and sensor applications. Chemphyschem 1:18–52
67. Silvert P, Herrera-Urbina R, Duvauchelle N, Vijayakrishnan V, Elhsissen KT (1996) Preparation of colloidal silver dispersions by the polyol process. Part 1—Synthesis and characterization. J Mater Chem 6:573–577
68. Silvert P, Herrera-Urbina R, Tekaia-Elhsissena K (1997) Preparation of colloidal silver dispersions by the polyol process. Part 1—Mechanism of particle formation. J Mater Chem 7:293–299
69. Smith LM, Sanders JZ, Kaiser RJ, Hughes P, Dodd C, Connell CR, Heiner C, Kent SB, Hood LE (1986) Fluorescence detection in automated DNA sequence analysis. Nature 321:674–679
70. Sokolov K, Chumanov G, Cotton TM (1998) Enhancement of molecular fluorescence near the surface of colloidal metal films. Anal Chem 70:3898–3905
71. Trevisiol E, Renard A, Defrancq E, Lhomme J (2000) Fluorescent labelling of oligodeoxyribonucleotides by the oxyamino-aldehyde coupling reaction. Nucleosides Nucleotides Nucleic Acids 19:1427–1439
72. Velikov KP, Zegers GE, van Blaaderen A (2003) Synthesis and characterization of large colloidal silver particles. Langmuir 19:1384–1389
73. Yu H, Chao J, Patek D, Mujumdar R, Mujumdar S, Waggoner AS (1994) Cyanine dye dUTP analogs for enzymatic labeling of DNA probes. Nucleic Acids Res 22:3226–3232
74. Zhang J, Matveeva E, Gryczynski I, Leonenko Z, Lakowicz JR (2005) Metal-enhanced fluoroimmunoassay on a silver film by vapor deposition. J Phys Chem B 109:7969–7975
75. Zynio SA, Samoylov AV, Surovtseva ER, Mirsky VM, Shirshov YM (2002) Bimetallic layers increase sensitivity of affinity sensors based on surface plasmon resonance. Sensors 2:62–70

Principal Component Global Analysis of Series of Fluorescence Spectra

Wajih Al-Soufi, Mercedes Novo, Manuel Mosquera, and Flor Rodríguez-Prieto

Abstract The analysis of series of molecular fluorescence or absorption spectra forms an integral part of innumerable investigations on the physicochemical properties of chemical or biological systems.

In many typical complex applications, such as photochemical systems with multiple interconversion processes in the ground and in the excited states or biochemical ligand binding studies with several possible binding sites, the number of species contributing to the spectral variation is not known a priori. Moreover, in the frequent case of strongly overlapping spectra of the species, their number cannot be estimated by simple inspection of the experimental spectra.

Principal Component Global Analysis (PCGA) is reviewed as an efficient and reliable way to determine how many species contribute to the observed spectral variation, to set up the correct mechanism and to estimate the values of the corresponding model-parameters. PCGA is applied to examples of host-guest interactions with two and three components and to systems showing complex ground and excited-state proton-transfer reactions with corresponding one and two acid-base equilibria.

1 Introduction

Analysis of series of molecular fluorescence or absorption spectra forms an integral part of innumerable investigations on the physicochemical properties of chemical or biological systems. The detailed analysis of the variation observed in these spectra by changing systematically an externally controllable variable such as pH, concentration, temperature, or time reveals the nature of the contributing species

W. Al-Soufi (✉)
Departamento de Química Física, Facultade de Ciencias, Universidade de Santiago de Compostela, E-27002 Lugo, Spain
e-mail: wajih.al-soufi@usc.es

C.D. Geddes (ed.), *Reviews in Fluorescence 2009*, Reviews in Fluorescence 2009,
DOI 10.1007/978-1-4419-9672-5_2,

and leads to a full understanding of the underlying reaction mechanisms and the corresponding equilibrium or kinetic constants.

In many typical complex applications, such as photochemical systems with multiple interconversion processes in the ground and in the excited states or biochemical ligand binding studies with several possible binding sites, the number of species contributing to the spectral variation is not known a priori. Moreover, in the frequent case of strongly overlapping spectra of the species, their number cannot be estimated by simple inspection of the experimental spectra.

Two main questions arise in these studies. First: how many species contribute to the observed spectral variation? And second: which is the correct mechanism and what are the values of the corresponding model parameters that best reproduce the spectra?

Classically both problems are tackled together making hypothesis on both the *number* of species and the underlying *model* and then testing the two hypothesis in one step, fitting the model to the experimental data at a single (or few) wavelengths with least-squares algorithms. On the basis of the quality of the resulting fit, the tested hypothesis is accepted or rejected. This can be a tedious task, since the answer to how many species contribute relies itself on the validity of the proposed model. Moreover, a model which works well at single wavelengths may fail to reproduce the full spectra. A correct model should describe the observed spectra in the whole accessible spectral interval with the same set of parameter values.

The answers to these questions are not trivial and in spite of innumerable publications relying on the interpretation of series of fluorescence spectra, there is no broad acceptance of a systematic solution.

In this contribution, we review principal component global analysis (PCGA) as an efficient and reliable way to get answers to these questions and present briefly the theory and typical applications.

PCGA separates the determination of the *number* of contributing components from the test and fit of different *models* to the data. In a first step, the series of experimental spectra is reduced by principal component analysis (PCA) to the *minimal* number of "abstract spectra" (eigenvectors), which reproduce all the systematic change in the original spectra. This yields the number of contributing components. In a second step, this information is used in a global analysis to test different model functions and to find estimates for the model parameters and the component spectra. In this step, the model functions are globally fitted to all spectra at *all* available wavelengths simultaneously. This approach allows one to estimate the emission spectra of contributing species from strongly overlapping spectra, which would be otherwise very difficult to determine if they cannot be experimentally isolated. Moreover, from the component spectra, important information can be derived, for example, the relative quantum efficiencies comparing spectral areas.

In the chemometric literature, many related analytical methods have been described, such as "window factor analysis" (WFA) [16], "self modeling curve resolution" [10, 25, 26, 28, 29], "second order global analysis" [9], and others. A comprehensive tutorial was published only recently on the analysis of reaction kinetic data [22].

In physical chemistry, the interest is focused mostly on the validation of different complex mechanisms which are generally not accessible to standardized solutions. Systems such as the lactim–lactam phototautomerization presented in Sect. 5.3 with multiple coupled equilibria and a complex dependence on the solvent proton concentration need careful and stepwise identification of the contributing species and a systematic resolution and validation of the underlying mechanism, combining information from different spectroscopic techniques and comparing results obtained in different solvents.

2 Example System

We illustrate the application of PCGA with typical physicochemical systems which depend on different external parameters: two examples of host–guest interactions with two and three components and two systems showing complex ground and excited-state proton-transfer reactions with corresponding one and two acid–base equilibria. The theory is accompanied with the results for the two component host–guest association equilibrium between the fluorescent dye Pyronine Y as guest and β-cyclodextrin as host. The results for the other example systems are explained in Chap. 5. More examples can be found in the literature [5, 15, 24].

2.1 Host–Guest Interaction Between Pyronine Y and β-Cyclodextrin

Small guests such as the fluorescent dye Pyronine Y (PY) can form inclusion complexes with β-cyclodextrin (CD), a toroidally shaped polysaccharide with a highly hydrophobic central cavity. Complexation induces significant changes in the physicochemical properties of PY and the stoichiometry of the complexes can be determined from titration experiments measuring series of absorption and emission spectra of the dye in the presence of different concentrations of CD. From these spectra the number of different species present in solution can be determined with PCA. GA confirms then the stoichiometry and yields the association equilibrium constants and the pure spectra of the species [23].

Figure 1 shows a series of $n_s = 22$ emission spectra from titration experiments. Red shift of the spectra together with a decrease in fluorescence intensity is observed as the concentration of CD is increased. The number of emitting species is not obvious from these data and has to be determined by further analysis such as PCGA as presented in the following section.

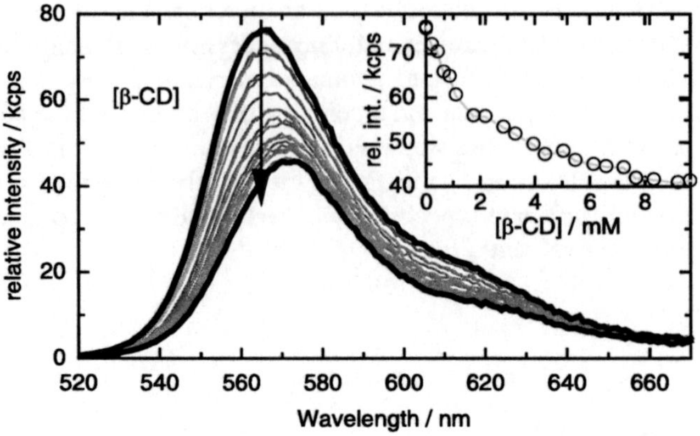

Fig. 1 Corrected fluorescence emission spectra of PY in the presence of different concentrations of β-CD in the range from 1×10^{-5} to 9.7×10^{-3} mol dm^{-3} ($\lambda_{exc} = 515$ nm, $n_w = 150$, $n_s = 22$). *Inset*: Intensity profile vs. β-CD concentration at 565 nm

3 Determination of the Number of Fluorescent Species by Principal Component Analysis

Principal component analysis (PCA) and the related factor analysis are well-known, validated statistical techniques, widely used in different areas of experimental science [12, 17]. Their application for dimensionality reduction, classification, regression, etc. is described extensively in the chemometric literature and PCA forms part of many statistical software packages. However, in spite of its widespread use in analytical chemistry, PCA has been rarely applied in physicochemical studies. This is especially surprising as the determination of the number of fluorescent species from series of emission spectra with PCA is straightforward and very efficient. The procedure is easily programmed in modern mathematical packages and can form part of routine data analysis.

We will give here a concise introduction to the key equations with the data from Sect. 2.1 as example.

PCA can be applied to any series of spectra which are *linear* combinations of a set of *constant* "component spectra" with relative contributions which depend on some "external" variable such as pH, concentration of a ligand or reaction partner, temperature, or time. In this case, the *i*th experimental spectrum, $\mathbf{y}_i$, is the sum of n_c component spectra $\mathbf{e}_j$, weighted with contributions (concentrations) d_{ij} plus some noise $\boldsymbol{\sigma}_i$:

$$\mathbf{y}_i = d_{i1}\mathbf{e}_1 + d_{i2}\mathbf{e}_2 + \cdots + d_{in_c}\mathbf{e}_{n_c} + \boldsymbol{\sigma}_i = \hat{\mathbf{y}}_i + \boldsymbol{\sigma}_i. \tag{1}$$

The spectral data can be arranged in an experimental data matrix $\mathbf{Y}$ with the rows containing the spectra and the columns reproducing the variation of the emission at one wavelength with the external variable. In matrix notation, the matrix of the theoretical spectra $\hat{\mathbf{Y}}$ is the product of the matrix of component spectra $\mathbf{E}$ with the matrix of concentrations $\mathbf{D}$ which depend on some model and on the "external" variable:

$$\underset{n_s \times n_w}{\hat{\mathbf{Y}}} = \begin{pmatrix} \hat{y}_{1,1} & \cdots & \hat{y}_{1,n_w} \\ \cdots & \cdots & \cdots \\ \hat{y}_{n_s,1} & \cdots & \hat{y}_{n_s,n_w} \end{pmatrix} = \underset{n_s \times n_c}{\mathbf{D}} \cdot \underset{n_c \times n_w}{\mathbf{E}}$$

$$= \begin{pmatrix} d_{1,1} & \vdots & d_{1,n_c} \\ \vdots & \vdots & \vdots \\ d_{n_s,1} & \vdots & d_{n_s,n_c} \end{pmatrix} \cdot \begin{pmatrix} e_{1,1} & \cdots & e_{1,n_w} \\ \cdots & \cdots & \cdots \\ e_{n_c,1} & \cdots & e_{n_c,n_w} \end{pmatrix}. \tag{2}$$

Here n_s denotes the number of different spectra (samples), each measured at n_w discrete wavelengths. Both n_w and n_s should be greater than n_c and we suppose for simplicity that $n_w > n_s$, since this is the common case.

Example PY + β-CD: Fig. 1 shows the series of spectra which form $\mathbf{Y}$ with $n_w = 150$ and $n_s = 22$.

If the component spectra $\mathbf{E}$ were known, $\mathbf{D}$ could be determined by linear regressions, $\mathbf{D} = \mathbf{Y} \cdot \mathbf{E}^+$, and vice versa. But in many cases, neither $\mathbf{E}$ nor $\mathbf{D}$ and not even the number of contributing species n_c are known and estimates have to be found for all of them.

Can we determine the number of component spectra n_c without knowing the component spectra $\mathbf{e}_j$ themselves?

Yes, taking into account that the experimental spectra $\mathbf{Y}$ are linear combinations of the n_c distinct (linearly independent) component spectra $\mathbf{e}_j$ (plus noise). The spectra $\mathbf{e}_j$ span a space of *dimension* n_c and each of the experimental spectra $\mathbf{y}_i$ is defined by a specific set of contributions d_{ij} which represent "coordinates" within this n_c-dimensional space. In matrix terms, we can say that $\text{rank}(\hat{\mathbf{Y}}) = n_c$. Thus, the minimal number of (any) linearly independent spectra needed to reproduce the systematic variation in the series of experimental spectra gives directly the number of observable components n_c. As we do not know the component spectra $\mathbf{E}$ themselves, we express $\mathbf{Y}$ by some other "abstract" spectra $\mathbf{V}$ with coefficients $\mathbf{Z}$:

$$\mathbf{y}_i = z_{i1}\mathbf{v}_1 + z_{i2}\mathbf{v}_2 + \cdots + z_{in_s}\mathbf{v}_{n_s}. \tag{3}$$

In PCA, the vectors $\mathbf{v}_l$ are the (left) *eigenvectors* of the quadratic and symmetric matrix $\mathbf{Y}^T\mathbf{Y}$, which has the same rank as $\mathbf{Y}$. The *eigenvectors* $\mathbf{v}_l$ are solutions of the eigenvalue equation (4) and are easily computed, for example, in Matlab (The MathWorks, MA, USA) with one line of code: `[V, L] = eig(Y'Y); V = V';`

$$\underset{n_s \times n_w}{\mathbf{V}} \; \underset{n_w \times n_w}{\mathbf{Y}^T\mathbf{Y}} = \underset{n_s \times n_s}{\Lambda} \; \underset{n_s \times n_w}{\mathbf{V}} \;. \tag{4}$$

The eigenvectors **V** reproduce all the variation in the experimental data **Y**, including the noise.

$$\underset{n_s \times n_w}{\mathbf{Y}} = \underset{n_s \times n_s}{\mathbf{Z}} \cdot \underset{n_s \times n_w}{\mathbf{V}} . \tag{5}$$

The product (z_{il} $\mathbf{v}_l$) defines the lth *principal component* (PC) of the ith experimental spectrum $\mathbf{y}_i$ (see (3)).

The eigenvectors $\mathbf{v}_l$ are orthogonal, that is $\mathbf{V}^\mathrm{T} = \mathbf{V}^{-1}$, so that the matrix of coefficients **Z** is easily calculated:

$$\mathbf{Z} = \mathbf{Y} \cdot \mathbf{V}^{-1} = \mathbf{Y} \cdot \mathbf{V}^\mathrm{T}. \tag{6}$$

The diagonal matrix **Λ** contains the eigenvalues λ_l, sorted so that the first has the highest value. The eigenvalues indicate directly how much of the *total* variance $\mathrm{SS}(\mathbf{Y}) = \sum_{i=1}^{n_s} \sum_{j=1}^{n_w} y_{ij}^2 = \sum_{l=1}^{n_s} \lambda_l$ in **Y** is contributed by the lth PC:

$$\lambda_l = \sum_{i=1}^{n_s} \sum_{j=1}^{n_w} (z_{il}\, v_{lj})^2 = \sum_{i=1}^{n_s} z_{il}^2. \tag{7}$$

In the absence of noise in **Y**, the first n_c eigenvectors would reproduce all the variation in **Y**. The eigenvalues corresponding to the remaining ($n_s - n_c$) eigenvectors would be zero. In practice, however, each of the n_s experimental spectra will have a unique, linearly independent noise pattern, and for a full reproduction of **Y**, n_s eigenvectors are needed, and in general rank(**Y**) $= n_s > n_c$. Nevertheless, n_c can be estimated by making the following assumption: The systematic (correlated) variation in the spectra due to n_c chemical components forms the main part of the total variation in **Y**, clearly separated from the (uncorrelated) variation due to noise or measurement imperfections. Thus, this "structural" variance is assumed to be reproduced by the first n_c PCs with the highest eigenvalues, whereas the remaining PCs represent the "residual" variance due to noise. The first n_c structural PCs will be represented by the truncated matrices $\tilde{\mathbf{V}}(n_c \times n_w)$, $\tilde{\mathbf{Z}}(n_s \times n_c)$, and $\tilde{\mathbf{\Lambda}}(n_c \times n_c)$. A superscript k at these matrices (e.g., $\tilde{\mathbf{V}}^k$) indicates that they are further cut down to only the first k PCs. The two sets of spectra **E** and **V** are then related by a "rotation" matrix **P**

$$\underset{n_c \times n_w}{\mathbf{E}} = \underset{n_c \times n_c}{\mathbf{P}} \cdot \underset{n_c \times n_w}{\tilde{\mathbf{V}}} \qquad \underset{n_s \times n_c}{\tilde{\mathbf{Z}}} = \underset{n_s \times n_c}{\mathbf{D}} \cdot \underset{n_c \times n_c}{\mathbf{P}} . \tag{8}$$

The following plots are used in order to discriminate between structural and residual PCs:

Eigenvector diagrams: Eigenvectors **V** and (normalized) coefficients $\mathbf{Z}^* = \mathbf{\Lambda}^{-1/2}\mathbf{Z}^\mathrm{T}$, which represent structural variation, can in general be well distinguished from those eigenvectors reproducing mainly noise or measurement imperfections and which change randomly with the variable.

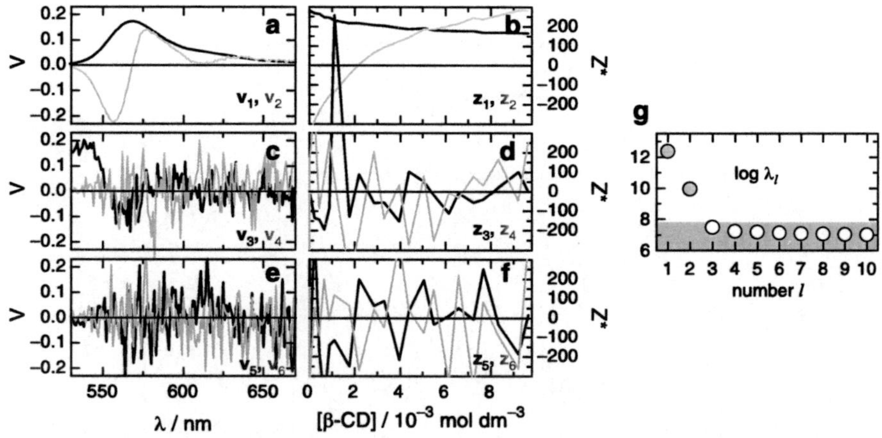

Fig. 2 Eigenvectors **V**, (normalized) coefficients $\mathbf{Z^*} = \mathbf{\Lambda}^{-1/2}\mathbf{Z}$, and eigenvalues of the emission spectra **Y** of the system PY + β-CD ($n_c = 2$) given in Fig. 1. **(a, b)** The structural components $\mathbf{v}_1$, $\mathbf{z}_1$ and $\mathbf{v}_2$, $\mathbf{z}_2$; (c)–(f) four residual components $\mathbf{v}_3$, $\mathbf{z}_3$ to $\mathbf{v}_6$, $\mathbf{z}_6$. Higher components are similar and not shown. (g) The first ten eigenvalues λ_1–λ_{10}. The first two eigenvalues (*gray*) belong to structural components, the other to residual components

Example PY + β-CD: The first two eigenvectors $\mathbf{v}_1$ and $\mathbf{v}_2$ shown in Fig. 2 vary systematically with wavelength. Comparing $\mathbf{v}_1$ with the spectra in Fig. 1, it is easily seen that it represents a mean spectrum reproducing the overall shape of the spectra. Its coefficient $\mathbf{z}_1$ drops with increasing host concentration following the decrease in overall emission. The second eigenvector $\mathbf{v}_2$ has just the form necessary to introduce the displacement of the spectrum to longer wavelength as its coefficient $\mathbf{z}_1$ increases from negative to positive values. These two first PCs reproduce already 96.4% of the total variance, the rest being mostly due to noise, as can be deduced from the uncorrelated variation of the following eigenvectors and coefficients $\mathbf{v}_3$, $\mathbf{z}_3$ to $\mathbf{v}_6$, $\mathbf{z}_6$.

Log-Eigenvalue Diagram: In the logarithmic plot of the eigenvalues λ_l vs. l, the residual eigenvalues define the level of "noise" variance. The number of PCs with an eigenvalue above this level give an upper limit for the number of structural components.

Example PY + β-CD: The first two of the eigenvalues shown in Fig. 2 are much higher than the following ones and represent the structural variance in the data reproduced by the first two PCs.

Residual Spectra: In order to detect systematic variation in **Y** not reproduced by the first k components, one can inspect residuals $\mathbf{R}^k$ between the experimental spectra **Y** and estimated spectra $\hat{\mathbf{Y}}^k$, which are calculated by including only the first k PCs. Residuals calculated with less than n_c PCs ($k < n_c$) show systematic deviations.

$$\mathbf{R}^k = \mathbf{Y} - \hat{\mathbf{Y}}^k = \mathbf{Y} - \tilde{\mathbf{Z}}^k \tilde{\mathbf{V}}^k. \tag{9}$$

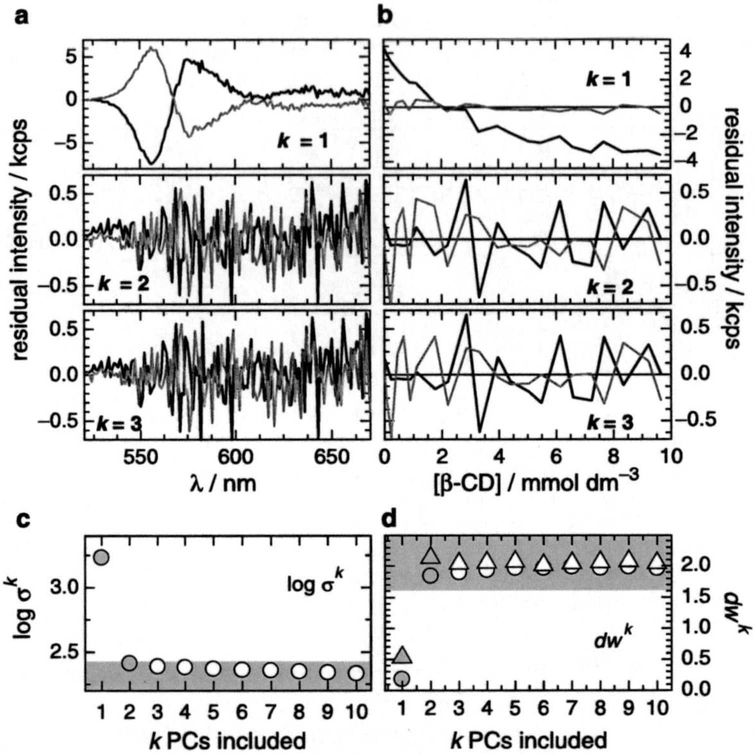

Fig. 3 Determination of the number of structural components from PCA of the emission spectra of Fig. 1. (**a**) Residual spectra $\mathbf{R}^k$ at two concentrations of β-CD: 0 mM (*black*) and 9.7 mM (*gray*) for $k = 1, 2$, and 3. (**b**) Residual profiles $(\mathbf{R}^k)^{\mathrm{T}}$ at two wavelengths 580 nm (*black*) and 610 nm (*gray*) for $k = 1, 2$, and 3. (**c**) Mean residual error σ^k vs. the number of included PCs k. (**d**) Mean Durbin–Watson Test values dw^k of residual spectra r_{ij} (*circles*) and of residual profiles r_{ji} (*triangles*) vs. the number of included PCs k. Values within the *gray shaded area* indicate uncorrelated residuals

The representation of these residual spectra r_{ij}^k vs. wavelength index j with successively increasing number of components k yields important information, revealing at which wavelengths and at which values of the external variable systematic deviations can be observed. Although variations in the PCs have no direct physical significance, typical influences of Raman bands, change-over wavelength of light sources, or strong baseline instabilities may be identified readily and be taken into account.

Example PY + β-CD: The residual spectra (Fig. 3) show random variation after inclusion of two components ($k = 2$).

Mean Residual Error Diagram: The mean residual error σ^k is estimated from the weighted root mean square of the residuals r_{ij}^k (9). Once all structural PCs are included, that is for $k \geq n_c$, the mean residual error reaches some constant level of instrumental noise.

$$\sigma^k = \left(\frac{\sum_{i=1}^{n_s}\sum_{j=1}^{n_w}\left(y_{ij}-\hat{y}_{ij}^k\right)^2}{(n_s-k)(n_w-k)}\right)^{1/2} = \left(\frac{\sum_{l=k+1}^{n_s}\lambda_l}{(n_s-k)(n_w-k)}\right)^{1/2}. \tag{10}$$

Example PY + β-CD: The inclusion of two components reduces the value of the mean residual error down to the noise level (Fig. 3c).

Durbin–Watson Test of the Residuals: The plot of mean Durbin–Watson Test values dw^k measures the correlation between consecutive values of the residuals r_{ij}^k. The residuals become uncorrelated for $k \geq n_c$. Uncorrelated residuals are indicated by the values of $dw^k > 1.5$–1.7 depending on n_w. [7, 8].

$$dw_i^k = \frac{\sum_{j=2}^{n_w}\left(r_{i,j}^k - r_{i,j-1}^k\right)^2}{\sum_{j=1}^{n_w}\left(r_{i,j}^k\right)^2}. \tag{11}$$

Example PY + β-CD: The mean Durbin–Watson Test values of the residual spectra (black circles in Fig. 3d) exceed the critical value of about 1.7 with the inclusion of the second component.

Residual Profiles: Series of experimental absorption or emission spectra are subject not only to random noise but also to erroneous variations of instrumental and experimental conditions such as baseline, detector sensitivity, source lamp intensity, or background counts. These may introduce systematic spectral variations which will appear in PCA as "spurious" components which are not easily distinguishable from structural components corresponding to chemical species. Structural and spurious components can be differentiated by analyzing the variation in the residuals as a function of the external variable, representing the *residual profiles* $(\mathbf{R}^k)^{\mathrm{T}}$. The contribution of the spurious components most probably does not follow in any systematic way the change in the external variable and should lead to random variation in the residual profiles. This can also be detected effectively in the Durbin–Watson Test values of the residual profiles.

Example PY + β-CD: In this system, the residual profiles show uncorrelated noise with two or more included PCs which is confirmed by their mean Durbin–Watson Test values (Fig. 3b and white circles in Fig. 3d). An example of spurious components can be seen in the case of HBI in Sect. 5.2. We conclude here that PCA of the emission spectra of PY in the presence of β-CD indicate that only two components are necessary to explain the observed systematic variation.

4 Model Validation and Parameter Estimation by Global Analysis

Nonlinear regression is a well established and widely applied technique for the estimation of model parameters. The simultaneous analysis of multiple series of experimental data sharing common model parameters, known as global analysis

(GA), had been introduced for fluorescence lifetime and anisotropy analysis [3, 6, 13, 27] and is extensively used in many areas of physical chemistry. Both specialized and general purpose programs for global nonlinear analysis can be found. Nevertheless, the great benefit of global analysis not only for parameter estimation but also for the much more valuable testing and validation of models is still strongly underestimated and has not been established as a standard tool in many laboratories. We will give in the following a short introduction to the theory of global nonlinear analysis of series of spectra and apply it to the example systems.

The basic procedure is straightforward. It comprises three steps: (a) postulation of a model defining the matrix of contributions (concentrations) **D** as a function of parameters and of the external variable; (b) estimation of the parameters in **D** and of the values of the component spectra **E** by iterative nonlinear least-squares fit of $\hat{\mathbf{Y}} = \mathbf{D} \cdot \mathbf{E}$ to the experimental data matrix **Y**; and (c) test of the adequacy of the used model on the basis of the goodness of the fit.

The elements of the matrix of contributions d_{il} are n_c (nonlinear) model functions h_l of the external variable x_i (pH, initial concentrations of some reactant, time, etc.) and depends on model parameters $a, b, \ldots$ such as equilibrium or rate constants:

$$\hat{d}_{il} = h_l(a, b, \ldots; x_i), \quad i = 1 \ldots n_s, \quad l = 1 \ldots n_c. \tag{12}$$

Nonlinear minimization then applies some iterative algorithm such as that of Marquardt [4, 21] in order to find values of the $a, b, \ldots$ and $\hat{\mathbf{E}}$ which minimize the sum of squares of the residuals χ^2 between **Y** and $\hat{\mathbf{D}}\hat{\mathbf{E}}$:

$$\chi^2 = \sum_{i=1}^{n_s} \sum_{j=1}^{n_w} \left((\mathbf{Y} - \hat{\mathbf{D}}\hat{\mathbf{E}})_{ij} \right)^2 = \sum_{i=1}^{n_s} \sum_{j=1}^{n_w} \left((\mathbf{Y} - \hat{\mathbf{D}}(\hat{\mathbf{D}}^{+}\mathbf{Y}))_{ij} \right)^2. \tag{13}$$

In this procedure, the model parameters $a, b, \ldots$ are estimated by iterative nonlinear minimization, whereas the $(n_c \cdot n_w)$ linear component spectra in $\hat{\mathbf{E}}$ are determined in each iteration directly by linear regression $\hat{\mathbf{E}} = \hat{\mathbf{D}}^{+}\mathbf{Y}$, that is by multiplying **Y** with the pseudo inverse of $\hat{\mathbf{D}}$ calculated with the given values of $a, b, \ldots$ in each iteration [14]. This way the number of unknown (linear) parameters is tremendously reduced.

In the case of a very big experimental matrix **Y**, it can be useful to reduce the size of the matrices from $(n_s \times n_w)$ to $(n_s \times n_c)$ using the information from PCA: $\hat{\mathbf{Y}} = \tilde{\mathbf{Z}}\tilde{\mathbf{V}}$ (5) and $\hat{\mathbf{E}} = \hat{\mathbf{P}}\tilde{\mathbf{V}}$ (8). The small rotation matrix $\hat{\mathbf{P}}$ can again be estimated by linear regression $\hat{\mathbf{P}} = \hat{\mathbf{D}}^{+}\tilde{\mathbf{Z}}$.

$$\chi^2 = \sum_{i=1}^{n_s} \sum_{j=1}^{n_w} \left((\tilde{\mathbf{Z}}\tilde{\mathbf{V}} - \hat{\mathbf{D}}\hat{\mathbf{P}}\tilde{\mathbf{V}})_{ij} \right)^2 = \sum_{i=1}^{n_s} \sum_{l=1}^{n_c} \left((\tilde{\mathbf{Z}} - \hat{\mathbf{D}}(\hat{\mathbf{D}}^{+}\tilde{\mathbf{Z}}))_{il} \right)^2. \tag{14}$$

The values of the parameters $a, b, \ldots$ and the component spectra $\hat{\mathbf{E}}$ are the result of the global analysis of all the $(n_s \times n_w)$ measurements available in the experiment. This has several important consequences: (1) The much higher number of data points leads in general to much smaller statistical errors in the parameters as compared to the standard analysis at one (or few) wavelength with only ($n_s \times 1$) data points. (2) Correlations between the parameters e_{ij}, a, b,... are reduced [11]. (3) A model and a set of parameter values which may be valid at a single wavelength or in some small wavelength range may fail to reproduce the data at the full wavelength interval. Checking a model against the full spectra is a much more demanding test and gives much more confidence in the validity of the model than single wavelength fits. (4) The component spectra obtained from this analysis constitute themselves an important result, since usually some of them cannot be determined directly.

Example PY + β-CD: From PCA, we know now that two components are responsible for the variation of the emission spectra of the PY with β-CD. If a 1:1 complexation equilibrium in the ground state is proposed, the two components are identified as free pyronine ($\mathbf{e}_1$) and its complex with β-CD ($\mathbf{e}_2$). Under the conditions of constant PY concentration and excess of β-CD, the following equations describe the relative concentrations of free pyronine (d_{i1}) and its complex (d_{i2}) as a function of the β-CD concentration and the association equilibrium constant K:

$$\hat{d}_{i1} = \frac{1}{1 + K[CD]_i}, \quad \hat{d}_{i2} = \frac{K[CD]_i}{1 + K[CD]_i}. \tag{15}$$

The results of the fit of (15) to the emission spectra of Fig. 1 are given in Table 1. The individual fits at three wavelengths give three values of K which span a wide interval with huge uncertainties. The global analysis at the same three wavelengths yields a common value of K with smaller uncertainty which assures a consistent fit at the three wavelengths. Finally, fitting all wavelengths simultaneously following (13), not only the smallest uncertainty in K is obtained but also the whole component spectra $\mathbf{e}_j$ as shown in Fig. 4. The spectrum of the free dye PY is known and coincides perfectly with $\mathbf{e}_1$, but that of the complex PY:β-CD, $\mathbf{e}_2$, cannot be measured directly, since it is not possible to get a concentration of β-CD high enough to have all the PY complexed.

Table 1 Results of nonlinear regression of (15) to the emission spectra of Fig. 1

	Individual analysis			Global analysis		
λ/nm	K/M^{-1}	$e_1/10^4$	$e_2/10^4$	K/M^{-1}	$e_1/10^4$	$e_2/10^4$
555	372 ± 18	5.7 ± 0.3	1.5 ± 0.6		5.7 ± 0.4	1.5 ± 0.6
565	411 ± 35	7.6 ± 0.6	3.3 ± 1.0	399 ± 19	7.6 ± 0.4	3.2 ± 0.7
575	438 ± 61	6.5 ± 0.5	4.0 ± 0.9		6.5 ± 0.4	4.0 ± 0.5
Global analysis (PCGA)						
$K = 397 \pm 4$ M^{-1}						

Individual and global analyses at three wavelengths and global analysis (PCGA) at all wavelengths between 521 and 670 nm ($n_w = 150$). The values of the component spectra $\mathbf{e}_j$ and the contributions $\mathbf{d}_j$ from PCGA are shown in Fig. 4. The standard errors are determined from the variance-covariance matrix calculated as part of the nonlinear minimization [4, 22]

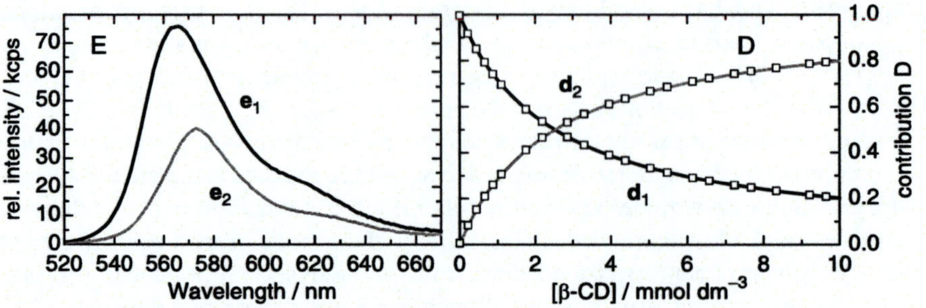

Fig. 4 Component spectra **E** and contributions **D** determined from PCGA of the emission spectra of the system PY + β-CD given in Fig. 1 and the value of K given in Table 1.

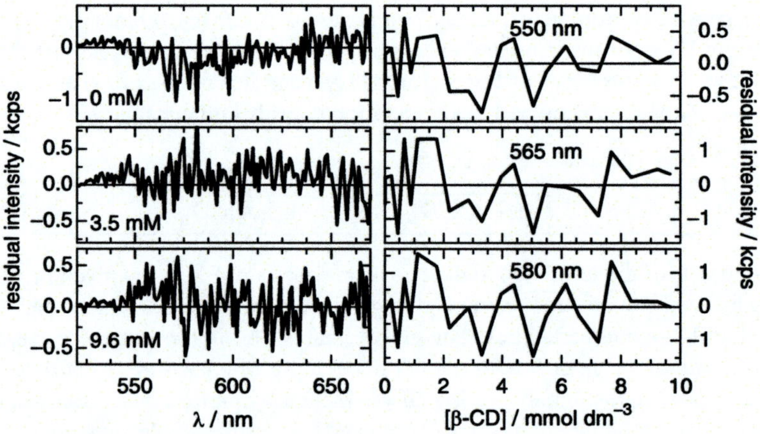

Fig. 5 Examples of the residuals $\mathbf{Y} - \hat{\mathbf{Y}}$ from the global analysis (PCGA) of the emission spectra **Y** of Fig. 1. *Left column*: residual spectra $(\mathbf{Y} - \hat{\mathbf{Y}})$ at three concentrations of β-CD as given in the plots. *Right column*: residual profiles $(\mathbf{Y} - \hat{\mathbf{Y}})^{\mathrm{T}}$ at three wavelengths as given in the plots

The goodness of the fits is checked on the basis of the residuals shown for certain wavelengths and host concentrations in Fig. 5. The randomly distributed residuals indicate good fits and validate the proposed model.

5 Results for the Example Systems

Here, we apply PCGA to some typical physicochemical systems. The results for the host–guest interaction between PY and CD have already been used to illustrate the method. The first example (Sect. 5.1) is that of a host–guest association with 1:1 and 1:2 host:guest stoichiometry and thus with three components. The following system (Sect. 5.2) shows two acid–base equilibria in the ground and the excited states and the third system (Sect. 5.3) a complex dependence of the concentration of different species coupled by excited-state proton-transfer reactions.

5.1 *Host–Guest Association Equilibria Between TNS and a Charged Cyclodextrin Derivative*

The complexation of 2-(*p*-toluidinyl)naphthalene-6-sulfonate (TNS) with the positively charged cyclodextrin, 6-deoxy-6-amino-β-cyclodextrin (CD) involves the simultaneous formation of two different 1:1 complexes, resulting from inclusion of the toluidin moiety or the naphthalenesulfonate moiety in a cyclodextrin unit, and a 1:2 complex, which is due to the complexation of the two aromatic moieties of TNS with two cyclodextrin molecules. The resolution of this complicated system, with four emitting species and four association equilibria, could only be achieved by careful and systematic analysis of both steady-state and time-resolved fluorescence data, together with direct structural information from NMR experiments [2].

Figure 6 shows the emission spectra ($n_s = 13$) of TNS at varying CD concentrations. PCA of these spectra indicates in all cases clearly three structural components as can be observed in Fig. 7. While the first three eigenvectors $\mathbf{v}_1$–$\mathbf{v}_3$ vary systematically with wavelength, the fourth eigenvector $\mathbf{v}_4$ shows already variations which are of the order of the noise itself and is not different from the following eigenvectors $\mathbf{v}_5$ or $\mathbf{v}_6$. The same is true for the coefficient $\mathbf{z}_1$–$\mathbf{z}_6$ and is confirmed by the plots of the eigenvalues λ_l, the mean residual error σ^k, and the Durbin–Watson Test values dw^k.

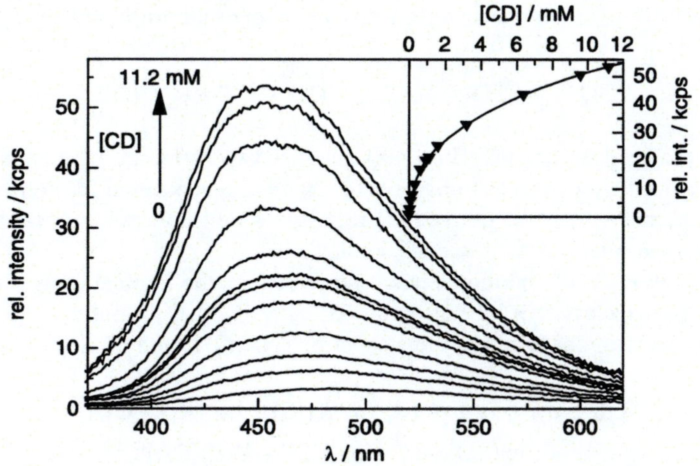

Fig. 6 Corrected fluorescence emission spectra ($\lambda_{exc} = 320$ nm) of TNS in the presence of different concentrations of CD in the range from 0 to 11.2×10^{-3} M. [TNS] = 2.0×10^{-6} M. *Inset*: Fluorescence intensity at 450 nm vs. CD concentration

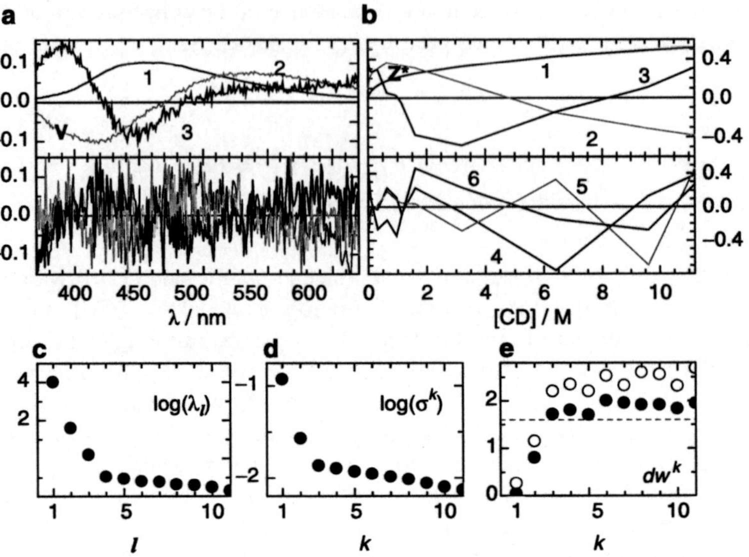

Fig. 7 Determination of the number of structural components n_c from PCA of the emission spectra of TNS (Fig. 6). (**a**) Eigenvectors $\mathbf{v}_1$–$\mathbf{v}_3$ (*above*) and $\mathbf{v}_4$–$\mathbf{v}_6$ (*below*). (**b**) (Normalized) coefficients $\mathbf{Z}^* = \mathbf{\Lambda}^{-1/2}\mathbf{Z}$, $\mathbf{z}_1$–$\mathbf{z}_3$ (*above*) and $\mathbf{z}_4$–$\mathbf{z}_6$ (*below*). (**c**) Eigenvalues λ_1–λ_{11}. (**d**) Mean residual error σ^k vs. the number of included PCs k. (**e**) Mean Durbin–Watson Test values dw^k of residual spectra r_{ij} (*black circles*) and of residual profiles r_{ji} (*open circles*) vs. the number of included PCs k

In order to account for the three structural components found by PCA, the components are assigned to the chemical species, free TNS, 1:1 complex (C11) and 1:2 complex (C12), and a mechanism with two complexation equilibria was proposed:

$$\mathrm{TNS} + \mathrm{CD} \overset{K_1}{\rightleftharpoons} \mathrm{C11}, \quad \mathrm{C11} + \mathrm{CD} \overset{K_1}{\rightleftharpoons} \mathrm{C12}. \tag{16}$$

This model does not take into account that two different 1:1 complexes are formed, since their concentrations have the same dependence on CD concentration. Their existence cannot be derived from steady-state fluorescence measurements, but from time-resolved fluorescence studies.

In the absence of interconversion processes in the excited state and under conditions of excess CD concentration ([CD] ≈ $[\mathrm{CD}]_0$), the following equations for the concentrations $\mathbf{D}$ of the three species, TNS, C11, and C12, are obtained.

$$\begin{pmatrix}\hat{d}_{i1} & \hat{d}_{i2} & \hat{d}_{i3}\end{pmatrix} = \begin{pmatrix} a & a \cdot K_1[\mathrm{CD}] & a \cdot K_1K_2[\mathrm{CD}]^2 \end{pmatrix}$$
$$\text{with } a = (1 + K_1[\mathrm{CD}] + K_1K_2[\mathrm{CD}]^2)^{-1}. \tag{17}$$

The fit of (17) to single emission intensity profiles as a function of [CD] allows for the determination of the association constants K_1 and K_2. Nevertheless, strong

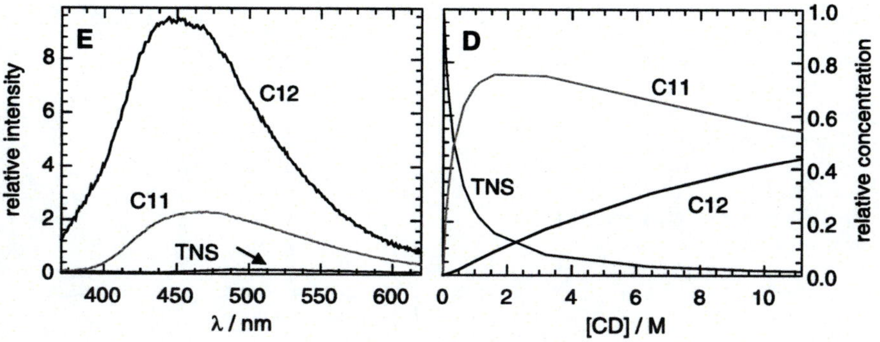

Fig. 8 Results of PCGA of the emission spectra of TNS (Fig. 6). *Left*: component spectra **E** corresponding to free TNS, 1:1 complex (C11), and 1:2 complex (C12). *Right*: coefficients **D** representing the relative concentrations of the species TNS, C11, and C12 ($K_1 = (2.99 \pm 0.03) \times 10^3\ M^{-1}$, $K_2 = 72 \pm 1\ M^{-1}$)

correlation is observed between the nonlinear parameter K_2 and the linear parameter e_{3j}, since only a small percentage of the complex C12 is present even at the highest [CD]. This correlation is efficiently reduced in the global analysis and precise values for the equilibrium constants and for the component spectra are obtained (Fig. 8). Note that the emission spectrum of complex C11 is a combination of the spectra of the two 1:1 complexes present. As these two species show the same dependence on [CD], they appear as an only component in PCA.

5.2 *Excited-State Proton Transfer of 2-(2′-Hydroxyphenyl) Benzimidazole*

In aqueous solutions, 2-(2′-hydroxyphenyl)benzimidazole (HBI) shows in the ground state two acid–base equilibria and a tautomeric equilibrium of three neutral species (Fig. 9). In the excited state, several photoinduced proton-transfer reactions were detected [1, 18]. The cationic form **C*** deprotonates very fast at the hydroxyl group, only emission from the keto form **K*** being detected at acidic pH. The neutral *cis*-enol form $\mathbf{E_c}$***** experiments an excited-state intramolecular proton transfer, leading also very fast to the neutral keto form **K***. The *trans*-enol form $\mathbf{E_t}$***** also deprotonates at the hydroxyl group, leading to the excited anion **A***, but some fluorescence from $\mathbf{E_t}$***** is also detected. The emission spectra of HBI in aqueous solutions change significantly with pH and show strong spectral overlap (Fig. 10).

The results of PCA of the emission series of HBI (Fig. 10) are shown in Fig. 11. All plots reveal the presence of three structural components. The variance of the fourth component (see λ_4) is still slightly higher than that of the following residual components. This additional variance is due to some systematic variation in the residual spectra (see $\mathbf{v}_4$ and residual spectra with $k = 3$) but not in the dependence

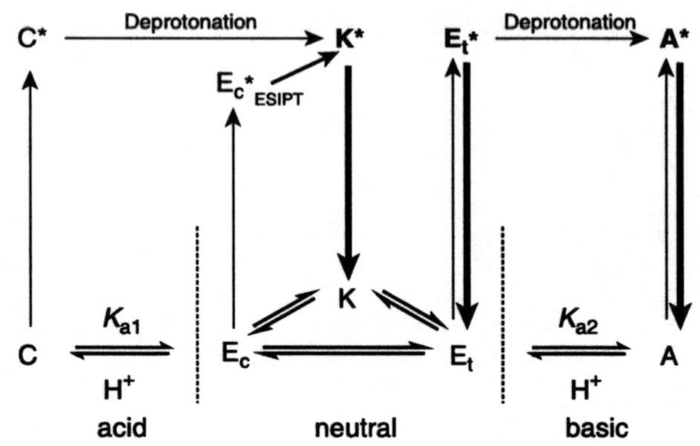

Fig. 9 Acid–base, tautomeric and conformational equilibria of HBI in aqueous solution. Molecular structures of HBI: **C** cationic form, **K** keto form, $\mathbf{E_t}$ *trans*-enol form, $\mathbf{E_c}$ *cis*-enol form, and **A** anionic form. Fluorescent species are set in *bold* letters

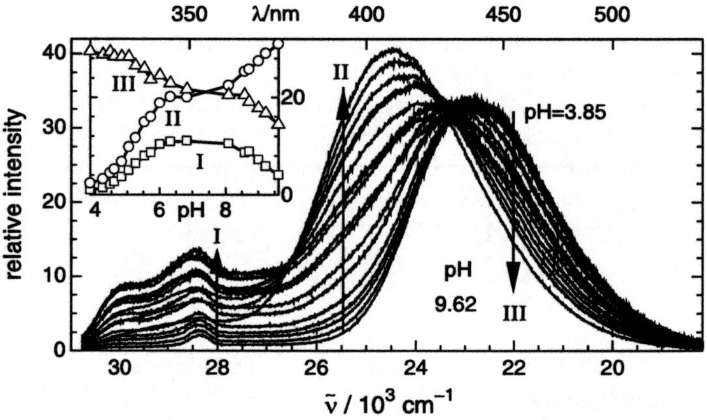

Fig. 10 Series of fluorescence emission spectra of HBI in aqueous solution with varying pH values between 3.85 and 9.62. [HBI] $= 2.4 \times 10^{-6}$ M, $\tilde{\nu}_{exc} = 31{,}650\ \mathrm{cm}^{-1}$. The count rates have been divided by 10^6 cps. Peaks at $28{,}350\ \mathrm{cm}^{-1}$ are due to Raman scattering. *Inset*: emission intensity profiles at the indicated wavenumbers

on pH (see $\mathbf{z}_4$ and residual profiles with $k = 3$). This is a typical pattern found for "spurious" components introduced by instrumental or experimental error during the measurement which does not depend on the external variable. We therefore confirm that three structural components are present, which according to the mechanism described above are assigned to the excited anion **A***and the neutral keto **K*** and *trans*-enol $\mathbf{E_t}$***** forms.

On the basis of the mechanism described, model functions for the component contributions d_{il} are proposed in (18), which were globally fitted to the series of spectra as a function of pH. d_{i1} represents the pH dependence of the anion emission, with contributing terms from the excitation of the ground-state anion and from the photoinduced emission of the anion at neutral pH, formed from the excited

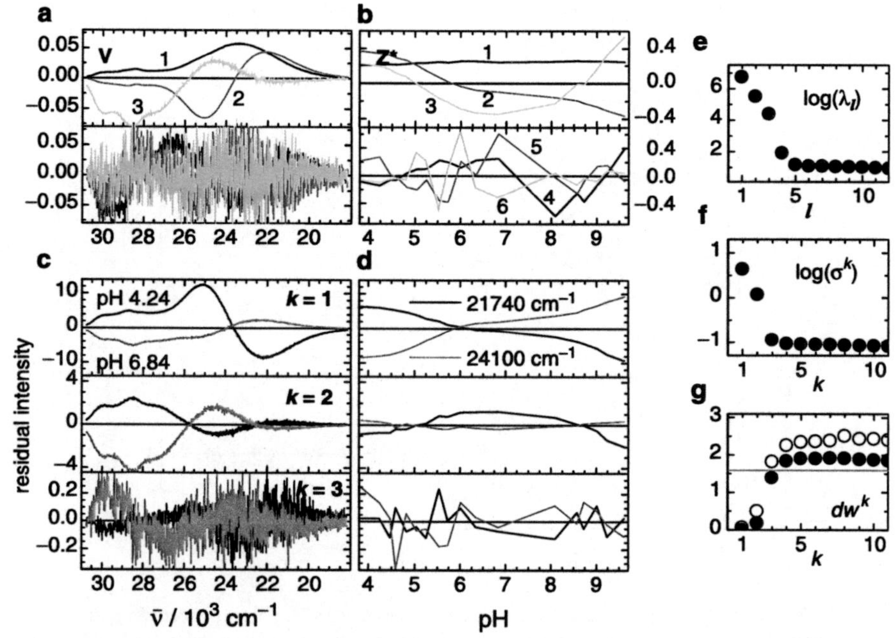

Fig. 11 PCA of the emission spectra of HBI (Fig. 10). (**a**) Eigenvectors $\mathbf{v}_1$–$\mathbf{v}_3$ (*above*) and $\mathbf{v}_4$–$\mathbf{v}_6$ (*below*). (**b**) (Normalized) coefficients $\mathbf{Z}^* = \mathbf{\Lambda}^{-1/2}\mathbf{Z}$, $\mathbf{z}_1$–$\mathbf{z}_3$ (*above*), and $\mathbf{z}_4$–$\mathbf{z}_6$ (*below*). (**c**) Residual spectra $\mathbf{R}^k$ at two pH: 4.24 (*black*) and 6.84 (*gray*) for $k = 1$, 2, and 3 included PCs. (**d**) Residual profiles $(\mathbf{R}^k)^\mathrm{T}$ at 21,740 cm^{-1} (*black*) and 24,100 cm^{-1} (*gray*) for $k = 1, 2$, and 3. (**e**) Eigenvalues λ_1–λ_{10}. (**f**) Mean residual error σ^k vs. the number of included PCs k. (**g**) Mean Durbin–Watson Test values dw^k of residual spectra r_{ij} (*black circles*) and of residual profiles r_{ji} (*open circles*) vs. the number of included PCs k

trans-enol form. d_{i2} represents the pH dependence of the *trans*-enol form emission. d_{i3} represents the pH dependence of the keto form emission, with contributing terms from the photoinduced formation of **K*** by deprotonation of the excited cation at acidic pH, and from the excitation of the neutral forms $\mathbf{E_c}$ and **K**. The values of γ and β have been obtained from comparison of excitation and absorption spectra and maintained constant in the fit. The resulting pK_a values, pure component spectra, and component contributions are given in Fig. 12.

$$\begin{aligned}(\hat{d}_{i1}\ \hat{d}_{i2}\ \hat{d}_{i3}) &= \left(a(K_{a1}K_{a2} + \gamma K_{a1}[\mathrm{H}^+]_i) \quad aK_{a1}[\mathrm{H}^+]_i \quad a([\mathrm{H}^+]_i^2 + \beta K_{a1}[\mathrm{H}^+]_i) \right)\\ &\text{with } a = (K_{a1}K_{a2} + K_{a1}[\mathrm{H}^+]_i + [\mathrm{H}^+]_i^2)^{-1}. \end{aligned} \tag{18}$$

5.3 Lactim–Lactam Phototautomerization

Ground-state tautomeric and acid–base equilibria, and excited-state proton-transfer processes of 2-(6′-hydroxy-2′-pyridyl)benzimidazole (HPyBI) in aqueous solution

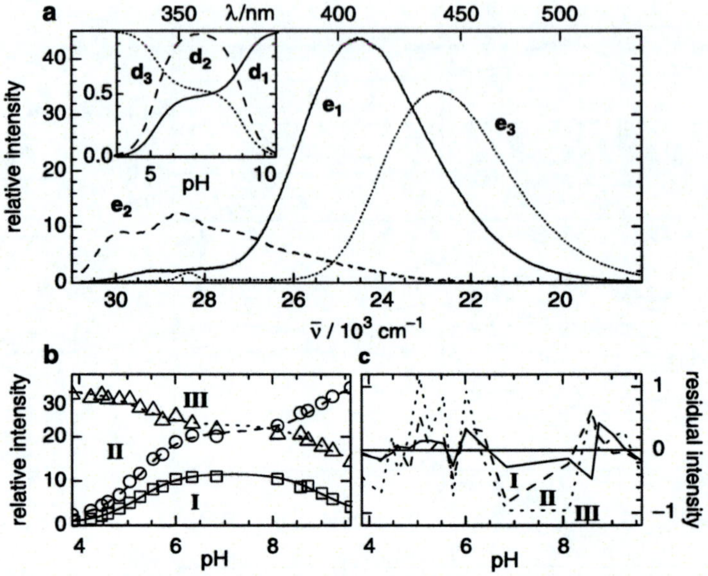

Fig. 12 Results of the global analysis of the emission spectra of HBI (Fig. 10) according to (18) and parameters $pK_{a1} = 5.246 \pm 0.002$ and $pK_{a2} = 8.996 \pm 0.004$, and fixed parameters $\beta = 0.53$ and $\gamma = 0.48$ obtained from excitation and absorption spectra. (**a**) Resolved component spectra $\mathbf{e}_1$ (*solid*), $\mathbf{e}_2$ (*dashed*), and $\mathbf{e}_3$ (*dotted*) corresponding to the species **A***, $\mathbf{E_t}$*****, and **K***, respectively. The peaks at 28,350 cm^{-1} are due to Raman scattering. *Inset*: Relative concentrations **D** vs. pH. (**b**) pH dependence of the experimental **Y** (*symbols*) and fitted $\hat{\mathbf{Y}}$ (*lines*) fluorescence intensity at three spectral positions as indicated in Fig 10. (**c**) Residual intensity profiles between the **Y** and $\hat{\mathbf{Y}}$ of (**b**)

have been studied by means of UV–vis absorption and fluorescence spectroscopy in both steady-state and time-resolved modes [19, 20]. There exist a lactim–lactam tautomeric equilibrium for both the neutral (**N** and **T**) and the protonated forms (**C** and **TC**) in the ground state (Fig. 13). An acid–base equilibrium between the protonated and neutral forms is established, with pK_a = 3.127. The excited-state behavior is summarized in Fig. 13. The increase of acidity of the OH group and basicity of the pyridine nitrogen upon excitation originates the phototaumerization of the neutral **N** and protonated **C** lactim species upon excitation. For the protonated form (Fig. 13b), the lactim–lactam phototaumerization process takes place via two competitive excited-state proton-transfer routes: a one-step water-assisted proton translocation (probably a double proton transfer) and a two-step pathway which involves first the dissociation of the lactim cation to form an emissive intermediate zwitterionic species and then the acid-catalyzed protonation at the pyridine nitrogen to give rise to the lactam tautomer. For the neutral lactim form, only the first route has been detected (Fig. 13a).

PCA was applied to the series of 28 fluorescence spectra of HPyBI recorded in water under excitation at 31,250 cm^{-1} (where only the lactim forms absorb) at different pH values (Fig. 14). The results of the analysis are shown in Fig. 15.

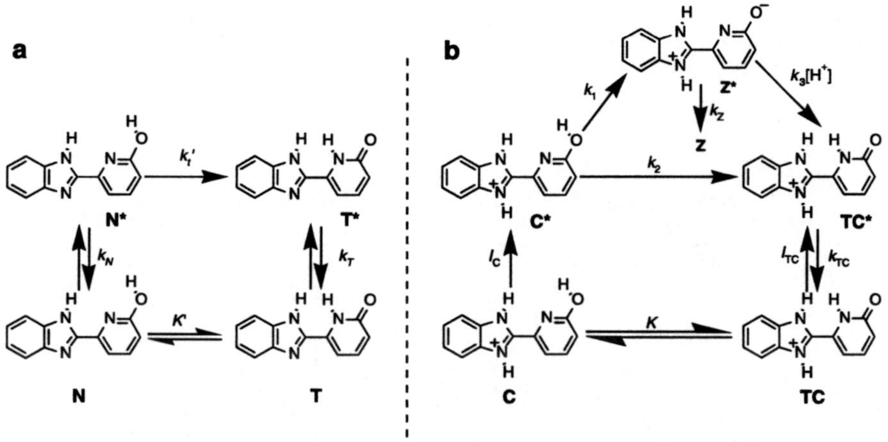

Fig. 13 Excitation and deactivation of (**a**) neutral and (**b**) protonated HPyBI in aqueous solution. The neutral (**N** and **T**) and protonated (**C** and **TC**) tautomeric forms are in equilibrium in the ground state, with $pK_a = 3.127$

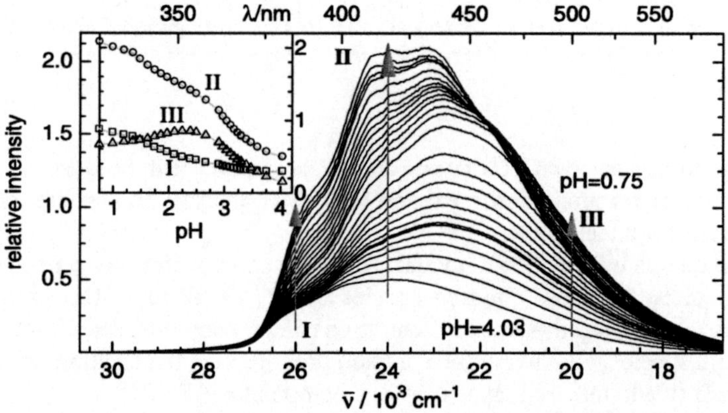

Fig. 14 Fluorescence spectra of HPyBI in aqueous solution with decreasing pH in the range 4.03–0.75. The *inset* shows the dependence on acidity of the fluorescence intensity at $\tilde{\nu}_{em} = 26{,}000\ cm^{-1}$ (*open square*), 24,010 cm^{-1} (*open circle*), and at 20,000 cm^{-1} (*open triangle*) ($n_s = 28$, $n_w = 550$, $\tilde{\nu}_{exc} = 31{,}250 cm^{-1}$, [HPyBI] $= 3 \times 10^{-6}$ mol dm^{-3})

The coefficients **Z***, the eigenvalues λ, the mean residual error σ^k, and the mean Durbin–Watson Test dw^k of residual profiles demonstrate that three independent spectral components are needed to reproduce all the experimental spectra at the different acidities. The eigenvectors **V** and the Durbin–Watson Test dw^k of the residual spectra (black circles in Fig. 15e) indicate some additional systematic spectral variation. The additional variance may be introduced by a very low

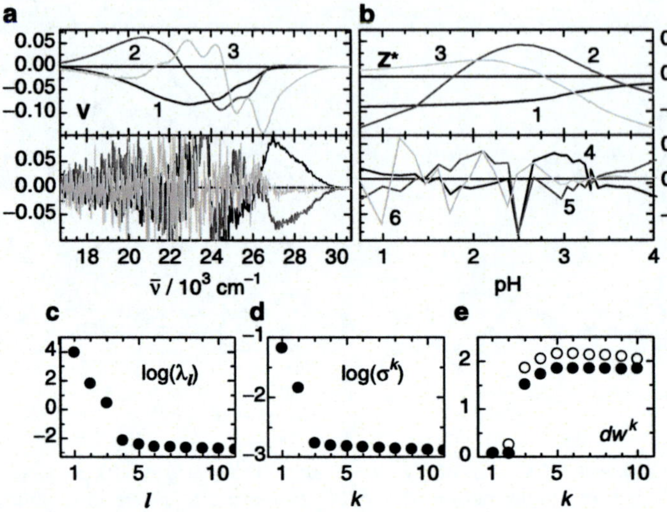

Fig. 15 Determination of the number of structural components n_c from PCA of the emission spectra of HPyBI (Fig. 14). **(a)** Eigenvectors $\mathbf{v}_1$–$\mathbf{v}_3$ *(above)* and $\mathbf{v}_4$–$\mathbf{v}_6$ *(below)*. **(b)** (Normalized) coefficients $\mathbf{Z}^* = \boldsymbol{\Lambda}^{-1/2}\mathbf{Z}$, $\mathbf{z}_1$–$\mathbf{z}_3$ *(above)* and $\mathbf{z}_4$–$\mathbf{z}_6$ *(below)*. **(c)** Eigenvalues λ_1–λ_{11}. **(d)** Mean residual error σ^k vs. the number of included PCs k. **(e)** Mean Durbin–Watson Test values dw^k of residual spectra r_{ij} *(black circles)* and of residual profiles r_{ji} *(open circles)* vs. the number of included PCs k

emission from species **C***, that is almost nonfluorescent because of the fast proton-transfer reactions in the excited state. The analysis shows that its fluorescence is almost undetectable.

Based on this information, a model with three components was proposed. In the working acidity range, the emissive species are **N***, **T***, **Z*** and **TC***. As the acidity dependence of the steady-state concentrations of the neutral forms **N*** and **T*** is the same, both species will always be in a fixed proportion at the exciting wavenumber at any acidity and they will appear as one component **NT***.

Taking into account the acidity dependence of the steady-state concentration of the respective excited species, the fluorescence emission spectra **F** from the series can then be written as a linear combination of the spectra of the emissive species (The ratio $\beta = 0.468$ is known from time-resolved fluorescence results):

$$\mathbf{F} = c_{NT}\,\mathbf{F_{NT}} + c_2\left((1-\beta)\,\mathbf{F_Z} + \beta\,\mathbf{F_{TC}}\right) + c_3\,\mathbf{F_{TC}}. \tag{19}$$

The model functions for the concentrations have been derived from the proposed mechanism:

$$\begin{pmatrix} c_{\mathrm{NT}} & c_2 & c_3 \end{pmatrix} = \begin{pmatrix} \frac{K_a}{K_a+[\mathrm{H}^+]} & a[\mathrm{H}^+] & a\alpha[\mathrm{H}^+]^2 \end{pmatrix}$$
$$\text{with } a = \left((1+\alpha[\mathrm{H}^+])(K_a+[\mathrm{H}^+])\right)^{-1}. \tag{20}$$

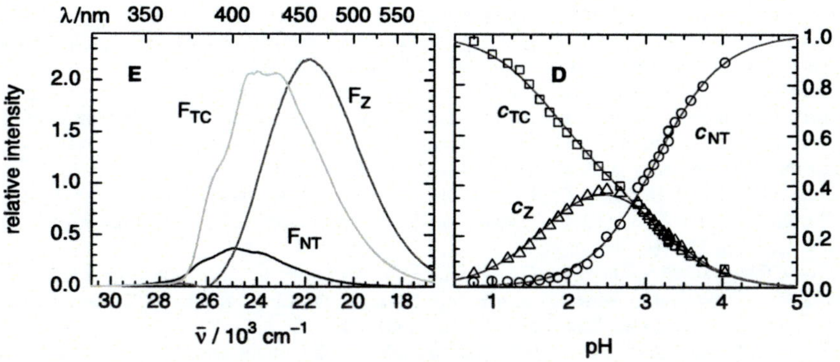

Fig. 16 Results of PCGA of the emission spectra of HPyBI (Fig. 14). *Left*: component spectra **E** corresponding to the neutral forms **N*** and **T*** (F_{NT}), the zwitterion **Z*** (F_Z), and the lactam cation **TC*** (F_{TC}). *Right*: coefficients **D** (*unfilled symbols*) representing the pH-dependent contributions of the component spectra of **N*** and **T*** (c_{NT}), **Z*** (c_Z), and **TC*** (c_{TC}). The *solid lines* represent the fit of the equations derived from the proposed model to these data. We used for these analyses the ratio $\beta = 0.468$ in H_2O provided from time-resolved fluorescence results

Global analysis with these model functions allowed us to obtain reliable estimations of the model parameters and the component spectra with $pK_a = 3.127 \pm 0.004$ and $\alpha = 57.3 \pm 1.2$ mol dm^{-3}. This allows one to obtain the pure spectrum of **Z*** and the coefficients c_{NT}, $c_Z = c_2(1 - \beta)$, and $c_{TC} = c_2\beta + c_3$ shown in Fig. 16. From the ratio of the areas beneath the species spectra, relative quantum efficiencies have been determined, which is especially interesting for the otherwise inaccessible zwitterion **Z***, which does not exist in the ground state [19].

Acknowledgments We thank the Spanish Ministry of Education and Science, the European Union ERDF, and the *Xunta de Galicia* for financial support.

References

1. Al-Soufi W, Novo M, Mosquera M (2001) Principal component global analysis of fluorescence and absorption spectra of 2-(2′-hydroxyphenyl)benzimidazole. Appl Spectrosc 55:630–636
2. Álvarez-Parrilla E, Al-Soufi W, Ramos Cabrer P, Novo M, Vázquez Tato J (2001) Resolution of the association equilibria of 2-(*p*-toluidinyl)-naphthalene-6-sulfonate (Tns) with beta-cyclodextrin and a charged derivative. J Phys Chem B 105:5994–6003
3. Beechem JM, Gratton E, Ameloot M, Knutson JR, Brand L (1991) The global analysis of fluorescence intensity and anisotropy decay data: second-generation theory and programs. In: Lakowicz JR (ed) Topics in fluorescence spectroscopy. Plenum Press, New York, pp 241–306
4. Bevington PR, Robinson KD (2003) Data reduction and error analysis for the physical sciences. McGraw-Hill Higher Education, USA

5. Carrazana J, Reija B, Ramos Cabrer P, Al-Soufi W, Novo M, Vázquez Tato J (2004) Complexation of methyl orange with beta-cyclodextrin: detailed analysis and application to quantification of polymer-bound cyclodextrin. Supramol Chem 16:549–559
6. Dommelen LV, Boens N, Ameloot M, De Schryver FC, Kowalczyk A (1993) Species-associated spectra and upper and lower bounds on the rate constants of reversible intramolecular two-state excited-state processes with added quencher. global compartmental analysis of the fluorescence decay surface. J Phys Chem 97:11738–11753
7. Draper NR, Smith H (1998) Applied regression analysis. Wiley, New York
8. Durbin J, Watson GS (1950) Testing for serial correlation in least squares regression. I. Biometrika 37:409
9. Dyson RM, Kaderli S, Lawrance GA, Maeder M, Zuberbühler AD (1997) Second order global analysis: the evaluation of series of spectrophotometric titrations for improved determination of equilibrium constants. Anal Chim Acta 353:381–393
10. Hamilton JC, Gemperline PJ (1990) Mixture analysis using factor analysis. II: self-modeling curve resolution. J Chemom 4:1–13
11. Johnson ML (2000) Parameter correlations while curve fitting. Methods Enzymol 321:424–446
12. Jolliffe IT (2002) Principal component analysis. Springer, New York
13. Knutson JR, Beechem JM, Brand L (1983) Simultaneous analysis of multiple fluorescence decay curves: a global approach. Chem Phys Lett 102:501–507
14. Lawton WH, Sylvestre EA (1971) Elimination of linear parameters in nonlinear regression. Technometrics 13:461–467
15. Lezcano M, Al-Soufi W, Novo M, Rodríguez-Núñez E, Vázquez Tato J (2002) Complexation of several benzimidazole-type fungicides with α- and β-cyclodextrins. J Agric Food Chem 50:108–112
16. Malinowski ER (1996) Automatic window factor analysis- a more efficient method for determining concentration profiles from evolutionary spectra. J Chemom 10:273–279
17. Malinowski ER (2002) Factor analysis in chemistry. Wiley, New York
18. Mosquera M, Penedo JC, Ríos Rodríguez MC, Rodríguez-Prieto F (1996) Photoinduced inter- and intramolecular proton transfer in aqueous and ethanolic solutions of 2-(2′-hydroxyphenyl) benzimidazole: evidence for tautomeric and conformational equilibria in the ground state. J Phys Chem 100:5398–5407
19. Penedo JC, Ríos Rodríguez MC, García Lema I, Pérez Lustres JL, Mosquera M, Rodríguez-Prieto F (2005) Two competitive routes in the lactim-lactam phototautomerization of a hydroxypyridine derivative cation in water: dissociative mechanism versus water-assisted proton transfer. J Phys Chem A 109:10189–10198
20. Penedo JC, Pérez-Lustres JL, Lema Garcia I, Ríos Rodríguez MC, Mosquera M, Rodríguez-Prieto F (2004) Solvent-dependent ground- and excited-state tautomerism in 2-(6′-hydroxy-2′-pyridyl)benzimidazole. J Phys Chem A 108:6117–6126
21. Press WH, Teukolsky SA, Vetterling WT, Flannery BP (2002) Numerical recipes in C++: the art of scientific computing. Cambridge University Press, New York
22. Puxty G, Maeder M, Hungerbühler K (2006) Tutorial on the fitting of kinetics models to multivariate spectroscopic measurements with non-linear least-squares regression. Chemom Intell Lab Syst 81:149–164
23. Reija B, Al-Soufi W, Novo M, Vázquez Tato J (2005) Specific interactions in the inclusion complexes of pyronines Y and B with β-cyclodextrin. J Phys Chem B 109:1364–1370
24. Rurack K, Hoffmann K, Al-Soufi W, Resch-Genger U (2002) 2,2′-Bipyridyl-3,3′-diol incorporated into AlPO4-5 crystals and its spectroscopic properties as related to aqueous liquid media. J Phys Chem B 106:9744–9752
25. Sun Y-P, Sears DF Jr, Saltiel J (1988) Principal component self-modeling analysis applied to conformational equilibration of 1,3-butadiene vapor. UV spectra and thermodynamic parameters of the two conformers. J Am Chem Soc 110:6277–6278

26. Sylvestre EA, Lawton WH, Maggio MS (1974) Curve resolution using a postulated chemical reaction. Technometrics 16:353–368
27. van Stokkum IHM, Larsen DS, van Grondelle R (2004) Global and target analysis of time-resolved spectra. Biochim Biophys Acta BBA Bioenerg 1657:82–104
28. Volkov VV (1996) Separation of additive mixture spectra by a self-modeling method. Appl Spectrosc 50:320–326
29. Windig W (1992) Self-modeling mixture analysis of spectral data with continuous concentration profiles. Chemom Intell Lab Syst 16:1–16

Hot Electron-Induced Electrogenerated Chemiluminescence

Johanna Suomi and Sakari Kulmala

Abstract In this chapter, we discuss the basics of cathodic hot electron-induced electrogenerated chemiluminescence (HECL). In the applications of HECL, we discuss, e.g., the usable electrode materials and their advantages as well as the applicable solution conditions in aqueous media. We also summarize the luminophore types excitable by this method and their usability as labels in practical bioaffinity assay applications.

1 Introduction

Electrochemiluminescence as an analytical tool is a relatively recent development. Electrochemiluminescence, i.e., electrogenerated chemiluminescence (ECL), is here considered a phenomenon in which light is generated at the surface and/or in the close proximity of an electrode as a result of a chemical reaction or a sequence of chemical reactions induced or initiated by a charge transfer between the electrode material and single or multiple solution species. Or more simply: in ECL one or more of the reagents is generated in situ in an electrolytic process. ECL shares many of the analytical advantages of chemiluminescence, the most important one being the low detection limit owing to the low background signal.

The ECL is normally initiated and controlled by the electrode potential of the working electrode and can be typically generated in nonaqueous electrolyte solutions and sometimes even in aqueous solutions by a potential step or sweep. Bard and his coworkers have quite recently described in detail the history and the fundamentals of traditional ECL and its applications in their excellent monographs and reviews [7, 78].

J. Suomi (✉)
School of Chemical Technology, Aalto University, Espoo, Finland
e-mail: johanna.suomi@aalto.fi

C.D. Geddes (ed.), *Reviews in Fluorescence 2009*, Reviews in Fluorescence 2009,
DOI 10.1007/978-1-4419-9672-5_3,

The main commercial applications of the ECL have so far been in bioaffinity assays. The use of $Ru(bpy)_3^{2+}$-based labels excitable in aqueous micellar electrolyte solutions in the presence of tripropyl amine or other suitable coreactants at active metal electrodes or carbon electrodes gives the basis for both immunoassays and DNA-probe assays [9, 12, 44]. Typically, a direct determination of an analytically interesting electrochemiluminescent molecule on the basis of its own ECL is very rarely applicable, at least in the aqueous world of bioanalysis. Thus, only labeling applications are important.

Generation of cathodic hot electron-induced electrogenerated chemiluminescence (HECL) in fully aqueous solution at thin insulating film-coated electrodes was reported for the first time about 15 years ago [3, 48, 49, 53]. Hot electron injection into aqueous solution was suggested to be the primary step of the reaction pathway leading to this cathodic ECL [48, 49, 54]. Since then, Bard and coworkers have also studied these phenomena but almost exclusively in nonaqueous solutions [82, 83].

The study of charge transport through insulating barriers upon application of voltage to metal/insulator/metal and metal/insulator/semiconductor (M/I/M and M/I/S) junctions, as well as the study of hot electrons and their transport in semiconductor/metal/semiconductor (S/M/S) structures was begun already in the 1960s [28, 86]. It was also very early observed that hot electrons can be transferred from these kinds of devices into vacuum, insulating liquids, [70] and later also into aqueous electrolyte solutions [13].

Hot electrons are usually defined as electrons having higher energy than the Fermi energy of the phase in question [72] or electrons having higher thermal energies than the thermal energy of the phase in question [79]. Hot electrons in semiconductors, metals, and metals covered with a thin insulating film can be generated by illuminating these electrodes with UV-light or by irradiating them with photons having even higher energy [20, 75, 77]. These hot electrons can be also emitted into the solution in contact with the electrode irradiated [77].

When electrons are photoemitted into aqueous electrolyte solution from the electrodes, these presolvated hot electrons (solution phase hot electrons) may either (1) escape back to the electrode material, (2) react directly with solutes, or (3) become solvated to form hydrated electrons (i.e., electrons solvated in water, e_{aq}^-). It has been reported that photoemission of electrons from oxide-coated aluminum electrodes results in injection of electrons to the conduction band of water and finally in the formation of hydrated electrons after solvation process when the insulating film is thinner than ca. 6 nm [75]. In this case, the very thin insulating film is sufficiently optically transparent to allow the excitation of the electrons in the Fermi level of the metal, and it is also partially preventing the escape of the presolvated hot electrons back to the metal behind the insulating film.

However, the most common way of generating hydrated electrons is pulse radiolysis of water, in which also hydroxyl radicals, hydrogen peroxide, atomic hydrogen, and molecular hydrogen are produced after a pulse of high-energy electrons or γ-rays. As a result of these studies, the chemistry of hydrated electron is quite well studied and its second-order reaction rate constants with a wide variety of compounds and ions are known [10]. However, very little is known about the reactivity and the reaction rates of presolvated hot electrons with different kind of solutes.

Photoemission and pulse radiolysis methods are not suited to producing solution phase hot electrons in a typical laboratory. Thus, the different kinds of tunnel junctions are much more promising, simple and low-cost methods for producing solution phase hot electrons. Instead of above-mentioned M/I/M junctions, we have mainly studied in our research another junction type to inject hot electrons into aqueous solutions: conductor/insulator/electrolyte solution (C/I/E) junctions.

We have proposed that during high-amplitude cathodic pulse polarization of this kind of C/I/E junctions, hot electrons can be tunnel emitted into an aqueous electrolyte solution, and we assume that also hydrated electron is an important mediating species in the reductions occurring at these type of electrons [3, 49, 52, 53]. However, we have not been able to show unequivocally that hydrated electrons are acting as mediating species in chemiluminescence generating pathways. Thus, only indirect evidence of the presence of fully hydrated electrons have been gained by first studying the hydrated electron-induced chemiluminescence and then the hot electron-induced electrochemiluminescence of a certain luminophore and comparing results, especially those obtained by using well-known hydrated electron scavengers. For these purposes, we have studied examples of luminescent transition metal chelates [4, 53], lanthanide chelates [45, 51, 58, 60], metal ions [45, 52], and organic luminophores [29, 50, 57, 69]. In all cases, the results support our theory of hydrated electrons having an important role in the reaction mechanisms, but the direct reactions of presolvated hot electrons may also be in response of the reduction reactions in the light generating pathways. However, it is very clear that hydrated electron in the presence of a suitable strongly oxidizing agent is able to induce chemiluminescence of these luminophores in fully aqueous solutions.

Typically, the luminophores are excited by two alternative or parallel pathways: (1) red-ox pathway in which the luminophore is first one-electron reduced by presolvated hot or hydrated electron and then one-electron oxidized by sufficiently strong oxidizing radical and/or (2) ox-red pathway in which the luminophore is first one-electron oxidized by reductively generated, sufficiently strongly oxidizing radical and then one-electron reduced to the excited state of the original oxidation by presolvated hot or hydrated electron. Typical coreactants yielding strongly oxidizing radicals by one-electron reduction are, e.g., peroxodisulfate ion, peroxodiphosphate ions, and hydrogen peroxide.

1.1 Generation of the Primary Species

Electrodes coated with a thin layer of insulating oxide are usable for the generation of HECL. For materials forming insulating oxides by anodic oxidation, in situ generation of suitable oxide film thickness is possible by anodic pulse [87] preceding the cathodic excitation pulse (Fig. 1).

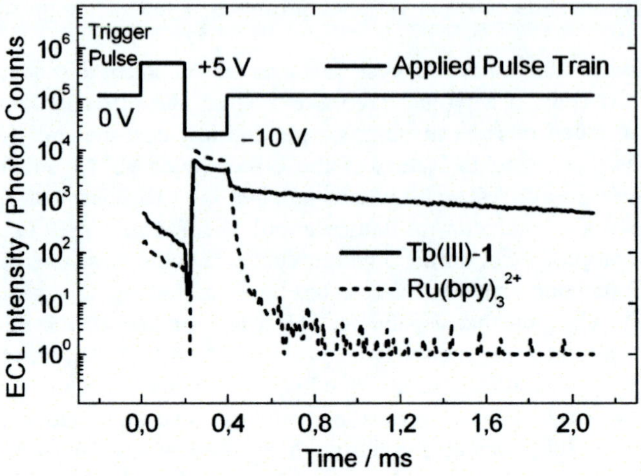

Fig. 1 ECL time profiles of $Ru(bpy)_3{}^{2+}$ and Tb(III)-1 chelates (Tb(III) chelated by 2,6-bis[*N*,*N*-bis(carboxymethyl)aminomethyl]-4-methylphenol). Conditions: 1 M Na_2SO_4 in 0.2 M borate buffer at pH 9.2, 1×10^{-3} M $K_2S_2O_8$, 1×10^{-6} M Tb(III)-1 or 1×10^{-7} M $Ru(bpy)_3{}^{2+}$ [49]

With other types of base conductor materials in C/I/E structures, the insulating film must be prepared in advance upon the conductor. For example, atomic layer deposition (ALD) is an excellent method of depositioning ultra thin insulating oxide films upon suitable sufficiently conducting materials [23].

Likewise in the experiments of Piazza and coworkers [75], the thickness of the insulating film in the C/I/E junction has a distinctive effect on electron emission into aqueous electrolyte solution. With all of the luminophores we have so far studied, the ultra thin insulating films (thinner than 4–6 nm) have provided the highest cathodic ECL intensity. In case of thicker insulating films, the ECL intensity has decreased exponentially as a function of oxide film thickness when thickness exceeds ca. 4 nm (Fig. 2) [48, 49, 51].

The reason for this has been revealed in solid-state studies using S/I/S junctions. DiMaria and coworkers have shown that if the thickness of SiO_2 film is at most ca. 4 nm in these devices, the hot electrons are transferred by direct tunnel emission through the barrier without a change in electron energy [14], while if the film is thicker than about 7 nm, the predominant tunneling mechanism is Fowler–Nordheim tunneling [14]. In Fowler–Nordheim tunneling in the S/I/S, M/I/S, and M/I/M junction, the electrons are first tunneled from the Fermi level of semiconductor or metal to the conduction band of the insulating film. After this, the electrons are transported by electric field in the conduction band.

Figure 3 illustrates the tunnel emission process in a C/I/E junction during high-amplitude cathodic pulse polarization. The electrons tunnel through thin insulator film (<4–5 nm) ballistically or near ballistically by direct field-assisted tunneling so that no considerable loss of energy of electrons occurs (denoted as hot *e*− type I in Fig. 3).

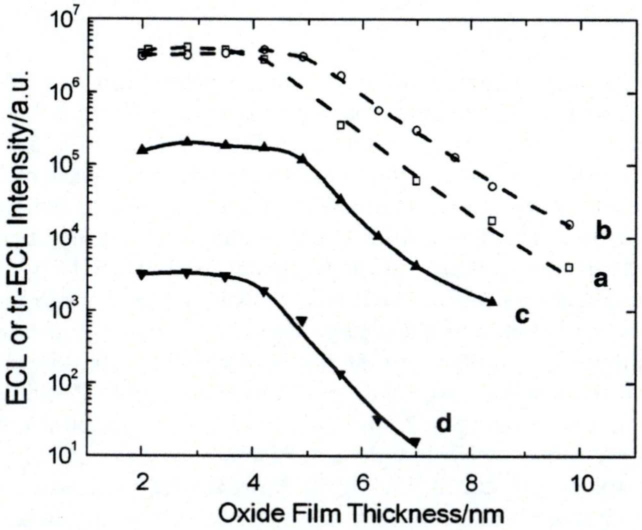

Fig. 2 The effect of oxide film thickness on HECL intensity at aluminum electrodes [49]. (a) 1.0 × 10^{-6} M Tb(III)-1 (=Tb(III) chelated by 2,6-bis[*N*,*N*-bis(carboxymethyl)aminomethyl]-4-methylphenol), TR-ECL, delay time 50 μs, time window 8.0 ms (*squares*), (b) 1.0 × 10^{-5} M Tb (III)-2 (=Tb(III) chelated by N^1-(4-isothiocyanatobenzyl)diethylenetriamine-N^1,N^2,N^3,N^3-tetraacetate), TR-ECL (*circles*), (c) 1.0 × 10^{-5} M $Ru(bpy)_3^{2+}$, cathodic ECL (*upward pointing triangles*), (d) 1.0 × 10^{-5} M $Ru(bpy)_3^{2+}$ TR-ECL (delay time 5 μs, time window 1.0 ms). Oxide films were manufactured by anodic oxidation

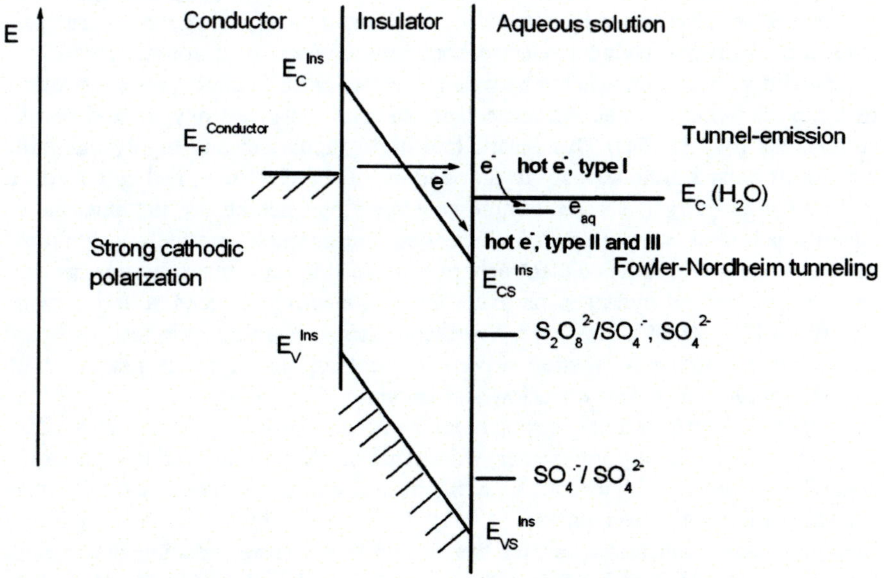

Fig. 3 Schematic energy diagram of tunnel emission and Fowler–Nordheim tunneling at an insulating film-coated electrode in contact with an aqueous electrolyte solution [48]. Generally, a direct tunnel emission can occur only if the insulating film is thinner than ca. 4–5 nm. In the FN-tunneling regime, the electrons are transferred into the electrolyte either from the bottom of the conduction band at the surface of the insulator (E_{CS}) or somewhere above it, if the FN-tunneling results in electron heating in the conduction band

However, with considerably thicker oxide films, Fowler–Nordheim (FN) tunneling predominates as an electron transportation mechanism. In FN-tunneling regime, the electrons are first tunneled to the conduction band of the insulating film and finally transferred into the solution from the bottom of the conduction band of the insulator (denoted as hot $e-$ type II) or from somewhat above the conduction band edge of oxide at the oxide/electrolyte solution interface due to the gaining of energy in the electric field while being transported in the conduction band [6, 14, 15]. During FN tunneling, once in the conduction band of the insulator, the electrons are partly gaining energy by the high electric field and partly losing the energy by inelastic scattering. Thus, the properties and thickness of the insulating film and the band bending phenomena determine the energy of the electrons reaching the I/E interface.

In practice, the conduction band edge of SiO_2 at oxide/electrolyte solution interface bends down well below the conduction band edge of water due to the cathodic polarization [3, 48, 49, 72] (Fig. 2). However, the hydrated electrons can be generated only when hot electrons are injected into the conduction band of water. Hence, with thicker oxide films, hydrated electrons cannot be generated and the electrons transferred to solution species are not sufficiently energetic to induce efficiently ECL [3, 4, 48, 51].

It is very likely that the cathodic reductions at these electrodes can be produced simultaneously by different types of electrons: (1) presolvated hot electrons, (2) hydrated electrons, (3) heterogeneously transferred electrons from the conduction band of the insulating film at insulating film/electrolyte interface, and (4) probably also from the surface states of the insulating film [3, 30, 48, 49, 53]. Normally, only presolvated hot electrons or hydrated electrons are sufficiently energetic to participate in reaction pathways leading to electrochemiluminescence in aqueous solutions.

In addition, presolvated hot or hydrated electrons can efficiently produce strongly oxidizing radicals from added coreactants such as dissolved oxygen and peroxo compounds [47, 51, 54]. This means that highly reducing and highly oxidizing conditions exist simultaneously in the vicinity of the electrode, and as a result it is possible to carry out chemiluminescent reactions, which are not possible in aqueous solutions at active metal electrodes. For example, sulfate and hydroxyl radicals cannot be produced electrochemically in fully aqueous solutions because peroxodisulfate and hydrogen peroxide are two-electron reduced at active metal electrodes [16, 30, 89]. However, hydrated electron is able to produce hydroxyl radicals in reaction with hydrogen peroxide and sulfate radical in reaction with peroxodisulfate ion at near diffusion controlled rate [10].

Hydrated electron is a very strong reducing species ($E(e_{aq}^{-}) = -2.9$ V vs. SHE) [10] having only slightly lower energy than that of conduction band edge of liquid water. This explains why they are able to induce chemiluminescence and ECL from a large variety of luminophores [4, 63]. In addition, hydrated electron is not following Marcus electron transfer theory and is therefore reducing fast species having not only moderately lower reduction potential than that of hydrated electron but also extremely strong oxidant as well. Thus, Marcus inverted region is not observed in its reactivity and it can easily carry out the last step of the above-mentioned ox-red excitation pathway of luminophores as well as very often also to initiate the red-ox excitation pathway.

Sulfate radical is one of the strongest one-electron oxidants known with standard reduction potential of about 3.4 V vs. SHE [71]. In the simultaneous presence of hydrated electrons and sulfate radicals, there is more than 6 eV of energy available in excitation reaction pathways. Figure 4 displays as an example the energetics of the HECL of $Ru(bpy)_3^{2+}$ and an aromatic Tb(III) chelate at oxide-coated n-silicon electrode in aqueous electrolyte solution.

Under these harsh conditions, it has been possible to simultaneously excite luminophores having very different redox properties and emitting either in the UV, VIS, or NIR range [1, 48, 56], which allows the development of multiparametric assays using wavelength discrimination in separation of the signals emerging from different labels. Figure 5 presents an example of a simultaneous HECL excitation of several different luminophores [48].

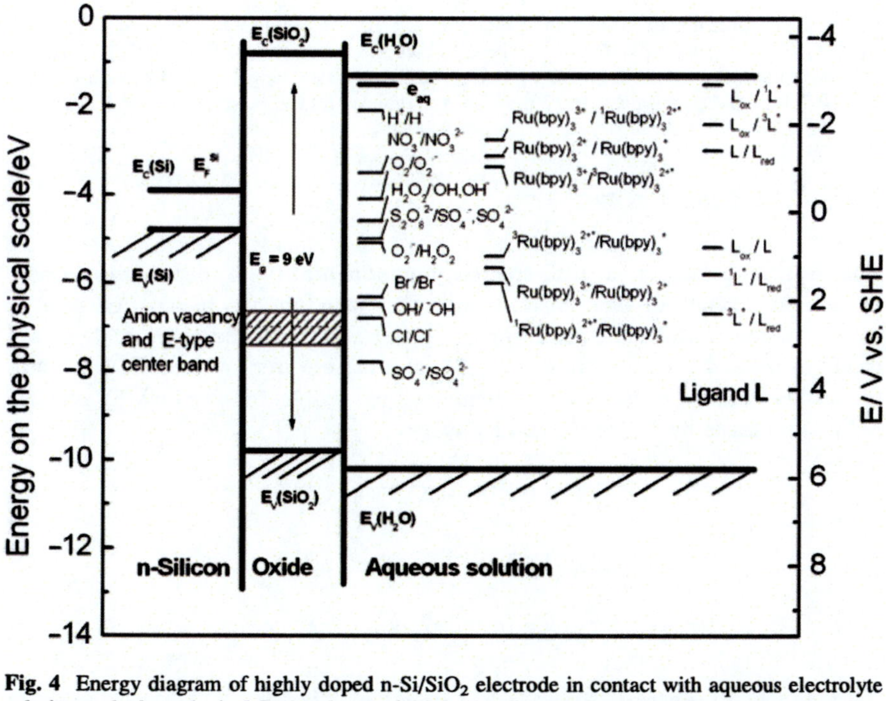

Fig. 4 Energy diagram of highly doped n-Si/SiO_2 electrode in contact with aqueous electrolyte solution under hypothetical flat band potential [3]. The diagram shows the formal redox potentials of the primary radicals and combined redox and luminescence properties of $Ru(bpy)_3^{3+}$ and an aromatic Tb(III)-chelate (Tb(III) chelated by 2,6-bis[*N*,*N*-bis(carboxymethyl)aminomethyl]-4-methylphenol). The diagram demonstrates clearly that on thermodynamical basis, both selected luminophores can be excited even to their singlet excited states in the simultaneous presence of hydrated electrons and sulfate radicals via both *ox-red* and *red-ox* pathways described in this chapter

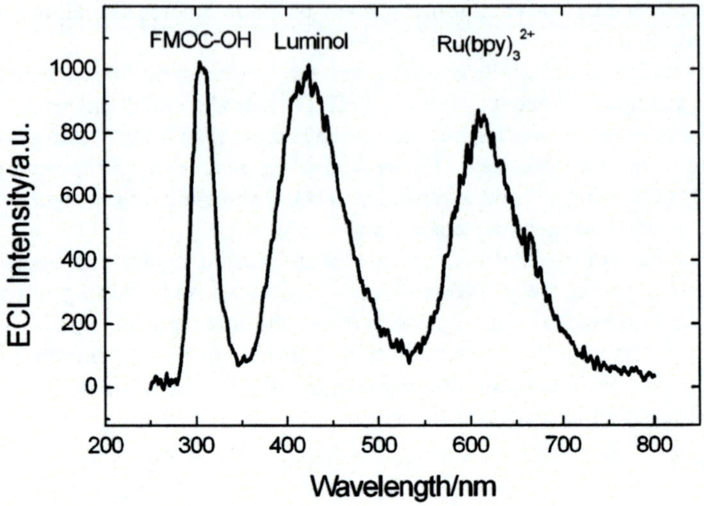

Fig. 5 ECL spectrum of a solution containing 1×10^{-6} mol/l FMOC-OH, 1×10^{-5} mol/l luminol, and 5×10^{-7} M $Ru(bpy)_3^{2+}$. The n-Si disk electrode was anodized at 5.7 V prior to ECL measurement. Conditions: 0.2 M borate buffer adjusted to pH 7.8 with sulfuric acid, 5×10^{-4} M $K_2S_2O_8$, pulse charge 120 μC, pulse frequency 80 Hz, pulse voltage 40 V [49]

1.2 Excitation Pathways

$Ru(bpy)_3^{2+}$ chelate is a well-studied luminophore (with derivatives usable as excellent labels in bioaffinity assays) and therefore we have often used it as one of the model compounds in our studies. Both in the cases of hydrated electron-induced chemiluminescence [54] and hot electron-induced ECL [4, 40], the results conform with the above-mentioned red-ox [(2) followed by (3)] and ox-red [(4) followed by (5)] excitation pathways:

$$e^-_{aq} + S_2O_8^{2-} \rightarrow SO_4^{\bullet -} + SO_4^{2-}, \quad (1)$$

$$Ru(bpy)_3^{2+} + e^-_{aq} \rightarrow Ru(bpy)_3^{+}, \quad (2)$$

$$Ru(bpy)_3^{+} + SO_4^{\bullet -} \rightarrow Ru(bpy)_3^{2+*} + SO_4^{2-}, \quad (3)$$

$$Ru(bpy)_3^{2+} + SO_4^{\bullet -} \rightarrow Ru(bpy)_3^{3+} + SO_4^{2-}, \quad (4)$$

$$Ru(bpy)_3^{2+} + e^-_{aq} \rightarrow Ru(bpy)_3^{2+*}. \quad (5)$$

After these excitation steps, the excited $Ru(bpy)_3^{2+*}$ then relaxes radiatively to the ground state, emitting at 620 nm. Both excitation pathways are energetically possible, but due to the much shorter lifetime of the one-electron reduced form of the chelate in comparison to that of one-electron oxidized chelate, the ox-red pathway is probably the predominating excitation route [4, 40].

A more general mechanism can be written for a luminophore L, which can be one of the luminescent organic luminophores or metal chelates we have studied:

$$e^-_{\text{hot}}(\text{electrode}) \rightarrow e^-_{\text{hot presolvated}}(\text{in the conduction band of water}), \quad (6)$$

$$e^-_{\text{hot presolvated}} \rightarrow e^-_{\text{aq}}, \quad (7)$$

$$L + e^-_{\text{aq}}(\text{or } e^-_{\text{hot presolvated}}) \rightarrow L^{\bullet -}, \quad (8a)$$

$$L^{\bullet -} + \text{Ox}^{\bullet} \rightarrow L^* + \text{Ox}^-, \quad (8b)$$

$$L + \text{Ox}^{\bullet} \rightarrow L^{\bullet} + \text{Ox}^-, \quad (9a)$$

$$L^{\bullet +} + e^-_{\text{aq}}(\text{or } e^-_{\text{hot presolvated}}) \rightarrow L^*, \quad (9b)$$

$$L^* \rightarrow L + h\nu. \quad (10)$$

Here, $\text{Ox}^{\bullet}$ is an one-electron oxidant, which can be cathodically produced from coreactants or even more simply by the F^+ or E_1' centers of the oxide film used as an insulating film of the electrode surface [3, 51, 60]. The reactions 8a and 8b present the oxidation-initiated reductive pathway (ox-red), while the reactions 9a and 9b show the reduction-initiated oxidative pathway (red-ox).

However, some luminophores such as luminol and its close analogues show chemiluminescence in which the cathodic generation of oxidizing radicals seems to be sufficient to trigger the luminescent pathway [58, 61, 68, 69].

1.3 Time-Resolved Detection

In addition to the usability of wavelength discrimination in multiparametric assays, HECL provides also the possibility to use time discrimination because luminophores with very different luminescence lifetimes can be simultaneously excited by cathodic pulses. The time constants of the electrolytic cell and excitation electronics set the limits of how short-lived HECL-displaying labels can be detected by time-resolved measurements. As already shown in Fig. 1, the HECL lifetime of some lanthanide chelates is of the order of milliseconds [50, 51], while that of $Ru(bpy)_3^{2+}$ and its derivatives is in the microseconds range [4, 40] and most of the organic luminophores show even much faster decay [30, 38, 85].

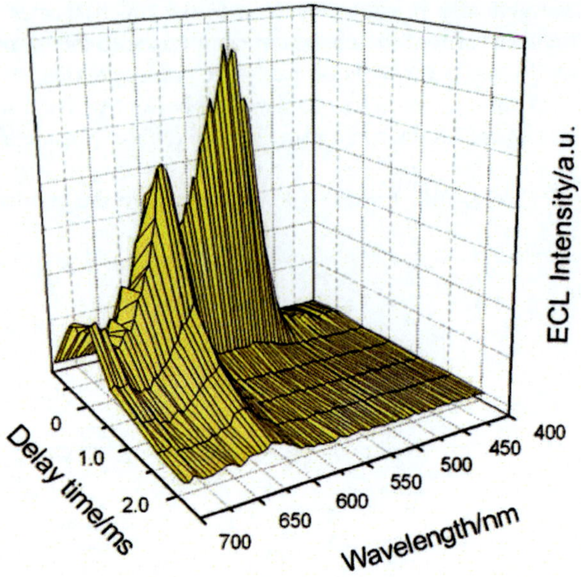

Fig. 6 Simultaneous excitation of FITC and Eu (III)-1-NCS at oxide-coated aluminum electrode [35]. Conditions: oxide-covered Al-strip working electrode, Pt-wire counter electrode, coulostatic pulse generator, applied pulse voltage −40 V, pulse frequency 20 Hz, pulse charge 120 μC, 0.2 M boric acid buffer at pH 9.2, 1×10^{-3} M $K_2S_2O_8$. Time-resolved spectra were measured with an instrument having relatively good sensitivity but poor resolution [76]

Figure 6 displays the time-resolved HECL spectrum of simultaneously excited aromatic Eu(III) chelate and fluorescein isothiocyanate (FITC). Aromatic Tb(III) chelates are best labels for utilization in bioanalytical HECL methods, due to their long luminescence lifetime and high HECL intensity [50, 51]. However, also some Eu(III) chelates seem to be usable although their HECL intensity is lower and luminescence lifetime is shorter in comparison to those of the corresponding Tb(III) chelates [21, 34]. Also some metalloporphyrins show relatively long-lived HECL emission but their use is somewhat problematic in solutions containing molecular oxygen [88].

HECL methods allow simultaneous excitation of some organic labels, $Ru(bpy)_3^{2+}$-based labels, and aromatic Eu(III) and Tb(III) chelates. Organic luminophores generally show a short-lived singlet state emission, but lanthanide chelates show long-lived ligand sensitized emission of the central ion, or in the cases of Gd(III) chelates, a relatively long-lived triplet state emission of the ligand, which allows the creation of multiparameter bioaffinity assays. Thus, novel types of immunoassays and DNA-probe assays can be developed in which a combination of time-resolution and wavelength discrimination can be used in the detection of the labels. The highest sensitivity-demanding parameter must be preferably labeled with Tb(III) chelates, and the parameters where less sensitivity is required should be labeled with organic luminophores, transition metal chelate labels, or with metalloporphyrins.

Time-resolved detection of the best so far developed Tb(III) chelates yield linear calibration curves spanning many orders of magnitude of concentration with detection limits close to picomolar level [50, 51].

2 Instrumentation

The apparatus for HECL generation need some kind of function generator because only pulsed excitation is possible, probably due to the indiffusion of protons during long-term cathodic polarization [51]. For research purposes, the function generator should preferably be programmable to allow easy generation of different voltage waveforms, e.g., if anodic oxide film are to be fabricated in situ during the measurement [3, 24].

Some kind of a potentiostat or a more simple pulsed voltage source is needed. The potential of the working electrode is normally controlled by means of a potentiostat. In many HECL applications, a real reference electrode is not necessary and it is sufficient to use a pseudo reference electrode, or even to use the counter electrode as a pseudo reference electrode. In the latter case, only the voltage between the working electrode and counter electrode is controlled, i.e., the real working electrode potential is reproducible only if the properties of the electrochemical cell remain unchanged between separate measurements. However, typical commercial potentiostats designed for electrochemists are often unable to generate sufficiently high voltages and current between the working and auxiliary electrode. In simple two-electrode cells, we have often used a coulostatic pulse generator, which gives always a fixed charge pulse before giving a trigger pulse for light detection unit [55].

The electronic circuitry used depends on the method of measurement. At low light levels, photon counting allows better discrimination against the noise than analogue recording. However, both of these methods are usable, although one can anticipate that with the continuous decrease in the cost of electronics photon counting will eventually win. The only instrumental advantage of analogue recording of the photomultiplier current is better linearity at high light levels owing to the pulse pile-up in photon counting. In the time-resolved measurements, the detection circuit is synchronized with the excitation pulses, allowing a fixed or variable time delay between the pulse and recording window.

As light detectors, we would at the moment recommend channel photomultiplier tubes (CPM), which are currently available at a quite reasonable price. CPM tubes are available with different cathode areas and sold separately or as nice photon counting heads, which are quite versatile, reliable, and easy to use.

In our earlier work, we have mainly used laboratory-made equipments consisting of either a laboratory-made coulostatic pulse generator or a commercial potentiostat with a laboratory-made pulse generator, a two-electrode cell, a photon counter unit, and an ordinary photomultiplier tube designed for photon counting. The HECL intensity has been measured with a photon counter unit consisting of a photomultiplier, a Stanford Research Systems photon counter, and a Nucleus MCS-II scale card attached to a computer. Measurements of HECL spectra have been performed with an ordinary Perkin-Elmer LS-5 or LS-50 B luminescence spectrometer with the electrodes put in the cuvette. However, better performance can presently be obtained with spectrometers containing sensitive CCD array or matrix chips allowing binning of the vertical rows.

Figure 7 shows a scheme of the laboratory-made ECL cell designed for metal band electrodes, and Fig. 8 that of another ECL cell type designed for the use of planar disk and slide electrodes, which is usable for silicon and coated glass electrodes.

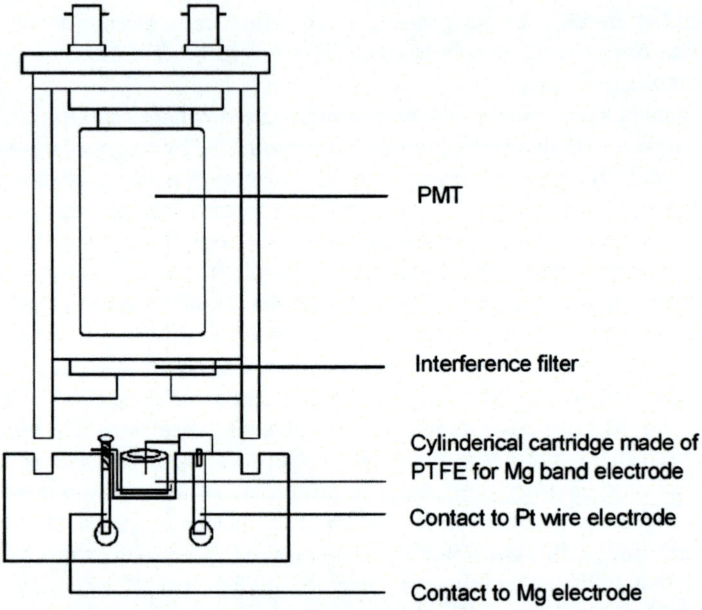

Fig. 7 The cell constructed for the use of Mg band electrodes. Mg band was first threaded through two slits close to the bottom of cylindrical PTFE cartridge. Then the cartridge was pushed into a cylindrical bowl also made of PTFE, and the end of Mg band was put under a screw, which was tightened against a brass bar giving an electrical contact to the negative polarity of a voltage source [24]

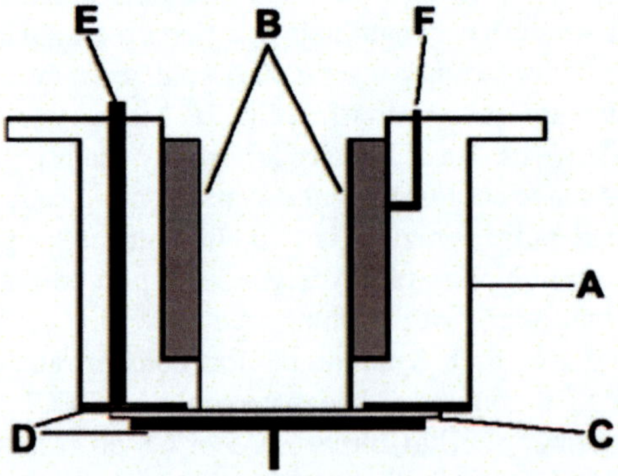

Fig. 8 Side view of ECL cell for measurements using silicon plate as working electrodes: (A) Teflon support; (B) Pt-counter electrode; (C) Al or Si plate working electrode; (D) working electrode contact; (E) counter electrode electrical contact wire. With this design, the electrical contact to the working electrode can be taken either from the bottom or from the upper surface sides of the working plate or disk electrode [31]

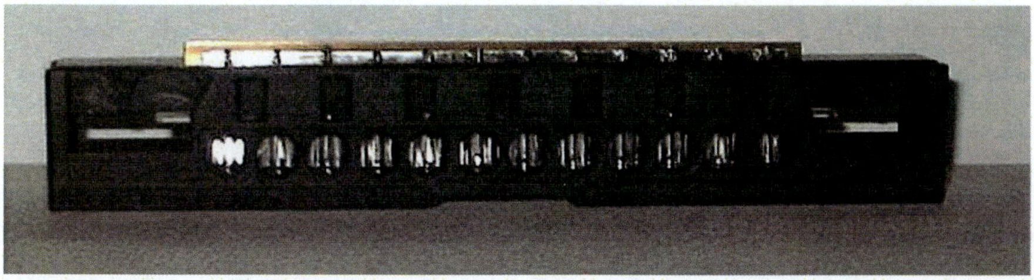

Fig. 9 Micro titer strip, Al rake working electrode, and Pt-counter electrodes assembled in a rack ready to be put on the conveyer track of the semiautomatic electrochemiluminometer [17]

Aluminum easily allows various electrode shape designs, e.g., the use of curved rakes, which fit in the standard 12-well microtiter strip and provide a nice light collection efficiency when HECL is measured through the walls of a microtiter strip [17] (Fig. 9).

2.1 Electrode Materials

The choice of counter electrode material in HECL cells seems not to be very critical. Use of platinum or gold is of course always safe, and thin layers of these metals may even be sufficiently optically transparent that the ECL can be measured through them [55]. Indium tin oxide (ITO) is also an excellent optically transparent counter electrode material. Often even stainless steel wire or tube is acceptable as counter electrode material in the time scale of HECL measurements.

The working electrode applied in HECL can have various shapes and structures. The conductor in the C/I/E junctions can either be a metal with a rather low work function [24, 51] or a strongly doped degenerate semiconductor [3, 23, 40].

Earlier, we have assumed that only very wide band gap (of the order of 8–9 eV) insulating oxides such as alumina, magnesia, and silica could be used as insulators in C/I/E junctions [3, 49]. However, it seems that much smaller band gaps are sufficient, whenever the oxide is still a true insulator (Fermi level in the mid-gap region) and is not behaving as an intrinsic semiconductor, which is common to many metal oxides [72].

It is rather interesting that also double barrier tunnel emission electrodes of type M/I/M/I work nicely [27], and thus possibly also different material layers could be applied.

For the actual HECL measurements, borate buffer seems to be the best buffer whenever its buffer capacity coincides in the right pH region. This is due to the very low reactivity of boric acid and borate ions with hydrated electron and many other radicals [10, 73].

2.2 Silicon Electrodes

The silicon wafers stored in contact with air have a natural oxide film, but for HECL applications the natural films are too thin (ca. 1 nm) [67] and usually not uniformly distributed on the silicon. Therefore, the film has to be manufactured on the silicon by repeatable techniques such as chemical vapor deposition [90], molecular beam epitaxy, or plasma and wet anodization and sputtering [8].

Both strongly doped n- and p-silicon allow a relatively easy fabrication of excellent Si/SiO_2 electrodes by thermal oxidation by rather standard procedures in clean rooms [31, 40]. However, quite good electrodes can be easily fabricated in any laboratory with normal laboratory ovens and even in room air [36].

2.3 Aluminum Electrodes

Generally, aluminum electrodes are less efficient than silicon electrodes in terms of signal intensity to cathodic background solid-state luminescence. However, they are also very low cost and easy to use. While the best quality silicon electrodes have to be oxidized under clean room conditions and cut to a specific size with special saw, aluminum electrodes at their simplest can be cut in any size from the Al band or sheet with ordinary scissors and just pressed flat or into the shape or cup before use [41]. In many cases, especially for purposes of disposable bioaffinity assay cartridges, a thin vacuum evaporated pure aluminum film that is just let to be oxidized in oxygen atmosphere or air seems to be the best choice, if low cost is one of the main issues.

The aluminum material we have been using a lot (nominally 99.9% pure Al band by Merck) has had a thick enough natural oxide film to produce strong HECL. Thus, deliberate oxidation of the film or anodic pulses before analysis have not been necessary. However, earlier batches of this aluminum worked much better than later batches. After some change of manufacture process of this aluminum band, the natural aluminum oxide film started to show a much stronger solid-state electroluminescence [47] during the cathodic pulse. In addition, this higher intensity solid-state electroluminescence contained also a slowly decaying component, which was very harmful for time-resolved measurements. This problem could be solved by coating the natural oxide layer of the electrode with ca. 100 nm thick layer of 99.999% Al and letting this oxidize in air at room temperature for 1 day. The background signal became very low and the actual analyte luminescence slightly increased [27], thus increasing hugely the sensitivity. The precision of the measurement was also increased.

Alumina is much less chemically resistant than silica and therefore Si/SiO_2 electrodes are usable in a wider pH range than Al/Al_2O_3 electrodes. In addition, if, e.g., bioaffinity assays are carried out on the surfaces of these electrodes and slight warming of the assay buffer (37°C) is applied, the alumina films seem to lose their insulating properties quite rapidly [17]. This kind of behavior does not occur on corresponding silica films.

2.4 Other Electrode Materials

Magnesia dissolves in acidic solutions [37, 66], but it is insoluble in alkaline solutions. Therefore, for applications in which extremely high pH is necessary for some reason, Mg/MgO electrodes are probably the best choice because aluminum electrodes are not usable above pH 10 [62] and also silica starts to dissolve under the more alkaline conditions.

Our research on using relatively impure (99.5%) oxide-covered Mg electrodes [24] showed that Mg/MgO electrodes can be used as disposable electrodes in an analogous way to Al/Al_2O_3 electrodes, even though we did not even use very alkaline electrolyte solutions.

Sometimes it would be beneficial to have an optically transparent working electrode. We have shown that both n-ZnO:Al/Y_2O_3 [23] and n-ZnO:Al/MgO [24] can be used, but they are quite expensive for disposable devices. If an optically transparent spacer made of PDMS or equivalent material is used between the n-ZnO:Al/insulator and ITO glasses, one could form a thin layer cell, which would be optically transparent in all directions. These kind of cells would not only allow HECL detection simultaneously by two detectors through both counter and working electrodes, but also alternatively time-resolved spectrophotometric observation of the rise and decay of different radicals generated in the cell by a cathodic pulse in both directions, parallel and perpendicular to the working electrode surface.

3 Applicable Luminophores in HECL-Based Detection

The tunnel emission of hot electrons through the insulating oxide film during cathodic pulse polarization of the electrode allows electrochemical excitation of a variety of luminophores emitting in a wide optical range. The following summarizes most of the luminophores we have used in HECL studies.

Table 1 shows the HECL lifetimes of different luminophores we have so far explored.

3.1 Metal Chelates

Many metal chelates are versatile luminophores and are often used as labels (markers, tags, and reporters) in bioanalysis and imaging [63, 80]. Our studies have mainly centered on rare earth and transition metal chelates, with the strongest interest in Tb(III) chelates [23, 25, 50, 51, 55, 59] and Ru(II) chelates [4, 33].

Typically multidentate lanthanide chelates are stable in a quite wide pH range but they can be fast dissociated in strongly acidic solutions [80] and many of them

Table 1 Comparison of HECL lifetimes

Compound	Lifetime (s)	Electrode, solution	Reference
Aromatic Tb(III) chelates	1.3×10^{-3} to 2.2×10^{-3}	Al/Al_2O_3; 1.0 M Na_2SO_4 in 0.2 M $H_3B_4O_3$ at pH 9.2	Kulmala et al. [51]
Eu(III) chelate	0.94×10^{-3}	Al/Al_2O_3; 1.0 mM $K_2S_2O_8$ in 0.05 M $Na_2B_4O_7$ at pH 7.8	Jiang et al. [35]
Pt(II) coproporphyrin	14.5×10^{-6} to 44.8×10^{-6} (three-step decay)	Al/Al_2O_3; 1.0 mM $K_2S_2O_8$ in 0.05 M $Na_2B_4O_7$ at pH 9.2	Canty et al. [11]
$Ru(bpy)_3{}^{2+}$	6.8×10^{-6} to 95×10^{-6} (three-step decay)	Al/Al_2O_3; 3.0 mM $K_2S_2O_8$ in 0.05 M $Na_2B_4O_7$ at pH 9.2	Jiang et al. [40]
$Ru(bpy)_2L_a{}^{2+}$	40×10^{-6}	Si/SiO_2; 1.0 mM $K_2S_2O_8$ in 0.05 M $Na_2B_4O_7$ at pH 9.2	Jiang et al. [39]
Luminol	3.0×10^{-3} and 7.2×10^{-3} (two-step decay)	Al/Al_2O_3; 1.0 mM $K_2S_2O_8$ in 0.05 M $Na_2B_4O_7$ at pH 9.2	Suomi et al. [84]
Lucigenin	18.4×10^{-6}	Al/Al_2O_3; 1.0 mM $K_2S_2O_8$ in 0.05 M $Na_2B_4O_7$ at pH 9.2	Jiang et al. [35]
Rhodamine B	19×10^{-6} (with $K_2S_2O_8$ ca. 12×10^{-6})	Al/Al_2O_3; 10.0 mM NaN_3 in 0.05 M $Na_2B_4O_7$ at pH 9.2	Jiang et al. [38]
Fluorescein	17×10^{-6} and 60×10^{-6} (two-step decay)	Al/Al_2O_3; 0.05 M $Na_2B_4O_7$ + 0.1 M Na_2SO_4 at pH 9.2	Ylinen et al. [91]
7-Hydroxy-4-methyl coumarin	6.8×10^{-6} and 25×10^{-6} (two-step decay)	Al/Al_2O_3; 0.05 M $Na_2B_4O_7$ + 0.1 M Na_2SO_4 at pH 9.2	Ylinen et al. [91]

start to lose their central ion as $Ln(III)OH_3$ precipitate during prolonged incubation under highly alkaline conditions [26].

Luminescent aromatic Tb(III) and Eu(III) chelates show HECL by ligand sensitized mechanism [35, 50, 51, 54]. The ligand is first excited by the ox-red or red-ox pathways, after which it transfers the excitation energy intramolecularly to the central ion exactly as in the photoluminescence of these chelates [63, 80] and finally the central ion emits light by its f-shell transitions. The requirements are that (a) the triplet state of the ligand must be higher in energy than the resonance level of the central lanthanide(III) ion, (b) the triplet state of the ligand must not react with the central ion, and (c) the excited central ion must not react with the ligand [63, 80].

When Tb(III) and Eu(III) are replaced by Gd(III) ion, the energetics of some ligands and central ions result in triplet state HECL emission of the ligand [46] and in singlet state emission of the ligand when Y(III) ion is used as central ion [50].

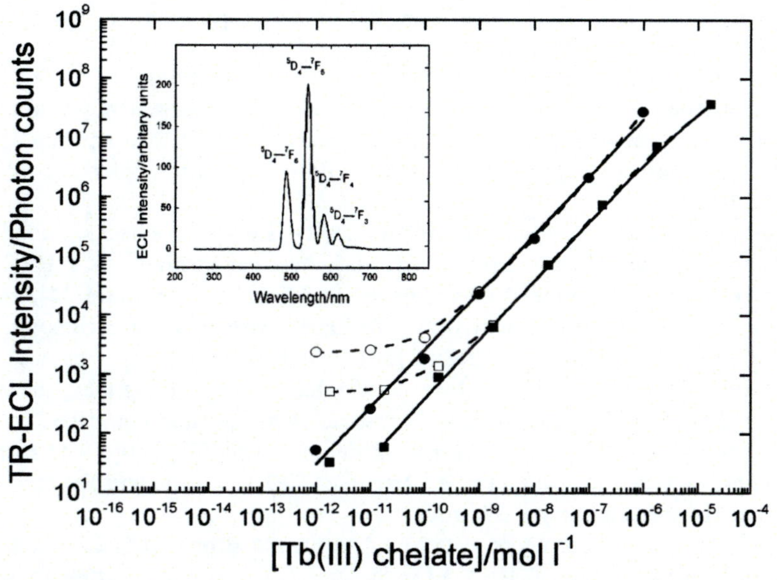

Fig. 10 Calibration curves of Tb(III) label chelates. (*Filled circles*) Tb^{3+}-2,6-bis[*N*,*N*-bis (carboxymethyl)aminomethyl]-4-methylphenol, (*filled squares*) Tb^{3+}-N^1-(*p*-isothiosyanatobenzyl)-diethylene triamine-N^1, N^2, N^3, tetra acetate. Conditions: 0.2 M borate buffer, pH 7.75, 0.1% NaN_3, pulse amplitude −10 V. *Open circles* and *squares* denote signal before subtracting the blank. Measured with semiautomatic electrochemiluminometer [17]

Tb(III) chelates can produce very intense HECL (Fig. 10), and therefore we have used them as labels in bioaffinity assays [5, 17] and as known good luminophores in the comparisons of different analysis conditions. Another strong point of Tb(III) chelates is their long-lived luminescence, 1.3–2.2 ms [51], which allows the use of time-resolved detection. Thus, the short-lived solid-state electroluminescence of the insulating films induced during the cathodic pulse can be discriminated, and as a result of the large signal-to-noise ratio, the detection limits can be very low [50, 51].

Since Tb(III) is very redox-inert, the above-mentioned requirements are normally easily fulfilled, but in the case of easily reducible Eu(III) the system is more complicated. In the studies, we found out [35] that one of the phenolic chelates (2,6-bis[*N*,*N*-bis(carboxymethyl)-aminomethyl]-4-methyl phenol-chelated Eu(III)) was not luminescent at all, although the corresponding Tb(III) chelate yields the highest HECL intensity obtained so far. A somewhat modified chelate, 2,6-bis [*N*,*N*-bis(carboxymethyl)-aminomethyl]-4-benzoylphenol-chelated Eu(III), gave a weak HECL emission, which could be strongly increased by the addition of peroxodisulfate. In the presence of azide, this chelate did not emit HECL. That gives one a possibility of yet another type of separation of label emissions: if both Tb(III) and Eu(III) are chelated with the latter ligand, in the presence of peroxodisulfate both HECL emissions will be triggered, while in the presence of azide only Tb(III) is triggered.

The synthesis of several phenolic Tb(III) chelates displaying strong HECL has been published [43], but as far as we know, these are not commercially available at the moment. However, some of the slightly less efficient HECL-emitting Tb(III) labels are commercially available from Perkin-Elmer Life Sciences, Wallac, Finland [17, 63].

$Ru(bpy)_3^{2+}$, tris(2,2′-bipyridine)ruthenium(II), is one of the best studied ECL-producing compounds [7, 18, 19, 33, 63]. We have shown that HECL of $Ru(bpy)_3^{2+}$ can be efficiently excited at oxide-coated aluminum [4], n-type silicon [31], and p-type silicon [40]. We have also studied a carboxylic acid derivative of the $Ru(bpy)_3^{2+}$ chelate [39], which can be used as a label with antibodies. This carboxylic acid derivative has a long enough luminescence lifetime for TR-ECL measurements for some purposes, and compared with an unmodified $Ru(bpy)_3^{2+}$ it has a bit red-shifted emission [38]. Figure 11 displays simultaneous HECL excitation of an aromatic Tb(III) chelate and $Ru(bpy)_3^{2+}$ at an oxide-covered aluminum electrode.

Platinum(II) coproporphyrin was found to have a strong ECL emission, with detection limit as low as ca. 10^{-11} M in the presence of peroxodisulfate ions [11]. The metalloporphyrin had previously been reported to show long-lived phosphorescence, and the electrochemiluminescence lifetime was sufficiently long [11] to allow time-resolved detection. However, with these labels, we recommend using azide as coreactant instead of peroxodisulfate, as the calibration curve with azide is much steeper although the detection limit is somewhat higher [11].

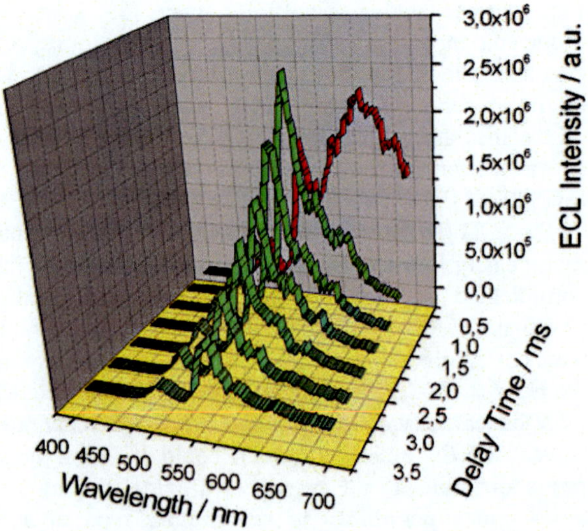

Fig. 11 Simultaneous ECL excitation of $Ru(bpy)_3^{2+}$ and a Tb(III) chelate at an oxide-coated aluminum electrode. Conditions: pulse voltage −10 V, pulse time 500 μs, pulse frequency 20 Hz, 0.05 M $Na_2B_4O_7$ at pH 9.2, 0.1 M Na_2SO_4 [4]

3.2 *Organic Dyes*

Many organic dyes used and synthesized for the purposes of photoluminescent labeling are also showing HECL [22, 30, 38, 64, 91]. However, it is necessary (1) that at least one of the primary radicals must be able to initiate the ox-red or re-ox excitation pathway and (2) the resulting luminophore radicals must not be too reactive with water [2]. Typically, the HECL and fluorescence emission spectra are similar which shows that the singlet state excitation of the luminophores is occurring also in HECL excitation pathways [5, 22, 30, 38, 64, 91].

For instance, fluorescein is a strongly emitting HECL-luminophore with detection limits in the subnanomolar level [91]. Also some coumarins show very high HECL intensity, with detection limit at approximately 0.1 nM [30].

Rhodamine B is a xanthene dye with many applications in photoluminescence. Its photostability has made it an attractive label in photoluminescent labeling and imaging [32], and it is also a strong HECL emitter [38]. The luminescence lifetime of Rhodamine B is relatively short, and therefore it could easily be used in time-resolved multilabelling experiments together with a label, e.g., Tb(III) chelate, with a very long luminescence lifetime [38].

SYBR Green I has been used since early 1990s as a fluorescent marker stain for double-stranded DNA and oligonucleotides. We were interested in finding out whether the dye would also be electrochemiluminescent [64]. SYBR Green I emitted strongly enough under hot electron-induced ECL conditions to give a detection limit of ca. 1 nM. However, it is by no means as efficient as the best electrochemiluminophores we have studied.

UV-emitting labels [48] as well as lanthanide chelates are attractive for HECL energy transfer assays [63] as donors, but so far we have not noticed that these kinds of experiments would have been published.

Derivatives of luminol [58, 84] and lucigenin [35] also show strong HECL, but their excitation probably occurs via several parallel routes and not just by simple ox-red or red-ox excitation pathways.

The drawback of using organic labels as HECL labels is that their luminescence lifetimes are typically very short, and low detection limits as those obtained by time-resolved detection of Tb(III) chelate labels cannot typically be reached. In the case of short-lived HECL-emitting labels, n-silicon doped with antimony and coated with thermal silica film seems to be the best electrode material studied so far due to its low solid-state electroluminescence background during the cathodic excitation pulse [40].

4 Choice of Coreactants

As mentioned above, borate buffer is the best choice for the actual HECL measuring stage because all the other buffers are more efficient free radical scavengers, and hence decrease the efficiency of the generation of HECL from the luminophores.

During other assay stages of a bioaffinity assay, the other commonly used buffers can freely be used if the washing steps are allowed in the assay [51].

Those luminophores for which hydroxyl radical is a good oxidant in the excitation pathway/pathways can normally be excited without any other additives than the dissolved oxygen that naturally exists in the aqueous excitation electrolyte solution [5, 38, 85]. In these cases, the HECL intensity can be enhanced by the addition of an optimal concentration of hydrogen peroxide to obtain the optimal amounts of the highly reducing and oxidizing radicals [38, 85].

For many luminophores, reaction with hydroxyl radical results in the disintegration of the molecule or metal chelate in aqueous solution rather than in HECL excitation via ox-red, red-ox, or other routes [5, 48]. The addition of a relatively high concentration of azide ions often solves this problem [55, 85] (Fig. 12). This is due to the fact that azide ion is an efficient hydroxyl radical scavenger producing fast azide radical [10], which typically only acts as one-electron oxidant [73]. However, azide radical has a relatively low oxidation power, and it has to be energetically sufficient in the excitation pathways of the luminophore or excitation cannot occur at all [5, 17, 51].

Some of the luminophores, which very efficiently disintegrate after reaction with hydroxyl radical, are very efficiently excited when the sulfate radical is used as an oxidant in the excitation pathways [2, 4, 5]. Typically about 1 mM solution of peroxodisulfate ions is optimal for the HECL excitation.

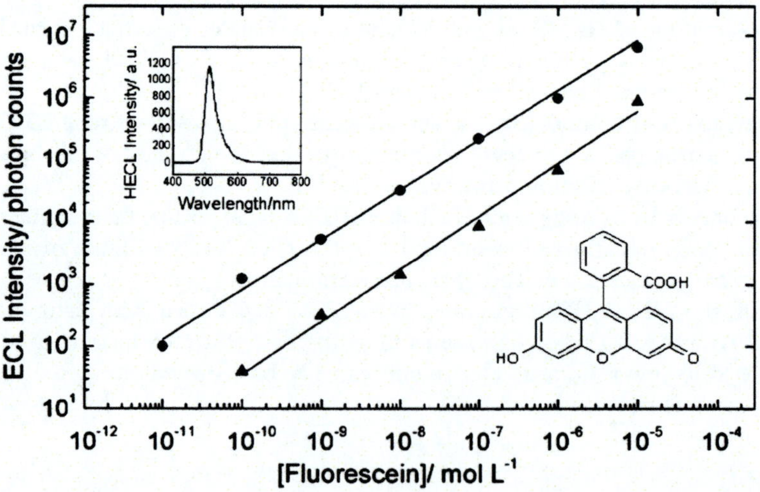

Fig. 12 Calibration curves of fluorescein at disposable oxide-coated aluminum electrodes (*triangles*) in the absence of azide ions and (*squares*) in the presence of 0.1 M NaN_3. Conditions: pulse voltage −10 V, pulse time 200 μs, pulse frequency 50 Hz, 0.05 M $Na_2B_4O_7$ at pH 9.2, 0.1 M Na_2SO_4, HECL was measured through a 515-nm interference filter [85]

Regardless of whether hydroxyl or sulfate radicals are primarily generated in the cell, they may "be converted" to other less oxidizing radicals by the addition of halide or pseudohalide salts [5, 10, 73], if the secondary radicals produced from halides or pseudo halides are more suitable oxidants for the system [55, 85].

5 Applications

HECL can be applied in detection of labels in various bioaffinity assay formats. Mainly Tb(III) chelate labels have so far been applied in heterogeneous [4, 5, 27, 31, 42, 51, 55] as well as homogeneous immunoassays [17, 42, 55] and recently also in DNA-probe assays [81].

Both immunometric (noncompetitive) [4, 5, 27, 31, 42, 51, 55] and competitive [17] immunoassay formats can be applied. The reason why homogeneous, i.e., separation free immunoassays [17, 42, 55] are possible, is that the presolvated hot and hydrated electrons are able to carry out the reductions only up to a distance of about 200 nm from the oxide film surface and, hence, only the labels within this range can be excited. In practice, this means that if the amount of the electrochemiluminescently labeled antibody is appropriately optimized for the immunometric assay, only the labels involved in the surface-antibody$_1$-antigen-labeled antibody$_2$ immunocomplexes are cathodically excited. Most of the unreacted labeled antibodies are too far from the electrode surface to be excited [63]. However, heterogeneous assays based on HECL detection have always lower detection limits than homogeneous assays, in which an optimal HECL measuring buffer can be used without the radical scavenging additives useful for the incubation buffer of the bioaffinity reaction stages [4, 5, 27, 55].

Although latex particles can be used as a solid phase of the assays [51], the detection limits are many orders of magnitude lower if the oxide surface of the electrode itself is being used as the solid phase of the assays [5].

We have mainly used human thyroid stimulating hormone (hTSH) and C-reactive protein (hCRP) as model analytes in our studies due to the great imporance of these analytes in clinical diagnostics and especially in rapid diagnostics [5, 27, 55]. hTSH is one of the analytes demanding extremely high sensitivity of the assays due to its low concentration in blood, whereas even the normal range of hCRP is very high in comparison to that of hTSH.

During the acute phase response to infection, inflammatory disease, surgery, trauma, and cancer, concentrations of hCRP in blood increase by many orders of magnitude and return to normal levels with resolution of the disease [65, 74]. Currently, there is coercive evidence that C-reactive protein (CRP) is also a sensitive marker for the development of cardiovascular disease in the general population, and recent studies suggest that CRP is not only a biomarker but also an active crucial mediator in the pathogenesis of atherosclerosis [65, 74].

CRP can provide prognostic information about the risk of future coronary events in apparently healthy persons. This application requires assays of higher sensitivity than has been available in the clinical laboratory, and therefore it seems to be well-suited for HECL detection-based assays [5]. On the basis of the work carried out in the Finnish universities [5], Labmaster Ltd, Turku, Finland, has further developed a rapid assay carried out using a disposable plastic cartridge containing an oxide-coated silicon electrode as working electrode and a small-sized HECL-based analyzer. The whole assay can be carried out in 6 min (Fig. 13).

The most interesting future developments seem to be truly homogeneous HECL energy transfer assays, in which none of the reactants of the bioaffinity assays are attached on the surface of the electrode prior to the assay, and possibly also the use of electroluminescent nanoparticles and quantum dots as labels [63].

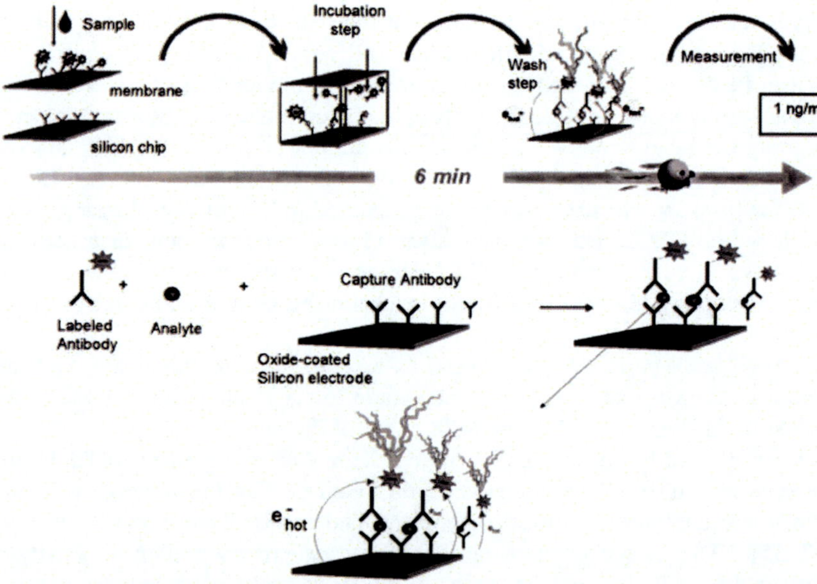

Fig. 13 Ultra sensitive immunometric immunoassays of hCRP. The immunoreaction reaches equilibrium in 5 min, and the total assay time is 6 min. The detection limit is 0.2 μg/L (blank + 3 SD). The linear assay range is about four orders of magnitude of concentration. The labeled secondary antibody and other reagents are dried in a porous membrane on which the sample is added. Sample (3.5 μL) dissolves the labeled antibody, and the analyte starts to form immunocomplexes with it while this liquid is filling the thin layer between the membrane and originally dry oxide-coated silicon electrode, which is also precoated with the primary (catching) antibody and suitable additives. After 5 min of incubation, the membrane is removed from top of the electrode and the electrode in the cartridge is washed with HECL measuring buffer, and the HECL intensity is measured

References

1. Ala-Kleme T, Haapakka K, Latva M (1999) Near-infrared electrogenerated chemiluminescence of ytterbium(III) chelates in aqueous electrolytes. Anal Chim Acta 395:205–211
2. Ala-Kleme T, Kulmala S, Jiang Q (2006) Generation of free radicals and electrochemiluminescence from simple aromatic molecules in aqueous solutions. Luminescence 21:118–125
3. Ala-Kleme T, Kulmala S, Latva M (1997) Generation of free radicals and electrochemiluminescence at pulse-polarized oxide-covered silicon electrodes in aqueous solutions. Acta Chem Scand 51:541–546
4. Ala-Kleme T, Kulmala S, Väre L, Juhala P, Helin M (1999) Hot electron-induced electrogenerated chemiluminescence of $Ru(bpy)_3^{2+}$ Chelate at oxide-covered aluminum electrodes. Anal Chem 71:5538–5543
5. Ala-Kleme T, Mäkinen P, Ylinen T, Väre L, Kulmala S, Ihalainen P, Peltonen J (2006) Rapid electrochemiluminoimmunoassay of human C-reactive protein at planar disposable oxide-coated silicon electrodes. Anal Chem 78:82–88
6. Arnold D, Cartier E, DiMaria DJ (1994) Theory of high-field electron transport and impact ionization in silicon dioxide. Phys Rev B 49:10278–10297
7. Bard AJ (ed) (2004) Electrogenerated chemiluminescence. Marcel Dekker, New York, pp 1–540
8. Barnett SA, Ferguson IT (1995) Introduction and general discussion. In: Glocker DA, Ismat Shah S (eds) Handbook of thin film process technology. Institute of Physics Publishing Ltd, Bristol, Chapter A2.0
9. Blackburn GF, Shah HP, Kenten JH, Leland J, Kamin RA, Link J, Peterman J, Powell MJ, Shah A, Talley DB (1991) Electrochemiluminescence detection for development of immunoassays and DNA probe assays for clinical diagnostics. Clin Chem 37:1534–1539
10. Buxton GV, Greenstock CL, Helman WP, Ross AB (1988) Critical review of rate constants for reactions of hydrated electrons, hydrogen atoms and hydroxyl radicals ($^{\bullet}OH/^{\bullet}O^{-}$) in aqueous solution. J Phys Chem Ref Data 17:513–886, and references therein
11. Canty P, Väre L, Håkansson M, Spehar A-M, Papkovsky D, Ala-Kleme T, Kankare J, Kulmala S (2002) Time-resolved electrochemiluminescence of platinum(II) coproporphyrin. Anal Chim Acta 453:269–279
12. Debad JD, Glezer EN, Wohlstadter J, Sigal GB, Leland JK (2004) Clinical and biological applications of ECL. In: Bard AJ (ed) Electrogenerated chemiluminescence. Marcel Dekker, New York, pp 359–396
13. Diesing D, Ruess S, Otto A, Lohrengel MM (1995) Time resolved charge transfer with hot electrons at the silver/electrolyte interface. Ber Bunsen Ges 99:1402–1405
14. DiMaria DJ, Cartier E (1995) Mechanism for stress-induced leakage currents in thin silicon dioxide films. J Appl Phys 78:3883–3894
15. DiMaria DJ, Fischetti MV (1988) Vacuum emission of hot electrons from silicon dioxide at low temperatures. J Appl Phys 64:4683–4691
16. Ege D, Becker WG, Bard AJ (1984) Electrogenerated chemiluminescent determination of Ru (bpy)3(2+) at low levels. Anal Chem 56:2413–2417
17. Eskola J, Mäkinen P, Oksa L, Loikas K, Nauma M, Jiang Q, Håkansson M, Suomi J, Kulmala S (2006) Competitive immunoassay by hot electron-induced electrochemiluminescence detection and using a semiautomatic electrochemiluminometer. J Lumin 118:238–244
18. Fähnrich KA, Pravda M, Guilbault GG (2001) Recent applications of electrogenerated chemiluminescence in chemical analysis. Talanta 54:531–559
19. Gerardi RD, Barnett NW, Lewis SW (1999) Analytical applications of Tris(2,2′-bipyridyl) ruthenium(III) as a chemiluminescent reagent. Anal Chim Acta 378:1–43, and references cited therein
20. Gleria M, Memming R (1976) Novel luminescence generation by electron transfer from semiconductor electrodes to ruthenium-bipyridil complexes. Z Phys Chem 101:171–179
21. Haapakka K, Kankare J, Kulmala S (1988) Cathodic europium(III) electroluminescence at an oxide-coated aluminum electrode. Anal Chim Acta 209:165–174

22. Haapakka K, Kankare J, Puhakka O (1988) Fluorophor-enhanced cathodic electroluminescence at an oxide-covered aluminum electrode. Anal Chim Acta 207:195–210
23. Håkansson M, Helin M, Putkonen M, Jiang Q, Kotiranta M, Suomi J, Ala-Kleme T, Kulmala S (2005) Electrochemiluminescence of Tb(III) chelates at optically transparent tunnel emission electrodes fabricated by atomic layer deposition. Anal Chim Acta 541:137–143
24. Håkansson M, Jiang Q, Helin M, Putkonen M, Niskanen AJ, Pahlberg S, Ala-Kleme T, Heikkilä L, Suomi J, Kulmala S (2005) Cathodic Tb(III) chelate electrochemiluminescence at oxide-covered magnesium and n-ZnO:Al/MgO composite electrodes. Electrochim Acta 51:289–296
25. Håkansson M, Jiang Q, Spehar A-M, Suomi J, Kotiranta M, Kulmala S (2005) Direct current-induced electrogenerated chemiluminescence of hydrated and chelated Tb(III) at aluminum cathodes. Anal Chim Acta 541:171–177
26. Håkansson M, Jiang Q, Spehar A-M, Suomi J, Kulmala S (2006) Extrinsic lyoluminescence of aluminium induced by lanthanide chelates in alkaline aqueous solution. J Lumin 118:272–282
27. Håkansson M, Jiang Q, Suomi J, Loikas K, Nauma M, Ala-Kleme T, Kankare J, Juhala P, Eskola JU, Kulmala S (2006) Cathodic electrochemiluminescence at double barrier $Al/Al_2O_3/Al/Al_2O_3$ tunnel emission electrodes. Anal Chim Acta 556:450–454
28. Handy RM (1966) Hot electron energy loss in tunnel cathode structures. J Appl Phys 37: 4620–4628
29. Helin M, Håkansson M, Canty P, Spehar A-M, Kulmala S (2002) Hot electron-induced electrogenerated chemiluminescence of 1-aminonaphthalene-4-sulfonate at oxide-covered aluminium electrodes in aqueous solution. Anal Chim Acta 454:193–201
30. Helin M, Jiang Q, Ketamo H, Håkansson M, Spehar A-M, Kulmala S, Ala-Kleme T (2005) Electrochemiluminescence of coumarin derivatives induced by injection of hot electrons into aqueous electrolyte solution. Electrochim Acta 51:725–730
31. Helin M, Väre L, Håkansson M, Canty P, Hedman H-P, Heikkilä L, Ala-Kleme T, Kankare J, Kulmala S (2002) Electrochemiluminoimmunoassay of hTSH at disposable oxide-coated n-silicon electrodes. J Electroanal Chem 524–525:176–183
32. Houghten RA, Dooley CT, Appel JR (2004) De novo identification of highly active fluorescent kappa opioid ligands from a rhodamine labeled tetrapeptide positional scanning library. Bioorg Med Chem Lett 14:1947–1951
33. Jiang, Q. (2006) Electrochemiluminescent and chemiluminescent labels for bioaffinity assays. Doctoral thesis, Helsinki University of Technology, Espoo, Finland
34. Jiang Q, Håkansson M, Spehar A-M, Ahonen J, Ala-Kleme T, Kulmala S (2006) Hot electron-induced time-resolved electrogenerated chemiluminescence of a europium(III) label in fully aqueous solutions. Anal Chim Acta 558:302–309
35. Jiang Q, Håkansson M, Suomi J, Ala-Kleme T, Kulmala S (2006) Cathodic electrochemiluminescence of lucigenin at disposable oxide-coated aluminum electrodes. J Electroanal Chem 591:85–92
36. Jiang Q, Ketamo H, Niskanen AJ, Suomi J, Håkansson M, Kulmala S (2006) Effects of thermal oxidation conditions of silicon electrodes on cathodic electrochemiluminescence of $Ru(bpy)_3^{2+}$ chelate. Electrochim Acta 51:3332–3337
37. Jiang Q, Kotiranta M, Langel K, Suomi J, Håkansson M, Spehar A-M, Ala-Kleme T, Eskola J, Kulmala S (2005) Ruthenium(II) tris(2,2′-bipyridine) chelate as a chemiluminophore in extrinsic lyoluminescences of aluminium and magnesium in aqueous solution. Anal Chim Acta 541:179–186
38. Jiang Q, Spehar A-M, Håkansson M, Suomi J, Ala-Kleme T, Kulmala S (2006) Hot electron-induced cathodic electrochemiluminescence of rhodamine B at disposable oxide-coated aluminum electrodes. Electrochim Acta 51:2706–2714
39. Jiang Q, Sun S, Håkansson M, Langel K, Ylinen T, Suomi J, Kulmala S (2006) Electrochemiluminescence and chemiluminescence of a carboxylic acid derivative of ruthenium(II) tris-(2,2′-bipyridine) chelate synthesized for labeling purposes. J Lumin 118:265–271

40. Jiang Q, Suomi J, Håkansson M, Niskanen AJ, Kotiranta M, Kulmala S (2005) Cathodic electrogenerated chemiluminescence of $Ru(bpy)_3^{2+}$ chelate at oxide-coated heavily doped silicon electrodes. Anal Chim Acta 541:159–165
41. Kankare J, Fälden K, Kulmala S, Haapakka K (1992) Cathodically induced time-resolved lanthanide(III) electroluminescence at stationary aluminum disk electrodes. Anal Chim Acta 256:17–28
42. Kankare J, Haapakka K, Kulmala S, Näntö V, Eskola J, Takalo H (1992) Immunoassay by time-resolved electrogenerated luminescence. Anal Chim Acta 266:205–212
43. Kankare J, Karppi A, Takalo H (1994) Novel labeling agents for immunoassay by time-resolved electrogenerated chemiluminescence. Anal Chim Acta 295:27–35
44. Kenten JH, Casadei J, Link J, Lupold S, Willey J, Powell M, Rees A, Massey R (1991) Rapid electrochemiluminescence assays of polymerase chain reaction products. Clin Chem 37: 1626–1632
45. Kulmala A, Kulmala S, Ala-Kleme T, Hakanen A, Korpela T (1999) Hydrated electron-induced chemiluminescence of thallium(I) ion. Anal Chim Acta 382:233–243
46. Kulmala S, Ala-Kleme T (1997) Room temperature phosphoresence emission in fully aqueous medium induced by injection of hot electrons into an electrolyte solution. Anal Chim Acta 355:1–5
47. Kulmala S, Ala-Kleme T, Hakanen A, Haapakka K (1997) F-centre luminescence from oxide-covered aluminium cathode induced by two step reduction of peroxydisulfate anions. J Chem Soc Faraday Trans 93:165–168
48. Kulmala S, Ala-Kleme T, Heikkilä L, Väre L (1997) Energetic electrochemiluminescence of (9-fluorenyl)methanol induced by injection of hot electrons into aqueous electrolyte solution. J Chem Soc Faraday Trans 93:3107–3113
49. Kulmala S, Ala-Kleme T, Joela H, Kulmala A (1998) Hot electron injection into aqueous electrolyte solution from thin insulating film-coated electrodes. J Radioanal Nucl Chem 232:91–95
50. Kulmala S, Ala-Kleme T, Kulmala A, Papkovsky D, Loikas K (1998) Cathodic electrogenerated chemiluminescence of luminol at disposable oxide-covered aluminum electrodes. Anal Chem 6:1112–1118
51. Kulmala S, Ala-Kleme T, Latva M, Loikas K, Takalo H (1998) Hot electron-induced electrogenerated chemiluminescence of rare earth(III) chelates at oxide-covered aluminum electrodes. J Fluor 8:59–65
52. Kulmala S, Ala-Kleme T, Väre L, Helin M, Lehtinen T (1999) Hot electron-induced electrogenerated luminescence of Tl(I) at disposable oxide-covered aluminum electrodes. Anal Chim Acta 398:41–47
53. Kulmala S, Haapakka K (1995) Mechanism of electrogenerated luminescence of terbium(III)-{2,6-bin[N, N-bis(carboxymethyl)aminomethyl]-4-benzoylphenol} chelate at an oxide-covered aluminum electrode. J Alloy Compd 225:502–506
54. Kulmala S, Hakanen A, Raerinne P, Kulmala A, Haapakka K (1995) Ruthenium(II) tris-(2,2′-bipyridine)-specific extrinsic lyoluminescences of x-ray irradiation colored and electrolytically colored alkali halides. Anal Chim Acta 309:197–210
55. Kulmala S, Håkansson M, Spehar A-M, Nyman A, Kankare J, Loikas K, Ala-Kleme T, Eskola J (2002) Heterogeneous and homogeneous electrochemiluminoimmunoassays of hTSH at disposable oxide-covered aluminum electrodes. Anal Chim Acta 458:271–280
56. Kulmala S, Helin M, Ala-Kleme T, Väre L, Papkovsky D, Kulmala A (1999) Electrochemiluminescent labels for applications in fully aqueous solutions at oxide-covered aluminium electrodes. Anal Chim Acta 386:1–6
57. Kulmala S, Kulmala A, Ala-Kleme T, Hakanen A, Haapakka K (1997) Intrinsic and 1-aminonaphthalene-4-sulphonate-specific extrinsic lyoluminescence of X-ray irradiated sodium chloride. Anal Chim Acta 340:245–256

58. Kulmala S, Kulmala A, Ala-Kleme T, Pihlaja J (1998) Primary cathodic steps of electrogenerated chemiluminescence of lanthanide(III) chelates at oxide-covered aluminum electrodes in aqueous solution. Anal Chim Acta 367:17–31
59. Kulmala S, Kulmala A, Helin M, Hyppänen I (1998) Hot electron-induced time-resolved electrogenerated luminescence of Tb(III) ions at stationary oxide-covered aluminum electrodes. Anal Chim Acta 359:71–86
60. Kulmala S, Kulmala A, Latva M (1997) X-ray irradiated sodium chloride as an excitation source for the sensitized terbium(III)-specific chemiluminescence of aromatic Tb(III) chelates in aqueous solutions. Anal Chim Acta 347:333–350
61. Kulmala S, Matachescu C, Joela H, Lilius E-M, Kupila E-L (1997) UV-irradiated potassium peroxodiphosphate and nitrate as excitation sources for luminol chemiluminescence. J Chem Soc Faraday Trans 93:3497–3504
62. Kulmala S, Matachescu C, Kulmala A, Papkovsky D, Håkansson M, Ketamo H, Canty P (2002) Chemiluminescence of luminol induced by dissolution of oxide-covered aluminum in alkaline aqueous solution. Anal Chim Acta 453:253–267
63. Kulmala S, Suomi J (2003) Current status of modern analytical luminescence methods. Anal Chim Acta 500:21–69, and references cited therein
64. Laakso P, Anttila H, Kairisto V, Eskola J, Kulmala S, Ala-Kleme T (2005) Hot electron-induced electrogenerated chemiluminescence of SYBR Green I. Anal Chim Acta 541:85–89, and references therein
65. Labarrere CA, Zaloga GP (2004) C-reactive protein: from innocent bystander to pivotal mediator of atherosclerosis. Am J Med 117:499–507
66. Mach R (1993) Thin film electroluminescence devices. In: Kitai AH (ed) Solid state luminescence. Chapman & Hall, London, pp 229–262
67. Mackintosh WD, Plattner HH (1977) The identification of the mobile ion during the anodic oxidation of silicon. J Electrochem Soc 124:396–400
68. Matachescu C, Kulmala S, Ala-Kleme T, Joela H (1997) Mechanism and analytical applicability of luminol-specific extrinsic lyoluminescence of UV-irradiated potassium peroxodisulfate. Anal Chem 69:3385–3390
69. Matachescu C, Kulmala S, Laine E, Raerinne P (1997) Luminol-specific extrinsic lyoluminescence of x-ray irradiated sodium chloride. Anal Chim Acta 349:1–10
70. Mead C (1961) Operation of tunnel-emission devices. J Appl Phys 32:646–652
71. Memming R (1969) Mechanism of the electrochemical reduction of persulfates and hydrogen peroxide. J Electrochem Soc 116:785–790
72. Morrison SR (1980) Electrochemistry at semiconductor and oxidized metal electrodes. Plenum Press, New York
73. Neta P, Huie RE, Ross AB (1988) Rate constants for reactions of inorganic radicals in aqueous solution. J Phys Chem Ref Data 17:1027–1284
74. Pepys MB, Hirschfield GM (2003) C-reactive protein. A critical update. J Clin Invest 111: 1805–1812
75. Piazza S, Splendore A, DiPaola A, Sunseri C, DiQuarto F (1993) Cathodic photoemission processes at the aluminum-electrolyte interface in the initial stage of passive film formation. J Electrochem Soc 140:3146–3152
76. Pihlajamäki SV, Kankare JJ (1986) Photon counter system with a rotating interference filter for measuring time-resolved electroluminescence spectra. Anal Instr 15:171–192
77. Pleskov YuV, Rotenberg ZA (1978) Photoemission of electrons from metals into electrolyte solutions as a method for investigation double-layer and electrode kinetics. In: Gerischer H, Tobias C (eds) Advances in Electrochemistry and Electrochemical Engineering. Wiley, New York, pp 1–124
78. Richter MM (2004) Electrochemiluminescence (ECL). Chem Rev 104:3003–3036
79. Schmidt WF (1997) Liquid-state electronics of insulating liquids. CRC, Boca Raton, Florida, p 260

80. Soini E, Lövgren T (1987) Time-resolved fluorescence of lanthanide probes and applications in biotechnology. Crit Rev Anal Chem 18:105–154, and references cited therein
81. Spehar-Délèze A-M, Suomi J, Jiang Q, de Rooij N, Koudelka-Hep M, Kulmala S (2006) Heterogeneous oligonucleotide-hybridization assay based on hot electron-induced electrochemiluminescence of a rhodamine label at oxide-coated aluminum and silicon electrodes. Electrochim Acta 51:5438–5444
82. Sung Y-E, Bard AJ (1998) Enhancement of electrochemical hot electron injection into electrolyte solutions at oxide-covered tantalum electrodes by thin platinum films. J Phys Chem B 102:9806–9811
83. Sung Y-E, Gaillard F, Bard AJ (1998) Demonstration of electrochemical generation of solution-phase hot electrons at oxide-covered tantalum electrodes by direct electrogenerated chemiluminescence. J Phys Chem B 102:9797–9805
84. Suomi J, Håkansson M, Jiang Q, Kotiranta M, Helin M, Niskanen AJ, Kulmala S (2005) Time-resolved detection of electrochemiluminescence of luminol. Anal Chim Acta 541:165–167
85. Suomi J, Ylinen T, Håkansson M, Helin M, Jiang Q, Ala-Kleme T, Kulmala S (2006) Hot electron-induced electrochemiluminescence of fluorescein in aqueous solution. J Electroanal Chem 586:49–55
86. Sze SM, Crowell CR, Carey GP, LaBate EE (1966) Hot-electron transport in semiconductor-metal-semiconductor structures. J Appl Phys 37:2690–2695
87. Tajima S (1977) Luminescence, breakdown and coloring of anodic oxide films on aluminum. Electrochim Acta 22:995–1011
88. Väre L, Canty P, Håkansson M, Spehar A-M, Papkovsky D, Ala-Kleme T, Kankare J, Kulmala S (2002) Time-resolved electrochemiluminescence of metalloporphyrins. Anal Chim Acta 453:269–279
89. White HS, Bard AJ (1982) Electrogenerated chemiluminescence. 41. Electrogenerated chemiluminescence and chemiluminescence of the tris(2,2′-bipyridine)ruthenium(2+)-peroxydisulfate(2-) system in acetonitrile-water solutions. J Am Chem Soc 104:6891–6895
90. Wolf S, Tauber RN (1986) Silicon Processing for the VLSI Era, vol 1. Lattice Press, Sunset Beach, USA
91. Ylinen T, Suomi J, Helin M, Ala-Kleme T, Kulmala S (2006) Time-resolved detection of hot electron-induced electrochemiluminescence of fluorescein in aqueous solution. J Fluoresc 16:27–33

Extension of Fluorescence Response to the Near-IR Region

Tarek A. Fayed

Abstract Development of efficient and photostable fluorophores that absorb and emit in the red and near-infrared (NIR, between 650 and 900 nm) spectral regions are of interest in many fields as optical engineering, analytical chemistry, biology, and medicine as new revolutionary tools for noninvasive and simple in vivo optical imaging. NIR fluorescent dyes have been often spotlighted owing to the advantages of this region as significant reduction of the background signal due to the lowest autoabsorption and autofluorescence of biomolecules in the NIR region, low-light scattering and deep penetration of NIR light, and the possibility to use low-cost excitation light sources. The present chapter provides an up-to-date overview on the different classes of NIR organic fluorophores, including cyanine and squaraine dyes, conformationally restricted and long chain-substituted boron-dipyrromethene (BODIPY) dyes, thiazines and oxazines, and pyrelene bisimides. The molecular structures, strategies to shift the absorption and emission bands of such dyes into the NIR region of the spectrum, and spectroscopic properties are comprehensively presented with hint to their applications.

1 Introduction

Many of the modern and important advances in physical chemistry, analytical chemistry, materials sciences, biology, and medicine have been made possible because of the availability of sophisticated extrinsic fluorescent reporters. These are molecular sensors whose fluorescence properties are transformed into signals useful for analytical, biochemical, or medical applications upon chemical or

---

T.A. Fayed (✉)
Chemistry Department, Faculty of Science, Tanta University, 31527 Tanta, Egypt
e-mail: tarek.fayed@science.tanta.edu.eg; tfayed2003@yahoo.co.uk

C.D. Geddes (ed.), *Reviews in Fluorescence 2009*, Reviews in Fluorescence 2009,
DOI 10.1007/978-1-4419-9672-5_4,

biochemical recognition of specific guests, and can be grouped into two major classes [1, 2]:

1. Synthetic systems such as organic dyes, inorganic nanoparticles, and lanthanide coordination complexes.
2. Biological systems such as fluorescent and bioluminescent proteins.

It is unlikely that any one of such luminescent systems will emerge as the universal solution for all optical applications, and each class warrants further development. Organic fluorescent dyes are most frequently used as reporters in optical chemical sensing. Their advantages are not only easy availability and low price, but also their versatility [2]. Organic dyes provide means for the direct determination of the location, activities, or concentrations of selected ions or electrically neutral species like biomolecules, without pretreatment of the sample. Although the use of organic dyes as fluorescent probes is not a new technique, it remains a dynamic and growing technology that continues to find new and more demanding applications. Therefore, the challenge for chemists is to develop new photoactive organic molecules that can probe specific molecular events with great sensitivity and selectivity.

Currently, there is considerable interest in developing of efficient and photostable fluorophores that absorb and emit in the red and near-infrared (NIR) spectral regions [3–7]. Herein, we focus on fluorescent organic dyes that emit in the NIR region (optical window 650–900 nm).

2 Characteristics of NIR Fluorescent Organic Dyes

NIR fluorescent dyes found wide applications in several areas in analytical chemistry [8] and have been applied for various devices such as laser dyes and organic light-emitting diodes (OLEDs) [9]. They are also attractive as probes for contrast agents for fluorescence imaging of biomedical samples [10]. There are several major advantages of using NIR fluorescent dyes over those emitting at shorter wavelengths [1]. The most important of these advantages is the reduction in background that ultimately improves the sensitivity achievable. Likewise, the number of fluorescent impurities is significantly reduced with excitation and detection at longer wavelengths. Besides reduced background, a further advantage is that low-cost, energy efficient, rugged diode lasers can be used in place of the more expensive and shorter lived gas lasers [8]. A variety of highly sensitive detectors in the visible-near-IR region are now available. Because there is very little undesired absorption and autofluorescence by common biomolecules at wavelengths greater than 650 nm, individual antigen and antibody molecules can be detected in serum samples [11]. Indeed, wavelengths above 650 nm can penetrate through skin and tissues, therefore, fluorescent NIR imaging probes are increasingly used for in vivo optical imaging of live animals [12]. Additional benefits associated with the use of NIR absorbing fluorophores include expanding the palette of multicolor imaging systems, such as those used in DNA sequencing and intracellular organelle labeling [13].

With the advent of both semiconductor and optical fiber detectors as well as laser diodes excitation sources, instruments working within the NIR region began to emerge, which opened up opportunities for researchers that had not previously been available [8]. However, challenges with many NIR fluorescent dyes are that the excitation and emission bands significantly overlap, and the fluorescence quantum yield is low due to conformational flexibility which leads to diminished fluorescence intensity.

The molecular structures of NIR fluorescent dyes are highly conjugated and have a lower energy gap between the ground and excited states compared to the visible region fluorescent dyes. The physical and chemical properties of NIR dyes are adjustable for different applications through chemical modification of the structure of the dye molecules. These properties include solubility of the dye in aqueous solutions, tuning of excitation and emission wavelengths, molecular aggregation, the biocompatibility in a given matrix, the binding ability of the dye to the target for a single analyte, etc. [1]. The structural modifications provide the dye molecules with much broader applications, particularly in the biological and analytical fields. Modifications may change emission wavelength and reaction ability of the dye molecule. Thus, the potential changes in the activity of the molecules must be considered. For example, as larger ligands are employed, more steric interactions may occur between the dye and the target, which reduce the efficiency.

In addition to the parameters characterizing organic fluorescent dyes and necessary for optical sensing applications [2], some other factors are important for the selection of NIR dyes as fluorescent reporters, and are going to be summarized below [1].

2.1 *Improving Water Solubility and Reducing Aggregation*

To effectively use NIR fluorescent dyes for sensing applications particularly for biological samples, a hydrophilic nature is usually essential. However, a number of NIR dyes are not water soluble due to their highly π-conjugated structures. Thus, suitable modification of such dyes is needed prior to their bioapplications. Gaining solubility in aqueous solutions is often achieved by linking of polar groups like sulfonate ($-SO_3^-$) groups to the dye structure [14], which bring negative charges to the dye molecules. However, the large number of negative charges may hinder the binding between the dye and the negatively charged bioanalytes. Alternatively, the dye molecules can be assembled inside a hydrophilic shell that has a hydrophobic inner layer such as phospholipids monolayer [15].

In addition to the hydrophobicity problem, aggregation is another major drawback of NIR fluorescent dyes applications. The dye molecules can easily aggregate in aqueous solutions, resulting in low fluorescence intensities and blue-shifted absorption peak. Two effective methods can prevent the aggregation. One is to mix a certain amount of water-soluble organic solvents into the aqueous solution to dissolve the dyes. The second way to prevent aggregation is to conjugate the dye molecule with other large biomolecules like a long “tail” of nucleotides or human serum albumin [16].

2.2 Separation of Excitation and Emission Bands

For many NIR fluorescent dyes, the excitation and emission bands are significantly overlapped. Only when the two bands are discrete, the dye can be effectively used for the detection of targets. A longer wavelength gap between the two maxima indicates a greater sensitivity. So far, two approaches can separate the emission and excitation bands effectively [1]. The first approach is to change the molecular structure of the dye slightly, keeping the binding properties with no change. In fact, modification of dye molecular structures is not always feasible in its spectra. Alternatively, two dyes can be used as one effective NIR dye if one dye emission band overlaps with the excitation of the other [17]. Based on fluorescence resonance energy transfer (FRET), one dye can be a donor and the other is an acceptor.

2.3 Fluorescence Lifetime of NIR Dye Molecules

Fluorescence lifetime is an important property of dye molecules. A longer lifetime gives excited electrons a greater possibility to release energy through nonradiative transitions. The lifetime of NIR fluorescent dyes can be changed upon modifications of dye molecular structures. It has been reported that the NIR dye molecules containing different heavy atoms (halogen) can alter their fluorescence lifetime. The lifetimes of these dyes vary with the identity of the halogen substitution near the center of the dye. The heavier halogen atom gives the longer lifetime [18]. The heavier atom shows a larger rate of intersystem crossing. Thus, the effect of modification of molecular structures on NIR dye lifetime should be considered prior to any modifications. Increased lifetime results in lower fluorescence quantum yield, and thus lower signal intensity.

So far three main classes of NIR dyes including cyanine dyes, squaraine dyes, thiazine, and oxazine dyes are well known and widely used in fluorescence sensing. Recently, a novel class of conformationally restricted boron-dipyrromethene (BODIPY) dyes has been reported. This class shows high photo- and chemical stabilities and become a promising NIR reagent. This chapter will provide an up-to-date overview on the different classes of long-wavelength organic fluorophores. The molecular structures, strategies to shift the absorption and emission bands into the NIR region of the spectrum, and spectroscopic properties are comprehensively presented with hint to their applications.

3 NIR Fluorescent Organic Dyes

3.1 Cyanine Dyes

NIR fluorescent cyanine dyes are increasingly employed as fluorescent reporters for interrogating biological processes such as enzyme activity [19], calcium and zinc [14, 20–22], nitric oxide [23] and proton concentration [24, 25], as well as

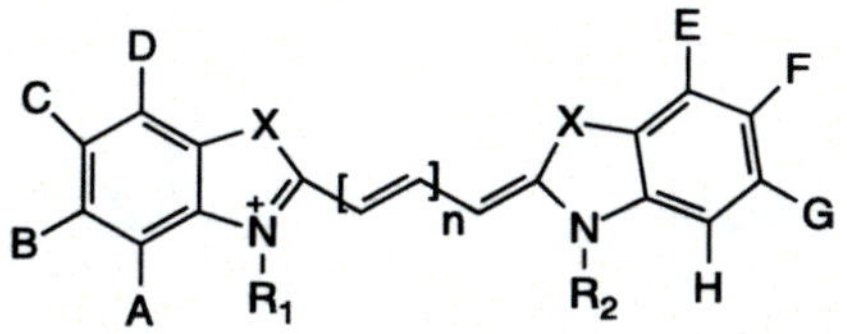

Fig. 1 Generic structure of a cyanine dye. A–H are hydrogens or functionalities including SO_3H or may be a labeling position. X = $C(CH_3)_2$, O, or S, R_1 and R_2 = alkyl groups and can include a labeling position, $n = 1–3$

in vivo imaging [16]. The appeal to this class of fluorophores derives from their straightforward syntheses, broad wavelength tunability, and particularly their high NIR absorption and emission wavelengths, large molar extinction coefficients, and moderate fluorescence quantum yields.

The molecules of cyanine dyes (Cy) are composed of a polymethine chain with two heterocyclic units at its terminals (Fig. 1). One of the nitrogen atoms is a positively charged iminium and the other is an amine. The length of the polymethine chain determines the spectral range of absorption and emission, with its increase by one vinylene unit (CH = CH), the spectra shift to longer wavelengths by about 100 nm [8].

3.1.1 Bioconjugate NIR Fluorescent Cyanine Dyes

Indocyanines, featuring two indole or benzindole rings linked by a polymethine chain, are the most important type of cyanine dyes used for bioconjugation. The absorption and emission wavelengths can be readily adjusted by varying the length of the polymethine chain and selecting indole rather than benzindole. The reactive group for conjugation to the probe molecule is typically NHS ester or isothiocyanate present on R1 and/or R2 group [26].

Most cyanine dyes self-aggregate in aqueous solutions [2], however, substitution by sulfonate groups on the aromatic rings of the dyes is very effective on increasing solubility and decreasing aggregation in water. Two examples of these sulfoindocyanine dyes are Cy7 (absorption/emission: 750 nm/777 nm) and Cy5 (absorption/emission: 650 nm/667 nm) which are highly soluble in water and fluoresce brightly [27]. The Cy5.5 and Cy7 NHS esters are important NIR dyes which all feature the sulfoindolocyanine structures. The dyes Cy5.18 and Cy7.18 have a linear polymethine chain structure. When a long chain ($n > 3$) is present, the double-bond conjugation system becomes less rigid, thus reducing the extinction coefficient and fluorescence yield. A bridged polymethine chain can improve this situation and make the long link less flexible. Two examples of dyes with this feature are IRDye 78 and IRDye 800 phosphoramidite which have a bridged cyclohexene ring [26].

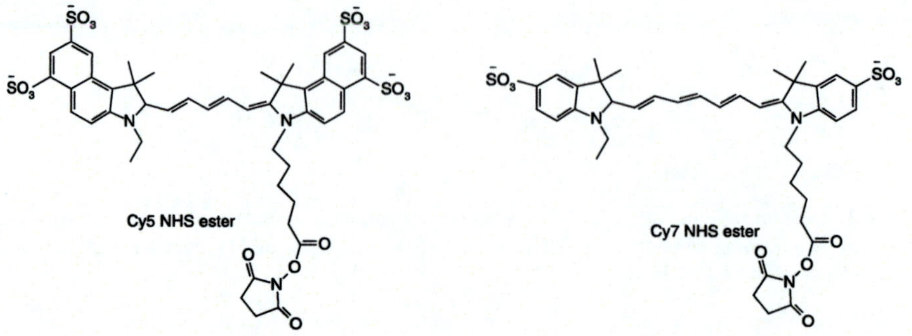

Although the arylsulfonate structure is an important development for cyanine dyes used in bioconjugation applications, further significant improvements have also been made by bonding the reactive groups directly to the aromatic ring [26] rather than the indole nitrogens. A longer wavelength dye in this class is the NIR820 NHS ester which has a maximum absorption at 790 nm and a maximum emission at 820 nm [26]. NIR820 dye, with a cyclohexene bridged heptamethine chain, has one NHS ester group directly attached to the indole ring. Another alternative attachment of the reactive group, namely attachment to the aromatic ring via a sulfamidoalkyl chain [26], has been described as, for example, IRD78-NHS ester. This group emphasizes that a space between the reactive group and the aromatic ring is useful to avoid a negative effect on the fluorescence efficiency of the dye.

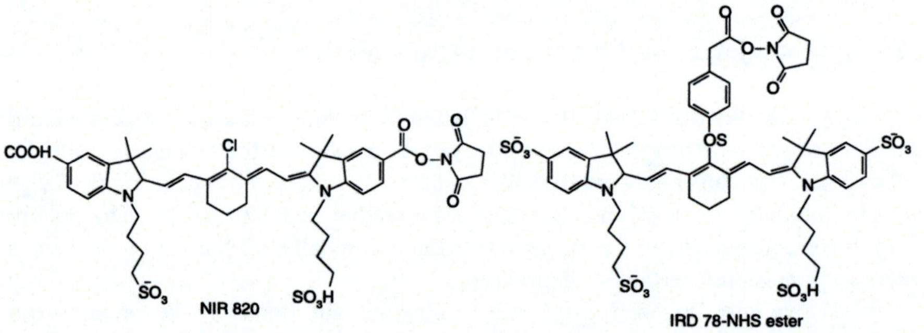

A set of NIR emitting heptamethine cyanine dyes (HCDs), **1–3**, have designed and synthesized for biological applications via NHS ester conjugation [27]. The dyes are water soluble and unaffected by pH variation, and show high photostability compared to commercial dyes with similar features. Among the synthesized dyes, HCDN, **3**, shows a broad absorption band covering the visible range (400–700) and a very broad emission band centered at 755 nm with a Stokes shift of more than 100 nm. While the position of the absorption and emission bands is not affected by the interaction between dye and Goat protein, increasing dye/protein ratio induces a linear increase of fluorescence performances of HCDs up to 30-fold excess. These spectroscopic features make HCDs suitable for multitasking fluorescence analysis in the NIR, IgG labeling, in vivo imaging, fluorescence microscopy, and FRET applications.

A conjugate between a NIR indocyanine dye and an organic polyamine polymer (polyethylenimine, PEI), **IR820-PEI**, [10] led to increased photostability and better optical properties (high fluorescence quantum yield and long emission maximum). **IR820-PEI** absorbs at 665 nm and emits at 780 nm displaying a large Stokes shift (115 nm). Also, it provides a useful reactive site that is able to bind DNA and can act as a multifunctional system for targeted gene delivery by conjugation.

With selected molecules, the delivery process can be monitored in vivo with noninvasive optical imaging techniques.

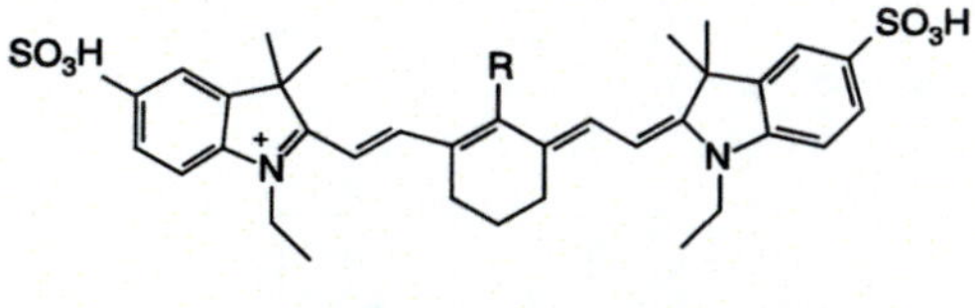

1 HCDS ; R = S-Ph-NHCO(CH2)3COOH
2 HCDO; R = NHCO(CH2)3COOH
3 HCDN; R = NH(CH2)3COOH

Heptamethine cyanine dyes (HCDs) **1-3**

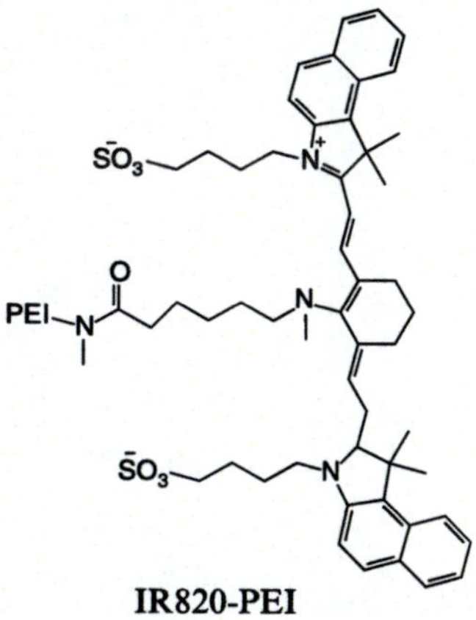

IR820-PEI

Two NIR cyanine fluorophores **NIR5.5-2** and **NIR7.0-2** which are analogues of commercially available **Cy 5.5** ($\lambda_{abs} = 674$ nm, $\lambda_{em} = 694$ nm) and **Cy 7.0** ($\lambda_{abs} = 750$ nm, $\lambda_{em} = 777$ nm), respectively, have been synthesized [28, 29]. The two dyes display absorption and emission maxima in the NIR region (**NIR5.5-2** $\lambda_{abs} = 688$ nm, $\lambda_{em} = 714$ nm; **NIR7.0-2** $\lambda_{abs} = 802$ nm, $\lambda_{em} = 818$ nm), large extinction coefficients (147,000 M^{-1} cm^{-1}), excellent solubility, and no tendency to self-aggregate in aqueous solutions. The fluorescence efficiency is compatible with their use in optical imaging applications and NIR fluorescence studies. The utility of these fluorescent labeling reagents is illustrated by the preparation of an internally quenched FRET-based fluorescent probe of caspase-3 protease, where efficient quenching effect between **NIR5.5-2** and **NIR7.0-2** at 705 and 798 nm was demonstrated. The observation of an efficient formation of an intramolecular heterodimer yielding a nonfluorescent ground state complex between the two NIR cyanine dyes in aqueous solutions opens the door for the use of this efficiently quenched fluorophore pair using coil-containing longer peptidic backbones (Fig. 2).

3.1.2 NIR Fluorescent Cyanine-Based Chemosensors

The accurate measurement of intracellular concentration of proton and metal cations like potassium, calcium, magnesium, zinc, and mercury, with penetration to multiple layers of cell tissue and high sensitivity, remains as an innovative analytical

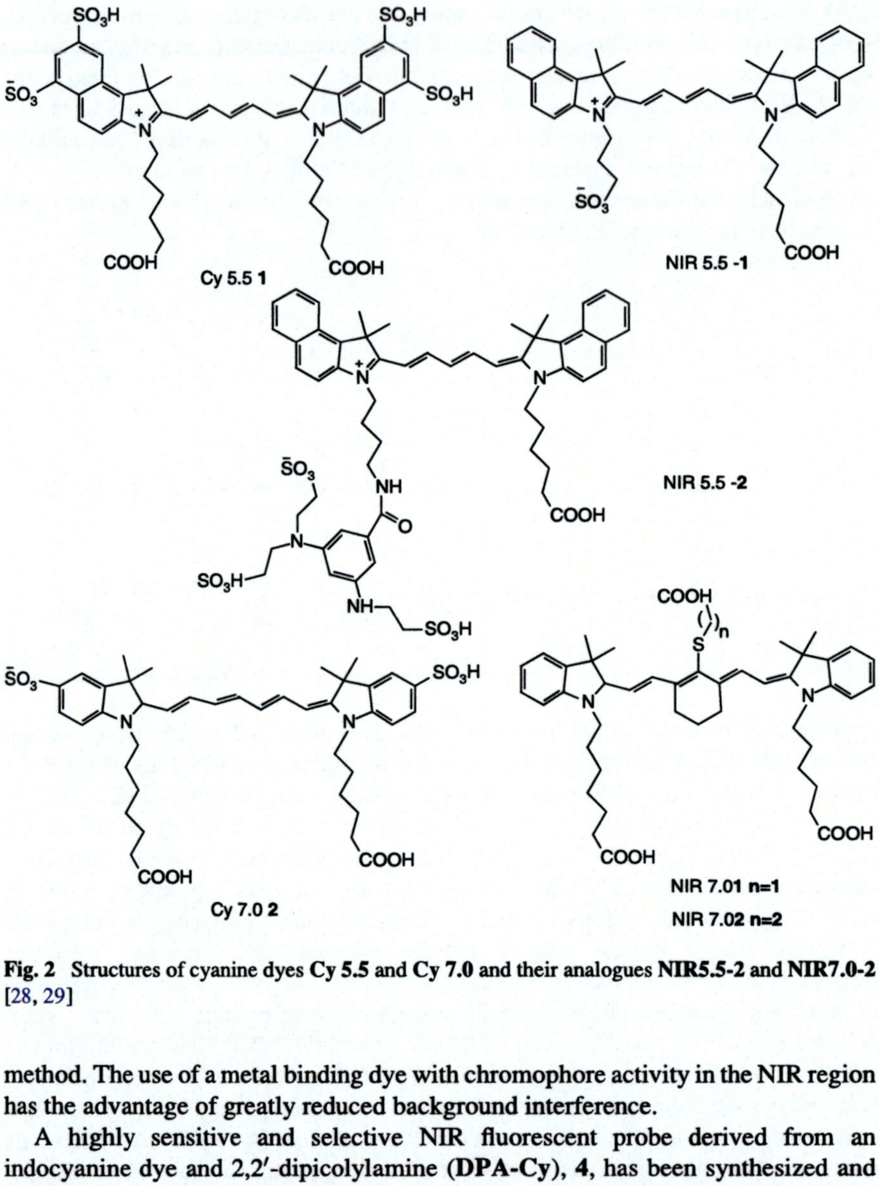

Fig. 2 Structures of cyanine dyes **Cy 5.5** and **Cy 7.0** and their analogues **NIR5.5-2** and **NIR7.0-2** [28, 29]

method. The use of a metal binding dye with chromophore activity in the NIR region has the advantage of greatly reduced background interference.

A highly sensitive and selective NIR fluorescent probe derived from an indocyanine dye and 2,2′-dipicolylamine (**DPA-Cy**), **4**, has been synthesized and applied for quantification and imaging of cellular zinc ion [20]. The probe shows absorption at 606 nm and emission at 800 nm in acetonitrile with a Stokes shift of 194 nm. Most common classes of fluorescent probe for metal ions are based on photoinduced electron transfer (PET) mechanisms. Disturbance of the electron cloud may affect the fluorescence intensity as the metal ion forms a complex with the NIR dye, which is most likely quenching of the fluorescence. In contrast, coordination of d^{10} transition metals or protons to the electron donor sites like amines is a commonly observed mechanism for fluorescence enhancement due to inhibition of PET. Based on this strategy, **DPA-Cy** is highly sensitive and selective to zinc, and gives a 20-fold turn-on response for detecting zinc. In addition, **DPA-Cy** is cell-permeable and responds to zinc quickly.

Another ratiometric fluorescent probe for Zn^{2+} in the NIR region, based on a tricarbocyanine chromophore, **5**, has been designed and evaluated [14]. Upon addition of Zn^{2+}, a 44-nm red shift of the absorption maximum and fluorescence modulation was observed. This change is due to the difference in the electron-donating ability of the amine substituent before and after reaction with Zn^{2+}.

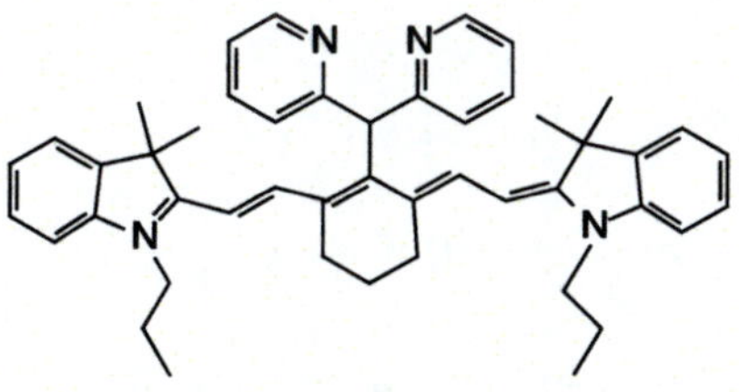

4

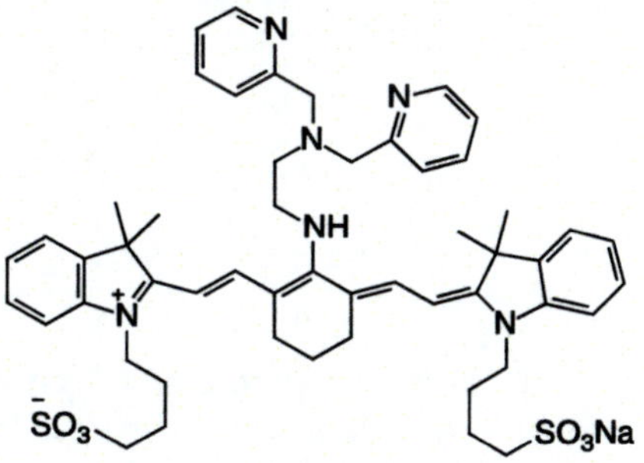

5

3,9-Dithia-6-mono-azaundecane-tricarbocyanine, **DMA-Cy**, **6**, has been synthesized and applied for imaging mercuric ions in biological systems [21]. This probe responds to mercuric ions stoichiometrically, rapidly, and reversibly at room temperature and produces strongly fluorescent product based on suppression of the PET quenching mechanism. The excellent sensitivity and selectivity for mercuric ions is demonstrated by its use in monitoring the real-time uptake of mercuric ions in HepG2 cells and 5-day-old zebrafish. In addition, the effect of other metal ions was examined individually. K^{+}, Na^{+}, Ca^{2+}, and Mg^{2+}, which exist at high levels in living cells, do not interfere even at high concentration. Besides, **DMA-Cy** was also selective for Hg^{2+} over Mn^{2+}, Co^{2+}, Ni^{2+}, Cu^{2+}, Zn^{2+}, Cd^{2+}, Pb^{2+}, and Ag^{+}. This indicated that the **DMA-Cy** was highly selective for Hg^{2+} over competing metal ion analytes in aqueous solution because of its thioether-rich receptor.

DMA-Cy

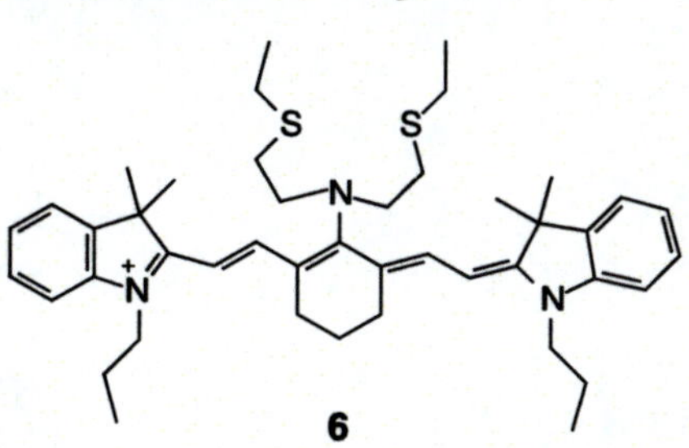

6

The PET and hence the NIR fluorescence of HCDs [22] with thio-substituents in the central position of the polymethine chain, **7**, is tuned by changing the electron-donating ability of the substituent. 4-Aminophenylthio-substitution led to an efficient PET and the lowest fluorescence quantum yield ($\phi_f = 0.0065$ compared to 0.038 for the other derivatives). Acetylation, protonation, or coordination of the amino group with transition metal cations like Zn^{2+}, Fe^{3+}, and Ce^{3+} could recover fluorescence greatly via suppressing the PET.

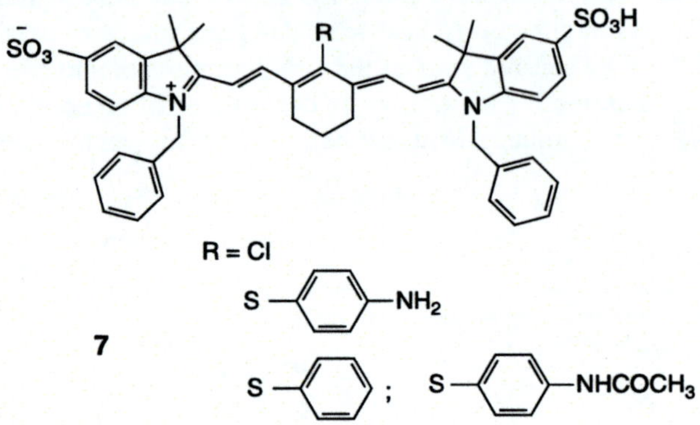

7

NIR fluorescent probes based on diaminocyanines **8** were designed for NO sensing, and their spectral properties have been investigated [23]. The probes are based on PET mechanism. The detection of NO is based on the change of the electron-donating ability of *o*-phenylenediamine upon selective NO-mediated transfomation of diamine into triazole under aerobic conditions (Fig. 3). The *o*-phenylenediamine quenches the fluorescence of the tricarbocyanine because of the electron transfer from *o*-phenylenediamine to the excited fluorophore. On the other hand, the formation of the triazole recovers its NIR fluorescence since the triazole ring have less electron-donating ability for such PET to occur. Comparison of the reaction rate with that of a widely used NO probe, diaminofluorescein, indicates that probe **8** can be applied to not only cellular but also in vivo NO imaging.

A class of norcarbocyanine that is sensitive to pH changes at a physiologically relevant range has been synthesized by the removal of a hydroxylsulfonylbutyl arm from indocyanine green dye (ICG), (Fig. 4) and their absorption and emission spectra have been studied in buffer solutions [24]. The simple structural design,

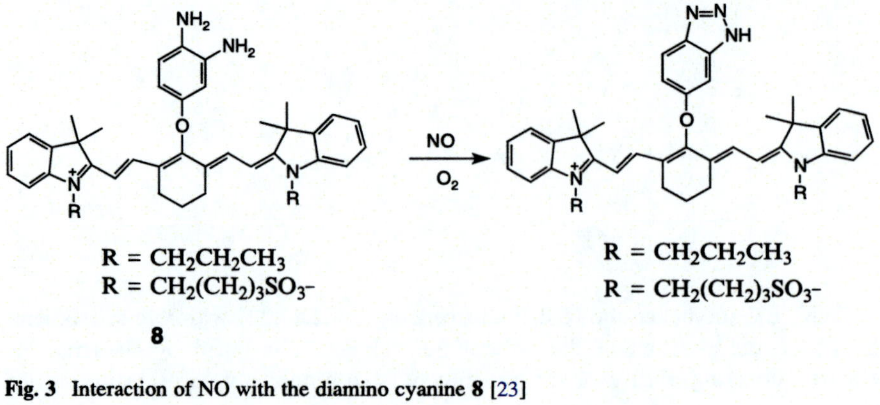

8

Fig. 3 Interaction of NO with the diamino cyanine **8** [23]

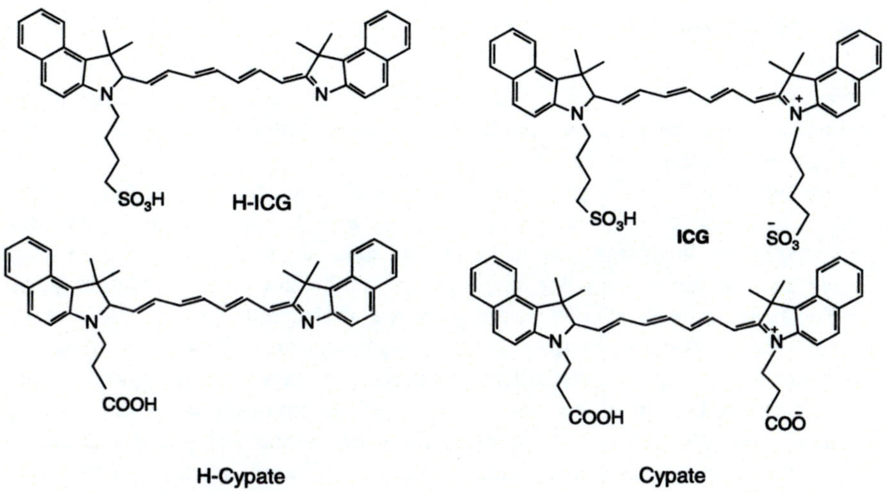

Fig. 4 Structure of H-ICG, H-cypate, ICG, and cypate [24]

ease of synthesis, and solubility are the characteristic features of such good NIR pH indicators for biomedical studies. The fluorescence emission peaks of these dyes (around 800 nm) showed slight but noticeable shifts at different pH values and significant intensity fluctuation at pH range close to the pK_a of the indole N atom in aqueous medium (pK_a = 7.2).

A HCD with an alkylamino group at the central position, **9**, was found to exhibit a large Stokes shift (>140 nm) and strong fluorescence [25]. This is due to excited-state intramolecular charge transfer (ICT) which was sensitive to the change in the viscosity and pH of the media.

An interesting neutral pH probe utilizing a fluorophore-spacer-receptor molecular framework based on tricarbocyanine, as a fluorophore, and 4′-4′-(aminomethylphenyl)-2,2′:6′,2″-terpyridine as a receptor, **10**, was developed [30]. The pH titration indicated that the probe can monitor the minor physiological pH fluctuations (pK_a of ~7.10), which is valuable for intracellular pH researches. By using this probe, the real-time imaging of cellular pH and the detection of pH in situ was achieved successfully in living HepG2 and HL-7702 cells.

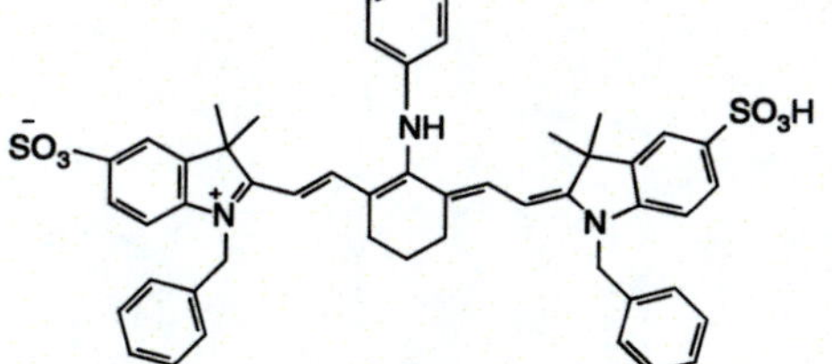

9

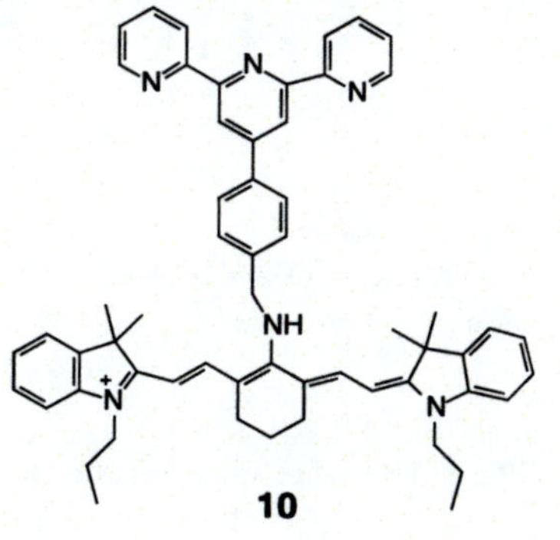

10

3.2 *Squaraine Dyes*

Squaraines belong to the class of polymethine dyes having resonance-stabilized zwitterionic structures with a cyclobutenediyliumdiolate core, classical examples of squaraine are shown in Fig. 5.

Squaraine dyes are among the most commonly used far-red and NIR fluorescent probes, due to their high extinction coefficients, fluorescence yields, and photostabilities. In addition, some squaraine dyes show high sensitivity toward the polarity of the surrounding environment, with fluorescence quantum yields tending to increase drastically in the presence of biomolecules, such as proteins and antibodies. Because of these features, squaraine dyes have been applied in many fields as in copiers, solar cells, optical discs, biological imaging, and fluorescent sensors in biological analysis. Coupling of electron-rich aromatic or heterocyclic compounds such as *N,N*-dialkylanilines, benzothiazoles, phenols, azulenes, and pryrroles with squaric acid yields a variety of symmetrical and unsymmetrical squaraines with tunable optical properties. Optical absorption and emission of squaraines can be pushed toward the long-wavelength region either by the use of strong electron donors or by the extension of conjugation. Using this approach, a large number of squaraines with near-IR absorption and emission has been synthesized [31–36]. Hence, the development of NIR fluorescent squaraine dyes along with description of their optical properties is going to be reviewed in the present section.

3.2.1 Symmetrical and Unsymmetrical NIR Fluorescent Squaraines

Replacing of the dialkylaniline moiety of bis[4-(*N,N*-dibutylamino)phenyl] squaraine, **DBAS**, by dialkylanthracene yields squaraine dyes **11–13** [31], with their absorption and emission spectra strongly shifted to the NIR region. This is due

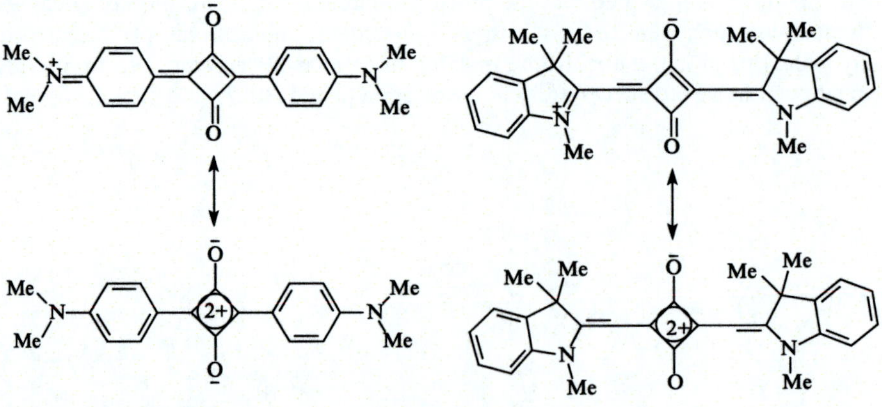

Fig. 5 Resonance structures of typical squaraine dyes

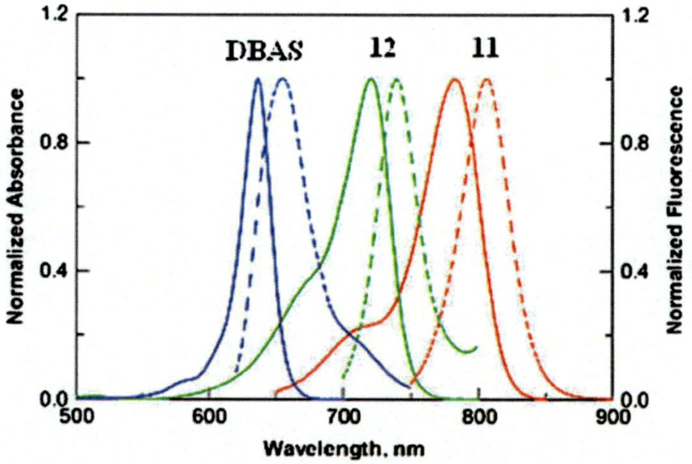

Fig. 6 Normalized absorption (*solid lines*) and emission spectra (*dashed lines*) of **11**, **12**, and **DBAS** in toluene, adopted from ref. [31]

to reduction in the HOMO–LUMO gap resulting from enhanced hydrogen bonding between the carbonyl group of the cyclobutane ring and the neighboring aromatic hydrogen in the dyes containing the anthracene ring. Figure 6 shows the absorption and emission spectra of **11 (Sq-1)** and **12 (Sq-2)** along with that of **DBAS**. The water compatibility and substantial enhancement of fluorescence of **13** in micellar media of SDS, CTAB, and TX-100 suggest that these dyes can be potentially useful for imaging of hydrophobic domains such as cell membranes.

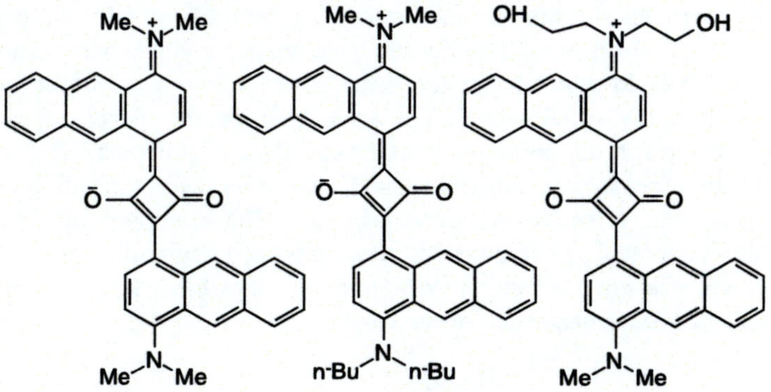

11 **12** **13**

The spectral, photophysical properties (quantum yields and fluorescence lifetimes) and photostabilities of a synthesized series of symmetrical, **14**, and unsymmetrical, **15**, ring-substituted squaraines absorbing and emitting in the red and NIR spectral region were measured and compared to the commonly used fluorescent label Cy5 [32]. The absorption maxima in aqueous media were found to be between 628 and 667 nm while the emission maxima are between 642 and 685 nm. The dyes exhibit high extinction coefficients (16.3–26.5 × 10^4 M^{-1} cm^{-1}) and lower quantum yields (2–7%) in aqueous buffer but high quantum yields (up to 45%) and long fluorescence lifetimes (up to 3.3 ns) in the presence of bovine serum albumin. Dicyanomethylene- and thio-substituted squaraines exhibit an additional absorption around 400 nm with extinction coefficients between 21.5 and 44.5 × 10^3 M^{-1} cm^{-1}. So, these dyes could be excited not only with red but also with blue laser diodes. Due to these favorable spectral and photophysical properties, these dyes can be used as bioconjugate fluorescent probes and labels for fluorescence intensity- and lifetime-based biomedical applications.

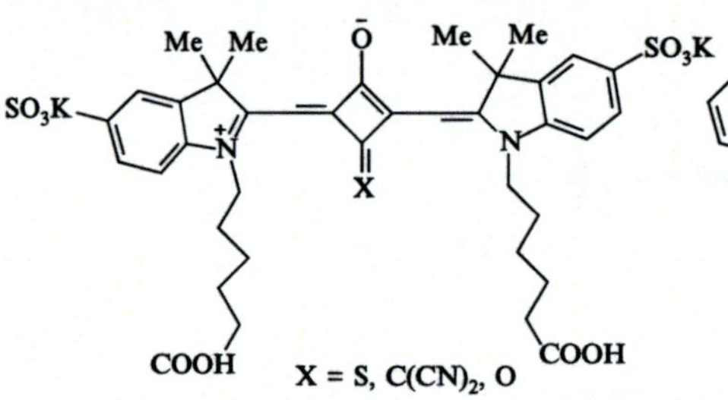

14

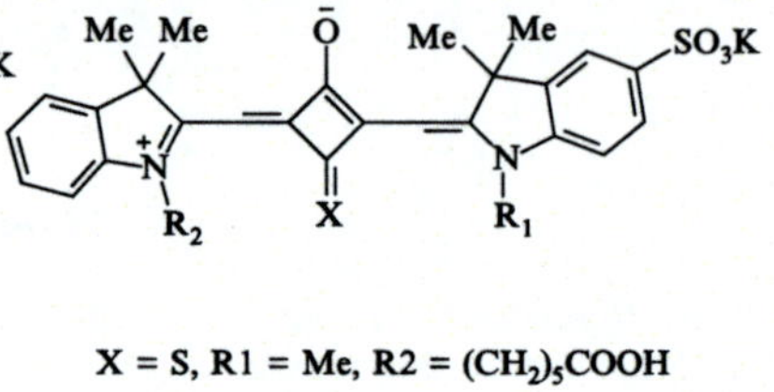

X = S, R1 = Me, R2 = $(CH_2)_5COOH$
X = $C(CN)_2$, R1 = Me, R2 = $(CH_2)_5COOH$
X = O, R1 = Me, R2 = $(CH_2)_5COOH$

15

The photophysical properties of some squaraine dyes containing amphiphilic substituents, **16a–16d**, have been investigated in the presence and absence of organized media [33]. The dyes exhibited absorption in the range 630–650 nm, with significant absorption coefficients in the aqueous medium. The fluorescence spectra of these dyes showed emission maximum from 660 to 675 nm, depending on the nature of substituents. The fluorescence quantum yields were in the range from 0.15 to 0.21 in ethanol, but ten times lower values were observed in the aqueous medium ($\phi_f = 0.01$–0.02). The absorption spectra of these dyes showed negligible changes in the presence of micelles (SDS, CTAB, and TX-100), however, their fluorescence quantum yields and lifetimes increase, indicating that these molecules undergo effective microencapsulation. These amphiphilic squaraines, which exhibit favorable photophysical properties, good solubility in the aqueous medium, and efficient interaction with micelles, can have potential biological applications as NIR fluorescence sensors.

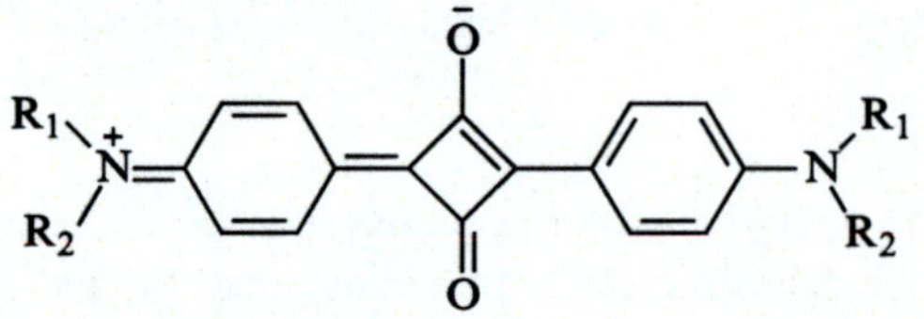

16 **a** R1 = R2 = CH_2-CH_2-OH **c** R1 = Me, R2 = $(CH_2$-CH_2-$O)_4CH_3$
b R1 = Me, R2 = $(CH_2)_3$-COOH **d** R1 = R2 = $(CH_2$-CH_2-$O)_4CH_3$

Water-soluble NIR fluorescent squaraines, **17**, namely, **KSQ-3** and **-4**, were synthesized and applied to biological labeling [34]. The large, planar, and hydrophobic squaraine dye becomes fully soluble in aqueous solution by the introduction of several alkyl substituents terminated with sulfo-group. **KSQ-4**, which is substituted with four sulfo-groups, exhibited perfect water solubility and significant fluorescence emission at 817 nm in the presence of bovine serum albumin ($\phi_f = 0.08$ when excited at 787 nm). These water-soluble NIR fluorescent probes enable much easier handling for biological analyses that have to be undertaken in aqueous solution.

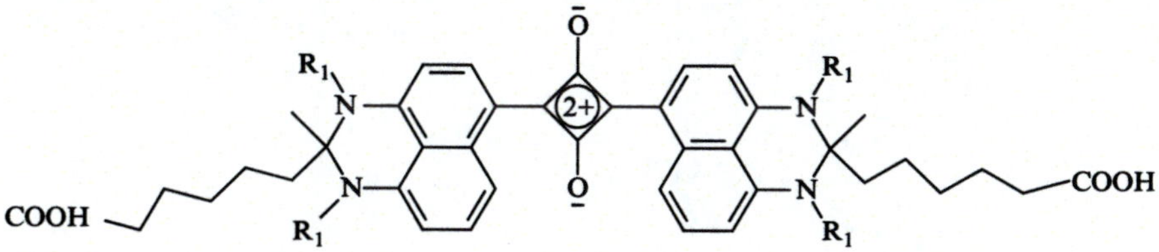

KSQ-3; R1=Me KSQ-4; R1 = $(CH_2)_3SO_3Na$

17

The influence of N-substitution on the spectral properties of 2, 3-dihydroperimidine-substituted squaraines, **18**, was investigated [35]. The absorption and fluorescence maxima of **KSQ-1** in cyclohexane were observed at 802 and 811 nm, respectively. However, the quantum yield was very low ($\phi_f = 0{:}074$ in any solvents), which implies limited applications of the already known **KSQ-1**. It is assumed that the presence of N–H bonds can favor nonradiative pathways for relaxation from the excited state, such as hydrogen-bond-induced intersystem crossing. On the other hand, **KSQ-2** (*N*-methyl-substituted dye) was found to have a higher extinction coefficient ($2.0 \times 10^5\ M^{-1}\ cm^{-1}$) and noticeably blue-shifted sharp absorption and fluorescence spectra ($\lambda_{max} = 737$ and 751 nm, respectively). A quantum yield more than eightfold higher compared to that of **KSQ-1** was observed ($\phi_f = 0{:}56$ and 0:47, in cyclohexane and toluene, respectively). Therefore, with just a simple alkylation, the optical properties are remarkably improved. In addition, a linear positive solvatochromic properties with the solvent polarizability in the spectral region around 780 nm was observed in the case of **KSQ-2**.

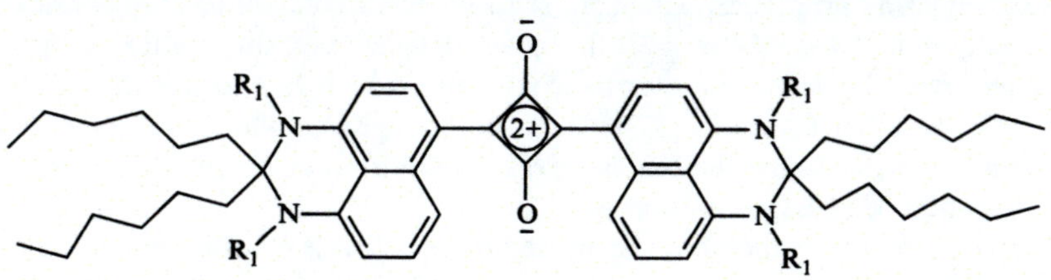

KSQ-1; R1=H KSQ-2; R1=CH_3

18

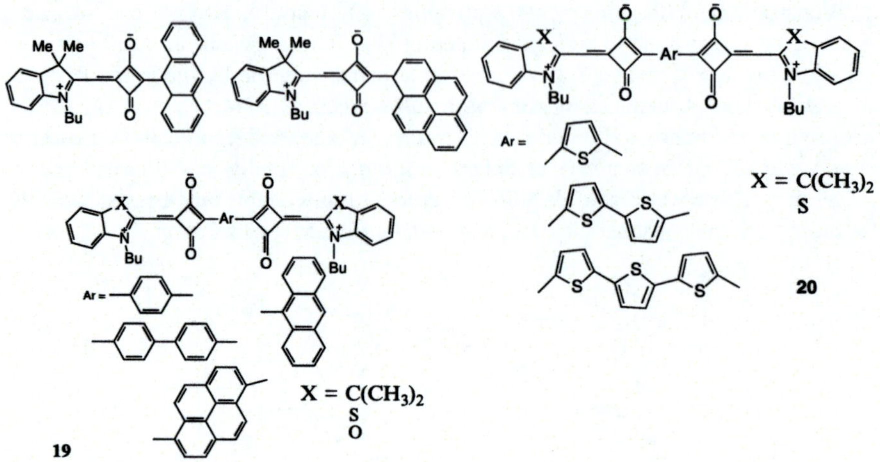

19

20

Fig. 7 Examples of some arenas and thiophene-bridged bis-squaraines [36]

Different arene- and thiophene-bridged bis-squarylium dyes **19** and **20** were synthesized and their spectral properties have been investigated [36]. Although the fluorescence intensities were much lower than those of the unbridged squarylium dyes, arene- and thiophene-bridged dyes, employing extensively conjugated spacers such as anthracene and pyrene or thiophene, bithiophene, and terthiophene, respectively, allowed to obtain NIR fluorescent dyes, especially in combination with the benzindolium heterocycles (Fig. 7).

3.2.2 NIR Fluorescent Squaraine-Based Chemosensors

Chemosensors, the usual configuration of which consists of an ionophore unit, integrated to a chromophore, is a topic of considerable interest because of their importance in specific detection of analytes in different fields such as chemistry, biology, medicine, and environmental studies [8, 37]. Organic dyes are extensively used as the signaling units in chemosensor design because of their intense absorption and emission properties, which are sensitive to external inputs. Squaraine dyes are ideally suited for this purpose because of their favorable optical properties associated with the peculiar zwitterionic structure, which gets perturbed with metal ions, pH, and other additives [37]. Therefore, ionophores which are integrated to squaraine dyes can signal the binding event in the form of measurable changes in the absorption or emission properties.

Sensor molecules which can report micromolar levels of calcium (in the presence of 1,000-fold excess of magnesium) are useful [38]. Intracellular Ca^{2+} signals are responsible for the initiation or regulation of a number of biological processes and there is a great interest in visualizing such Ca^{2+} spikes in real times. A red to NIR emitting squaraine-based water-soluble chemosensor **21** has been applied for selective detection of Ca^{2+} ions in the micromolar range as indicated by decreasing the fluorescence intensity around 730 nm, whereas large excess of Mg^{2+} has no effect on the fluorescence intensity.

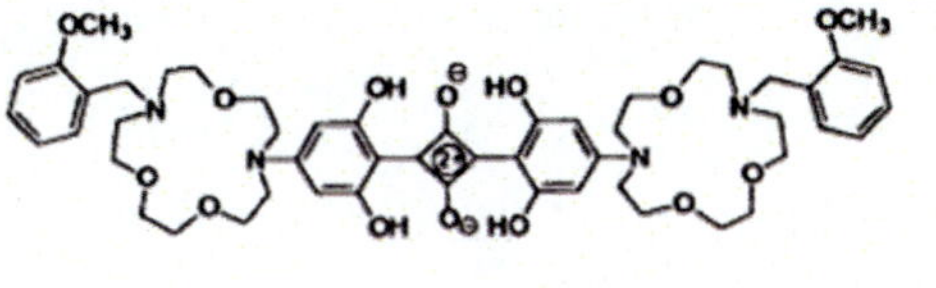

22

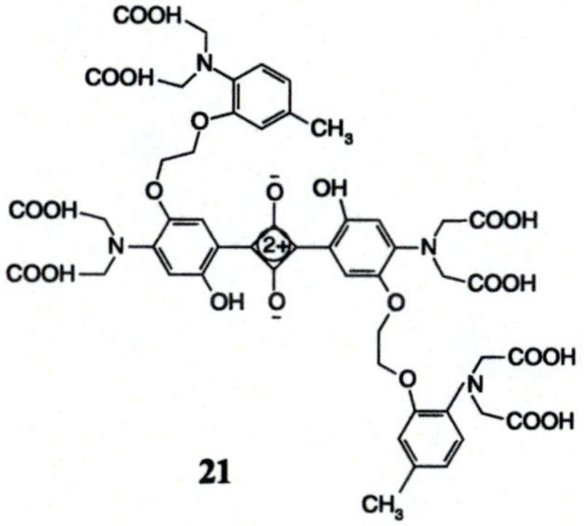

21

A squaraine with a lariat-crown ether cation-receptor unit **22** was synthesized and shown to selectively respond to Na^+ ions in polar protic solvents [39]. The selective binding of this ligand to Na^+ ions is accompanied by a decrease in the fluorescence intensity around 650 nm.

The performance of squaraine-based sensors as cation probes has been improved by applying the Swager's concept [37, 40] of molecular wire-based signal amplification in pyrrole-based NIR fluorescent polysquaraines. This was illustrated using polysquaraines **23** as a selective and specific probe for Li^+ over Na^+ and K^+ as reflected by the fluorescence enhancement on adding the former cation. The bichromophore **24** containing flexible oxyethylene bridge showed strong perturbation in the NIR absorption and emission spectra, noticeable fluorescence quenching, with high selectivity toward alkaline earth metal cations, particularly to Mg^{2+} and Ca^{2+} ions, whereas no optical response was noticed against alkali metal ions. Although the bichromophore **24** showed similar changes in the absorption spectra with Mg^{2+} and Ca^{2+}, it revealed preferential formation of a 1:1-folded complex with Mg^{2+} and 1:2 sandwich dimer with Ca^{2+} as shown in Fig. 8.

Electron donor-substituted squaraines with tetrahydroquinoxaline as the terminal group (**25a–25c**) [41] interact with different metal ions in THF, particularly **25c**, and it was found that copper ion can be detected selectively. The degree of red shift in the NIR emission of these dyes depends on the nature and position of the substituent group. A relatively large red shift in the emission of **25b** ($\lambda_f = 790, 823$, and 774 for **25a–25c**, respectively) indicates that the donor group directly linked to central four-membered (C_2O_4) ring heavily affects excited state energies of these molecules.

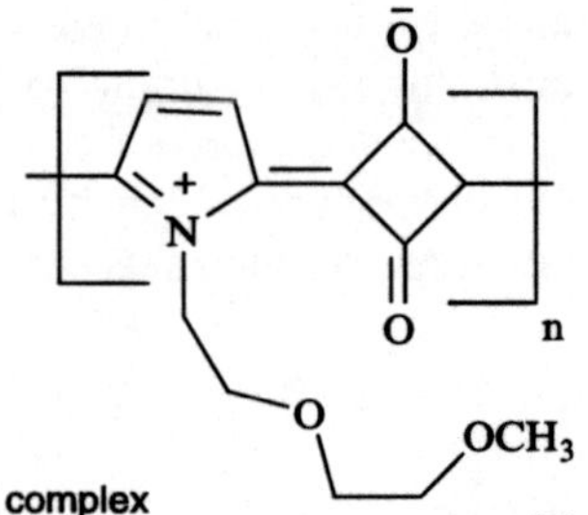

23

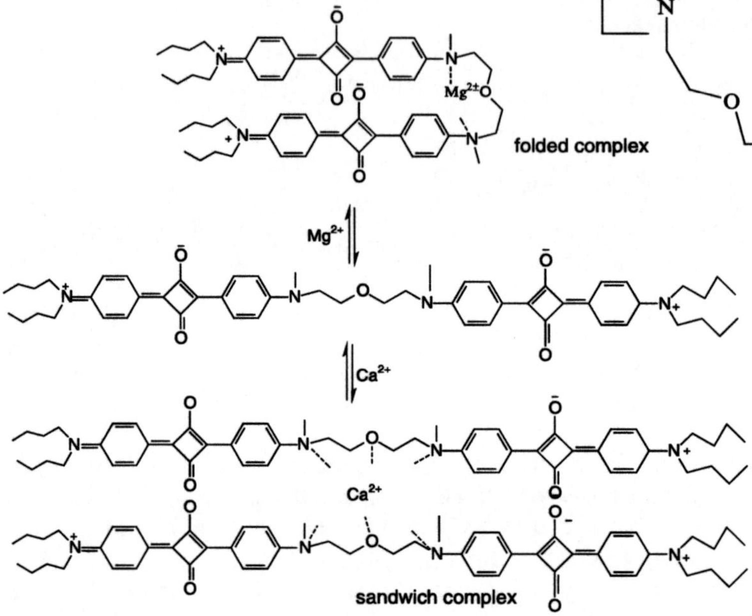

24

Fig. 8 Cation-induced regidization of bis-squaraine **24** [40]

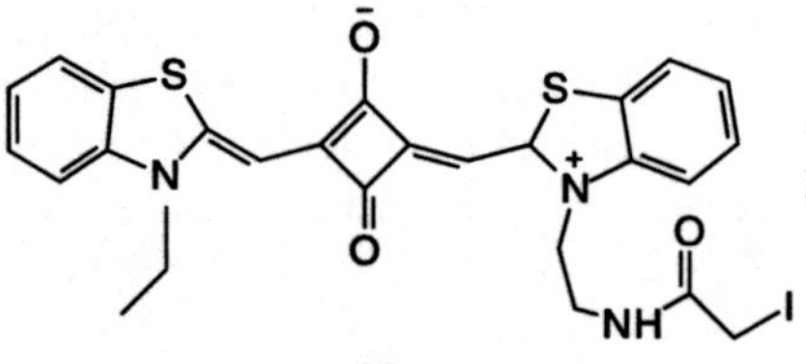

26

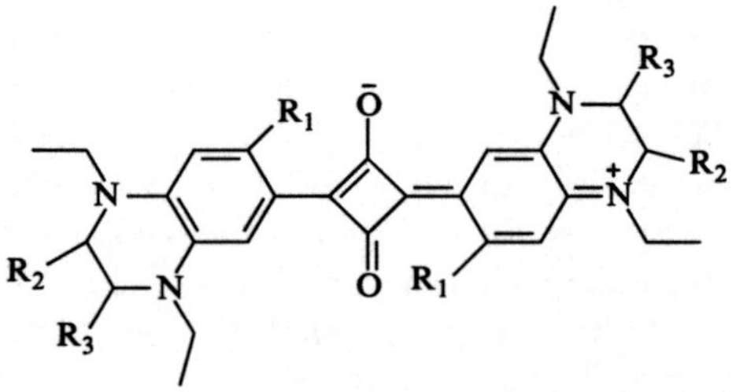

25

a R_1=H, R_2=H, R_3=H

b R_1=MeO, R_2=H, R_3=H

c R_1=H, R_2=ph, R_3=ph

An environmental sensitive NIR fluorescent squaraine dye with a thiol-reactive linker, **26**, has been applied for biosensing of glucose [42]. The absorption and fluorescence maxima in aqueous medium were above 650 and 665 nm, respectively, with Stokes shifts of 10 nm. The fluorescence intensity at 665 nm show

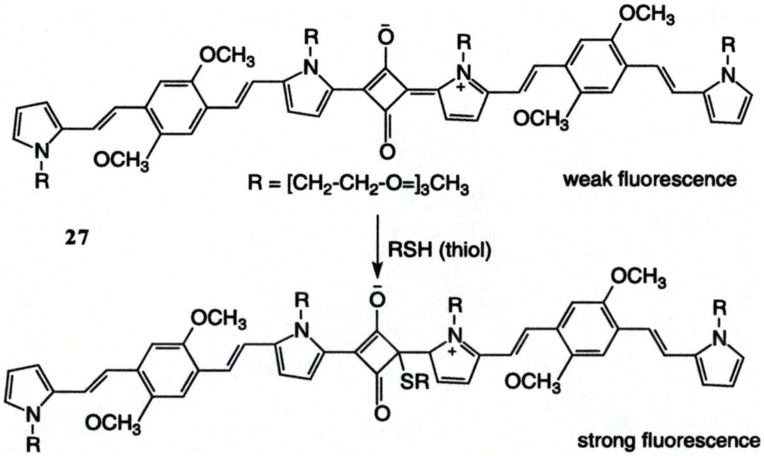

Fig. 9 Chemical activation of the weak fluorophore **27** to an active fluorophore through thiol group attack [43]

remarkable enhancement in the presence of glucose as well as when conjugated to glucose/galactose binding protein (GGBP).

A π-extended squaraine dye **27** that exhibits remarkable absorption and emission changes in the presence of aliphatic thiols has been developed [43] and used for the detection of low molecular-weight amino thiols such as cysteine and homocysteine in human blood plasma. The weakly fluorescent NIR **27** is chemically activated through a conjugation break upon nucleophilic attack by a thiol such that a strongly emitting fluorophore is generated (Fig. 9). The emission spectrum of **27** exhibits a weak band at 800 nm ($\lambda_{ex} = 730$ nm) in acetonitrile/water (1:1) solution and showed a dramatic change upon addition of cysteine. When excited at 410 nm, the emission spectrum of the **27**-cysteine adduct exhibited a bright orange fluorescence band with a maximum at 592 nm, whereas a decrease in the intensity of the weak NIR emission at 800 nm ($\lambda_{ex} = 730$ nm) with increasing concentration of cysteine was observed.

Since monitoring of pH changes is an important way to study signal transduction pathways and ligand interactions with G-protein-coupled receptors, a NIR pH label (**square-650-pH-NHS, K8-1407**) that shows ratiometric behavior in both the excitation and emission modes of measurement has been characterized [44]. The free label has a pK_a of 7.11, and the antibody-conjugated label has a pK_a of 6.28. Because the fluorescence of **Square-650-pH** dramatically increases as the pH of its environment decreases from neutral to acidic (Fig. 10), it is an ideal tool for studying phagocytic events from nonspecific binding to cells and their regulation by drugs and/or environmental factors. Another potential application is acidity regulation in mammalian cells.

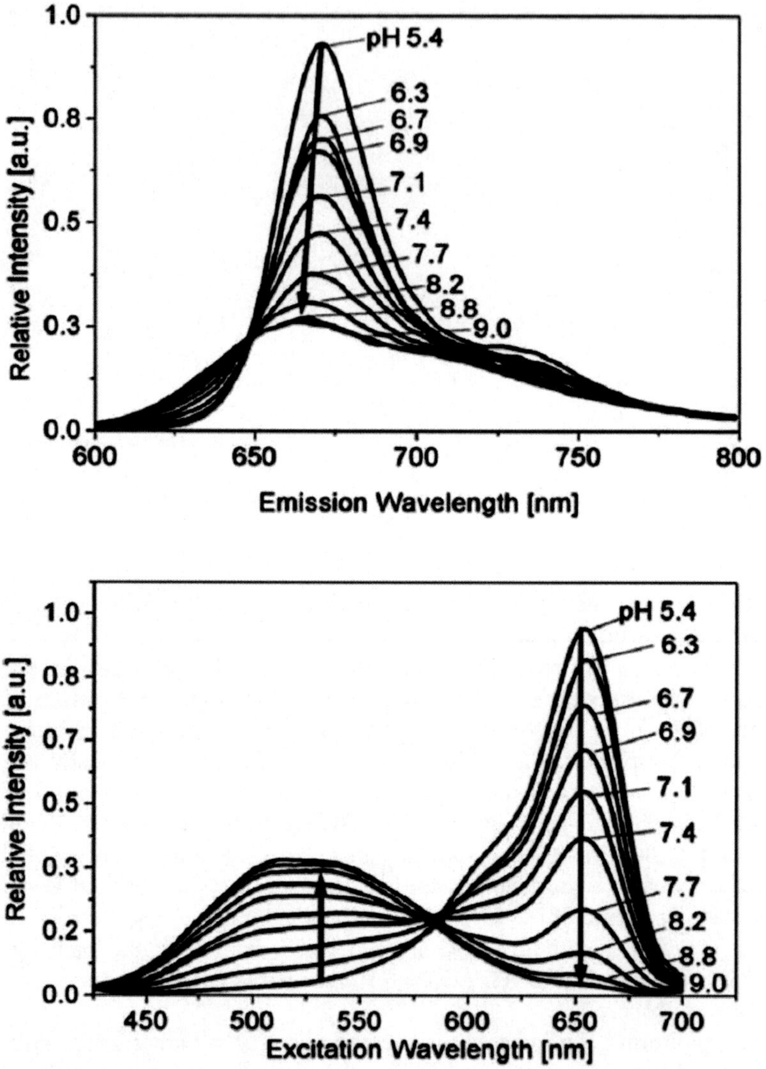

Fig. 10 pH-dependent emission ($\lambda_{exc} = 589$ nm) and excitation ($\lambda_{em} = 710$ nm) spectra of Squaraine-650-pH [44]

3.3 BODIPYs Dyes

Boron-dipyrromethene (4,4-difluoro-4-bora-3a,4a-diaza-s-indacene, BODIPY) fluorescent dyes have been the subject of intense investigations [6, 45–48] due to their high extinction coefficients, intense fluorescence, high photostability, and insensitivity to the polarity of the medium as well as to pH. However, typical BODIPYs have some drawbacks, such as relatively short fluorescence emission maxima (λ_{max} around 500 nm) and low extinction coefficients (ε around 80 × 10^3 M^{-1} cm^{-1}). Currently, one of the key goals in the field of fluorescence sensing research is to develop methods for shifting the major absorption and emission to the NIR region by modifying the BODIPY core. This includes attachment of electron-donating groups at the periphery, extension of the π-conjugation system, and creation of a rigid structures [6], see Fig. 11.

With these facts in mind, the present section would explore the modification of the simple BODIPY core to bathochromically shift the spectral bands while keeping the high emission efficiency and the possibility of further functionalization for the construction of fluorescent labels.

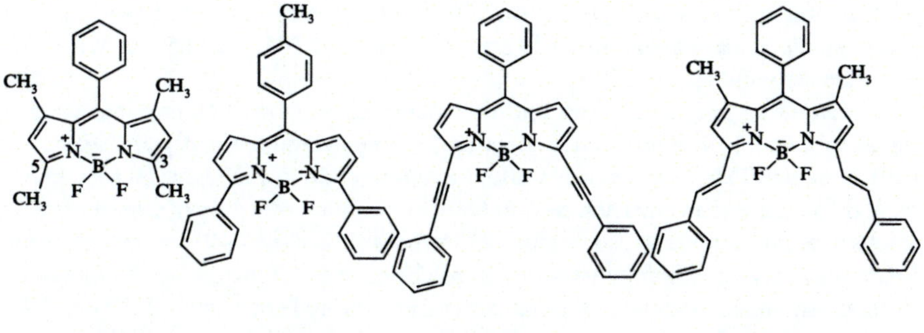

λ_{abs} = 479 nm, λ_f = 505 nm, ϕ_f = 0.60 (CH_3CN) **28**

λ_{abs} = 547 nm, λ_f = 582 nm, ϕ_f = 0.21 (MeOH) **29**

λ_{abs} = 605 nm, λ_f = 622 nm, ϕ_f = 1.0 (MeOH) **30**

λ_{abs} = 628 nm, λ_f = 642 nm, ϕ_f = 0.84 (CH_3CN) **31**

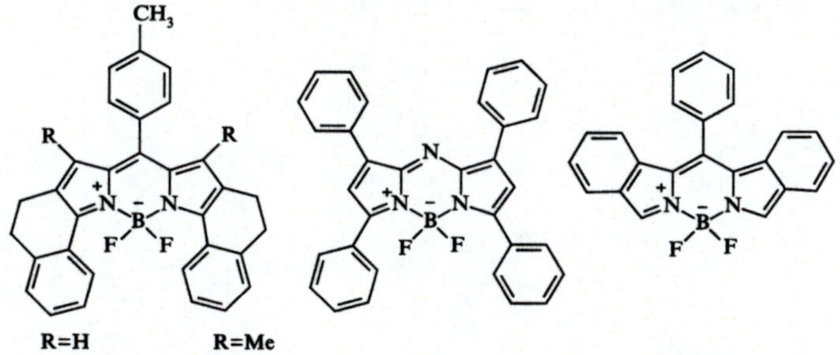

R=H: λ_{abs} = 634 nm, λ_f = 647 nm, ϕ_f = 0.38; R=Me: λ_{abs} = 619 nm, λ_f = 639 nm, ϕ_f = 0.72 ($CHCl_3$) **32**

λ_{abs} = 647 nm, λ_f = 669 nm, ϕ_f = 0.34 (EtOH) **33**

λ_{abs} = 597 nm, λ_f = 606 nm, ϕ_f = 0.91 (CH_3CN) **34**

Fig. 11 Examples of NIR fluorescent BODIPY and their spectral properties [6]

3.3.1 NIR Fluorescent BODIPY-Based Chemosensors

The introduction of certain substituents, usually amino groups, makes it possible to switch the fluorescence or shift absorption and emission wavelengths of BODIPYs upon ion coordination, change of pH, or environmental polarity. Dyes with such responsive features have been used, for example, for the sensing of biologically, toxicologically, or environmentally relevant species, including saxtoxin [45], Zn^{2+}, Fe^{3+} [46], and Hg^{2+} [47].

One of the first NIR fluorescent pH-responsive BODIPY derivatives was realized by introducing a styryl branch [48] that was equipped with an aniline group through the 3 position of the core, **35**. The fluorescence maximum of **35** is shifted to the NIR in polar solvents (from 611 nm in hexane to 731 nm in MeCN, $\lambda_{exc} = 595$ nm) because of a photoinduced charge transfer (CT) process in the excited state, involving the dimethylamino group as an electron donor and the BODIPY core as the electron acceptor. Protonation of the dimethylamino group switches off the CT process and recovers the typical BODIPY-like narrow, structured, and solvent polarity-independent absorption and emission bands at 555 and 565 nm, with high fluorescence yield ($\phi_f = 0.97$).

The emission properties of an acid-switchable fluorophore with two independent signaling units [8-(4-hydroxyphenyl) and 3-(*p*-dimethylaminostyryl)], BODIPY, **36**, were studied in THF [49]. Optically dilute solutions of **36** in THF have an absorbance peak at 565 nm and an emission peak at 660 nm. Addition of a drop of perchloric acid results in a considerable hypsochromic shift in both absorption (~40 nm) and emission (~100 nm) spectra, which is expected considering the ICT nature of the emission. Without any added modulator, the quantum yield of the red emission in THF was 0.25, whereas upon addition of acid, a bright yellow emission is observed with increased quantum yield ($\phi_{em} = 0.84$). However, addition of a base (potassium *tert*butoxide) yields a significantly quenched emission ($\phi_{em} = 0.032$) at the same wavelength (660 nm) due to deprotonation of the phenolic hydroxyl group. This system functions as a combinatorial logic circuit for half-subtractor.

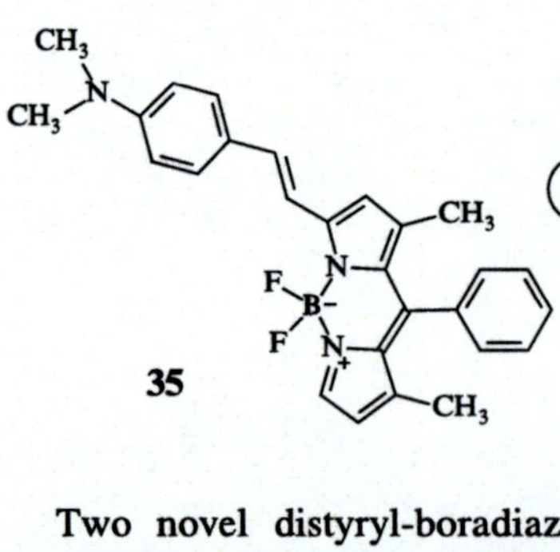

35

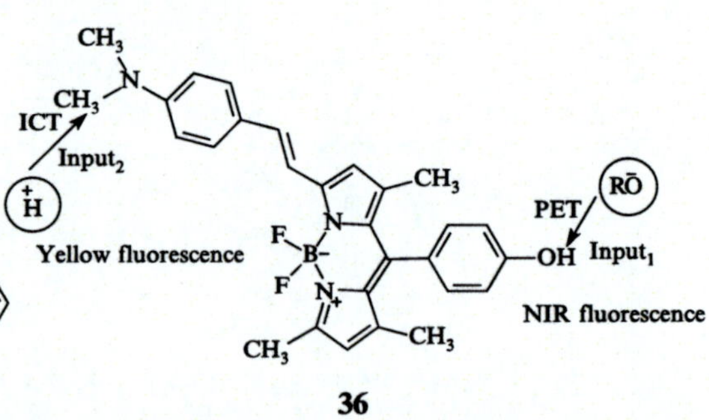

36

Two novel distyryl-boradiazaindacene dyes with dimethylaminostyryl and pyridylethenyl substituents, **37** and **38**, have been synthesized [50]. The dyes display opposite spectral shifts on protonation with trifluoroacetic acid (TFA) in organic solvents. This bidirectional switching of the dyes was shown to be directly related to ICT donor and acceptor characteristics of the substituents attached to the

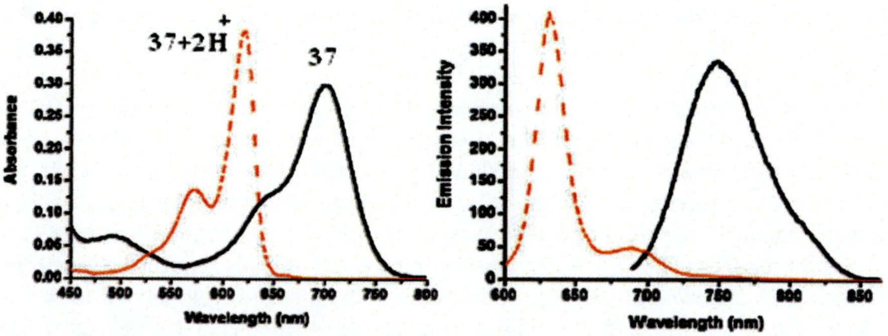

Fig. 12 Absorption (*left*) and emission (*right*) spectra of compound **37** in response to the addition of a small aliquot of TFA, adopted from ref. [50]

BODIPY core. The presence of two different protonation sites yields two different sets of absorption and emission spectra. For **37**, **37**-2H$^+$, the emission peaks appear at 753 and 630 nm, respectively. A blue shift on adding TFA was observed due to the functionality of ICT donor dimethylamino group which is considerably less effective on protonation. The pyridyl-substituted dye **38** shows a different behavior where gradual addition of a small amount of TFA results in a single distinct but red-shifted spectrum, with absorption peak at 660 nm. Similarly, the emission spectrum shows just one red-shifted peak with maximum at 677 nm. It is known that the pyridyl groups become a stronger electron acceptor on protonation, so producing a bathochromic shift, Fig. 12.

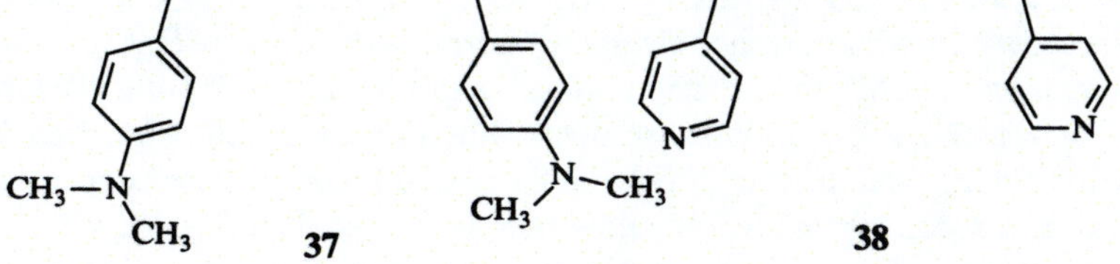

37 **38**

To examine the PET sensing properties of BODIPY, the photophysical properties of the benzylamine-substituted analogues, **39a** and **39b**, have been examined in various solvents [51]. Incorporation of diethylamine and morpholine receptors facilitates off/on microenvironment polarity and pH sensing in the red region of the spectrum. While little change in the spectral properties was observed on substrate recognition, large changes in the fluorescence intensity act as the sole signaling event. The emission maxima of **39a** is only 12 nm hypsochromically shifted from

DMF (680 nm) to cyclohexane (668 nm). Identical behavior was recorded for **39b**. In contrast to the invariance of the emission wavelength maxima in various solvents, the emission intensity shows a ninefold enhancement on comparing the extremes of DMF and cyclohexane. In more polar solvents, little fluorescence is observed since the excited state is quenched efficiently by the PET process (sensor is off). Whereas in nonpolar solvents, the PET process is an ineffective competing process for fluorescence emission and the sensor is switched on. Also, the fluorescence spectra show a marked response to the acid analyte. For example, the nonprotonated **39a** is weakly emissive in DMF solutions but upon protonation with aqueous HCl a strong emission is observed at 682 nm with greater than eightfold from the off to the on state. The ability of **39a** to localize and be readily detected in vitro gives an indication of the potential of this fluorophores as markers for specific cellular events in conjunction with other receptor units and fine-tuning of the switching responses.

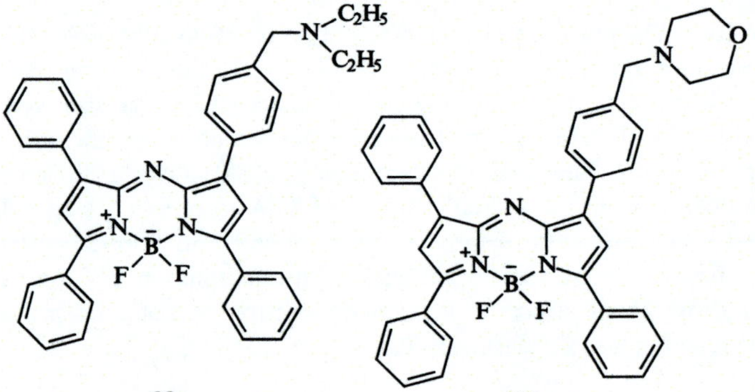

39a **39b**

Although, there are several examples of BODIPY-based chemosensors, little is reported about boratriazaindacenes acting as chemosensors for metal cations. The boratriazaindacene derivative **40** shows spectacular metal ion selectivity [47]. This is in part due to the rigid nature of the ligand and the selectivity imposed by the type of the donor atoms. Among the studied metal ions, Ba(II), Ca(II), Cd(II), Co(II), Cu(II), Fe(II), Hg(II), K(I), Li(I), Mn(II), Pb(II), Zn(II), cupric, and zinc cations show minor fluorescence changes, whereas on addition of Hg(II) at the same concentration (20 μM), the emission ratio changes more than 90-fold. The fluorescence sensor **40** has 2-pyridyl substituents which create a well-defined pocket for metal ion binding and its interaction with the mercuric ions creates large changes in the emission spectrum that would allow ratiometric sensing, Fig. 13.

Incorporation of an aza crown by means of a methylene spacer to an aza-BODIPY chromophore leads to PET-based chemosensor, **41** [45]. Fluorescence of the free ligand is quenched by the PET process from the crown's amino nitrogen to the BODIPY. When a cationic species, such as saxitoxin, binds to the aza crown of **41**, the electron transfer is inhibited and the emission at 680 nm is greater than 100% switched on.

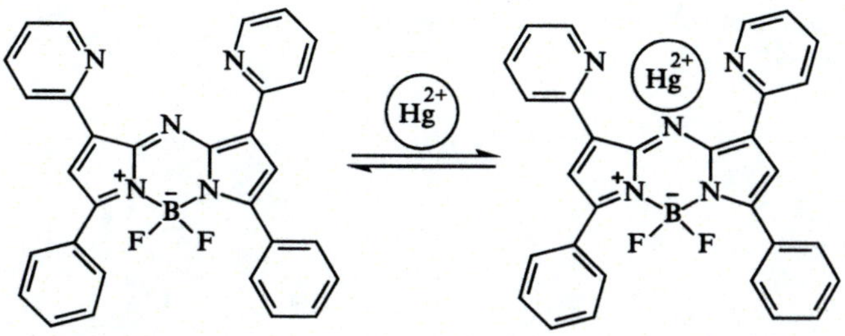

40

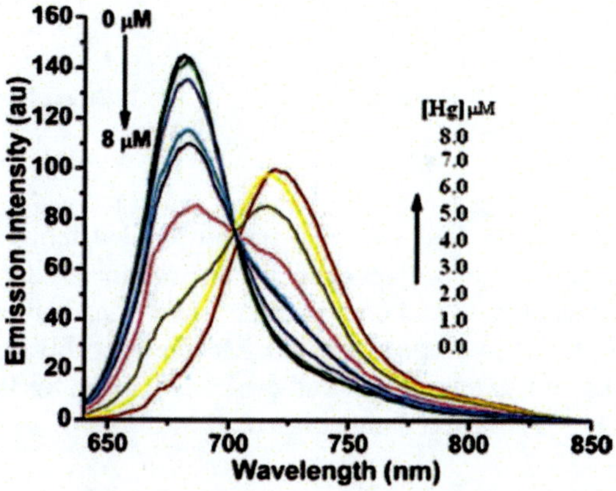

Fig. 13 Interaction of the probe **40** with Hg(II) ions and the corresponding emission response to increasing concentrations of Hg(II) ions in the form of perchlorate salt. Excitation was at 660 nm, adopted from ref. [47]

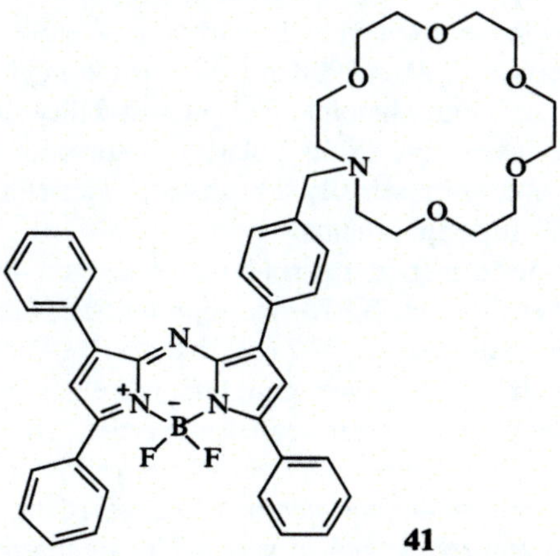

41

Incorporation of a size-restricted dithia-aza-oxa macrocycle, via a phenyl linker into a *meso*-substituted boron-dipyrromethene, **42** (Fig. 14), yields amplified fluorescence in the red-visible spectral range upon binding of Fe^{3+} ions [46]. The response to Fe^{3+} and potentially interfering metal ions is studied in highly

Fig. 14 Interaction of Fe^{3+} with probe **42** [46]

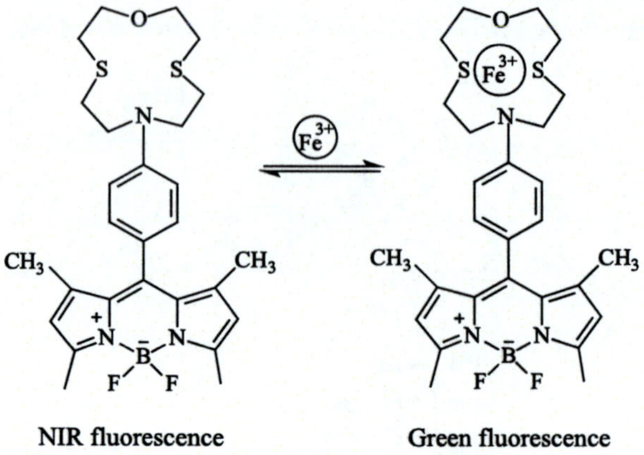

42

polar aprotic and protic solvents as well as in neat and buffered aqueous solution. In organic solvents, the fluorescence is quenched by an intramolecular charge or electron transfer in the excited state and coordination of Fe^{3+} leads to a revival of the fluorescence without pronounced spectral shifts. Most remarkably, the unbound BDP derivative shows dual emission in water and can be employed for the selective ratiometric signaling of Fe^{3+} in buffered aqueous solutions.

3.3.2 Conformationally Restricted NIR Fluorescent BODIPY

With appropriate substitution and restriction, different aza-BODIPY dyes have been developed [52, 53], Fig. 15. These fluorescent aza-dipyrromethene dyes have a number of advantages in comparison to the carbon analogues, such as peak fluorescence can be tuned between 700 and 900 nm, high quantum yield, narrow excitation and emission bands, and high chemical and photostability. In addition, the sharp fluorescent peaks are insensitive to the polarity of solvents. Efforts are currently underway to develop nonsymmetrically substituted, water-soluble versions to allow conjugation for biosensing experiments.

A series of high-performance fluorophores based on boron-dipyrromethene (BODIPY), named Keio Fluors (**KFL**), are reported [54]. The KFL dyes cover a wide spectral range from the yellow (547 nm) to the NIR (738 nm) region. Based on simple molecular modifications, their emission wavelength could be easily and subtly controlled without losing their optical properties. The emission wavelength of the **KFL** dyes is tuned with emission colors from yellow, orange, red, far-red, to NIR. Figure 16 shows representative spectra. Moreover, these **KFL** dyes also retain their excellent optical properties, such as spectral bands sharper than quantum dots, high extinction coefficients, and high quantum yields (0.56–0.98), without any critical solvent polarity dependent decrease of their brightness. These advantageous characteristics make the **KFL** dyes potentially useful as new candidates of fluorescent standards to substitute or to complement existing long-wavelength fluorescent dyes, such as cyanines, oxazines, rhodamines, or other BODIPY dyes.

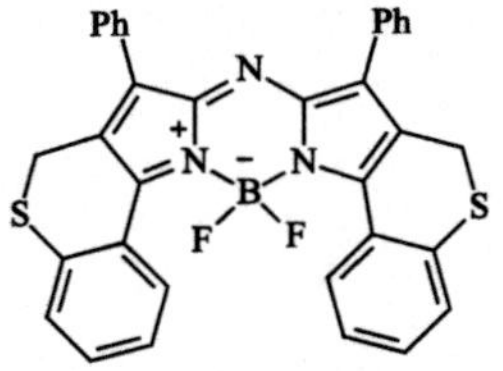

$\lambda_{abs} = 706$ nm
$\lambda_f = 730$ nm
$\phi_f = 0.11$
($CHCl_3$)

43

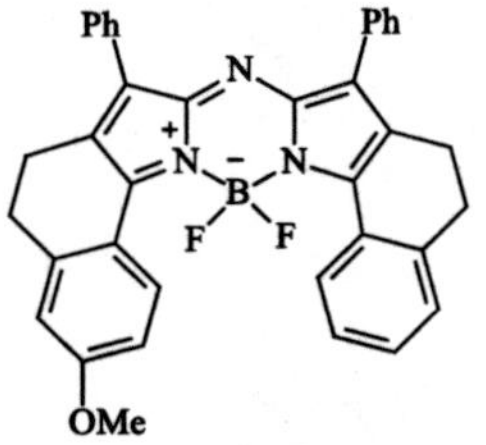

$\lambda_{abs} = 723$ nm
$\lambda_f = 734$ nm
$\phi_f = 0.32$
($CHCl_3$)

44

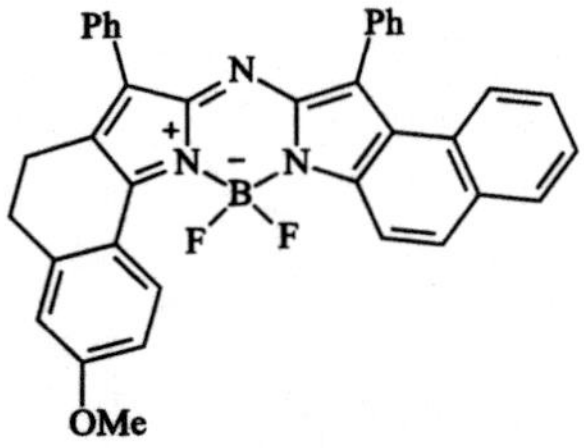

$\lambda_{abs} = 715$ nm
$\lambda_f = 730$ nm
$\phi_f = 0.11$
($CHCl_3$)

45

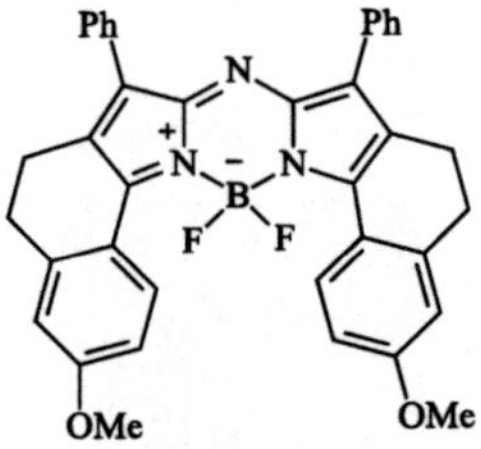

$\lambda_{abs} = 740$ nm
$\lambda_f = 752$ nm
$\phi_f = 0.28$
HCl_3)

46

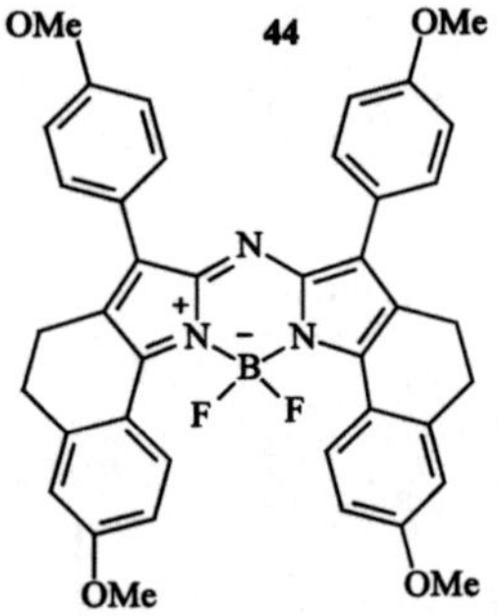

$\lambda_{abs} = 736$ nm
$\lambda_f = 748$ nm
$\phi_f = 0.29$
($CHCl_3$)

47

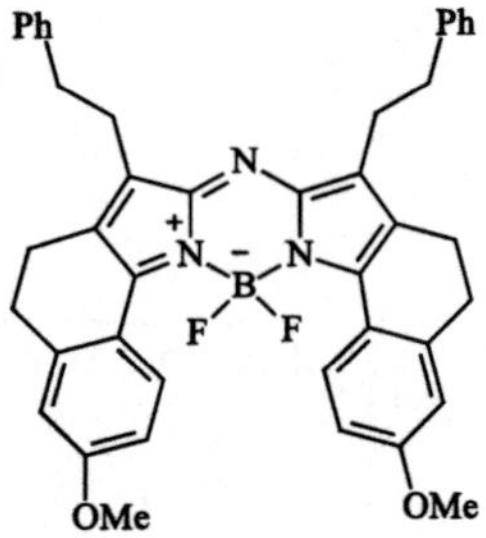

$\lambda_{abs} = 721$ nm
$\lambda_f = 732$ nm
$\phi_f = 0.31$
($CHCl_3$)

48

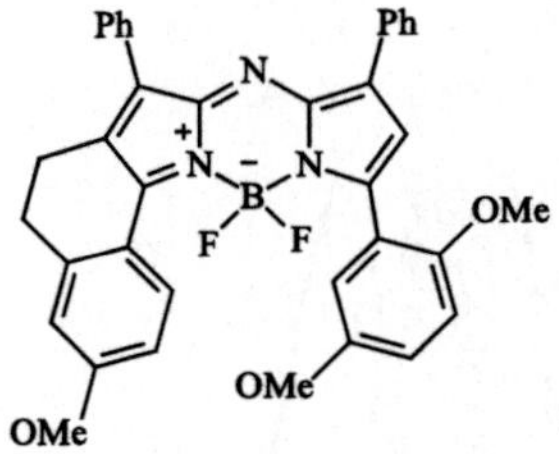

$\lambda_{abs} = 681$ nm
$\lambda_f = 728$ nm
$\phi_f = 0.24$
($CHCl_3$)

49

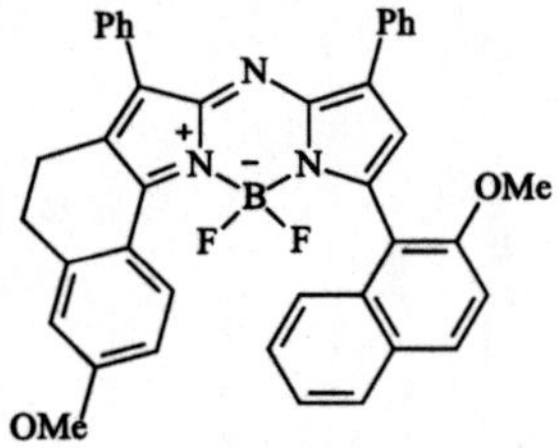

$\lambda_{abs} = 668$ nm
$\lambda_f = 692$ nm
$\phi_f = 0.46$
($CHCl_3$)

50

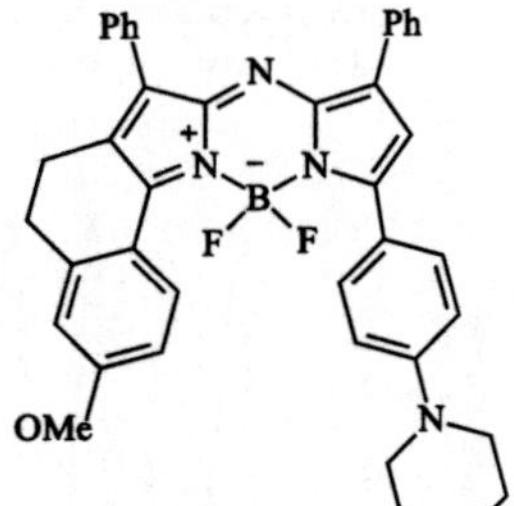

$\lambda_{abs} = 774$ nm
$\lambda_f = 815$ nm
$\phi_f = 0.05$
($CHCl_3$)

51

Fig. 15 Examples of substituted and conformationally restricted BODIPY fluorephores and their spectral data [52, 53]

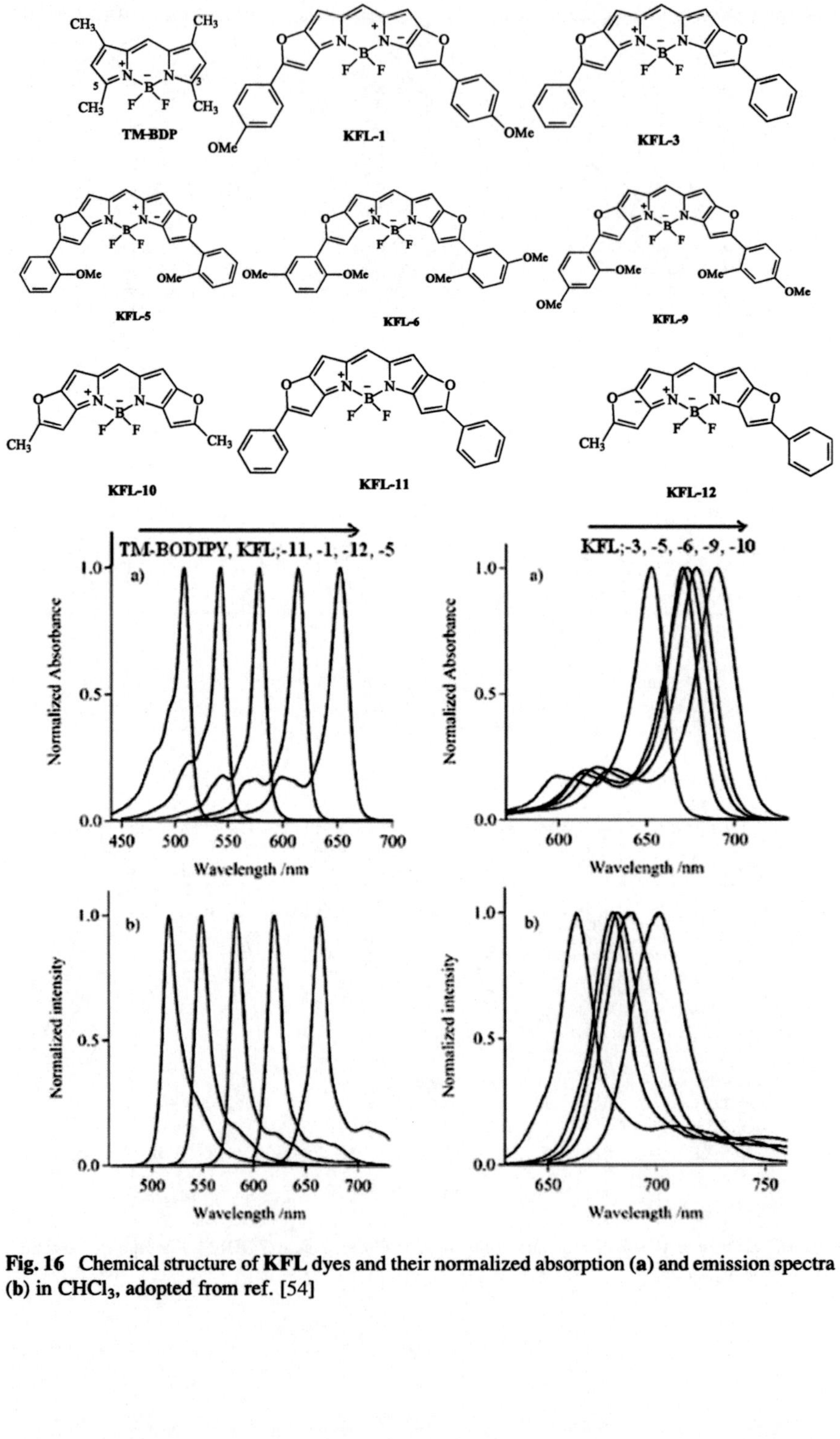

Fig. 16 Chemical structure of **KFL** dyes and their normalized absorption (**a**) and emission spectra (**b**) in $CHCl_3$, adopted from ref. [54]

Introduction of phenathrene rings, **53** and **54**, shifts the absorption and emission bands 130 nm with respect to the simplest BODIPY core, **52**, while keeping high emission efficiency [55]. The dyes exhibit narrow and structured absorption and emission bands, high extinction coefficients, and low dependence of spectroscopic properties on environment polarity. The absorption and emission wavelengths are blue-shifted upon increasing the electron donor strength of substituents. In addition, fluorescence of NMe_2-derivative can be switched on by protonation ($pK_a = 1.9$ in water).

For a series of alkoxy substituted mono- and distyryl-BODIPY NIR fluorescent dyes, **55** and **56**, respectively, a small degree of solvatochromism was observed [56]. This is most likely due to alkoxy group being a weakly electron donor. However, there is no doubt that this series of boradiazaindacenes will be attractive candidates for practical applications.

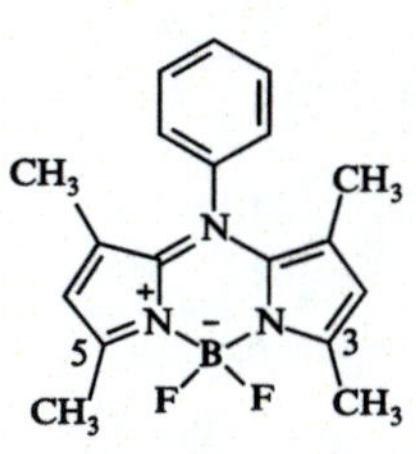

λ_{abs} = 498 nm
λ_f = 510 nm
ϕ_f = 0.59
(Me_2O)

52

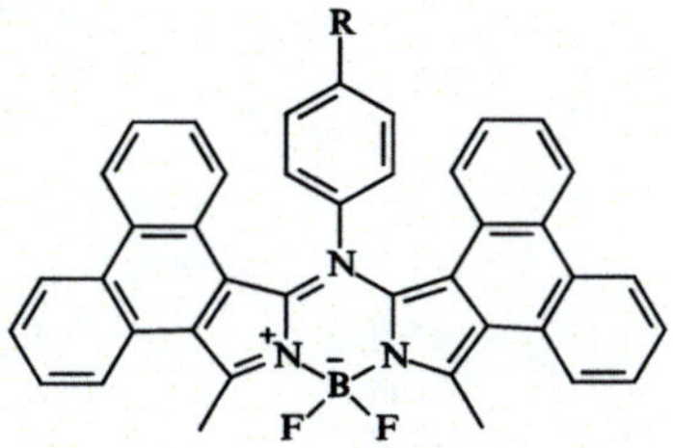

R = H; λ_{abs} = 627 nm, λ_f = 647 nm, ϕ_f = 1.0
R = I; λ_{abs} = 630 nm, λ_f = 652 nm, ϕ_f = 0.92
R = NMe_2; λ_{abs} = 618 nm, λf = 634 nm, ϕ_f = 0.94
(Me_2O)

53

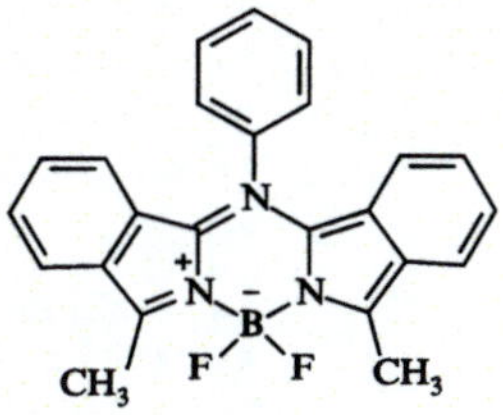

λ_{abs} = 598 nm
λ_f = 605 nm
ϕ_f = 0.98
(Me_2O)

54

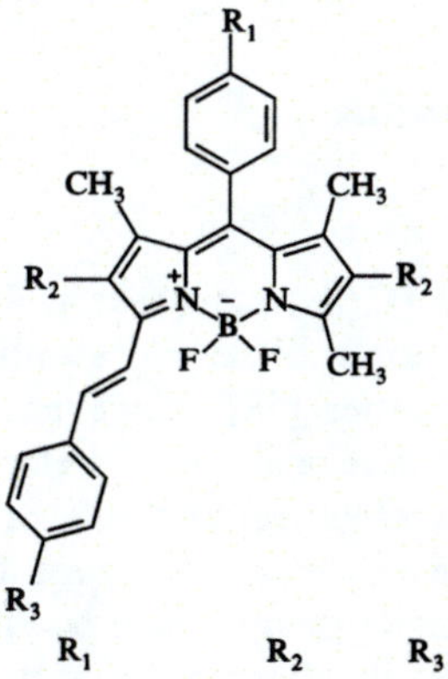

R_1	R_2	R_3
H	Br	$OCH_2COOtBu$
$OCH_2COOtBu$	Br	$OCH_2COOtBu$
$OCH_2COOtBu$	Et	$OCH_2COOtBu$

λ_{abs} = 657 nm
λ_f = 679 nm
ϕ_f = 0.42

λ_{abs} = 646 nm
λ_f = 668 nm
ϕ_f = 0.44

55

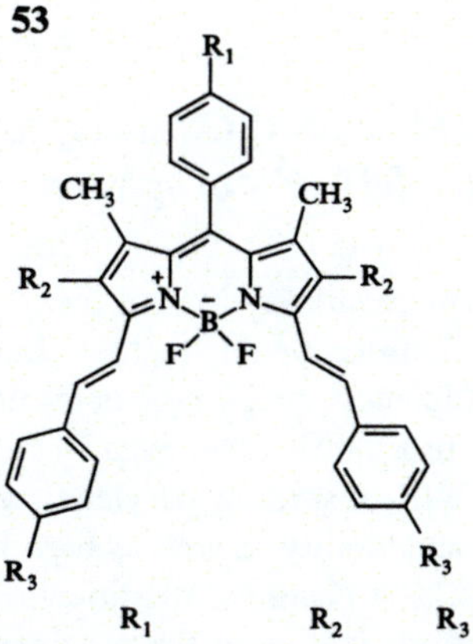

R_1	R_2	R_3
H	Br	$OCH_2COOtBu$
$OCH_2COOtBu$	Br	$OCH_2COOtBu$
$OCH_2COOtBu$	Et	$OCH_2COOtBu$

λ_{abs} = 656 nm
λ_f = 678 nm
ϕ_f = 0.40

(isopropanol)

56

Bis-ethynyl-substituted BODIPY dyes with long-wavelength absorption and emission, **57**, have been synthesized [57]. The compounds have two photon absorption active materials with good linear optical properties. The preliminary fluorescence imaging experiments indicated their cell permeability and nontoxicity.

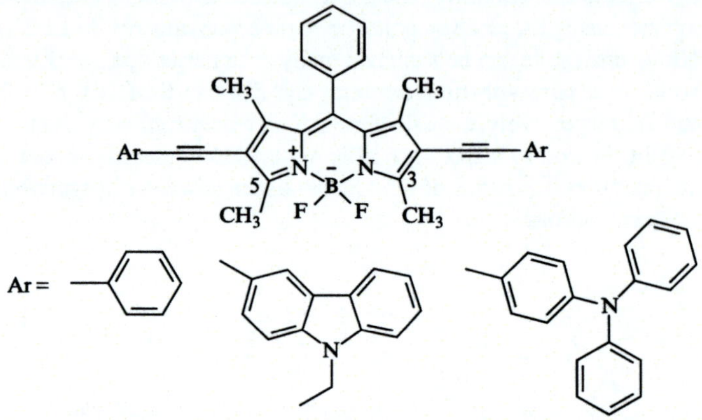

λ_{abs} = 573 nm
λ_f = 618 nm
ϕ_f = 0.60

λ_{abs} = 597nm
λ_f = 670nm
ϕ_f = 0.36
(THF)

λ_{abs} = 593nm
λ_f = 687 nm
ϕ_f = 0.31

57

3.4 Pyrelenebisimides, Oxazine, Rhodamines, and Other NIR Fluorophores

3,4:9,10-Perylenetetracarboxylic acid bisimides (PTCAIs) have been widely used in practical applications owing to their high photo- and thermal stability, high luminescence efficiency, and optoelectronic properties [58]. However, most of PTCAIs cannot absorb NIR light, even in the solid state, and are not water soluble. Therefore, new NIR absorbing perylene, with excellent properties such as high photo- and thermal stability as well as high luminescence quantum yield, has been strongly required. The chemical modification at imide groups could be drastically changed by functionalization of the perylene core with electron donor or acceptor groups [58–60]. Monofunctionalization of the imide structure allows terrylenediimides now to be coupled with a variety of compounds, for example, by cross-coupling, which can lead to an array of terrylenediimides with new functional groups such as hydroxy, amino, or carboxy groups needed to link up with other molecules. Such core-functionalized perylene bisimides are usually synthesized from the corresponding halogenated, in particular, brominated derivatives [58–60]. The substituted terrylenediimides offer, depending on the substituents used, exciting features such as good solubility in common organic solvents, water solubility [58], or NIR absorption and emission.

Introduction of dialkylamino groups into the bay regions leads to a large bathochromic shift of the absorption bands [58]. The reaction of tetrabromoterrylenediimide **58** with the electron-rich secondary cyclic amine piperidine gives NIR-fluorophore terrylenediimide **59d**. The absorption maximum of the longest wavelength absorption band of tetrapiperidinyl-substituted terrylenediimide **59d** at 819 nm is shifted bathochromically compared to that of **58** by 168 nm and emits around 800 nm, due to the electron-donating effect of the piperidinyl groups. Terrylenediimide **59a** with four phenoxyl substituents and good solubility in common organic solvents and water-soluble terrylenediimide **59c** were synthesized.

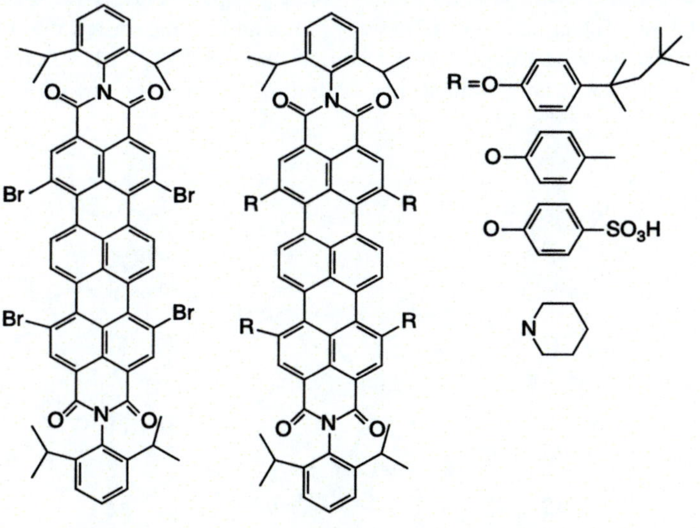

a **b** **c** **d**

58 **59**

Incorporation of piperidinyl group to the mixture of dibrominated perylene bisimides **60**, the regioisomeric pure 1,6-Piper-Pery **61** was separated and characterized [59]. Introduction of the electron-donating amines into the perylene core extends the π-system of push–pull aromatic of the perylene molecules and resulting in the bathochromic shifts of their absorption and emission with fluorescence peak at 760 nm. Efforts are currently underway to develop nonsymmetrically substituted, water-soluble versions to allow conjugation for biosensing experiments.

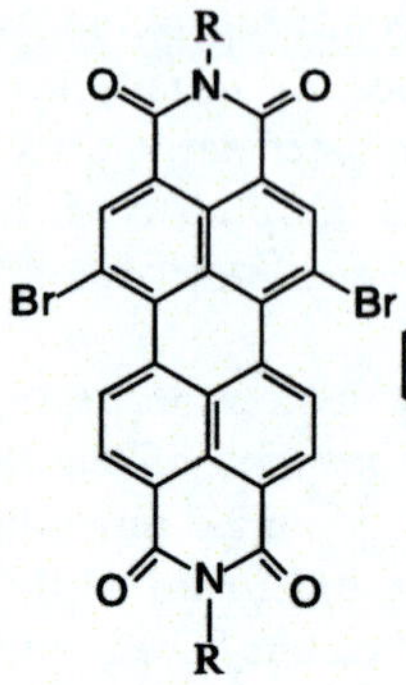

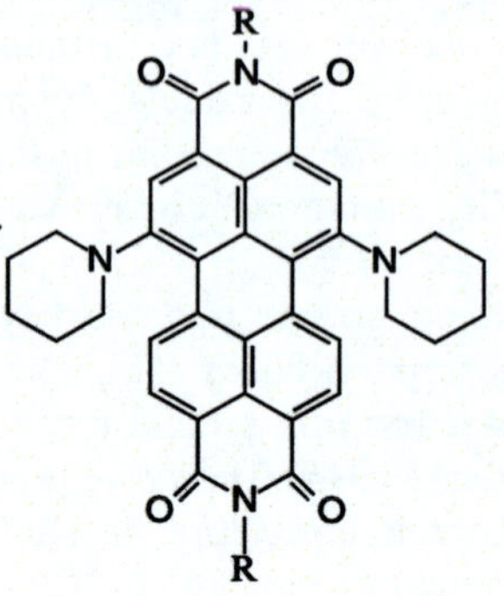

R = C_6H_{11}

n–C_4H_9

2,6 C_3H_7–C_6H_3

60 **61**

Strongly fluorescent NIR dyes have been obtained [60] by the core-substitution of perylene bisimides with joined nitrogen donor groups, 4-phenyl-1,2,4-triazolin-3,5-dione **63**, **64**. The emission of **63a** is at 775 nm and is thus shifted into the NIR. The emission of **64a** is strongly further shifted into the NIR, with a fluorescence maximum at 873 nm, and fluorescence emission can even be recorded beyond 1,100 nm.

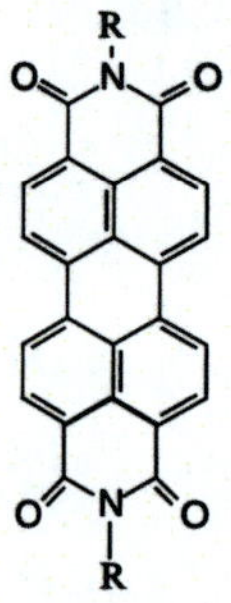

62

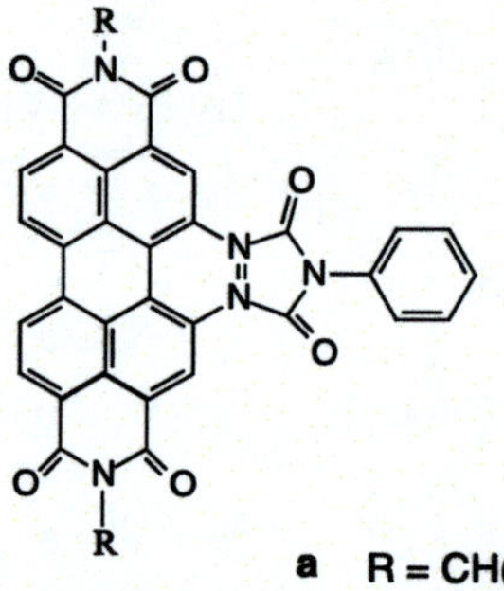

63

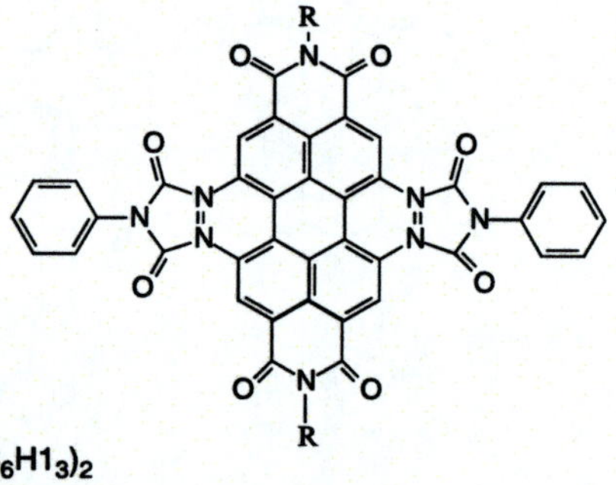

64

a R = $CH(C_6H_{13})_2$

b $CH(C_8H_{17})_2$

c $CH(CH_3)_2$

Oxazine and rhodamine dyes are mainly visible fluorophores, but some have been extended into the NIR region and used as probes for proteins [3, 8], particularly the amine derivatives. Oxazine and rhodamine dyes are structurally similar compounds with two nitrogen atoms forming the push–pull conjugated system, with oxazine having an additional oxygen atom to bridge the conjugation. They are structurally more compact than cyanine dyes. Some long-wavelength examples of these dye families are: Rhodamine 800 (absorption/emission: 680 nm/700 nm), Nile Blue (absorption/emission: 638 nm/660 nm), and oxazine 750 (absorption/emission: 673/691 nm).

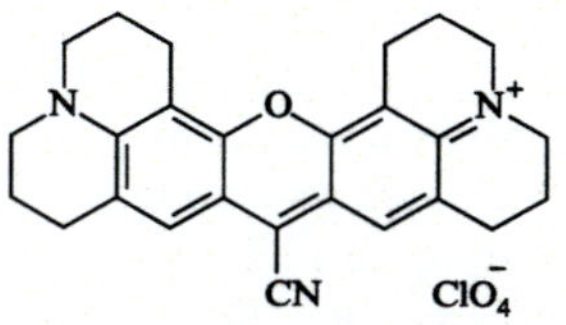

Rhodamine 800

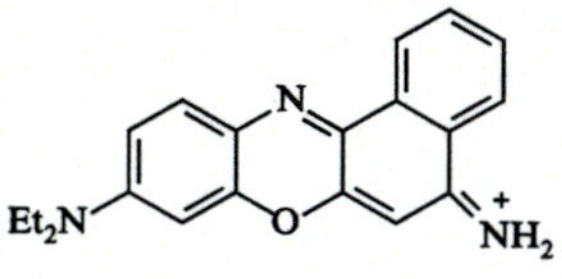

Nile Blue

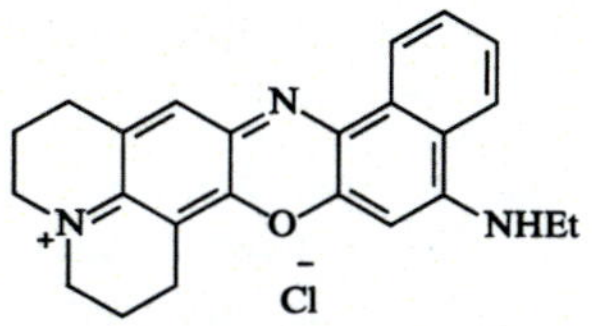

Oxazine 750

NIR fluorophores containing benzo[c]heterocycle subunits have been synthesized [61]. Some of these probes (**65**, **66**, and **67**) emit characteristic fluorescence signal in the NIR (between 612 and 720 nm) only when bound to amyloid-beta peptide (Aβ) [62]. The compounds exhibited Aβ-dependent changes in fluorescence quantum yield, lifetime, and emission spectra that may be imaged microscopically or in vivo using new lifetime and spectral fluorescence imaging techniques. The turn-on of the probes when bound to Aβ enable quantitative molecular imaging in vivo.

Molecular probes based on the Δ-pyrazoline chromophore **58** were synthesized [63]. The probes emit at the NIR region (around 680 nm) and show a cation-induced switching on of the fluorescence upon binding to main group elements (Sr, Ca, and Ba) and/or heavy metal ions (Hg, Pb, and Ag). In these substituted 1,3,5-triaryl-D2-pyrazolines, not only does the spacer-separated receptor at the 5-position acts as an electron donor but also the main chromophore itself. Thus, upon combining intramolecular charge and electron transfer processes in a simple fluorophore-spacer-receptor ionophore with a small but rigid spacer, an efficient cation-triggered switching on of the ICT fluorescence can selectively be achieved, with advantageous emission features such as broad and largely Stokes-shifted bands and considerably high fluorescence quantum yields in the red/NIR spectral region.

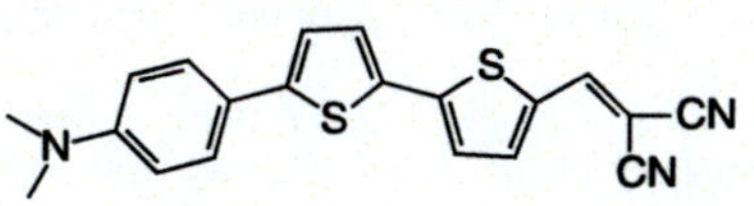

65

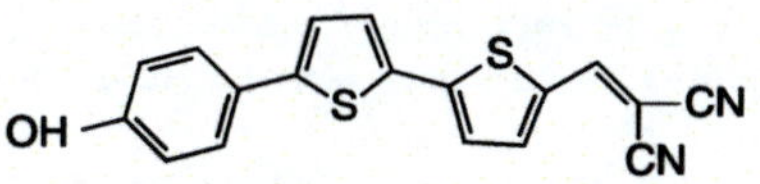

66

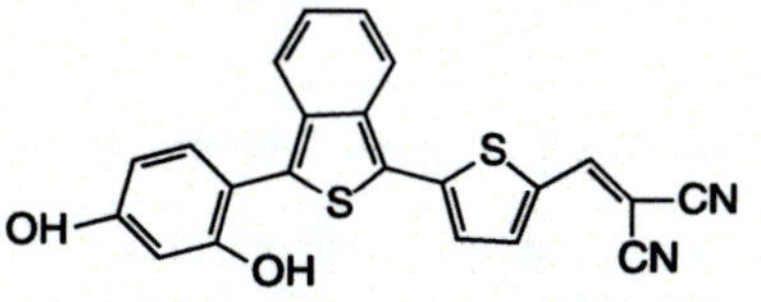

67

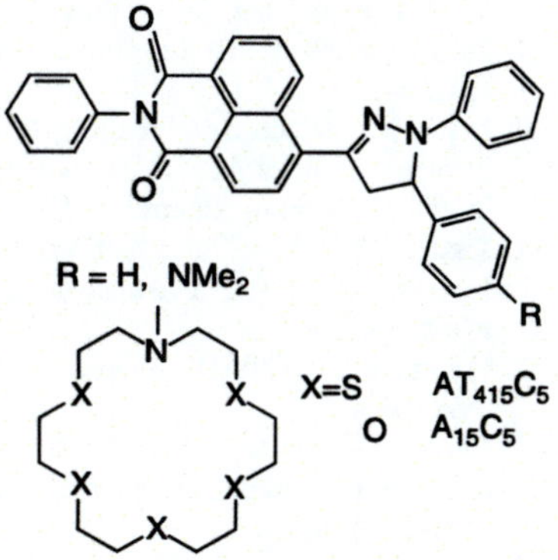

68

4 Concluding Remarks

Various types of organic dyes are NIR fluorophores and can be used in analytical chemistry and biological systems as fluorescent probes as well as in vivo imaging of biosamples in medicine. Active research in this field is aiming to improve probe brightness and solubility for highly sensitive assays and applications. The number of fluorophores in the near-IR region is growing as a consequence of the ongoing dye synthesis research efforts. Since the fluorophore properties needed for a particular application may vary, the growing number of near-IR fluorophores will greatly benefit the development of technological, analytical, and biological applications using long wavelength emitting dyes.

References

1. Carrie L, Amiot SX, Song L, Lingyun P, Julia XZ (2008) Near-infrared fluorescent materials for sensing of biological targets. Sensors 8:3082–3105
2. Demchenko AP (2009) Introduction to fluorescence sensing. Springer, Amsterdam
3. Changxia S, Jinghe Y, Lei L, Xia W, Yang L, Shufang L (2004) Advances in the study of luminescence probes for proteins. J Chromatogr B 803:173–190
4. Pa S-Y, Kun J, Sea-Wha OB (2005) The novel functional chromophores based on squarylium dyes. Korean Chem Soc 26:428–432
5. Kiyose K, Aizawa S, Sasaki E, Kojima H, Hanaoka K, Terai T, Urano Y, Nagano T (2009) Molecular design strategies for near-infrared ratiometric fluorescent probes based on the unique spectral properties of aminocyanines. Chem Eur 15:9191–9200
6. Loudet A, Burgess K (2007) BODIPY dyes and their derivatives: syntheses and spectroscopic properties. Chem Rev 107:4891–4932
7. Pham W, Cassell L, Gillman A, Koktysh D, Gore JC (2008) A near-infrared dye for multichannel imaging. Chem Commun 16:1895–1897
8. Lakowicz JR (1994) Topics in fluorescence spectroscopy: probe design and chemical sensing, vol 4. Plenum Press, New York
9. Drexhage KH (1990) Structure and properties of laser dyes in dye lasers. Springer, New York, pp 155–200
10. Masotti A, Vicennati P, Boschi F, Calderan L, Sbarbati A, Ortaggi G (2008) A novel near-infrared indocyanine dye – polyethylenimine conjugate allows DNA delivery imaging in vivo. Bioconjug Chem 19:983–987
11. Kasten FH (1993) Introduction to fluorescent probes: properties, history and applications. In: Mason WT (ed) Fluorescent and luminescent probes for biological activity. Academic Press, pp. 12–33
12. Frangioni JV (2003) In vivo near-infrared fluorescence imaging. Curr Opin Chem Biol 7:626–634
13. Cheng Z, Wu Z, Xiong Z, Gambhir SS, Chen X (2005) Near-infrared fluorescent RGD peptides for optical imaging of integrin avb3 expression in living mice. Bioconjug Chem 16:1433–1441
14. Kiyose K, Kojima H, Urano Y, Nagano T (2006) Development of a ratiometric fluorescent zinc ion probe in near-infrared region, based on tricarbocyanine chromophore. J Am Chem Soc 128:6548–6549

15. Chen J, Corbin IR, Li H, Cao W, Glickson JD, Zheng G (2007) Ligand conjugated low-density lipoprotein nanoparticles for enhanced optical cancer imaging in vivo. J Am Chem Soc 129:5798–5799
16. Malick J, Gryczynski I, Geddes CD, Lakowicz JR (2003) Metal-enhanced emission from indocyanine green; a new approach to in vivo imaging. J Biomed Opt 8:472–478
17. Pham W, Choi Y, Weissleder R, Tung CH (2004) Developing a peptide-based near-infrared molecular probe for protease sensing. Bioconjug Chem 15:1403–1407
18. Flanagan JH, Owens CV, Romero SE, Waddell E, Kahn SH, Hammer RP, Soper SA (1998) Near-infrared heavy-atom-modified fluorescent dyes for base-calling in DNA-sequencing applications using temporal discrimination. Anal Chem 70:2676–2684
19. Ching-Hsuan T, Sebastian B, Umar M, Ralph W (1999) Preparation of a cathepsin D sensitive near-infrared fluorescence probe for imaging. Bioconjug Chem 10:892–896
20. Tang B, Huang H, Xu K, Tong L, Yang G, Liu X, An L (2006) Highly sensitive and selective near-infrared fluorescent probe for zinc and its application to macrophage cells. Chem Commun 34:3609–3611
21. Tang B, Cui JL, Xu HK, Tong LL, Yang WG, An GL (2008) A sensitive and selective near-infrared fluorescent probe for mercuric ions and its biological imaging applications. Chem BioChem 9:1159–1164
22. Song F, Peng X, Lu Erhu WY, Zhou W, Fan J (2005) Tuning the photoinduced electron transfer in near-infrared heptamethine cyanine dyes. Tetrahedron Lett 46:4817–4820
23. Sasaki E, Kojima H, Nishimatsu H, Urano Y, Kikuchi K, Hirata Y, Nagano T (2005) Highly sensitive near-infrared fluorescent probes for nitric oxide and their application to isolated organs. J Am Chem Soc 127:3684–3685
24. Zhang Z, Achilefu S (2005) Design, synthesis and evaluation of near-infrared fluorescent pH indicators in a physiologically relevant range. Chem Commun 47:5887–5889
25. Tang B, Fabiao Y, Li P, Tong L, Duan X, Xie T, Wang X (2009) A near-infrared neutral pH fluorescent probe for monitoring minor pH changes: imaging in living HepG2 and HL-7702 cells. J Am Chem Soc 131:3016–3023
26. Patsenker L, Tatarets A, Kolosova O, Obukhova PY, Fedyunyayeva I, Yermolenko I, Eewald T (2008) Fluorescent probes and labels for biomedical applications. Ann NY Acad Sci 1130:179–187
27. Chiara AB, Giuseppe C, Barolo C, Viscardi G, Coluccia S (2006) Novel heptamethine cyanine dyes with large stokes' shift for biological applications in the near infrared. J Fluoresc 16:221–225
28. Bouteiller C, Clave G, Bernardin A, Chipon B, Massonneau M, Renard PY, Romieu A (2007) Novel water-soluble near-infrared cyanine dyes: synthesis, spectral properties, and use in the preparation of internally quenched fluorescent probes. Bioconjug Chem 18:1303–1317
29. Curino A, Bugge HT, Weissleder R, Tung C-H (2004) Design, synthesis, and characterization of urokinase plasminogen-activator-sensitive near-infrared reporter Benedict law. Chem Biol 11:99–106
30. Peng X, Song F, Lu E, Wang Y, Zhou W, Fan J, Gao Y (2005) Heptamethine cyanine dyes with a large stokes shift and strong fluorescence: a paradigm for excited-state intramolecular charge transfer. J Am Chem Soc 127:4170–4171
31. Basheer MC, Santhosh U, Alex S, Thomas KG, Cherumuttathu HS, Suresh D (2007) Design and synthesis of squaraine based near infrared fluorescent probes. Tetrahedron 63:1617–1623
32. Anatoliy LT, Irina AF, Tatyana SD, Yevgeniy AP, Andrey OD, Ewald AT, Leonid DP (2006) Synthesis of water-soluble, ring-substituted squaraine dyes and their evaluation as fluorescent probes and labels. Anal Chim Acta 570:214–223
33. Kalliat TA, Danaboyina R (2005) Near-infrared fluorescent probes: synthesis and spectroscopic investigations of a few amphiphilic squaraine dyes. J Phys Chem A 109:5571–5578
34. Keitaro U, Daniel C, Koji S (2008) Water-soluble NIR fluorescent probes based on squaraine and their application for protein labeling. Anal Sci 24:213–217

35. Keitaro U, Daniel C, Koji S (2007) A squaraine-based near-infrared dye with bright fluorescence and solvatochromic property. Chem Lett 36:1424–1425
36. Shigeyuki Y, Taishi O, Nobuo A, Hiroyuki N (2008) The synthesis and optical properties of bis-squarylium dyes bearing arene and thiophene spacers. Dyes Pigm 77:525–536
37. Shigeyuki Y, Hiroyuki N (2008) Squarylium dyes and related compounds. Top Heterocycl Chem 14:133–181
38. Akkaya EU, Turkyilmaz S (1997) A squaraine-based NIR fluorescent chemosensor for calcium. Tetrahedron Lett 38:4513–4516
39. Umut O, Engin UA (1998) A squaraine-based sodium selective fluorescent chemosensor. Tetrahedron Lett 39:5857–5860
40. Ayyappanpillai A (2005) Chemistry of squaraine-derived materials: near-IR dyes, low band gap systems, and cation sensors. Acc Chem Res 38:449–459
41. Yogesh C, Gitish KD, Raja BK, Satish P (2009) Tetrahydroquinoxaline based squaraines: synthesis and photophysical properties. Dyes Pigm 83:162–167
42. Joseph T, Douglas BS, Terry JA, Sandra AA, Bruce PJ (2007) Synthesis and biosensor performance of a near-IR thiol-reactive fluorophore based on benzothiazolium squaraine. Bioconjug Chem 18:1841–1846
43. Sivaramapanicker S, Kizhumuri PD, Ayyappanpillai A (2008) A near-infrared squaraine dye as a latent ratiometric fluorophore for the detection of aminothiol content in blood plasma. Angew Chem Int Ed 47:7883–7887
44. Yevgen AP, Larysa IM, Anatoliy LT, Vadim IS, Ewald AT, Leonid DP (2009) Near-infrared, dual-ratiometric fluorescent label for measurement of pH. Anal Biochem 390:136–140
45. Robert EG, Hua M, Mahbubul HM, John BT, Jennifer SP (2007) Visible fluorescence chemosensor for saxitoxin. J Org Chem 72:2187–2191
46. Bricks JL, Kovalchuk A, Trieflinger Ch, Nofz M, Büschel M, Tolmachev AI, Daub J, Rurack K (2005) On the development of sensor molecules that display Fe^{III}-amplified fluorescence. J Am Chem Soc 127:13522–13529
47. Coskun A, Yilmaz MD, Akkaya EU (2007) Bis(2-pyridyl)-substituted boratriazaindacene as an NIR-emitting chemosensor for Hg(II). Org Lett 9:607–609
48. Rurack K, Kollmannsberger M, Daub J (2001) Molecular switching in the near infrared (NIR) with a functionalized boron-dipyrromethene dye. Angew Chem Int Ed 40:385–387
49. Coskun A, Deniz E, Akkaya EU (2005) Effective PET and ICT switching of boradiazaindacene emission: a unimolecular, emission-mode, molecular half-subtractor with reconfigurable logic gates. Org Lett 7:5187–5189
50. Deniz E, Isbasar GC, Bozdemir OA, Yildirim LT, Siemiarczuk A, Akkaya EU (2008) Bidirectional switching of near IR emitting boradiazaindacene fluorophores. Org Lett 10:3401–3403
51. Hall MJ, Allen LT, O'Shea DF (2006) PET modulated fluorescent sensing from the BF2 chelated azadipyrromethene platform. Org Biomol Chem 4:776–780
52. Zhao W, Carreira EM (2006) Conformationally restricted Aza-BODIPY: highly fluorescent, stable near-infrared absorbing dyes. Chem Eur J 12:7254–7263
53. Zhao W, Carreira EM (2005) Conformationally restricted Aza-Bodipy: a highly fluorescent, stable, near-infrared-absorbing dye. Angew Chem Int Ed 44:1677–1679
54. Umezawa K, Matsui A, Nakamura Y, Citterio D, Suzuki K (2009) Bright, color-tunable fluorescent dyes in the Vis/NIR region: establishment of new "tailor-made" multicolor fluorophores based on borondipyrromethene. Chem Eur J 15:1096–1106
55. Descalzo AB, Xu H, Xue Z, Hoffmann K, Shen Z, Weller MG, You X, Rurack K (2008) Phenanthrene-fused boron–dipyrromethenes as bright long-wavelength fluorophores. Org Lett 10:1581–1584
56. Dost Z, Atilgan S, Akkaya EU (2006) Distyryl-boradiazaindacenes: facile synthesis of novel near IR emitting fluorophores. Tetrahedron 62:8484–8488

57. Zhang D, Wang Y, Xiao Y, Qian S, Qian X (2009) Long-wavelength boradiazaindacene derivatives with two-photon absorption activity and strong emission: biological imaging applications. Tetrahedron 65(39):8099–8103
58. Nolde F, Qu J, Kohl C, Pschirer NG, Reuther E, Mullen K (2005) Synthesis and modification of terrylenediimides as high-performance fluorescent dyes. Chem Eur J 11:3959–3967
59. Fan L, Xu Y, Tian H (2005) 1,6-Disubstituted perylene bisimides: concise synthesis and characterization as near-infrared fluorescent dyes. Tetrahedron Lett 46:4443–4447
60. Heinz L, Patrick B (2003) An approach to novel NIR dyes utilizing α-effect donor groups. Dyes Pigm 59:109–116
61. Meek ST, Nesterov EE, Swager TM (2008) Near-infrared fluorophores containing benzo[c] heterocycle subunits. Org Lett 10:2991–2993
62. Raymond SB, Skoch J, Hills ID, Nesterov EE, Swager TM, Bacskai BJ (2008) Smart optical probes for near-infrared fluorescence imaging of Alzheimer's disease pathology. Eur J Nucl Med Mol Imaging 35:93–98
63. Rurack K, Resch-Genger U, Bricks JL, Spieles M (2000) Cation-triggered 'switching on' of the red/near infra-red (NIR) fluorescence of rigid fluorophore–spacer–receptor ionophores. Chem Commun 2:2103–2104

Luminescence Analysis of $Bi_4(TiO_4)_3$ and $LiZnVO_4$ Ceramic Powders

Bhaskar Kumar Grandhe and Buddhudu Srinivasa

Abstract A couple of novel ceramic luminescent materials based on $Bi_4(TiO_4)_3$ and $LiZnVO_4$ matrices have been developed by a conventional solid-state reaction method. An intense *blue* emission at 480 nm has been measured with an excitation at 418 nm from the Bismuth Titanate $Bi_4(TiO_4)_3$ ceramic powder. A bright green emission at 533 nm has been measured upon with $\lambda_{exci} = 359$ nm. The mechanisms involved in the observance of such prominent visible color emissions from these ceramic powders have been explained. Along side the measurement of their luminescence spectra, structural details of these ceramic powders have been carried out based on the recorded profiles of X-ray diffraction (XRD), scanning electron microscopy (SEM), energy dispersive X-ray spectroscopy (EDS), Fourier-transform IR spectroscopy (FTIR), and Raman spectra. With regard to thermal properties of these novel optical materials, their comprehensive differential thermo gravimetric analysis (TG–DTA) has also been undertaken and the important and interesting results are thus reported here.

1 Introduction

Production of more efficient luminescent materials has continually been carried out for different applications such as compact light sources, display screens and photonics devices, etc. [1]. Oxide-based inorganic ceramic phosphors have been investigated for their optical applications like high-resolution devices, namely, cathode-ray tubes, electroluminescent devices, plasma display panels, and field emission displays. Oxide phosphors have been investigated because of their higher chemical stability over the other phosphors such as sulfide phosphors [2, 3].

B. Srinivasa (✉)
Department of Physics, Sri Venkateswara University, Tirupati 517502, Andhra Pradesh, India
e-mail: drsb99@hotmail.com

C.D. Geddes (ed.), *Reviews in Fluorescence 2009*, Reviews in Fluorescence 2009,
DOI 10.1007/978-1-4419-9672-5_5,

During the past few years, Titanate- and Vanadate-based materials have attracted special attention due to their significant applications [4–6]. In recent times, more attention has been focused on visible photoluminescence (PL) at room temperature because of its potential technological application on the development of new luminescent materials. The broad band luminescence in the visible light range is considered to be more effective to obtain a good color rendering property from the lighting devices [7]. Mixed-metal oxides play an important role in the development of several newer materials [8]. Literature survey in this direction of work indicates that no systematic study has so far been made on Bismuth Titanate ($Bi_4(TiO_4)_3$) and Lithium Zinc Vanadate ($LiZnVO_4$) ceramic powders; therefore, in the present investigation, we have prepared $Bi_4Ti_3O_{12}$ and Lithium Zinc Vanadate ($LiZnVO_4$) ceramic powders by a conventional solid-state route and characterized by employing different techniques to understand their structural, luminescence, and thermal properties.

2 Experimental Details

Bismuth Titanate ($Bi_4(TiO_4)_3$) and Lithium Zinc Vanadate ($LiZnVO_4$) ceramic powders were synthesized by using a solid-state reaction method. Highly pure and analytical reagent grade chemicals such as Bi_2O_3, TiO_2, Li_2CO_3, ZnO, and NH_3VO_4 were used as the starting materials. Those chemicals were weighed based on the calculated composition and each of them was finely powdered using mortar and a pestle for 2 h to obtain homogeneous precursor. These precursors were then transferred each separately into silica crucibles and were heated in an electrical furnace from the room temperature to 1,100°C (5 h) for Bismuth Titanate ceramic powder and up to 850°C (5 h) for Lithium Zinc Vanadate, respectively.

The structures of the prepared ceramic powders were characterized on a XRD 3003 TT Seifert diffractometer with CuK_α radiation ($\lambda = 1.5406$ Å) at 40 kV and 20 mA. The morphologies of the ceramic powders were examined on a Zeiss EVO MA 15 scanning electron microscope. The elemental analysis of the synthesized products was carried out using the Oxford EDAX equipment (INCA-Penta Fet X3) attached to the scanning electron microscopy (SEM) system. Fourier-transform IR (FTIR) spectra of the samples were recorded on a Nicolet IR-200 spectrometer using KBr pellet technique from 4,000 to 400 cm^{-1}. Raman Spectra of the prepared ceramics were recorded by using a high-resolution Jobin Yvon Model HR800UV system attached with a He-Ne laser (633 nm) as the excitation source with an output power of 15 mW having a laser beam spot size of 100 μm using an appropriate lens system. The photoluminescence spectra of these powders were recorded on a Jobin Yvon Fluorolog-3 Fluorimeter with a Xe-arc lamp (450 W) as an excitation source. Thermogravimetry (TG) and differential thermal analysis (DTA) were performed for the precursors in N_2 atmosphere at a heating rate of 10°C/min by using Netzsch STA 409 Simultaneous Thermal Analyzer.

3 Results and Discussion

3.1 *Section A: Analysis of Blue Luminescent $Bi_4(TiO_4)_3$ Ceramic Powder*

The X-ray diffraction (XRD) profile of $Bi_4(TiO_4)_3$ ceramic powder is shown in Fig. 1 and the patterns are indexed and are found to be in good agreement with the orthorhombic structure following JCPDS Card No. 12-0213. The crystallite size of the prepared ceramic powders has been estimated from the Scherer's equation, $D = 0.9\lambda/\cos\theta$, where D is the crystallite size, λ is the wavelength of X-ray (0.15405 nm), θ and β are the diffraction angle and full width at half maximum (FWHM) of an observed peak, respectively. Intense diffraction peaks have been selected to compute the crystallite size and it is found to be in an average size of 66 nm.

SEM micrograph of the $Bi_4(TiO_4)_3$ ceramic powder is shown in Fig. 2. The obtained micrograph shows that the particles are agglomerated and irregular in shapes. Therefore, the average diameter of the grain size has approximately been measured and it is in the range of 1 μm. The narrow widths of the diffraction peaks also indicate that the prepared ceramic powder possess large-sized grains [9]. It may be mentioned that crystalline powders in micrometer dimension could display encouraging and significance luminescent intensities [10]. Energy dispersive X-ray spectroscopy (EDS) spectrum has been measured to carry out the elemental analysis of synthesized $Bi_4(TiO_4)_3$ ceramic powder. The obtained elemental characteristic X-ray radiation for Bi is M_α 2.4 keV, L_α 10.8 keV, and L_β 13.0 keV, for Ti is

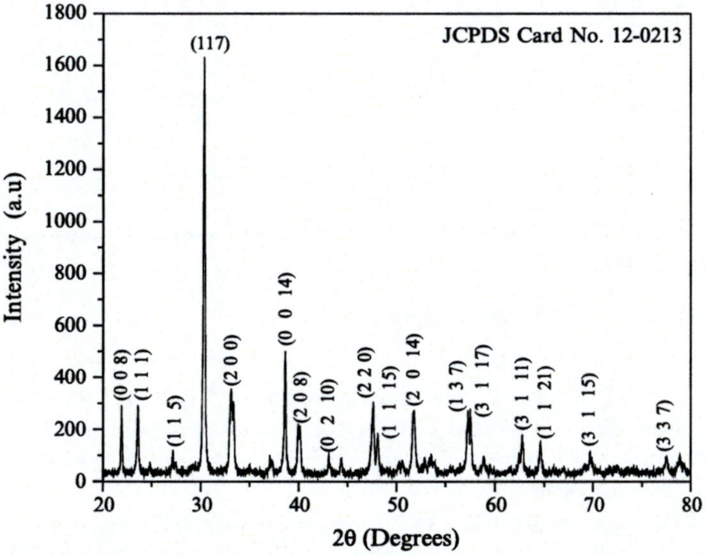

Fig. 1 XRD profile of $Bi_4(TiO_4)_3$ ceramic powder

Fig. 2 SEM image of $Bi_4(TiO_4)_3$ ceramic powder

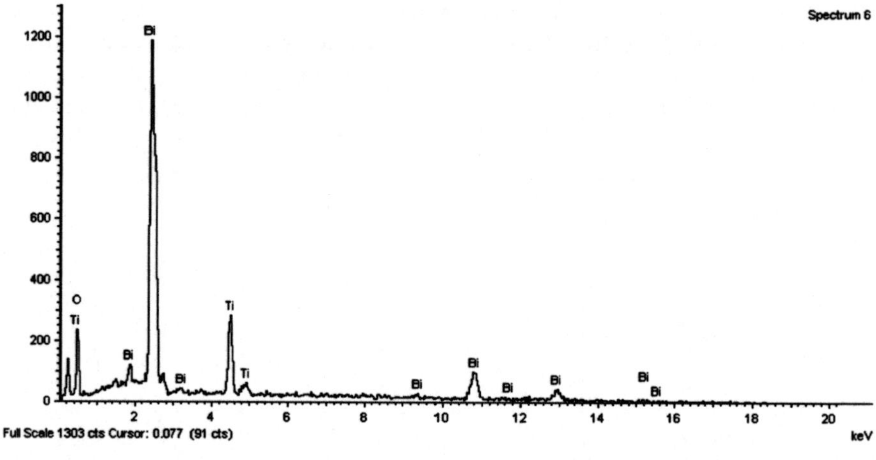

Fig. 3 EDS profile of $Bi_4(TiO_4)_3$ ceramic powder

K_α 4.5 keV and K_{ab} 4.9 keV, and for oxygen is K_α 0.5 keV [11]. The existence of the unassigned carbon peak at 0.2 keV in the EDS spectrum does not belong to the prepared sample, it is from the carbon tape that was used to hold the specimen during the measurement (Fig. 3).

The FTIR spectrum of the $Bi_4(TiO_4)_3$ ceramic powder is shown in Fig. 4 and from the figure, three sharp bands at 815, 580, and 400 cm^{-1} are observed. The former two bands are ascribed to the Ti–O stretching vibrations, while the latter one to the

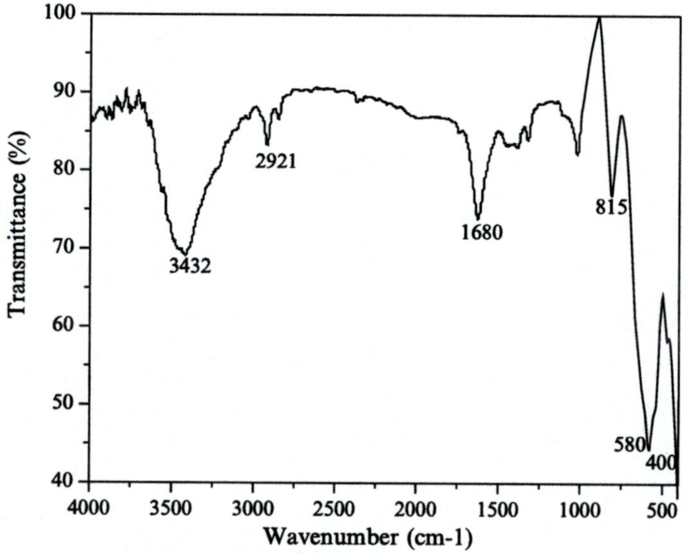

Fig. 4 FTIR spectrum of $Bi_4(TiO_4)_3$ ceramic powder

Ti–O bending vibrations [12]. Due to the absence of a C=O vibration at around 1,450 cm^{-1}, the powder is free from carbonates. This result is satisfactory to strengthen the technological point of view since most of the properties are dependent on the quality of the raw powder used [13]. Bands at 2,921 and 3,432 cm^{-1} could arise from the antisymmetric and symmetric stretching band of H_2O and OH groups, while a band at 1,680 cm^{-1} originates due to the bending vibrations of H_2O. These three bands are the characteristic vibrations of moisture present in the sample [14].

Figure 5 shows the Raman spectrum of $Bi_4(TiO_4)_3$ ceramic powder with intense Raman peaks implying the strong interactions among the atoms, which mainly arise from the stretching and bending of the shorter metal–oxygen bonds within the anionic groups. The TiO_6 octahedra in the $Bi_4(TiO_4)_3$, accordingly, plays an important role in the lattice vibrations. In accordance with Raman data of $Bi_4Ti_3O_{12}$, $BaTiO_3$, and $PbTiO_3$ [15], a shorter bond length of Ti–O compared to Bi–O suggests that the Raman phonon mode at 849 cm^{-1} originates mainly from the vibrations of atoms inside the TiO_6 octahedra. It is attributed to the symmetric Ti–O stretching vibration. The modes at 540 cm^{-1} (TO) and 565 cm^{-1} (LO) indicate the doubly degenerate symmetric O–Ti–O stretching vibrations. They split into a longitudinal (LO) and transverse (TO) components due to the long-range electrostatic forces that might be associated with lattice ionicity. The peaks located at 268 and 227 cm^{-1} are the modes ascribed to the O–Ti–O bending vibrations. Although the mode at 227 cm^{-1} is a Raman inactive according to the O_h symmetry of TiO_6, it has often been observed because of the distortion of octahedron. The mode at 324 cm^{-1} s due to a combination of the stretching and bending vibrations. Two other modes at 538 and 558 cm^{-1} correspond to the opposing excursions of the

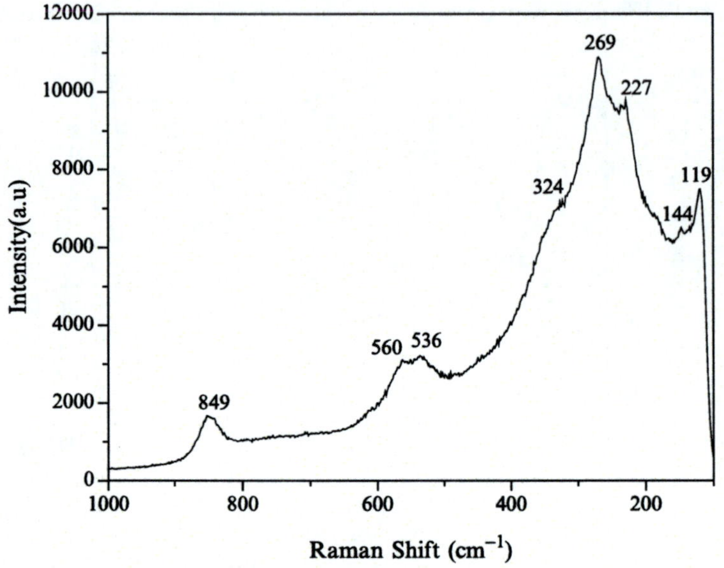

Fig. 5 Raman spectrum of $Bi_4(TiO_4)_3$ ceramic powder

external apical oxygen atoms of the TiO_6 octahedra. The TiO_6 octahedra have shown a considerable distortion at the room temperature so that some phonon modes at 324, 538, and 849 cm^{-1} appear wider in nature but weak in strength. The Raman modes at 118 and 144 cm^{-1} originate due to the vibrations between Bi and O atoms. All the assignments have been made and found tallying with the literature reports [15–17].

Figure 6 shows the comprehensive TG–DTA profiles of $Bi_4(TiO_4)_3$ precursor which is mixed in a stoichiometric situation. A sharp exothermic peak at 1,100°C is attributed to the heat loss during the crystallization of $Bi_4(TiO_4)_3$ ceramic powder. The homogenous mixture of Bi_2O_3 and Ti_2O_3, when it was heated at 1,100°C for 5 h, shows an XRD pattern corresponding to $Bi_4(TiO_4)_3$ (JCPDS Card No. 12–0213) which supports the assignment of the exothermic peak at 1,100°C in the DTA curve due to the crystallization of $Bi_4(TiO_4)_3$. No significant weight loss was found in the TG curve up to 1,100°C temperature and beyond that the precursor becomes gradually decomposing.

Figure 7a shows an excitation spectrum of $Bi_4(TiO_4)_3$ ceramic powder upon monitoring the emission at $\lambda_{emis} = 480$ nm and it has resulted a strong excitation band at 418 nm. Figure 7b shows the emission spectrum of $Bi_4(TiO_4)_3$ ceramic powder in the range of 450–750 nm with an excitation at 418 nm. We have noticed a strong blue emission at 480 nm which is ascribed to the electron–hole recombination of localized exciton. Electrons in the valence band are excited to some localized levels and thus form small polarons, and the electron polarons interact with holes possibly trapped near oxygen vacancies to form localized exciton, such

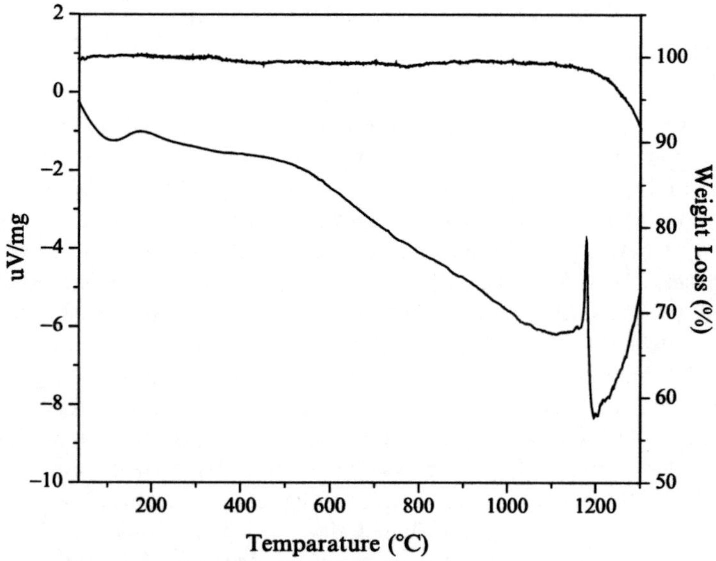

Fig. 6 TG–DTA profiles of $Bi_4(TiO_4)_3$ ceramic powder

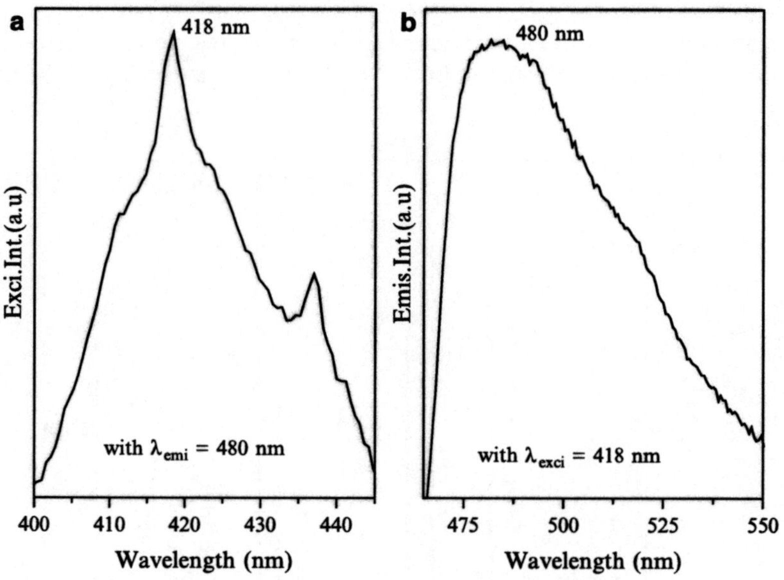

Fig. 7 (**a**) Excitation and (**b**) emission spectra of $Bi_4(TiO_4)_3$ ceramic powder

as self-trapped exciton. The recombination of the localized excitons could be resulting in blue emission as has been reported in the literature [18]. Electron–hole recombination of localized excitons associated with oxygen vacancies could therefore be a possible reason for this blue emission.

3.2 *Section B: Analysis of Green Luminescent $LiZnVO_4$ Ceramic Powder*

The XRD profile of $LiZnVO_4$ ceramic powder is shown in Fig. 8 and from the figure it is confirmed that the sample structure is in the form of rhombohedral, following the usage of the JCPDS Card No. 38–1332. The crystallite size of the prepared ceramic powder has been estimated using the Scherer's equation, $D = [(0.9\lambda)/(\beta \cos \theta)]$, where D is the crystallite size, λ is the wavelength of X-ray (0.15406 nm), θ and β are the diffraction angle and FWHM of an observed peak, respectively. Intense diffraction peaks have been selected to compute the mean crystallite size and it is found to be in 96 nm.

The SEM image of the $LiZnVO_4$ ceramic powder is shown in Fig. 9 which reveals that the particles are densely agglomerated and the average diameter of the grain size is in the range of 1–2 μm. The narrow widths of the diffraction peaks also indicate that the prepared ceramic powders possess large-sized grains and such sizes are found to be more encouraging for different applications [9, 10]. To verify the elements present in the optical material studied, an EDS profile has been recorded as shown in Fig. 10. The EDS profile does not show the presence of lithium because of instrument limitation [19]. The existence of the carbon peak at 0.2 keV in the EDS spectrum is due to the carbon tape that was employed in holding the specimen.

Figure 11 shows the FTIR spectrum of $LiZnVO_4$ ceramic powder. Bands in the region of 1,100–400 cm^{-1} are due to vanadium–oxygen stretching vibrations of

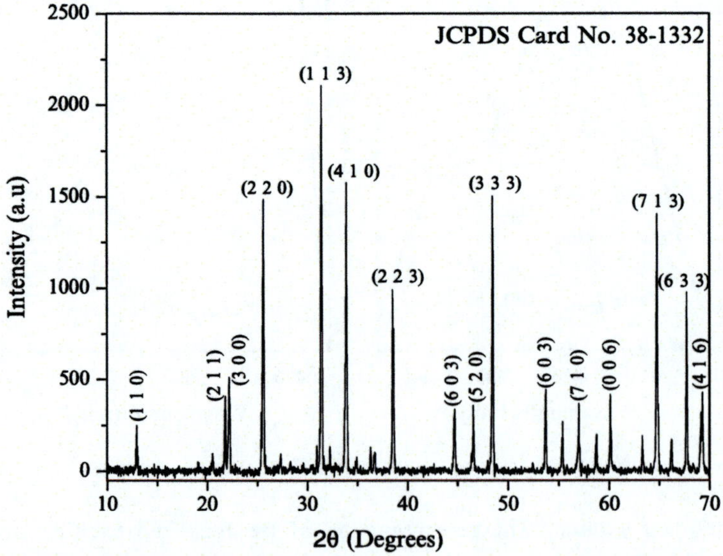

Fig. 8 XRD profile of $LiZnVO_4$ ceramic powder

Fig. 9 SEM image of $LiZnVO_4$ ceramic powder

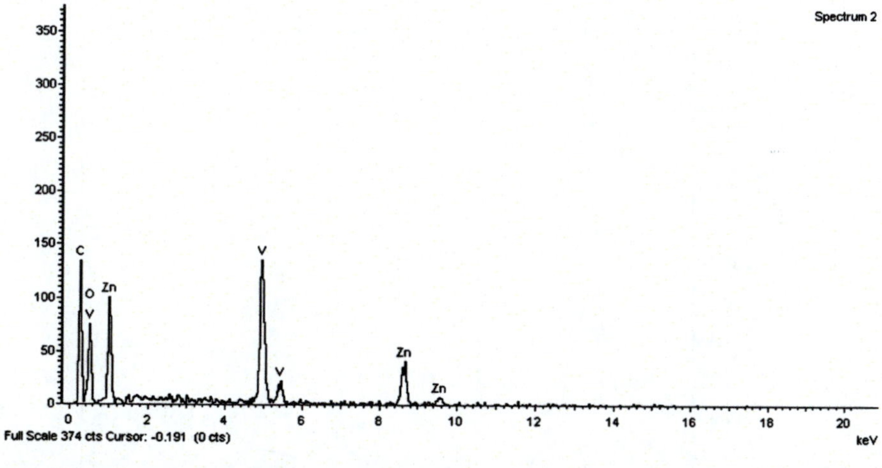

Fig. 10 EDS profile of $LiZnVO_4$ ceramic powder

VO_4 and the other metal–oxygen bonds present in the material [8, 20]. A low intensity band at 414 cm^{-1} corresponds to the stretching vibration of Zn–O [5]. Another peak at 490 cm^{-1} is due to the presence of γ-phase of Lithium Vanadate [20]. The bands observed at 647, 712, 790, and 930 cm^{-1} are all due to V–O vibration modes of $LiZnVO_4$ ceramic powder [21, 22]. The bands at 3,432 and 1,642 cm^{-1} are assigned to O–H stretching and H–O–H bending vibrations, respectively, and these two bands are the characteristic vibrations of moisture present in the sample [14]. Figure 12 shows the Raman spectrum of $LiZnVO_4$

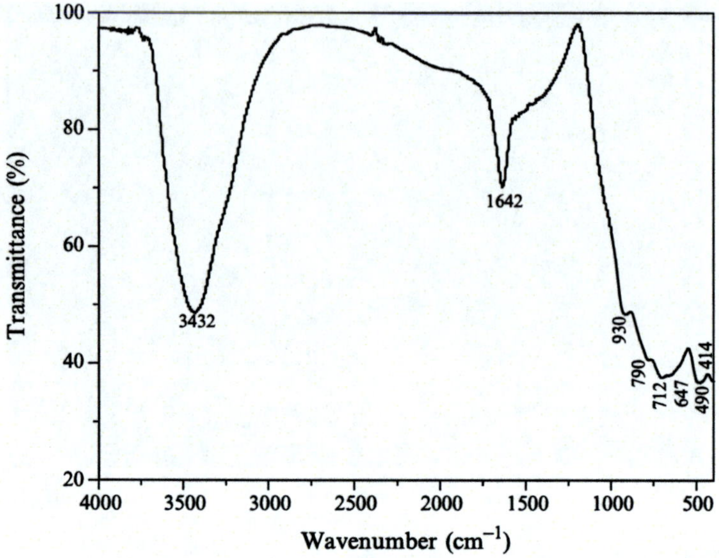

Fig. 11 FTIR spectrum of $LiZnVO_4$ ceramic powder

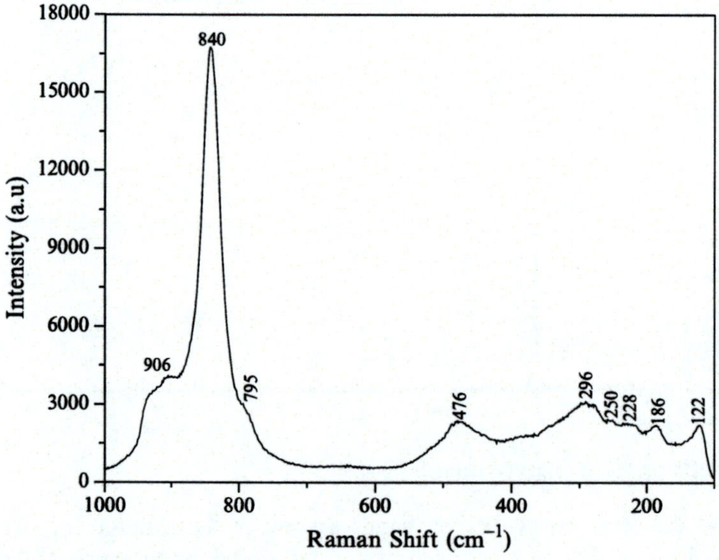

Fig. 12 Raman spectrum of $LiZnVO_4$ ceramic powder

ceramic powder with intense Raman peaks which imply the stronger interactions among the atoms, which mainly arise from the stretching and bending of the shorter metal–oxygen bonds [15]. The Raman peaks located at 186, 250, and 476 cm^{-1} are ascribed to the Li–O vibrations [21] and the peaks at 122, 296,

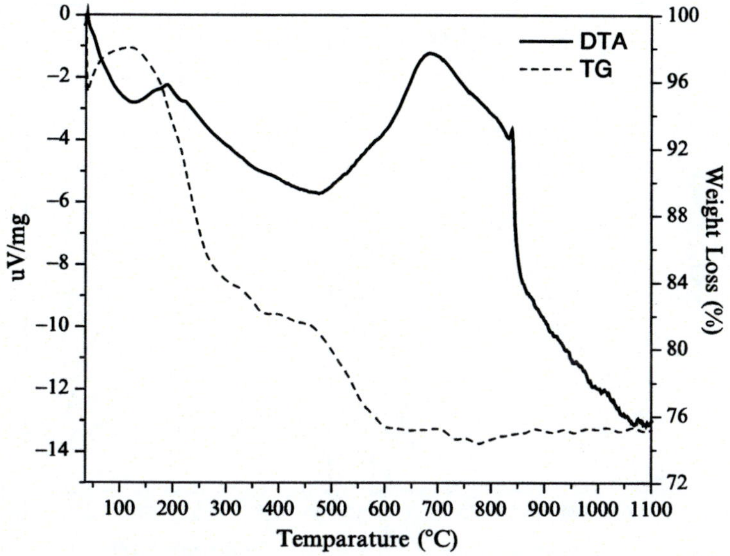

Fig. 13 TG–DTA profiles of $LiZnVO_4$ ceramic powder

795, and 906 cm^{-1} are attributed to V–O vibrations. Intense peak at 840 cm^{-1} is due to the asymmetric stretching υ_{as} (VO_4) and symmetric stretching υ_s (VO_4) vibration modes of the VO_4 groups [21, 23, 24]. The Raman band at 228 cm^{-1} is attributed to the Zn–O vibration [5].

Figure 13 shows the comprehensive TG and DTA profiles of the $LiZnVO_4$ precursor containing the chemicals mixture (Li_2CO_3, ZnO, and NH_3VO_4) in stoichiometric ratio in the temperature range upto 1,100°C. In the temperature range from 35 to 300°C, the sample shows both exothermic and endothermic peaks in the DTA curve, which is in accordance with the first weight loss. These observations can be attributed to the decomposition of the organic species which are used during the grinding process of raw materials for homogeneity [25]. The second weight loss in the range of 300 and 700°C is due to the degradation of organic material from the synthesized material which could arise due to the elimination of CO_2 and H_2O contents in the precursor. The exothermic peak observed at around 650°C is due to the elimination of CO_2 and also because of the formation of constituent oxides [26]. Upon increasing the temperature, a solid-state reaction occurs among the chemicals of Li_2CO_3, ZnO, and NH_3VO_4 and the peaks observed in the DTA curve thus confirm such a reaction. An exothermic peak at 840°C is attributed to the heat loss during the crystallization of $LiZnVO_4$ ceramic powder. The well-powdered mixture of Li_2CO_3, ZnO, and NH_3VO_4, upon its sintering at 850°C for 5 h, shows an XRD pattern corresponding to $LiZnVO_4$ (JCPDS Card No. 38–1332) validating the assignment of the exothermic peak at 840°C in the DTA curve due to the crystallization of $LiZnVO_4$. TG curve indicates a weight loss of nearly 25% when the temperature is raised from 35 to 700°C because of the degradation of the material

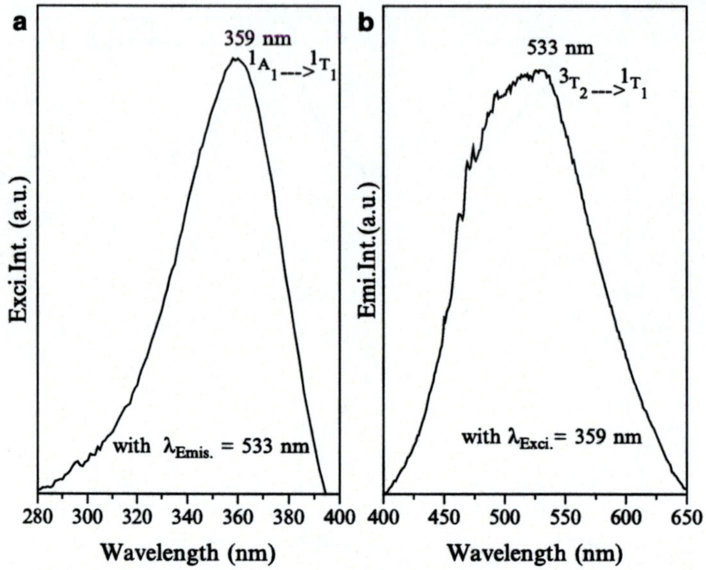

Fig. 14 (a) Excitation and (b) emission spectra of $LiZnVO_4$ ceramic powder

by loosing the existing moisture, organic species, and other gases present in the precursor. No significant weight loss has been observed in the TG curve beyond 700°C.

Figure 14a shows an excitation spectrum of $LiZnVO_4$ ceramic powder by monitoring with a green emission at $\lambda_{emis} = 533$ nm and from the figure, an intense excitation band at 359 nm ($^1A_1 \rightarrow {}^1T_1$) has been noticed which is similar to the band reported in the literature [27]. This absorption band (excitation peak) of $VO_4{}^{3-}$ group is because of a charge transfer from the oxygen ligands (O^{2-}) to the central vanadium atom (V^{5+}) [28]. It is encouraging because of the fact that this material could be excited in the UV region. Figure 8b shows an emission spectrum of $LiZnVO_4$ ceramic powder with an excitation at 359 nm. Vanadates, like tungstates, show luminescence without any addition of dopant ion into the matrix. Local defects such as atom vacancies or interstitials may induce new energy levels in the band gap and these vacancies could be arising due to the possibility of the observance bright green emission at 533 nm ($^3T_2 \rightarrow {}^1T_1$) from the prepared $LiZnVO_4$ ceramic powder [27, 29].

4 Conclusion

In summary, it could be concluded that prominently *blue* color emitting ferroelectric $Bi_4(TiO_4)_3$ and brightly *green* luminescent $LiZnVO_4$ ceramic powders have successfully been made by a solid-state reaction method. XRD and SEM analysis of the optical materials indicate that these are in orthorhombic and rhombohedral structures, respectively, and the particles of those are approximately in 1 μm in size. FTIR and Raman spectra have been employed to identify the functional groups and Raman modes of the ceramic powder are studied. TG–DTA profiles have substantiated their optimum sintering temperatures. $Bi_4(TiO_4)_3$ ceramic powder has revealed a strong blue emission at 480 nm and it arises due to electron–hole recombinations of localized excitons that associate with the oxygen vacancies. Existence of luminescence centers with oxygen O^- defects in the vanadium–oxygen clusters could be a possible reason for the observance of bright green emission from the $LiZnVO_4$ ceramic powders. Based on the results reported in the present work, we suggest that these ceramic powders could be used as novel optical materials for visible color display applications.

References

1. Yan B, Su XQ (2006) Chemical co-precipitation synthesis of luminescent BixY1-xVO_4: RE (RE = Eu^{3+}, Dy^{3+}, Er^{3+}) phosphors from hybrid precursors. J Non-Cryst Solids 352:3275–3276
2. Yu CC, Liu XM, Yu M, Lin CK, Li CX, Wang H, Lin J (2007) Enhanced photoluminescence of Ba_2GdNbO_6: Eu^{3+}/Dy^{3+} phosphors by Li^+ doping. J Solid State Chem 180:3058–3065
3. Chang YS, Lin HJ, Chai YL, Li YC (2008) Preparation and luminescent properties of europium-activated $YInGe_2O_7$ phosphors. J Alloy Comp 460:421–425
4. Pazik R, Hreniak D, Strek W, Speghini A, Bettinelli M (2006) Structural and luminescence properties of Eu^{3+} doped $Ba_xSr_{1-x}TiO_3$ (BST) nanocrystalline powder prepared by different methods. Opt Mater 28:1284–1288
5. Ni S, Wang X, Zhou G, Yang F, Wang J (2010) Crystallized $Zn_3(VO_4)_2$: synthesis, characterization and optical property. J Alloy Comp 491:378–381
6. Sivakumar T, Chang HY, Halasyamani PS (2007) Synthesis, structure, and characterization of a new two-dimensional lead (II) vanadate, $Ba_3PbV_4O_{14}$. Solid State Sci 9:370–375
7. Chen R, Song F, Chen D, Peng Y (2009) Improvement of the luminescence properties of $CaTiO_3$: Pr obtained by modified solid-state reaction. Powder Technol 194:252–255
8. Bosacka M (2006) The synthesis and selected properties of new double vanadates $M_2InV_3O_{11}$, where M = Zn, Mg. Mater Res Bull 41:2181–2186
9. Li W, Su D, Zhu J, Wang Y (2004) Mechanical and dielectric relaxation in neodymium-modified bismuth titanate ceramics. Solid State Commun 131:189–193
10. Xiao X, Yan B (2007) Chemical co-precipitation synthesis and photoluminescence of Eu^{3+} or Dy^{3+} doped $Zn_3Nb_2O_8$ microcrystalline phosphors from hybrid precursors. Mater Sci Eng B 136:154–158
11. Pookmanee P, Boonphayak P, Phanichphant S (2004) Chemical synthesis of bismuth titanate microparticles. Ceram Int 30:1917–1919
12. Pavlovic N, Kancko D, Szecsenyi KM, Srdic VV (2009) Synthesis and characterization of Ce and La modified bismuth Titanate. Process Appl Ceram 3(1–2):88–95
13. Simoes Z, Aguiar EC, Ries A, Longo E, Varela JA (2007) Niobium doped $Bi_4Ti_3O_{12}$ ceramics obtained by the polymeric precursor method. Mater Lett 61:588–591
14. Lei F, Yan B (2008) Hydrothermal synthesis and luminescence of $CaMO_4$:RE^{3+} (M = W, Mo; RE = Eu, Tb) submicro-phosphors. J Solid State Chem 181:855–862

15. Ling ZC, Xia HR, Liu WL, Han H, Wang XQ, Sun SQ, Ran DG, Yu LL (2006) Lattice vibration of bismuth titanate nanocrystals prepared by metalorganic decomposition. Mater Sci Eng B128:156–160
16. Liu WL, Xia HR, Han H, Wang XQ (2004) Structural and dielectrical properties of bismuth Titanate nanoparticles prepared by metalorganic decomposition method. J Cryst Growth 269:499–504
17. Hu Z, Gu H, Hu Y, Zou Y, Zhou D (2009) Microstructural, Raman and XPS properties of single-crystalline $Bi_{3.15}Nd_{0.85}Ti_3O_{12}$ nanorods. Mater Chem Phys 113:42–45
18. Gu H, Hu Z, Hu Y, Yuan Y, You J, Zou W (2008) The structure and photoluminescence of $Bi_4Ti_3O_{12}$ nanoplates synthesized by hydrothermal method. Colloids Surf A Physicochem Eng Aspects 315:294–298
19. Kojima Y, Suzuki K, Kawai Y (2006) Hydrogen generation from lithium borohydride solution over nano-sized platinum dispersed on $LiCoO_2$. J Power Sources 155:325–328
20. Vijayakumar M, Selvasekarapandian S, Kesavamoorthy R, Nakamura K, Kanashiro T (2003) Vibrational and impedance spectroscopic studies on lithium vanadate prepared by solid-state reaction. Mater Lett 57:3618–3622
21. Julien C, Massot M, Vicente CP (2000) Structural and vibrational studies of $LiNi_{1-y}Co_yVO_4$ $(0 \leq y \leq 1)$ cathodes materials for Li-ion batteries. Mater Sci Eng B 75:6–12
22. Shenouda AY (2006) Structure and electrochemical behavior of lithium vanadate materials for lithium batteries. Electrochim Acta 51:5973–5981
23. Chakroun-Ouadhour E, Ternane R, Ben Hassen-Chehimi D, Trabelsi-Ayadi M (2008) Synthesis, characterization and electrical properties of a lead sodium vanadate apatite. Mater Res Bull 43:2451–2456
24. Gotic M, Music S, Ivanda M, Soufek M, Popovic S (2005) Synthesis and characterisation of bismuth (III) Vanadate. J Mol Struct 744–747:535–540
25. Ngamjarurojana A, Khamman O, Yimnirun R, Ananta S (2006) Effect of calcination conditions on phase formation and particle size of zinc niobate powders synthesized by solid-state reaction. Mater Lett 60:2867–2872
26. Zhao J, Hua Quan M, Zhang L (2006) Preparation of potassium lithium niobate powders with low Li content via the Pechini method. Ceram Int 32:843–846
27. Nakajima T, Isobe M, Tsuchiya T, Ueda Y, Kumagai T (2009) A revisit of photoluminescence property for vanadate oxides AVO_3 (A:K, Rb and Cs) and $M_3V_2O_8$ (M:Mg and Zn). J Lumin 129(12):1598–1601
28. Wu X, Huang Y, Shi L, Jin Seo H (2009) Spectroscopy characteristics of vanadate $Ca_9Dy(VO_4)_7$ for application of white-light-emitting diodes. Mater Chem Phys 116:449–452
29. Zubkov VG, Tyutyunnik AP, Tarakina NV, Berger Ludmila IF, Surat L (2009) Synthesis, crystal structure and luminescent properties of pyrovanadates $A_2CaV_2O_7$ (A = Rb, Cs). Solid State Sci 11:726–732

Time-Correlated, Single-Photon Counting Methods in Endothelial Cell Mechanobiology

Peter J. Butler, Ramachandra R. Gullapalli, Tristan Tabouillot, and Michael C. Ferko

1 Overview

While mechanical forces are known to guide the development of nearly all biological tissues including bone, cartilage, and many soft tissues, much attention has focused on endothelial cell mechanobiology and the role of blood flow-induced forces in regulating the health of blood vessels. It is now well accepted that modulation of endothelial cell physiology and pathophysiology by fluid mechanical forces is a principal reason why atherosclerotic lesions are located at areas of disturbed flow including at arterial branch points and areas of high arterial curvature. However, the molecular identity of endothelial cell mechanosensors remains elusive largely due to the complexity of cell mechanics and to the difficulty in identifying when and where a candidate mechanosensor has been perturbed. Thus, new methods of cell-specific mechanical modeling along with molecular-scale readouts of perturbation by force are needed to help unravel the magnitude-, time-, and position-dependent responses of endothelial cells to mechanical forces.

We outline in this chapter our use of pulsed lasers, time-correlated single-photon counting (TCSPC) instrumentation, modeling software, and multimodal fluorescence microscopy, to identify the spatial and temporal aspects of molecular-scale mechanosensing in live intact endothelial cells subjected to well-defined forces. It is anticipated that cell-specific modeling and experimentation combined with new tools in fluorescence spectroscopy will reveal the identity, location, and temporal aspects of endothelial cell mechanosensors, thus providing insight into the mechanical origins of vascular health and disease.

P.J. Butler (✉)
Department of Bioengineering, The Pennsylvania State University,
205 Hallowell Building, University Park, PA 16802, USA
e-mail: pbutler@psu.edu

C.D. Geddes (ed.), *Reviews in Fluorescence 2009*, Reviews in Fluorescence 2009,
DOI 10.1007/978-1-4419-9672-5_6,

2 Introduction: Hemodynamics, Endothelial Cell Responses, and Molecular Dynamics Measurements

The endothelium is a primary regulator of vascular health and derangement of this system can lead to diseases such as atherosclerosis. It is subject to hemodynamic forces (e.g., shear stress, τ) which depend on the location in the vasculature, the heart rate, and the metabolic demands of tissues. Endothelial cells (ECs) are sensitive to temporal shear gradients [1–3] and spatial shear gradients [4], the nature of which is thought to determine whether ECs exhibit an atherogenic or atheroprotective phenotype [5]. Shear stress also plays a major role in the coordination of blood flow in the microvasculature [1, 6] and, hence, in the maintenance of capillary blood pressure, and the delivery of oxygen, nutrients, and immunity-related leucocytes to the tissue.

Biological cells convert forces to intracellular biochemical signaling cascades by mechanotransduction, a process that is responsible for diverse physiological phenomena including bone and vascular wall remodeling, vascular caliber control, and embryonic development [7]. In contrast, atherosclerosis, sensory dysfunction in diabetes mellitus, and hair cell damage in the inner ear leading to vertigo are examples of cellular mechanical regulatory processes gone awry [8]. Physiological forces arise from blood flow-induced shear stresses, hydrostatic pressure, extracellular matrix deformations, intracellular contractions, and osmotic swelling. The length-scale of these forces spans single molecules, cellular organelles, cells, tissues, organs, and physiological systems; the timescales span nanoseconds to days [9].

The essential ingredients to a comprehensive understanding of mechanotransduction are a well-defined force, cellular mechanical properties, and a physiological readout directly related to the force. Well-defined forces arise from fluid flow [10], atomic force microscopes [11, 12], optical traps [13], and magnetic beads [14]. Cellular deformations to these forces can be modeled using continuum mechanics [14–16], or ultrastructural characterization of load-bearing structures [17–19]. Examples of readouts of force-induced physiological responses include calcium signaling [20], ion channel activity [21], phosphorylation of proteins [22], and transcription of new RNA [23]. A major challenge in mechanotransduction is to differentiate mechanosensors (cellular structures which are perturbed by physical forces), mechanotransducers (molecules which undergo biochemical changes in response to force and initiate signaling cascades), and mechanobiology (the resultant changes in cellular structure and function).

Analysis of single molecules in cells subjected to force can provide the link between force and mechanotransduction. Major technical and computational advances in the analysis of fluorescence fluctuations provide new methods to assess single-molecule dynamics in model and cellular systems [24, 25]. For example, TCSPC instrumentation is available on a single compact PCI-card which, when integrated with pulsed and continuous wave-lasers, can be used to assess nanosecond- to second-scale dynamics of single fluorescent molecules. Analysis methods including fluorescence correlation spectroscopy (FCS) and fluorescence lifetime can then be used to gain information on diffusion, conformational changes, aggregation, chemical kinetics, and other important biomolecular phenomena [24, 25].

We have developed cell-specific computational methods [26] and novel integrated microscopy and spectroscopy techniques [27] with the goal of identifying the location and molecular identity of mechanosensors of shear stress. The intended application of this system is to assess force-induced changes in dynamics of molecules occurring on a time scale of nanoseconds, while addressing long-term adaptive responses of cells on the order of hours. It is anticipated that these new tools will enable a comprehensive analysis of cellular mechanobiology and lead to major clinical advances in treating or preventing diseases such as atherosclerosis which have their origins in mechanotransduction [8].

In Sect. 3, we begin with an overview of recent methods in cell-specific computational modeling of fluid dynamics and cellular mechanics. This work helps focus attention on areas of stress concentration. In Sect. 4, we outline the application of TCSPC to FCS, measurements of fluorescence lifetime, and prospects for detection of membrane mechanosensing. In Sects. 4 and 5, we discuss the promise and limitations of these techniques in endothelial cell mechanobiology.

3 Advances in Cell-Specific Modeling[1]

The precise mechanisms by which apical shear stress leads to localized intracellular signaling remains unknown but may involve three possible general mechanisms. First, shear stress may directly perturb a cellular structure which is directly linked to a diffusible factor leading to downstream intracellular biochemical signaling. For example, shear stress activates G-proteins which are coupled directly to the membrane [28]. Second, cellular architecture may redistribute apically applied forces to intracellular organelles where forces induce signaling locally. For example, stress fibers connected to the apical membrane may distribute stress to remote cellular locations and activate stress-fiber-associated proteins there [19]. Third, shear stress may induce a combination of these effects in which biochemical signals are initiated at the apical surface, a diffusible activating factor is generated, and propagation and focusing of this signal is facilitated by local forces. There is strong evidence for this third model. First, shear-induced nitric oxide production depends on an intact glycocalyx [29, 30] and shear causes increases in EC-membrane lipid lateral diffusion [31] and free volume [32], suggesting shear stresses are directly "felt" by the apical surface of the cell which includes the glycocalyx and plasma membrane. In support of decentralized

[1]Portions of Sect. 3 have appeared previously in reference [26] (used with permission).

forces, shear stress caused deformation of intermediate filaments [33], strain focusing at focal adhesions (FAs) [34], and stress focusing in the cell interior [35]. But shear stress also activates PECAM-1, a protein in cell junctions near the cell surface. This activation may lead to production of a diffusible factor which induces activation of integrins in FAs, where stresses may be concentrated [36]. Local potentiation of cell signaling by forces are suggested by studies in which endothelial cells actively reorganize their points of attachment (e.g., FAs), and align them in the direction of flow [37, 38]. Directional reorientation of FAs may be due to a combination of integrin activation and forces which bias the location of new integrin-extracellular matrix bonds. In support of this idea, shear-induced shc-integrin association (thought to be responsible for activation of various mitogen-activated protein kinases) was dependent on new integrin binding to the extracellular matrix [39]. Shear stress also leads to polarized adaptive changes in cell mechanics [40, 41]. Taken together, such studies support a model of cellular mechanotransduction in which the global activation of signaling pathways by shear-forces are converted to local signaling events in discrete locations in cells by force amplification and force-induced directional biasing of signal propagation. Thus, there is a need for models which quantify how the unique architecture of endothelial cells can amplify forces from shear stress at discrete cellular locations.

In response to this challenge, we recently developed new integrated methods in fluorescence imaging and image processing for the development of solid models with cell-specific topographies and subcellular organelles [16]. The goal of this research was to use these methodologies along with quantitative total internal reflection fluorescence microscopy (qTIRFM) to create a cell-specific, multicomponent, three-dimensional (3-D) solid elastic continuum model of an endothelial cell in a confluent monolayer with experimentally determined FAs [26]. Finite element analysis was used to compute stress transmission throughout the endothelial cell due to fluid flow or magnetic bead twisting applied at the apical surface. It is possible that juxtaposition of high and low moduli organelles and constrained and unconstrained regions are two mechanisms of stress amplification in cells. Thus, we quantified the effects of FAs and material inhomogeneity on deformation, strain, and stress distributions in the cell interior. This type of cell-specific modeling based on experimentally determined topographies and boundary conditions may help identify potential sites of force-induced potentiation and directional biasing of cell signaling.

The main contributions of this study were to provide the first quantitative predictions of stress distributions in focally adhered ECs resulting from apically applied fluid flow. These results arise from finite element analysis of a cell-specific model in which surface topography and FA location and area were experimentally determined. The model predicts that shear-induced stresses were generally small but significantly amplified and focused near FAs and the high-modulus nucleus. Some of the main results are reported in the next sections. Other analysis of deformations, strains, and other mechanical parameters can be found in ref. [26].

3.1 Effects of FAs on Distributions of Stress, Strains, and Displacements

Inclusion of FAs as attachment locations in a homogeneous linear elastic continuum model resulted in heterogeneous internal stresses, strains, and displacements. Stresses near FAs were nearly 40-fold larger (Fig. 1) than surface shear stresses (Fig. 2), thus supporting the widely held contention that FAs are a means of force amplification of shear stress.

3.2 Effects of Material Inhomogeneities on Stress Distributions

Composite materials made of stiff and soft structures, when stressed, exhibit stress concentrations in and around the stiff structure. Consistent with this principle, we quantified stress amplification near the high-modulus nucleus (Fig. 3). While overall strains were small (~0.5%), inclusion of the nucleus in the model increased extra-nuclear strains that were twofold greater than the case when the nucleus was not included in the model. Similarly, stresses in the nucleus were large (~50 dynes/cm^2) although strains were low (~0.05%).

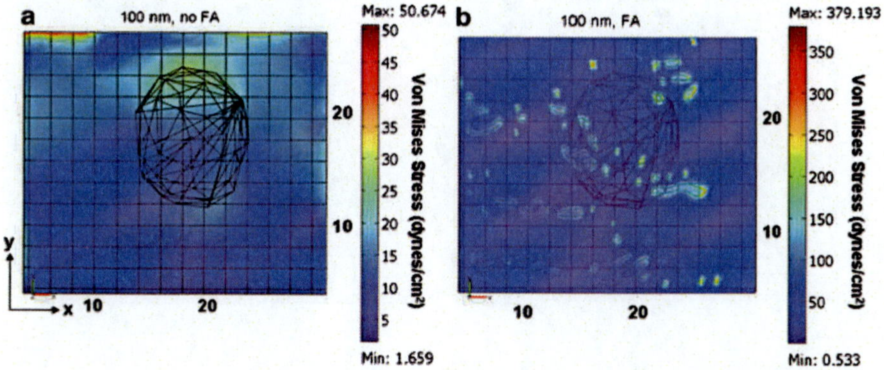

Fig. 1 Effects of focal adhesion on shear-induced stresses. Von Mises stress distributions were evaluated at $z = 0.1$ μm above the coverslip for (**a**) model computed without focal adhesions and (**b**) model solved with focal adhesions. Young's moduli were 775 and 5,100 Pa for the cytoplasm and nucleus, respectively. Poisson's ratio was 0.33. Extracellular fluid viscosity was 0.0084 poise

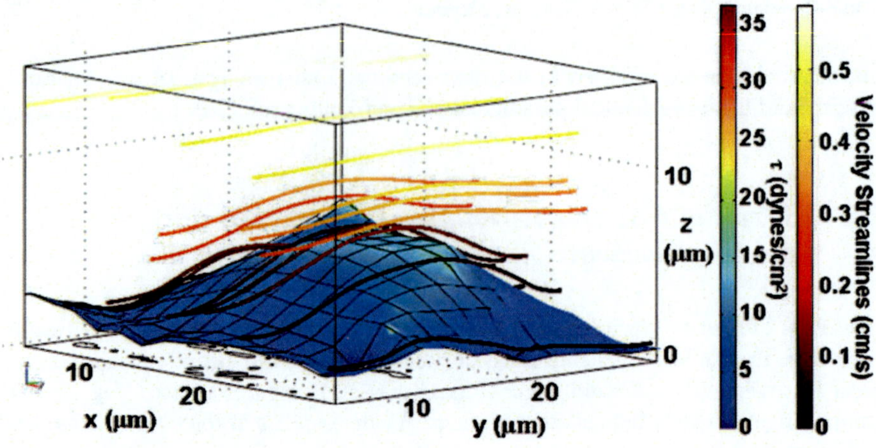

Fig. 2 Representative subcellular shear stress distribution: a nominal shear stress of 10 dynes/cm^2 was simulated over the solid cell monolayer model in the positive *y*-direction. Stress distributions show stress peaks at the apical region over the nucleus while stress is minimum in the valleys between cells. Simulated cell represents a single cell in a monolayer. Simulated velocity field is shown using streamlines. Color plot of shear stress in dynes/cm^2; streamline color corresponds to fluid velocity (cm/s). Axes in μm

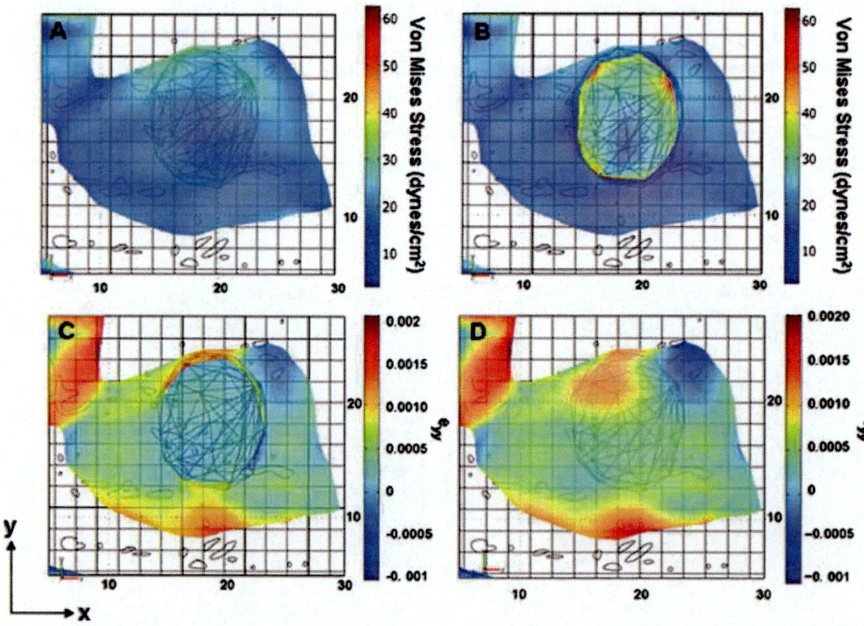

Fig. 3 Effects of nucleus on shear-induced stresses and strains. **(a)** Von Mises stress distribution for a model in which the nuclear modulus was set equal to the cytoplasmic modulus (774 Pa) in order to neglect the effects of the nucleus without altering mesh conditions. Note uniformity of stresses in the vicinity of the nucleus. **(b)** Von Mises stress distribution for a model solved with the nuclear modulus equal to 5,100 Pa (actual modulus). Note that the stiffer nucleus induced local high stress concentrations in the nucleus–cytoplasm interface. **(c)** e_{yy} strains without nucleus. **(d)** e_{yy} strains with nucleus. e_{yy} strains were larger than those in *x* and *z* directions and larger than shear strains

3.3 Relationship Between Stress and Endothelial Cell Mechanotransduction

Mechanotransduction is the process by which cells convert mechanical stimuli into intracellular chemical signaling cascades. Mechanotransduction is known to be involved in the regulation of homeostasis of many tissues including the vascular endothelium [42]. Interest in the precise mechanisms by which endothelial cells sense and respond to mechanical signals arises from a now well-established correlation between spatial and temporal variations in hemodynamic shear stress and the focal nature of atherosclerotic lesions. Candidate cellular structures responsible for mechanotransduction in ECs include FAs, plasma membrane subdomains [31, 43], cell–cell junctions [36], and others.

It has not yet been shown, however, that physiological shear stress results in sufficient stress in these regions to directly activate signaling. In order to better understand force transmission in single cells, finite element mechanical approaches deal effectively with complex geometries and force impositions [15, 44] and provide good approximations of force levels in idealized cells and in cells in which topography and mechanical properties have been measured. In support of the use of continuum theory in mechanotransduction, Mack and co-authors used FEA of an idealized cell subject to point deformation from a magnetic bead to correlate the stresses in FA to FA translocation as determined by fluorescence images of GFP-paxillin, a protein that binds to FAK in mature FAs. Similarly, Charras and colleagues demonstrated that FEA-predicted stresses correlated well with ion channel opening in bone cells [45].

While it is difficult to couple information from a continuum model to forces on individual molecules, Charras and Horton recently showed that stresses and strains resulting from fluid shear stress are below the threshold needed to activate candidate shear-sensitive molecules [15]. However, because their model did not include the nucleus or FAs, they could not predict the stress-amplification mechanisms suggested in our study. In our model, strains on the order of 1–4% are predicted. If these strains were applied to the membrane near FAs, it is conceivable that transmembrane proteins could be activated there.

It is possible that cell–cell junctions [46] are additional mechanisms of force amplification. A recent report by Tzima and colleagues suggests that PECAM-1 along with adapter molecules may constitute a force-sensitive system which indirectly activates integrin molecules [36]. In that study, both shear stress and bead pulling were used as forcing functions to activate integrins in endothelial cells. Since cell junctions were not explicitly included in the present model, it is not yet clear that bead pulling and shear stress elicit comparable stresses at cell junctions. Our study focused on cellular stresses in response to apically applied fluid shear stress in order to begin to evaluate extant theories on focal-adhesion-mediated mechanotransduction of fluid shear stress. It is possible that other modes of force (e.g., cell contraction, substrate stretching, and bead pulling) may stimulate different molecular signaling pathways than those activated by fluid shear stress [15].

3.4 Cell-Specific Stress Analysis and Mechanotransduction

In summary, this study presents the first estimates of heterogeneous displacements and stress–strain fields in sheared and focally adhered endothelial cells. These heterogeneous stress responses are due to the inclusion in the model of distributed attachment locations (i.e., FAs) and cellular components with different elastic moduli. The location and directions of upstream tensile and downstream compressive stresses computed in the vicinity of individual FAs were consistent with the observation of FA growth in the downstream direction of flow and FA retraction in the upstream side, thus providing needed quantitative information to elucidate mechanisms of mechanotaxis of ECs [37, 38, 47]. Further work is necessary to extract stresses from individual FAs and to precisely compute forces on integrin-extracellular matrix bonds to evaluate integrin-mediated force sensing. If such modeling methods are followed by cellular probing for physiological changes and molecular signaling events on the same cells, direct correlations could be made between subcellular stresses and signaling (i.e., cell-specific mechanotransduction) to assess which structures are the dominant mechanotransducers under physiological and pathophysiological conditions. Elucidating molecular mechanisms of alterations in EC phenotypes from atheroprotective to atherogenic in response to prevailing wall shear stress is an important component of the development of therapeutic interventions in vascular disease.

In order to determine the precise molecular mechanisms which couple force and biochemical signaling in ECs, it is useful to combine engineering analysis of intact endothelial cells with real-time experimental readouts of the same cell using comparable spatial and temporal scales; in other words, cell-specific models constructed from accurate input parameters of cell topographies and boundary conditions followed by micron-scale interrogation of molecular changes in areas of stress or strain focusing. Such analysis would lead to direct, same-cell correlation of experimental measurements and stress distributions. While atomic force microscopy (AFM) and electron microscopy (EM) can yield ultra-high-resolution topographies for solid models, fluorescence microscopy can yield high-resolution images of surface and internal cellular features of viable cells. In addition to yielding insight into the role of surface topography, nuclei, and FAs in shear-induced stress distributions, these methodologies set the stage for cell-specific modeling and experimentation to elucidate the fundamental mechanisms of mechanotransduction.

4 Time-Correlated Single-Photon Counting[2]

4.1 Background

We hypothesize that mechanosensation will be accompanied by changes in molecular dynamics in areas of the cell reaching a threshold of stress. Thus, on the same microscope used for modeling, we have integrated TCSPC instrumentation, which

[2]Portions of Sect. 4 were published recently in [27] (used with permission).

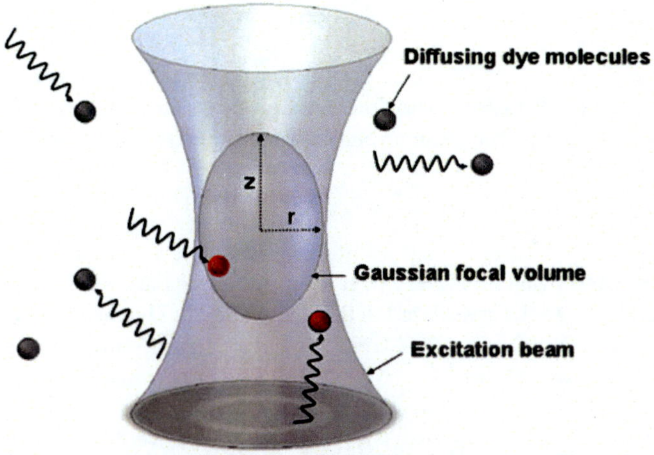

Fig. 4 Confocal volume for fluorescence correlation spectroscopy. The radial dimension, *r*, of the confocal volume is close to the diffraction limit of the objective and is determined by keeping the (*r*/*z*) parameter constant (<10) in (6) when fitting the autocorrelation curve from R6G molecules in water. The radius thus obtained from experiment is 326 ± 10 nm. Fluorescent molecules are excited by the entire laser beam and fluorescence emission is only collected in the confocal volume. When particles move into and out of the confocal volume in *x*, *y*, and *z* directions, 3-D diffusion is considered. When particles only move in *x*–*y* plane, 2-D diffusion is considered

is capable of measuring changes in lateral diffusion, rotational diffusion, and fluorescence lifetime of fluorescent molecules targeted to submicron location of adherent endothelial cells subjected to hemodynamic stress.

4.2 Fluorescence Correlation Spectroscopy

The detection of single molecules in femtoliter volumes was made possible by the development of confocal optics, high-sensitivity detectors, and robust fluorophores [48]. Probe molecules move into and out of a confocal volume yielding fluorescence fluctuations (Fig. 4). Autocorrelation analysis of these fluctuations can provide diffusion coefficients and other single-molecule information [49, 50].

In the following analysis, we consider the relationship between autocorrelation of fluorescence fluctuations and molecular diffusion. We start with the Stokes–Einstein relationship for the diffusion coefficient, D, of a sphere with a hydrodynamics radius, R, in a solvent of viscosity, η:

$$D = \frac{k_B T}{6\pi\eta R}, \tag{1}$$

where k_B is the Boltzmann constant and T is the absolute temperature. We define a characteristic molecular diffusion transit time, τ_D, across a small area of radius, r, such that:

$$r^2 = 4D\tau_D. \quad (2)$$

Under conditions of equilibrium, the fluorescence fluctuates around an average value $\langle F \rangle$, due to diffusion. The intensity of these fluctuations, $\delta F(t)$, can be autocorrelated to obtain the autocorrelation function, $G(\tau)$, given by,

$$G(\tau) = \frac{\langle \delta F(t+\tau)\delta F(t) \rangle}{\langle F(t) \rangle}, \quad (3)$$

where t (time) and τ (lag time) vary over all times of the data collection period. The fluorescence fluctuations are related to the instantaneous change in concentration in the observation volume, $\delta C(r, t)$. To relate the autocorrelation function to diffusion, we use the diffusion equation:

$$D\,\nabla^2\,\delta C(r,t) = \frac{\partial \delta C(r,t)}{\partial t} \quad (4)$$

and the relationship between fluorescence fluctuations and concentration of the fluorescent molecules in the confocally defined optical probe volume:

$$\delta F(t) = K \int_{\text{Volume}} \delta C(r,t) I(r) \theta(r), \quad (5)$$

where $I(r)$ is the excitation intensity profile, $\theta(r)$ is the collection efficiency profile, and K is a proportionality constant. The confocal probe volume is created by placing a small aperture (pinhole or fiber optic) in an image plane that is conjugate to the focus of a high numerical-aperture, infinity-corrected objective. The laser beam has a Gaussian intensity profile that fills the back aperture of the objective, leading to a Gaussian illumination profile, $I(r)$, such that $I(r) = I_0 e^{(-r^2/z^2)}$ where r and z are the probe volume radius and half-height, respectively, and are defined as the point where the intensity falls off to $1/e^2$ of the maximum (center) intensity (Fig. 3). A single constant factor known as the structure factor, ω, is defined as the ratio of z/r. The autocorrelation obtained from the experiment and its relation to the theoretical Gaussian confocal volume and the characteristic diffusion time is [49–51]:

$$G(\tau) = \frac{1}{N}\left(\frac{1}{1+(\tau/\tau_D)}\right)\left(\frac{1}{1+(1/\omega)^2(\tau/\tau_D)}\right)^{1/2}, \quad (6)$$

where N is the average number of diffusing fluorophores in the confocal volume. The number of fluorescent molecules present in the sample volume is the inverse of the term $G(0)$.

The relationship between the autocorrelation function and diffusion on two-dimensional (2-D) structures (e.g., plasma membrane) is given by:

$$G(\tau) = \frac{1}{N}\left(\frac{1}{1+(\tau/\tau_D)}\right). \quad (7)$$

This measurement in two dimensions can be extended to include multicomponent diffusion of multiple noninteracting species of fluorescent molecules, according to the equation:

$$G(\tau) = \sum_{i=1}^{n} b_i \left(\frac{1}{1 + (\tau / \tau_{Di})} \right), \tag{8}$$

where b_i is the relative proportion of the noninteracting diffusing molecules.

In the case of 2-D anomalous diffusion (e.g., in a cellular membrane), the time-dependence of the mean square displacements is not linear but of the form $\langle \Delta r^2 \rangle = \Gamma\tau^{\alpha}$, and the autocorrelation is given by [52]:

$$G(\tau) = \frac{1}{N} \frac{1}{1 + \Gamma\tau^{\alpha}/r^2}, \tag{9}$$

where Γ is the transport coefficient and α is the temporal exponent with values between zero and one.

4.3 Fluorescence Lifetime

Time-resolved fluorescence lifetime spectroscopy enables analysis of subtle changes in photophysics of fluorescent molecules [53]. When a fluorescent molecule is excited to a higher energy state using a picosecond pulse of laser light, it remains in the excited state for a finite time before it decays to the ground level energy state. Using a high-frequency pulsed laser, histograms of photon emission times relative to excitation times can be generated and fit with a negative exponential (or multiple exponentials) with a characteristic decay time (or lifetime), τ (different than the characteristic diffusion time, τ_D, of FCS). Fluorescence lifetime depends on local molecular microenvironmental factors including ionic strength, hydration, oxygen concentration, binding to macromolecules, and the proximity to other molecules that can deplete the excited state by resonance energy transfer [53]. The fluorescence lifetime and quantum yield are related to intrinsic photophysical characteristics of a fluorescent molecule such as radiative and nonradiative decay mechanisms. The fluorescence quantum yield, Q, is the ratio of the number of photons emitted to the number of photons absorbed, according to:

$$Q = \frac{k_r}{k_r + k_{nr}}, \tag{10}$$

where k_r and k_{nr} are the radiative and nonradiative decay rates of the molecule, respectively. Fluorescence lifetime is given by:

$$\tau = \frac{1}{k_r + k_{nr}}. \tag{11}$$

The value for k_{nr} depends on the mode of the nonradiative decay, such as collisional quenching, hydration, and vibrational relaxation. Thus, any alteration of k_{nr} also leads to a detectable change in the value of the fluorescence lifetime.

The value for fluorescence lifetime is obtained by an iterative reconvolution of an instrument response function (IRF) with the fluorescence intensity using an assumed decay law, which can be approximated by a sum of exponentials [53]:

$$I(t) = \sum_i \alpha_i \exp(-t/\tau_i), \tag{12}$$

where α_i is the fraction of molecules with lifetime τ_i, normalized to unity. Fluorescence lifetimes are independent of fluorescence probe concentrations and can provide information not obtainable from intensity variations alone. When polarized light is used to excite a molecule whose excitation dipole is oriented parallel to the polarization of the pulse, it is possible to separate the parallel and perpendicular components of the emitted fluorescence signal and to extract rotational diffusion constants [53] (Fig. 5).

4.4 Effects of Dye Concentration on FCS-Determined Diffusion Coefficients

Rhodamine 6G (R6G) was used to calibrate FCS because of its known diffusion coefficient (2.8×10^{-6} cm^2/s in water), high quantum yield (~0.95), and the relatively large absorption cross section (1.7×10^{-16} cm^2 at 514.4 nm) [54, 55]. Using these known values for rhodamine dye diffusion, we computed the radius of the focal volume to be 0.326 ± 0.010 μm. This value compares favorably with the theoretically computed values of the radius of the focal volume of 0.264 μm. This experimental radius of the confocal volume was used in all subsequent experiments on cells and vesicles to compute the respective diffusion coefficients. In order to test whether dye concentration would affect the resulting diffusion coefficients (e.g., due to aggregation), we measured the dye diffusion coefficient at progressively increasing concentrations of R6G of 1, 2, 5, 8, and 10 nM in distilled water at room temperature. For each concentration level, ten autocorrelation curves were obtained and the resulting diffusion coefficients were averaged (Fig. 6). The value of the autocorrelation curve at $\tau = 0$ s, $G(0)$, is inversely proportional to the average number of molecules present in the confocal volume (6). $G(0)$ increased from 0.3 to 2.49 corresponding to a decreasing average particle concentrations of 3.3–0.40 which are consistent with the number of molecules calculated from the known concentration of diffusing dye (data not shown). The values of τ_D for the five concentrations of R6G measured was approximately 95 μs (inset of Fig. 8) indicating that diffusion of the R6G dye for very dilute concentrations is independent of the number of molecules in the confocal volume.

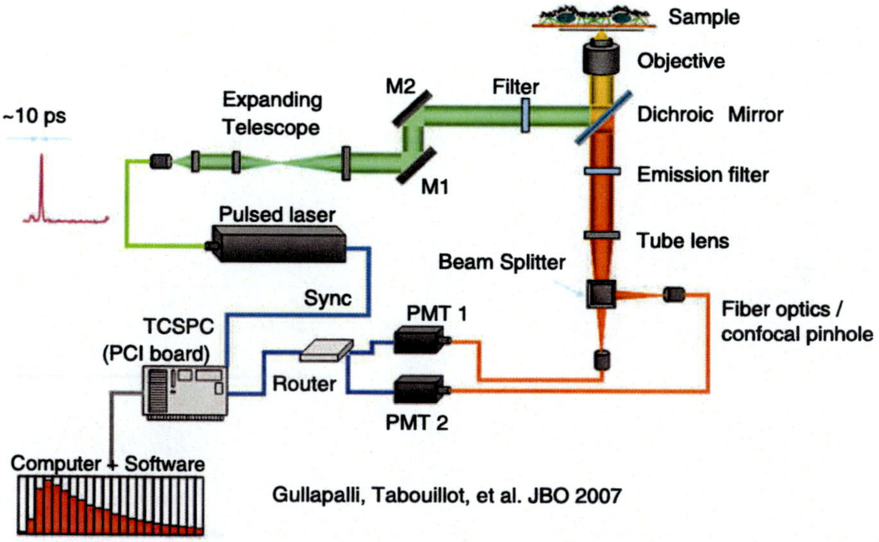

Fig. 5 Optical setup. The Kr/Ar-ion, diode, or pulsed Nd:YAG laser beam is transmitted via fiber coupling to the TIRF or confocal port. For confocal illumination, upon exiting the fiber, the beam is collimated with lens L1, expanded by L2 and L3, steered by the mirrors M1 and M2, reflected off the dichroic mirror (DM1), and enters the right side port of the microscope (note that the tube lens for the side port has been removed). After excitation of the sample, the fluorescence emission signal is collimated by the objective and exits the side port, passes through the dichroic mirror and is focused – using lens L4 – onto the optical fiber which is connected to the PMT. A polarizing beam splitter (PBS) can be introduced before the fiber to separate light with polarization that is parallel or perpendicular to that of the excitation light. The PMTs convert single photons to electrical pulses which are routed to the TCSPC board. Laser light from the TIRF system shares the back port of the microscope with the epifluorescence tube (Epi). Lens L5 and L6 collimate the epifluorescence and TIRF light, respectively. The TIRF illumination is focused at the objective back aperture by the lens L7. When the sliding mirror, Mslide1, is removed from the light path, the right side port is closed and the emission signal can be collected by the camera via the tube lens (TL). In addition, the optical trap can be inserted above the fluorescence cube turret with an infrared dichroic mirror (DMslide) mounted on a custom-built slider (data not shown)

The effective volume (V_{eff}) of the confocal volume measured in this study can be approximated using the equation $V_{\text{eff}} = \pi^{3/2} r_0^2 z_0$[51]. Where the radius of the confocal volume, r_0, is calibrated from the experiment (326 nm) and z_0 is the value of the axial half-height obtained from the structure factor. The confocal volume of our one-photon FCS setup calculated in this way is 0.889 ± 0.09 fl when averaged over all the 50 measurements of the R6G concentrations. The term "one photon" indicates that fluorescence excitation is accomplished using a one-photon mechanism rather than two photons and that confocality is achieved using a pinhole placed in a conjugate image plane.

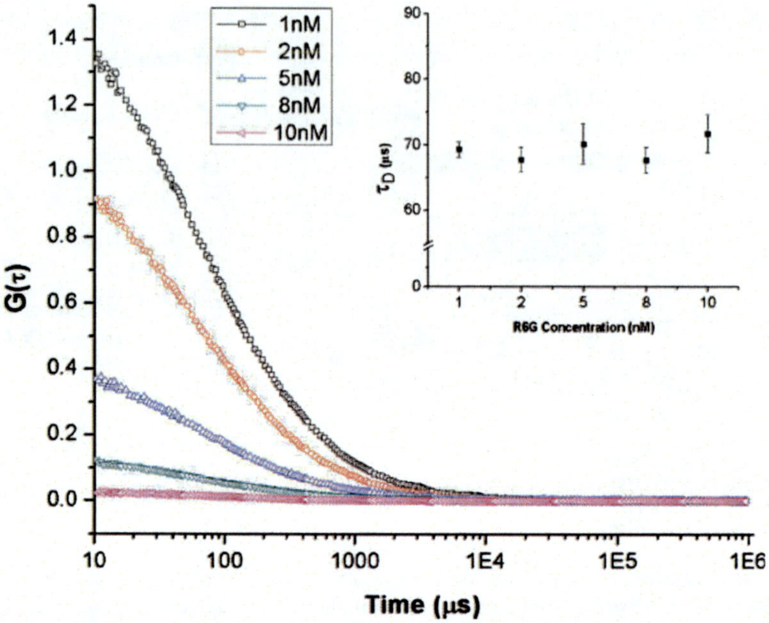

Fig. 6 Sample autocorrelation curves for various concentrations of Rhodamine 6G ranging from 1 to 10 nM. *G*(0) (inversely proportional to the average number of particles in the observation volume) was determined by fitting autocorrelation functions with (6). (*Inset*) The average diffusion times, τ_D, of R6G molecules obtained from fitting autocorrelation curves with (6) (*error bars* indicate the standard deviation of ten measurements)

4.5 Effects of Solvent Viscosity on FCS-Determined Diffusion Coefficients

In order to assess the sensitivity of FCS-determined diffusion coefficients to solvent viscosity, fluorescence fluctuation analysis was performed on R6G dye dissolved in aqueous solutions of glycerol concentrations ranging from 10 to 80% glycerol. The solution bulk viscosities were measured in a cone-and-plate viscometer. Ten FCS measurements were collected for each solution, autocorrelation curves were computed, and diffusion coefficients were averaged (Fig. 7a). All the experiments were performed at room temperature. The average τ_D ranged from a value of 0.1 ms for 10% glycerol to 6.5 ms for 80% glycerol. Data in Fig. 7b show that FCS-determined diffusion coefficients decrease with increasing viscosity in a manner consistent with (1). In addition, values of viscosity obtained from FCS-determined diffusion coefficients using (1) differed from bulk viscosity measurements obtained by cone-and-plate viscometry by 4–20% indicating a good correspondence between the two methods (data not shown).

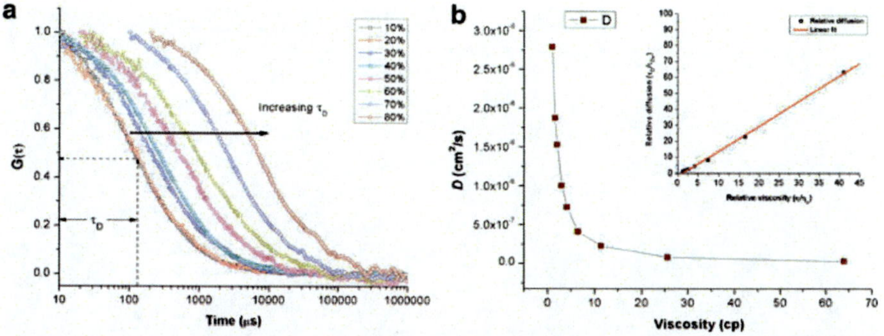

Fig. 7 (**a**) Dependence of the autocorrelation curve of R6G on viscosity of glycerol/water solution. Increase of viscosity (corresponding to increasing percentage of glycerol, v/v) leads to longer characteristic diffusion times, τ_D. (**b**) Diffusion coefficients (*D*) of 5 nM R6G obtained by fitting autocorrelation curves in (a) with (6). The *x*-axis indicates the values of viscosity of the dye solution measured with a cone-and-plate viscometer. (*Inset*) The relationship between normalized viscosity (η/η_0) and normalized characteristic diffusion time (τ_D/τ_{D0}) was fit with a line. η_0 and τ_{D0} are viscosity and characteristic diffusion time, respectively, of dye in pure water

4.6 FCS Measurements of DMPC Giant Unilamellar Vesicles and Endothelial Cell Membranes

FCS measurements were performed on DMPC vesicles and endothelial cell membranes stained with DiI-C_{18} dye excited with 520 nm light from the CW laser or 532 nm light from the pulsed solid-state laser. Fluorescence light with wavelength of 545 nm and longer was collected for analysis. For cells and vesicles, 5 nM DiI was found to best ensure sufficient fluorescence signal and ease-of-fit of autocorrelation curves. About 40,000 photons per second were collected and the number of fluorescent molecules in the probe volume at any time was between 1 and 8. The laser beam position in the *x* and *y* plane, relative to the imaging system, was assessed by preparing a single monolayer of DiI on a glass coverslip and bleaching it with the laser. The coordinates of the laser spot were then recorded using the camera. During FCS measurements, the position of the focus in the *z*-axis was adjusted to correspond to the membrane by moving the microscope objective such that fluorescence intensity was maximized.

For DMPC vesicles, 20 FCS measurements were taken at the GUV apical region, and no more than five measurements were obtained on a single vesicle. The temperature in the chamber was maintained at 26°C, which is above the phase-transition temperature (24°C) for DMPC. The results were fit using the equation for 2-D diffusion (7) to obtain a diffusion coefficient for DiIC_{18} in DMPC vesicles of $5.7 \pm 1.4 \times 10^{-8}$ cm^2/s.

In BAECs, moving the stage such that the confocal volume intersected the desired location on the membrane apical surface-enabled FCS measurements on spatially distinct points on a cell (Fig. 8). FCS measurements were performed on

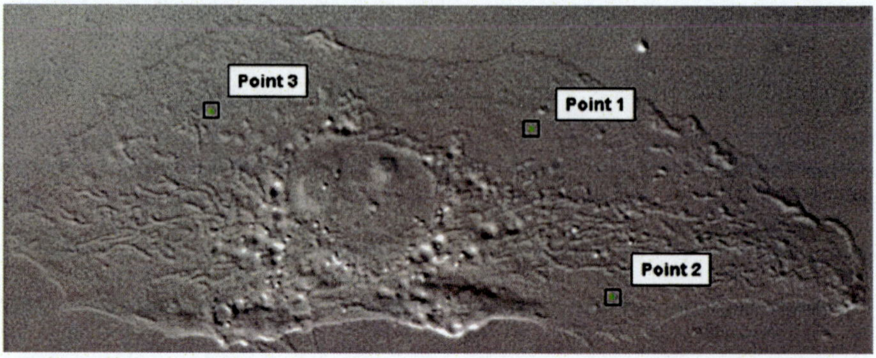

Fig. 8 Bovine aortic endothelial cell imaged with DIC microscopy. Superimposed points in the image indicate location of confocal volume on the membrane surface of a representative cell from which fluorescence fluctuation data were collected. The cell in Fig. 8 corresponds to cell 1 in Table 2

5 cells at 3 points in each cell, and 5 measurements were recorded at each point for a total of 75 measurements. Schwille et al. showed that the diffusion of DiI in rat basophilic leukemia cell membrane can be assessed using models for 2-D, 2-species fit or anomalous diffusion [52]. In another recent study, Gielen et al. reported that the diffusion on a nerve cell was well described using a 2-D, 2-species fit [56]. Consistent with this finding, we observed that a model for 2-D, 2-species diffusion fit the autocorrelation curves better than models for either anomalous subdiffusion or 2-D single species. Thus, (8) was used to determine diffusion coefficients in BAEC membranes. Values given in Table 1 represent averages of five FCS measurements taken on the cell and point indicated. Analysis of a 2-D, 2-species fit gives a fast fraction (considered to be unbound dye) and a slow fraction (which is considered to be the membrane mobile fraction). The proportion of dye in the membrane, therefore, ranged from 78 to 93%, with an average value of 83% and exhibited diffusion coefficients ranging from $1.75 \pm 0.3 \times 10^{-8}$ to $7.03 \pm 1.5 \times 10^{-8}$ cm^2/s. The value of the fast fraction (or unbound dye) ranged from 7 to 22% with an average value of 17% with a diffusion coefficient ranging from $2.11 \pm 1.1 \times 10^{-6}$ to $6.27 \pm 3.5 \times 10^{-6}$ cm^2/s. Two sample *t*-test statistics were used to analyze the differences between diffusion coefficients at different points on a given cell with $p < 0.05$. As an indication of heterogeneity in diffusion in living cells, statistically significant differences in slow-component diffusion between at least two points on a given cell were found in every cell tested (see Table 1). Representative plots of the autocorrelation function, curve fits, and residuals from cell and model membranes are shown in Fig. 9.

4.7 Fluorescence Lifetime Measurements

We measured the fluorescence lifetime of DiD dissolved in the polar protic solvents, ethanol and methanol, and polar aprotic solvents, dimethyl sulfoxide (DMSO), and

Table 1 Diffusion coefficients of FCS measurements obtained from BAECs

	Bovine aortic endothelial cells – diffusion coefficient using 2-D, 2-species fit								
	Point 1			Point 2			Point 3		
	D_1 ($\times 10^{-6}$ cm^2/s)	b_2	D_2 ($\times 10^{-8}$ cm^2/s)	D_1 ($\times 10^{-6}$ cm^2/s)	b_2	D_2 ($\times 10^{-8}$ cm^2/s)	D_1 ($\times 10^{-6}$ cm^2/s)	b_2	D_2 ($\times 10^{-8}$ cm^2/s)
Cell 1	2.58 ± 0.4	0.84 ± 0.01	2.54 ± 0.1	3.01 ± 0.6	0.79 ± 0.79	3.95 ± 0.3[a]	3.15 ± 1.6	0.82 ± 0.02	2.74 ± 0.7[b]
Cell 2	2.32 ± 1.1	0.82 ± 0.02	1.75 ± 0.3	4.29 ± 0.9	0.79 ± 0.01	4.50 ± 0.3[a]	4.13 ± 0.9	0.82 ± 0.02	4.48 ± 1.3[a]
Cell 3	2.85 ± 1.4	0.82 ± 0.02	3.62 ± 0.3	5.29 ± 2	0.85 ± 0.01	5.20 ± 0.4[a]	6.27 ± 3.5	0.83 ± 0.01	5.94 ± 0.4[a,b]
Cell 4	2.86 ± 2.7	0.78 ± 0.06	5.17 ± 3.0	4.08 ± 3.3	0.88 ± 0.01	3.99 ± 1.1	4.91 ± 1.1	0.85 ± 0.02	7.03 ± 1.5[b]
Cell 5	3.58 ± 2.4	0.86 ± 0.01	5.21 ± 1.7	5.47 ± 0.8	0.85 ± 0.01	6.91 ± 0.7	2.11 ± 1.1	0.85 ± 0.03	2.19 ± 0.1[a,b]

2-D, 2-species fit was performed on 75 measurements, obtained from five different cells. Three different spatial points were chosen for each individual cell (see Fig. 8). D_1 and D_2 are diffusion coefficients of the fast and slow moving DiI molecules, respectively. b_2 and $1 - b_2$ are their respective relative fractions. Two sample *t*-test analysis was used to determine whether differences in diffusion coefficients between points were statistically significant. The superscript "a" indicates significant difference between point 1 and points 2 or 3 while significant difference between point 2 and point 3 is indicated by the superscript "b" ($p < 0.05$). Slow-component diffusion values ranged from 1.75×10^{-8} to 7.03×10^{-8} cm^2/s and averaged 4.38 ± 1.62 ($\times 10^{-8}$) cm^2/s (mean ± SD)

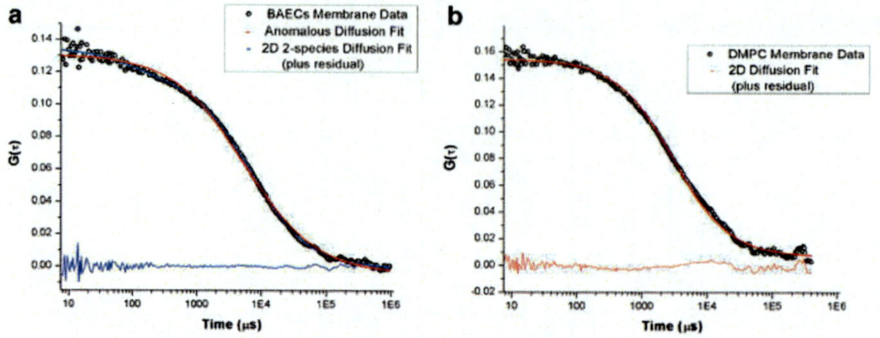

Fig. 9 (**a**) Representative plots of the autocorrelation of fluorescence intensity fluctuations arising from BAEC membranes stained with 5 nM DiI-C_{18}. Autocorrelation functions were fit with (7)–(9) describing 2-D diffusion (1-species), 2-D diffusion (2-species), and anomalous diffusion, respectively. The residual of the 2-D, 2-species fit is shown at the bottom of the graph (*blue line*). (**b**) Representative autocorrelation curve and 2-D diffusion fit of DiI-C_{18} in a DMPC membrane. The residual of the fit is shown at the bottom of the graph

Table 2 Fluorescence lifetimes of DiD in polar protic (ethanol and methanol) and polar aprotic (DMSO and DMF) solvents

Solvent	Diffusion time (μs)	Lifetime (ns), Experiment (τ_1,τ_2)	χ^2	Dielectric constant[a]	Dipole moment[a]
Ethanol	NA	0.48, 1.22	1.18	24.3	1.69
Methanol	NA	0.40, 1.02	1.14	33	1.70
Dimethyl sulfoxide (DMSO)	51.8	1.10, 1.93	1.30	47.2	3.96
Dimethyl formamide (DMF)	47.2	0.74, 1.57	1.50	38.3	3.82

Pulsed-laser excitation (650 nm) and TCSPC were used to determine fluorescence lifetime. Photon arrival time histograms were fit with a biexponential decay (12). Diffusion of DiD in the polar aprotic solvents, DMF and DMSO, was determined using FCS and (6). Fluorescence emission of DiD in polar protic solvents, ethanol and methanol, was sufficient for lifetime analysis but too weak to generate an autocorrelation curve which could be fit

[a] From the Handbook of chemistry and physics tables

dimethyl formamide (DMF). In contrast to polar aprotic solvents, polar protic solvents have dissociable H^+ and can form hydrogen bonds. They also have lower dielectric constants and lower dipole moments than polar aprotic solvents. Since fluorescence lifetime depends on the electronic cloud configuration of the dye molecule, these solvents are expected to affect fluorescence lifetime by altering electron distribution and intramolecular charge transfer. In each solvent, fluorescence lifetimes were best fit using biexponential decays (12) consistent with earlier results for DiI [57]. Each value reported in Table 2 represents the average of ten measurements. Lifetime values for ethanol and methanol were (τ_1, τ_2) 0.48, 1.22 and 0.40, 1.02 ns, respectively, suggesting that fluorescence lifetime can be different even in closely related solvents. Buschmann et al. obtained the fluorescence lifetime

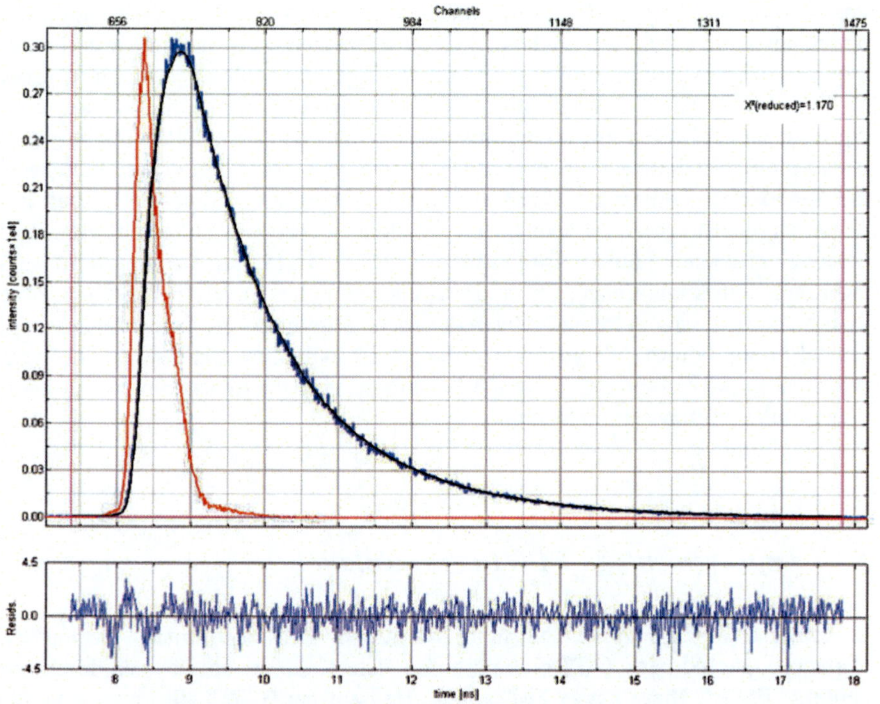

Fig. 10 A fluorescence decay (*blue line*) and instrument response function (*red line*) for DiD fluorescence in ethanol. The decay histogram is fit with a double exponential (*black line*) to obtain the fluorescence lifetime. The χ^2 value of the fit is 1.17. The full width at half maximum (FWHM) of the instrument response function is ~300 ps

values of DiD in various solvents [58]. Using a single exponential to fit the lifetime curve, they obtained a value of 1.12 ns for the decay of DiD in ethanol. In our experiments, the characteristic diffusion time, τ_D, of DiD in DMSO and DMF were 51.2 and 47.8 μs, respectively. However, the fluorescence lifetimes of the dye (using a biexponential decay) were 0.74, 1.57 and 1.10, 1.93 ns, in DMSO and DMF, respectively (Table 2). Thus, molecules with similar diffusion characteristics can be readily distinguished based on lifetime. Fluorescence emission of DiD in ethanol and methanol is intrinsically weak and therefore autocorrelation curves were noisy and could not be fit with models for FCS. A representative curve showing the fluorescence decay with a fit along with the IRF is shown in Fig. 10.

5 Discussion

The integrated modeling-microscopy-spectroscopy setup described in this study was designed to address the following fundamental issues in mechanotransduction research: (a) forces are spatially distributed, (b) relationships between force application and cell perturbation requires the integration of continuum and molecular-scale predictive models, and (c) determination of molecular activation requires temporal and spatial registration of single-molecule events in live, intact cells under controlled conditions. To meet these design criteria, we have constructed and calibrated a multimodal microscope for cellular imaging and single-molecule fluorescence detection and analysis. First, collection volume of the fluorescence emission was reduced to <1 fl using a fiber-aperture placed in a plane confocal to the laser focus. Second, fluorescence was collected with high-detection efficiency photomultiplier tubes (PMTs). Third, noise arising from stray laser light, autofluorescence, and fluorescence from glass coverslips and impurities in the solution was systematically reduced or eliminated by using high-quality optical filters. Fourth, all systems (confocal and trap) were spatially correlated to positions on the cell through careful 3-D calibration of the focus point and imaging plane of the multimodal microscope.

5.1 Improvements in FCS Instrumentation

In contrast to majority of the more traditional FCS instruments, which use avalanche photodiodes (APDs) for photon detection, we employ single-photon counting PMTs. These newer PMTs (e.g., Hamamatsu H7422-40P) have a higher quantum efficiency than older generation PMTs making them suitable for photon correlation studies [59]. In addition, PMTs are easier to align [because their active area is larger (~mm^2 vs. ~μm^2)], less susceptible to photon-induced degradation, and have a superior response time (10's of picoseconds) compared to APDs (100's of picoseconds). These properties also make them optimal for fluorescence lifetime measurements. Thus, the use of PMTs enables fluorescence lifetime and FCS data to be obtained from a single set of measurements [60].

We use software to generate autocorrelation curves from FIFO data obtained from time-correlated single-photon counting instrumentation (i.e., TCSPC-FCS). Hardware autocorrelator cards can also generate autocorrelation curves in near real-time using current fluctuations from APDs or PMTs. However, when combined with pulsed-laser excitation, TCSPC instrumentation records all photon arrival times and saves them into memory directly. This feature makes all the raw photon data available for additional computational analysis necessary for generation of photon counting histograms (PCHs), multispectral fluorescence lifetime imaging (FLIM), burst-induced fluorescence lifetimes (BIFL), and higher-order correlations without the need for an additional, expensive, hardware correlator [59, 60]. Advances in TCSPC instrumentation and modern autocorrelation software algorithms (e.g., multi-tau time correlation method) ensure near real-time autocorrelation curves that were previously only possible with dedicated hardware correlators [61]. Recent studies support the increasing use of TCSPC for FCS and lifetime measurements [60].

5.2 FCS Can Detect Subtle Changes in Dye Concentrations

In order to test whether the system could detect subtle changes in concentration of diffusing species, we performed FCS analysis of autocorrelation curves arising from diffusion of rhodamine dye which varied in concentration from 1 to 10 nM. These calibration experiments support the use of FCS to determine the absolute number of molecules in the observation volume. Absolute molecular concentrations are essential in the study of the binding kinetics of molecules present on the surfaces of cell membranes and in solution [62]. In the context of mechanotransduction, it is thus possible to follow a surface signal transduction pathway by following the association and disassociation of a receptor complex in response to an externally imposed force from fluid flow. One of the main limitations of the technique of FCS is the requirement of extremely low concentrations of probe molecules. We observed degradation of correlation signal with dye concentrations greater than 10^{-8} M. While requirement for low concentrations of molecules in FCS makes imaging impractical, it is an advantage in studies of binding kinetics in cells, where the concentrations of aggregating molecules are typically low. The low concentration also means the molecular events in cells can be studied with relatively less perturbation of the biological system compared to methods which require high concentrations of fluorescent molecules such as fluorescence recovery after photobleaching (FRAP) and any imaging-based methods such as fluorescence resonance energy transfer (FRET).

5.3 FCS-Determined Diffusion Coefficients Are Sensitive to Small Changes in Viscosity

In our calibration experiment, we show that increases in glycerol concentration resulted in right-shifted autocorrelation curves indicating longer diffusion times for increasingly viscous environments (Fig. 7a). Thus, FCS is likely to be sufficiently sensitive to monitor subtle changes in the local microviscosity which may be biologically significant. In our recent study, we showed that the diffusion of DiI in cellular membranes is not uniform and varies substantially in response to applied shear over the cells. Combined with 3-D piezo-scanning, FCS can monitor such changes occurring in cells due to imposed fluid flow or other mechanical perturbations to determine spatially resolved mechanical effects on membranes.

5.4 Differences in DiI Diffusion in Cell and Model Membranes Determined by FCS

In order to test whether the CMDM could detect spatially varying diffusion of lipids in cell and model membranes, we measured the fluorescence fluctuations arising from DiI in model DMPC membranes and bovine aortic endothelial cells. Model membranes are useful in studying diffusion in lipid bilayers, phase segregation in lipids, and model protein diffusion in vesicles [31, 63, 64]. They are compositionally simple and homogenous in nature enabling one to fit the simple 2-D equation (7) of diffusion to the autocorrelation curves. In DMPC vesicles above the phase-transition temperature, we obtained diffusion coefficient values of $5.7 \pm 1.4 \times 10^{-8}$ cm^2/s. More importantly, the small differences of diffusion values obtained within measurements performed on one vesicle and between different individual vesicles support the assumption that the vesicles are homogenous in nature and that the FCS measurement is repeatable. Vaz et al. obtained values of the diffusion coefficients in DMPC/Chol bilayers at various temperatures using FRAP. At 303 K, the value of D_L for a pure DMPC bilayer was $\sim 6 \times 10^{-8}$ cm^2/s which agrees well with the value obtained in the current study [65, 66]. FRAP experiments in other model membranes of POPC (l-palmitoyl-2-oleoyl-sn-glycero-3-phosphocholine) show unrestricted translational diffusion $D = 4.4 \times 10^{-8}$ cm^2/s in fluid membrane and restricted diffusion $D = 0.7 \times 10^{-9}$ cm^2/s in polymer-stabilized membranes [67]. Previous FCS experiments on DLPC (dilauroyl phosphatidylcholine) revealed a translational diffusion coefficient of 3×10^{-8} cm^2/s [63].

FCS analysis of DiI fluorescence from stained endothelial cells revealed two important observations. First, the choice of diffusion models used to fit autocorrelation curves of DiI fluorescence in cell membranes is not straightforward, consistent with observations by others [52, 56]. To fit our cell lipid diffusion data, we used models for both 2-D, 2-species (8) and anomalous subdiffusion (9). In general, we found that fitting of autocorrelation curves with a model for 2-D, 2-species fit resulted in uncorrelated residuals and smaller χ^2 values. The spatially heterogeneous nature of the cell membrane necessitates the requirement for a complex fit such as that for 2-D, 2-species. In our study, the 2-D, 2-species fit reveals the existence of a fast moving component, whose origin is unclear but may reflect unbound dye, as has been suggested in other studies [56]. It has been speculated that the fast component of the diffusion might arise from the unwashed dye present in the focal volume [56]. Measurements of dye in solution away from the membrane support this hypothesis. However, the relatively high fraction (83%) of the mobile membrane component confirms that FCS is effective in measuring cell membrane lipid diffusion. The results obtained by the 2-D, 2-species fit compare favorably with membrane diffusion values in the literature using either FCS or FRAP [31, 52, 56].

The second important observation in this study is that translational diffusion coefficients for different measurement points on endothelial cells vary significantly (cells 1–5, Table 2). The values for the membrane diffusion on different points on the cells used in this study ranged from $1.75 \pm 0.3 \times 10^{-8}$ to $7.03 \pm 1.5 \times 10^{-8}$ cm^2/s. While all the values for membrane diffusion are within the same order of magnitude, the inter-point variation within a single cell or between individual cells used in this study is not trivial, but represents an example of the heterogeneous nature of the membrane surface. The variability in the values of the diffusion coefficient might arise from the complexity of the cellular membrane such as differences in lipid phase and density, local protein concentration and aggregation [68], and membrane interaction with the cytoskeleton and glycocalyx [69]. Thus, the ability to obtain FCS data from specific locations is an important step toward assessing the nature of spatial heterogeneity of molecular dynamics in live cells.

5.5 *Time-Resolved Fluorescence of DiD in Polar Protic and Aprotic Solvents*

We used our pulsed, red laser to generate data for fluorescence lifetime and FCS to test the ability of the system to deflect subtle differences in how DiD interacts with its immediate surroundings. The measurement of fluorescence lifetime is independent of dye concentration and is solely determined by changes in the dye's photophysics induced by inter- and intramolecular interactions. Data reported in Table 2 suggest that the same fluorescent molecule, DiD, has longer fluorescence lifetimes in aprotic solvents compared to protic solvents (Table 2). Fluorescence lifetime depends on nonradiative decay mechanisms. Thus, the shorter fluorescence lifetimes in protic solvents may be due to such nonradiative decay mechanisms as hydrogen bonding and intermolecular proton transfer. Newer modalities of FLIM exploit this ultrasensitivity of a fluorophore for changes in its immediate surrounding to generate spatial maps of lifetime [60]. It is expected that lifetime analysis will enable rapid mapping of mechanically induced, molecular-scale perturbation of the dye-lipid-water microenvironments in intact cell membranes.

5.6 *Multimodal Microscopy*

In order to correlate molecular-scale events with cellular structures and their locations, we have integrated TCSPC into a multimodal microscope consisting of differential interference contrast (DIC), total internal reflection fluorescence (TIRF), and epifluorescence microscopy followed by deconvolution using Autodeblur software (Autoquant, Albany, NY, USA) (Fig. 5). A 3-D piezoelectric stage provides high-resolution positioning, such that molecular-scale analysis can be associated with subcellular regions of live adherent cells. Much of this hardware is commercially available, making these techniques accessible to researchers in biology and mechanobiology. The engineering challenge is to integrate these modalities on a single platform. In response to this challenge, microscope manufactures have developed microscopes with multiple input and output ports and increased infinity space for the introduction of custom optics.

6 Conclusions

Integrated instrumentation and interdisciplinary approaches will help solve biological and medical problems originating from the interface of cellular signaling and cell mechanics, an emerging area of biology. Finite element modeling, Monte Carlo simulations, and molecular dynamics can provide predictive multiscale models of cell behavior. However, due to the level of complexity involved in the cellular signal transduction and inherent compartmentalization of cellular signaling in microenvironments, verification of models needs to be done at the single-molecule level on intact cells with high-precision spatial registration. Alterations in dynamics of molecules are assumed to accompany changes in their signaling (e.g., phosphorylation and dimerization). Such changes are detectable by FCS and fluorescence lifetime analysis, making single-molecule spectroscopy an ideal tool to address the precise molecular mechanism by which forces induce changes in cellular biology.

Some possible future avenues of investigation are monitoring the diffusion-related kinetics of transmembrane proteins such as G-protein-coupled receptors in response to externally applied forces due to fluid flow, micropipette aspiration, or an optical trap. One of the authors (P.J.B.) has performed earlier work showing changes in membrane fluidity in response to shear flow using FRAP [31]. The study can be extended to spatially map the changes occurring on a plasma membrane with single-molecule sensitivity using the current instrumental setup. The current setup can also potentially be used to follow the differential kinetics of protein molecules at the apical and basal surfaces of the cell by combining the techniques of one-photon FCS with TIR.

In conclusion, we have successfully built a multimodal microscope that combines cellular modeling with microscale monitoring of molecular dynamics at multiple timescales with prescribed macroscale force conditions. This system addresses an unmet need for integration of force application, analysis of cell mechanics, and molecular-scale mechanotransduction detection. It is now possible to forge new research directions in which cell-specific, multicomponent models of mechanotransduction are developed from 3-D live cell imaging and validated with molecular-scale biological readouts.

References

1. Butler PJ, Weinbaum S, Chien S, Lemons DE (2000) Endothelium-dependent, shear-induced vasodilation is rate-sensitive. Microcirculation 7:53–65
2. Butler PJ, Tsou TC, Li JY, Usami S, Chien S (2002) Rate sensitivity of shear-induced changes in the lateral diffusion of endothelial cell membrane lipids: a role for membrane perturbation in shear-induced MAPK activation. FASEB J 16:216–218
3. Frangos JA, Huang TY, Clark CB (1996) Steady shear and step changes in shear stimulate endothelium via independent mechanisms–superposition of transient and sustained nitric oxide production. Biochem Biophys Res Commun 224:660–665
4. DePaola N, Gimbrone MA Jr, Davies PF, Dewey CF Jr (1992) Vascular endothelium responds to fluid shear stress gradients. Arterioscler Thromb 12:1254–1257
5. Nerem RM, Levesque MJ, Cornhill JF (1981) Vascular endothelial morphology as an indicator of the pattern of blood flow. J Biomech Eng 103:172–176
6. Koller A, Kaley G (1991) Endothelial regulation of wall shear stress and blood flow in skeletal muscle microcirculation. Am J Physiol 260:H862–H868
7. Orr AW, Helmke BP, Blackman BR, Schwartz MA (2006) Mechanisms of mechanotransduction. Dev Cell 10:11–20

8. Ingber DE (2003) Mechanobiology and diseases of mechanotransduction. Ann Med 35:564–577
9. Davies PF (1997) Overview: temporal and spatial relationships in shear stress-mediated endothelial signalling. J Vasc Res 34:208–211
10. Barbee KA, Mundel T, Lal R, Davies PF (1995) Subcellular distribution of shear stress at the surface of flow-aligned and nonaligned endothelial monolayers. Am J Physiol 268: H1765–H1772
11. Mathur AB, Truskey GA, Reichert WM (2000) Atomic force and total internal reflection fluorescence microscopy for the study of force transmission in endothelial cells. Biophys J 78:1725–1735
12. Trache A, Meininger GA (2005) Atomic force-multi-optical imaging integrated microscope for monitoring molecular dynamics in live cells. J Biomed Opt 10:064023
13. Wang Y, Botvinick EL, Zhao Y, Berns MW, Usami S, Tsien RY, Chien S (2005) Visualizing the mechanical activation of Src. Nature 434:1040–1045
14. Huang H, Dong CY, Kwon HS, Sutin JD, Kamm RD, So PT (2002) Three-dimensional cellular deformation analysis with a two-photon magnetic manipulator workstation. Biophys J 82:2211–2223
15. Charras GT, Horton MA (2002) Determination of cellular strains by combined atomic force microscopy and finite element modeling. Biophys J 83:858–879
16. Ferko MC, Patterson BW, Butler PJ (2006) High-resolution solid modeling of biological samples imaged with 3D fluorescence microscopy. Microsc Res Tech 69:648–655
17. Osborn EA, Rabodzey A, Dewey CF Jr, Hartwig JH (2006) Endothelial actin cytoskeleton remodeling during mechanostimulation with fluid shear stress. Am J Physiol Cell Physiol 290: C444–C452
18. Sultan C, Stamenovic D, Ingber DE (2004) A computational tensegrity model predicts dynamic rheological behaviors in living cells. Ann Biomed Eng 32:520–530
19. Wang N, Suo Z (2005) Long-distance propagation of forces in a cell. Biochem Biophys Res Commun 328:1133–1138
20. Geiger RV, Berk BC, Alexander RW, Nerem RM (1992) Flow-induced calcium transients in single endothelial cells: spatial and temporal analysis. Am J Physiol 262:C1411–C1417
21. Levitan I, Christian AE, Tulenko TN, Rothblat GH (2000) Membrane cholesterol content modulates activation of volume-regulated anion current in bovine endothelial cells. J Gen Physiol 115:405–416
22. Li S, Kim M, Hu YL, Jalali S, Schlaepfer DD, Hunter T, Chien S, Shyy JY (1997) Fluid shear stress activation of focal adhesion kinase. Linking to mitogen-activated protein kinases. J Biol Chem 272:30455–30462
23. McCormick SM, Eskin SG, McIntire LV, Teng CL, Lu CM, Russell CG, Chittur KK (2001) DNA microarray reveals changes in gene expression of shear stressed human umbilical vein endothelial cells. Proc Natl Acad Sci USA 98:8955–8960
24. Hess ST, Huang S, Heikal AA, Webb WW (2002) Biological and chemical applications of fluorescence correlation spectroscopy: a review. Biochemistry 41:697–705
25. Vukojevic V, Pramanik A, Yakovleva T, Rigler R, Terenius L, Bakalkin G (2005) Study of molecular events in cells by fluorescence correlation spectroscopy. Cell Mol Life Sci 62:535–550
26. Ferko MC, Bhatnagar A, Garcia MB, Butler PJ (2006) Finite-element stress analysis of a multicomponent model of sheared and focally-adhered endothelial cells. Ann Biomed Eng 35 (2):208–223
27. Gullapalli RR, Tabouillot T, Mathura R, Dangaria J, Butler PJ (2007) Integrated multimodal microscopy, time resolved fluorescence, and optical-trap rheometry: toward single molecule mechanobiology. J Biomed Opt 12(1):014012
28. Gudi SR, Clark CB, Frangos JA (1996) Fluid flow rapidly activates G proteins in human endothelial cells. Involvement of G proteins in mechanochemical signal transduction. Circ Res 79:834–839

29. Florian JA, Kosky JR, Ainslie K, Pang Z, Dull RO, Tarbell JM (2003) Heparan sulfate proteoglycan is a mechanosensor on endothelial cells. Circ Res 93:e136–e142
30. Mochizuki S, Vink H, Hiramatsu O, Kajita T, Shigeto F, Spaan JA, Kajiya F (2003) Role of hyaluronic acid glycosaminoglycans in shear-induced endothelium-derived nitric oxide release. Am J Physiol Heart Circ Physiol 285:H722–H726
31. Butler PJ, Norwich G, Weinbaum S, Chien S (2001) Shear stress induces a time- and position-dependent increase in endothelial cell membrane fluidity. Am J Physiol Cell Physiol 280: C962–C969
32. Haidekker MA, L'Heureux N, Frangos JA (2000) Fluid shear stress increases membrane fluidity in endothelial cells: a study with DCVJ fluorescence. Am J Physiol Heart Circ Physiol 278:H1401–H1406
33. Helmke BP, Goldman RD, Davies PF (2000) Rapid displacement of vimentin intermediate filaments in living endothelial cells exposed to flow. Circ Res 86:745–752
34. Helmke BP, Rosen AB, Davies PF (2003) Mapping mechanical strain of an endogenous cytoskeletal network in living endothelial cells. Biophys J 84:2691–2699
35. Hu S, Chen J, Fabry B, Numaguchi Y, Gouldstone A, Ingber DE, Fredberg JJ, Butler JP, Wang N (2003) Intracellular stress tomography reveals stress focusing and structural anisotropy in cytoskeleton of living cells. Am J Physiol Cell Physiol 285:C1082–C1090
36. Tzima E, Irani-Tehrani M, Kiosses WB, Dejana E, Schultz DA, Engelhardt B, Cao G, DeLisser H, Schwartz MA (2005) A mechanosensory complex that mediates the endothelial cell response to fluid shear stress. Nature 437:426–431
37. Davies PF, Robotewskyj A, Griem ML (1994) Quantitative studies of endothelial cell adhesion. Directional remodeling of focal adhesion sites in response to flow forces. J Clin Invest 93:2031–2038
38. Zaidel-Bar R, Kam Z, Geiger B (2005) Polarized downregulation of the paxillin-p130CAS-Rac1 pathway induced by shear flow. J Cell Sci 118:3997–4007
39. Jalali S, del Pozo MA, Chen K, Miao H, Li Y, Schwartz MA, Shyy JY, Chien S (2001) Integrin-mediated mechanotransduction requires its dynamic interaction with specific extracellular matrix (ECM) ligands. Proc Natl Acad Sci USA 98:1042–1046
40. Sato M, Levesque MJ, Nerem RM (1987) An application of the micropipette technique to the measurement of mechanical properties of cultured bovine aortic endothelial cells. J Biomech Eng 109:27–34
41. Sato M, Nagayama K, Kataoka N, Sasaki M, Hane K (2000) Local mechanical properties measured by atomic force microscopy for cultured bovine endothelial cells exposed to shear stress. J Biomech 33:127–135
42. Davies PF (1995) Flow-mediated endothelial mechanotransduction [Review] [407 refs]. Physiol Rev 75:519–560
43. Rizzo V, Sung A, Oh P, Schnitzer JE (1998) Rapid mechanotransduction in situ at the luminal cell surface of vascular endothelium and its caveolae. J Biol Chem 273:26323–26329
44. Karcher H, Lammerding J, Huang H, Lee RT, Kamm RD, Kaazempur-Mofrad MR (2003) A three-dimensional viscoelastic model for cell deformation with experimental verification. Biophys J 85:3336–3349
45. Charras GT, Williams BA, Sims SM, Horton MA (2004) Estimating the sensitivity of mechanosensitive ion channels to membrane strain and tension. Biophys J 87:2870–2884
46. Fujiwara K, Masuda M, Osawa M, Kano Y, Katoh K (2001) Is PECAM-1 a mechanoresponsive molecule? Cell Struct Funct 26:11–17
47. Li S, Butler P, Wang Y, Hu Y, Han DC, Usami S, Guan JL, Chien S (2002) The role of the dynamics of focal adhesion kinase in the mechanotaxis of endothelial cells. Proc Natl Acad Sci USA 99:3546–3551
48. Haustein E, Schwille P (2003) Ultrasensitive investigations of biological systems by fluorescence correlation spectroscopy. Methods 29:153–166
49. Elson E, Magde D (1974) Fluorescence correlation spectroscopy. I. Conceptual basis and theory. Biopolymers 13:1–27

50. Magde D, Elson E, Webb WW (1974) Fluorescence correlation spectroscopy. II. An experimental realization. Biopolymers 13:29–61
51. Zander C, Enderlein J, Keller RA (2002) Single-molecule detection in solution methods and applications. Wiley-VCH, Berlin
52. Schwille P, Korlach J, Webb WW (1999) Fluorescence correlation spectroscopy with single-molecule sensitivity on cell and model membranes. Cytometry 36:176–182
53. Lakowicz JR (1999) Principles of fluorescence spectroscopy. Springer, New York
54. Widengren J, Mets U, Rigler R (1995) Fluorescence correlation spectroscopy of triplet states in solution: a theoretical and experimental study. J Phys Chem 99:13368–13379
55. Kubin RF, Fletcher AN (1982) Fluorescence quantum yields of some rhodamine dyes. J Lumin 27:455–462
56. Gielen E, Vercammen J, Sykora J, Humpolickova J, Vandeven M, Benda A, Hellings N, Hof M, Engelborghs Y, Steels P, Ameloot M (2005) Diffusion of sphingomyelin and myelin oligodendrocyte glycoprotein in the membrane of OLN-93 oligodendroglial cells studied by fluorescence correlation spectroscopy. C R Biol 328:1057–1064
57. Packard BS, Wolf DE (1985) Fluorescence lifetimes of carbocyanine lipid analogues in phospholipid bilayers. Biochemistry 24:5176–5181
58. Buschmann V, Weston KD, Sauer M (2003) Spectroscopic study and evaluation of red-absorbing fluorescent dyes. Bioconjug Chem 14:195–204
59. Becker W (2006) Advanced time-correlated single photon counting techniques. Springer, New York
60. Becker W, Bergmann A, Haustein E, Petrasek Z, Schwille P, Biskup C, Kelbauskas L, Benndorf K, Klocker N, Anhut T, Riemann I, Konig K (2006) Fluorescence lifetime images and correlation spectra obtained by multidimensional time-correlated single photon counting. Microsc Res Tech 69:186–195
61. Wahl M, Gregor I, Patting M, Enderlein J (2003) Fast calculation of fluorescence correlation data with asynchronous time-correlated single-photon counting. Opt Express 11:3583–3591
62. Pramanik A, Rigler R (2001) Ligand-receptor interactions in the membrane of cultured cells monitored by fluorescence correlation spectroscopy. Biol Chem 382:371–378
63. Korlach J, Schwille P, Webb WW, Feigenson GW (1999) Characterization of lipid bilayer phases by confocal microscopy and fluorescence correlation spectroscopy. Proc Natl Acad Sci USA 96:8461–8466
64. Bacia K, Scherfeld D, Kahya N, Schwille P (2004) Fluorescence correlation spectroscopy relates rafts in model and native membranes. Biophys J 87:1034–1043
65. Almeida PF, Vaz WL, Thompson TE (1992) Lateral diffusion in the liquid phases of dimyristoylphosphatidylcholine/cholesterol lipid bilayers: a free volume analysis. Biochemistry 31:6739–6747
66. Vaz WLC, Clegg RM, Hallmann D (1985) Translational diffusion of lipids in liquid-crystalline phase phosphatidylcholine multibilayers – a comparison of experiment with theory. Biochemistry 24:781–786
67. Schutz GJ, Schindler H, Schmidt T (1997) Single-molecule microscopy on model membranes reveals anomalous diffusion. Biophys J 73:1073–1080
68. Kusumi A, Suzuki K (2005) Toward understanding the dynamics of membrane-raft-based molecular interactions. Biochim Biophys Acta 1746:234–251
69. Tarbell JM, Pahakis MY (2006) Mechanotransduction and the glycocalyx. J Intern Med 259:339–350

Going Beyond Continuous Glucose Monitoring with Boronic Acid-Appended Bipyridinium Salts

Alexander Schiller, Boaz Vilozny, Ritchie A. Wessling, and Bakthan Singaram

Abstract A two-component sensing system comprising a fluorescent dye and a boronic acid-containing molecule that acts dually as a fluorescence quencher and a saccharide receptor was developed for continuous glucose monitoring in blood. Boronic acid-based bipyridinium salts as tunable receptors have been introduced to increase glucose selectivity over that of other boronate-forming analytes. Powerful solution-phase sensor arrays for neutral and anionic carbohydrates were created with these probes. They can be also used in new label-free fluorescent assays for carbohydrate-modifying enzymes.

1 Introduction

Boronic acids are known to reversibly bind diols with high affinity to form cyclic boronate esters [1–3]. As a result, boronic acid-containing molecular probes have been developed as chemosensors for the recognition of diol-containing analytes, such as carbohydrates, nucleotides, and hydroxy carboxylic acids [4–7]. Since 1990, much research has been focused on the design of boronic acid-containing fluorescent carbohydrate sensors that operate under physiological conditions [8]. By contrast, reports on colorimetric sugar probes using boronic acids are still limited [9–13].

Glucose in particular plays a central role in various metabolic processes. For example, control of blood glucose concentration is of critical importance for patients suffering from diabetes mellitus, a disorder affecting many millions of people worldwide. Therefore, many probes utilizing boronic acid-based receptors have been developed to selectively detect glucose at physiological concentrations.

A. Schiller (✉)
Institute for Inorganic and Analytical Chemistry, Friedrich-Schiller-University Jena, Humboldtstr. 8, 07743 Jena, Germany
e-mail: alexander.schiller@uni-jena.de

C.D. Geddes (ed.), *Reviews in Fluorescence 2009*, Reviews in Fluorescence 2009,
DOI 10.1007/978-1-4419-9672-5_7,

The ultimate goal is to develop implantable sensors that can be used to continuously monitor glucose concentrations in people suffering from diabetes [14–17]. A new application for this technology is to achieve tight glycemic control in the Intensive Care Unit [18, 19]. Determination of fructose or other saccharides in food and beverages and monitoring of fermentation processes are other important applications of interest.

To develop a successful in vivo boronic acid-based glucose sensor, an important criterion that must be met is the preferential binding of glucose over other physiologically significant monosaccharides, such as fructose. This remains a challenge because most organic boronic acids display higher binding affinities for fructose [8]. To increase glucose selectivity, the concept of bidentate binding [20] via diboronic acids has been utilized [21–25]. In this approach, the molecular probe contains two arylboronic acid moieties spatially disposed in a way that allows for cooperative binding to a single glucose molecule. For a given diboronic acid-containing scaffold, the spacing between the boronic acids can be adjusted through synthetic modifications to create a glucose-specific binding pocket. However, in many systems, the boronic acids are covalently linked to an indicator (a fluorophore or chromophore) [4–8], and such modifications to the receptor can prove difficult or impossible to achieve without altering the photophysical properties of the indicator as well. Further, these receptors may require elaborate syntheses.

This problem can be avoided by employing a modular approach to the sensor design, where the receptor and the indicator moieties exist as covalently discrete entities. This technique has been used in well-known indicator displacement assays (IDA). An indicator is first allowed to bind reversibly to a receptor. A competitive analyte is introduced into the system causing the displacement of the indicator from the host, which in turn modulates an optical signal [26, 27]. Signal modulation in an IDA is possible based on several mechanisms: photoinduced electron transfer (PET) [26, 28], fluorescence resonance energy transfer (FRET) [29], electronic energy transfer (EET) [30], or simple changes in local ionic strength or pH [31]. The common interactions between the indicator or analyte and the host are H-bonding, electrostatic interactions, and complexing with metal ions [32]. These interactions are dependent on the geometry of the guest, its charge, its hydrophobicity, and the solvent system. Innovative IDAs have been described by the Anslyn group [32–35] and Severin et al. [36–41].

In 1999, Singaram and Wessling began their study of a two-component sensing system comprising a fluorescent dye that serves as the indicator unit and a diboronic acid-appended viologen that serves a dual function as fluorescence quencher and saccharide receptor [42]. Progress in this endeavor is documented in a series of papers by the Singaram group detailing the chemistry of this system and the development of sensors based on immobilization of the sensing components in a hydrogel [43–60]. In contrast to a standard IDA, the indicator in this system is displaced by the analyte from an allosteric interaction (Fig. 1). This means that the analyte does not compete at the same binding site with the indicator. By contrast, it binds at another site (allos stereos *Greek* "other object"), thereby inducing a

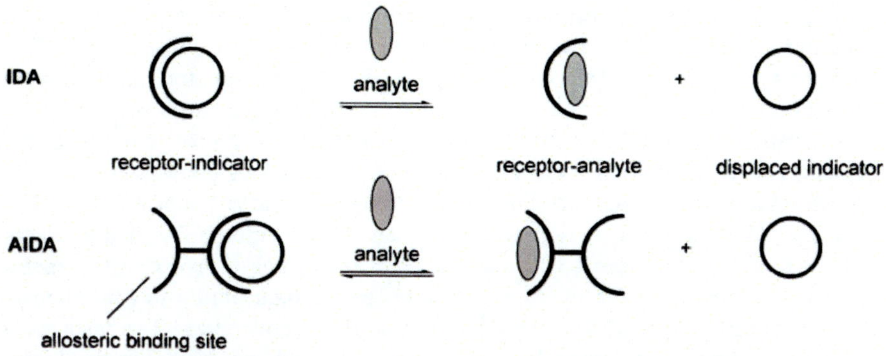

Fig. 1 Comparison between an indicator displacement assay (IDA) and an allosteric indicator displacement assay (AIDA). IDA: An indicator is first allowed to bind reversibly to a receptor. A competitive analyte is introduced into the system causing the displacement of the indicator from the host, which in turn modulates an optical signal. AIDA: The indicator is displaced by the analyte from an allosteric interaction. The analyte binds at another site at the receptor rather than the indicator, inducing a decrease in affinity of the indicator for the receptor

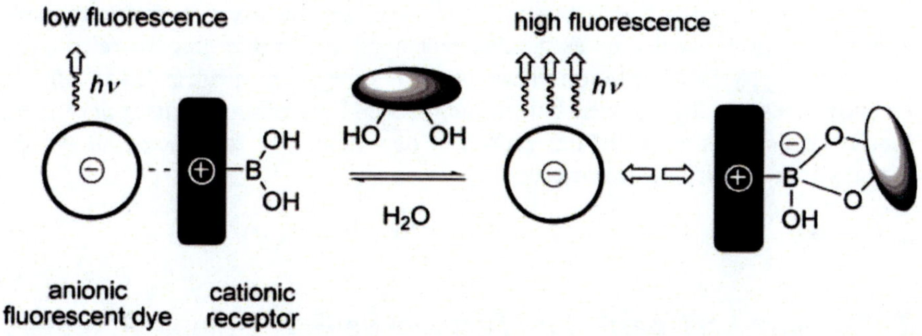

Fig. 2 AIDA in sugar sensing – proposed mechanism for sensing diol-containing analytes based on a cationic receptor (boronic acid-based bipyridinium salt) and an anionic fluorescent dye. Reprinted with permission from [48]. Copyright 2007 Wiley-VCH Verlag GmbH & Co KGaA

decrease in the affinity of the indicator for the receptor. We call this new type of assay an "allosteric indicator displacement assay" (AIDA).

The first use of an AIDA as a probe for saccharide sensing was reported by the Singaram group in 2002 [60]. The proposed signaling mechanism involves ground-state complex formation between an anionic fluorescent dye and a cationic receptor (boronic acid-based bipyridinium salt) that facilitates electron transfer from the dye to the bipyridinium salt (viologen), resulting in a diminution of the fluorescent emission of the dye (Fig. 2). When a diol-containing analyte interacts with the ground-state complex at physiological pH, the boronic acids are converted to anionic boronate esters partially neutralizing the charge on the cationic viologen. The reduced charge weakens the complex, leading to dissociation. Consequently, quenching is diminished, resulting in an increase in fluorescence intensity. To help substantiate the step of glucose-induced dissociation of dye and quencher, UV–Vis, fluorescence, and ^{11}B NMR spectroscopy were used to monitor the change in charge of boron from neutral to anionic upon addition of glucose at pH 7.4 [53].

In contrast to the soluble probes described previously, sensors can detect the presence of glucose in body tissues or in flowing blood. Sensors are usually prepared via immobilization of the sensing components in polymer matrices. The Singaram group developed a procedure to covalently immobilize a fluorescent anionic dye and a viologen-appended boronic acid into poly(2-hydroxyethyl methacrylate) hydrogels. These hydrogels were tested for their ability to continuously and reversibly detect glucose over the course of several hours. The tests were carried out using a cuvette-based system as well as a fiber-optic-based configuration. Under physiological conditions (0.1 M phosphate buffer, pH 7.4, 37°C), the fluorescent hydrogels displayed an excellent dynamic response to glucose concentrations within the biologically significant range (2.5–20 mM) [51, 52, 59]. A related system is currently being commercialized for clinical use in blood glucose monitoring by GluMetrics, Inc. (http://www.glumetrics.com) [61].

Going beyond continuous glucose monitoring, this review article summarizes the research carried out by the Singaram and Schiller group to address the problem of obtaining selectivity with boronic acid receptors by using bis-boronic acid-appended bipyridinium salts as tunable receptors to increase glucose selectivity. Further, a powerful solution-phase sensor array for neutral and anionic carbohydrates employing these boronic acid-based bipyridinium salts is described. Finally, we show the use of our probes in new label-free fluorescent assays for carbohydrate-modifying enzymes.

2 Boronic Acid-Based Bipyridinium Salts as Tunable Receptors

The Singaram group has reported a series of tunable receptors for the detection of saccharides with the two-component probe. Six cationic bis-boronic acid-appended benzyl viologens (3,3′-*o*/*m*-, 4,3′-*o*/*m*-, and 4,4′-*o*/*m*-BBV) have been synthesized and thoroughly investigated (Fig. 3). The anionic fluorescent dye, 8-hydroxypyrene-1,3,6-trisulfonic acid trisodium salt (HPTS) was used as an indicator. In addition, three corresponding benzyl viologens (3,3′-, 4,3′-, and 4,4′-BV) were used as controls.

The major structural elements of the viologens are the bipyridyl core and the benzyl boronic acid group. Established reaction schemes and readily available starting materials were used to quickly access the viologen structures. Structural variables include the position of nitrogens in bipyridyl rings and position of boronic acid moieties on the benzyl ring (Fig. 3). In addition to the commercially available 4,4′-bipyridine, gram quantities of 3,4′- and 3,3′-bipyridines have also made by cross-coupling reactions [49]. Bromomethyl phenylboronic acid or benzyl bromide

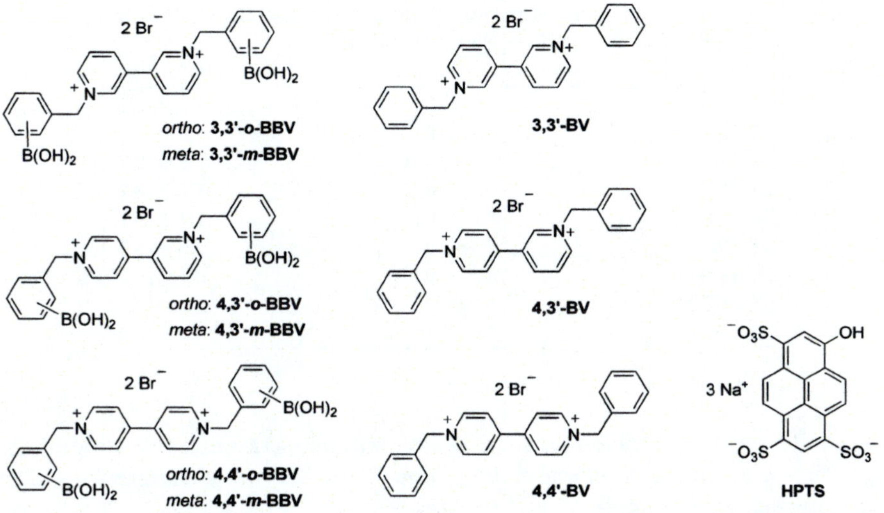

Fig. 3 Six carbohydrate receptors comprising cationic bis-boronic acid-appended benzyl viologens (3,3′-*o*/*m*-, 4,3′-*o*/*m*-, and 4,4′-*o*/*m*-BBV) and the anionic fluorescent dye, 8-hydroxypyrene-1,3,6-trisulfonic acid trisodium salt (HPTS). Three corresponding benzyl viologens (3,3′-, 4,3′-, and 4,4′-BV) were used as controls

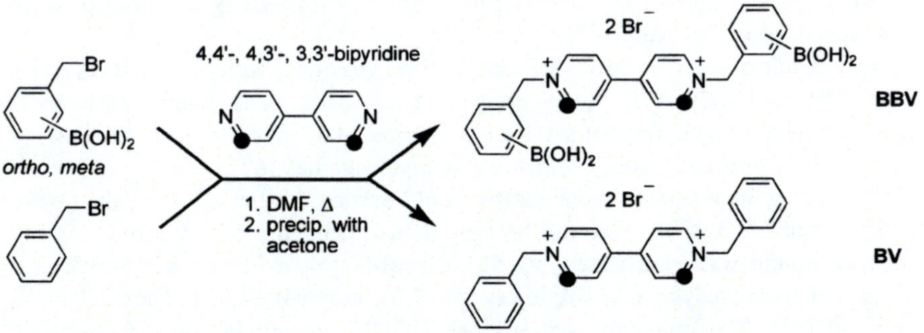

Fig. 4 Synthesis of BBVs and BVs as controls (*dots* indicate variable positions of nitrogens in bipyridyl rings)

was allowed to react in dimethylformamide with the corresponding bipyridines to obtain the desired viologens. Isolation of the products was achieved by precipitation from acetone (Fig. 4).

The quenching characteristics for each of the receptors with HPTS were established by titration of the fluorescent dye with the corresponding viologens. The calculated Stern–Volmer constants [29], K_s and V, indicating the degree of static and dynamic quenching, respectively, are summarized in Table 1. The quenching efficiencies of the boronic acid appended bipyridinium salts were found to be of the following order: 4,4′-BBVs > 3,4′-BBVs > 3,3′-BBVs.

Table 1 Static (K_s) and dynamic (V) quenching constants for BBVs with HPTS (4×10^{-6} M in pH 7.4 phosphate buffer)

Quencher	K_s (10^3 M^{-1})	V (10^3 M^{-1})
4,4′-*o*-BBV	8.9 ± 0.2	2.9 ± 0.1
4,4′-*m*-BBV	8.1 ± 0.1	3.0 ± 0.1
4,4′-BV	15.0 ± 1.0	2.3 ± 0.2
3,4′-*o*-BBV	6.6 ± 0.2	1.7 ± 0.1
3,4′-*m*-BBV	7.5 ± 0.2	1.7 ± 0.1
3,4′-BV	9.5 ± 0.5	2.8 ± 0.3
3,3′-*o*-BBV	4.3 ± 0.2	0.14 ± 0.03
3,3′-*m*-BBV	5.0 ± 0.1	0.70 ± 0.05
3,3′-BV	7.4 ± 0.3	2.2 ± 0.2

A similar trend was shown with the benzyl bipyridinium salts. Variation of the position of nitrogens in bipyridyl rings from 4,4′ to 3,3′ was accompanied with a loss in quenching strength.

Several other instances of viologens serving as quenchers for fluorescent organic dyes have been reported [62–67]; other papers have described the quenching of fluorescent polymers with viologens [68–70]. As mentioned previously, the Singaram group has also reported quenching of fluorophores by viologens and the formation of nonfluorescent ground-state complexes of HPTS with both 4,4′-*o*-BBV [60] and *N*,*N*′-bis(benzyl-2-boronic acid)-4,7-phenanthrolinium dibromide [58].

As a result of their electron-deficient nature, viologens function as acceptors in many charge-transfer (CT) complexes [71–76]. Baptista et al. maintain that HPTS and methyl viologen (paraquat) form a ground-state complex on the basis of UV–Vis, fluorescence, and laser flash photolysis studies [62].

Regarding the quenching mechanism in our system, further UV–Vis absorbance studies confirmed that HPTS and boronic acid-appended bipyridinium salts form distinct ground-state complexes [53, 56]. Corroborating evidence was provided by X-ray structure analysis of a single-crystal of the complex $(4,4'\text{-}m\text{-BBV})_3(\text{HPTS})_2$ (Fig. 5) [77]. The unit cell contains one HPTS^{3-} molecule, 1.5 $4,4'\text{-}m\text{-BBV}^{2+}$ molecules, one methanol, and one water molecule. One viologen cation is entirely contained within the asymmetric unit; one of the boronic acids has undergone esterification with methanol present in the crystallizing medium. The molecules are arranged in a slipped stacking motif that is commonly observed for a donor–acceptor CT complex of this type [78]. X-ray structures of CT complexes involving similar viologen acceptors have been reported [79–82].

The six BBV receptors (3,3′-*o*/*m*-, 4,3′-*o*/*m*-, and 4,4′-*o*/*m*-BBV) were tested against glucose (glc), galactose (gal) and fructose (fru). The apparent binding affinities of each for the three monosaccharides were determined by fluorescence measurements. Binding constants (K_b) were calculated from a 1:1 binding algorithm (1).

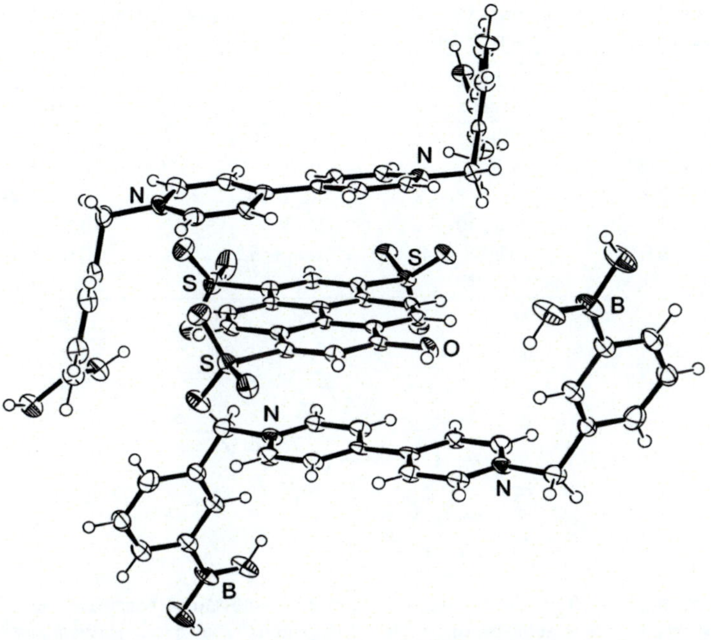

Fig. 5 ORTEP plot [77] of the partial unit cell of (4,4′-*m*-BBV^{2+})$_3$(HPTS^{3-})$_2$. The HPTS^{3-} species lies between these two cations with the sulfonate groups staggered with respect to the basic π-stacking motif. The 4,4′-*m*-BBV^{2+} molecule that is entirely contained within the asymmetric unit has one boronic acid that has undergone an esterification with methanol. Additional solvent molecules (water, methanol) are omitted for clarity [49]

$$\frac{F}{F_0} = \frac{\left(1 + \frac{F_{max}}{F_0}\right) K_b[A]}{1 + K_b[A]} \tag{1}$$

where F is measured fluorescence intensity, F_0 is fluorescence intensity with no addition of analyte, F_{max} is maximum fluorescence intensity, K_b is the apparent binding constant, and $[A]$ is the analyte concentration. The values are listed in Table 2.

The 4,4′-bipyridyl-based receptors (4,4′-*o*- and 4,4′-*m*-BBV) display monosaccharide selectivity similar to that reported for monoboronic acids. This is expected since for both structures, the intramolecular distance between the two boronic acids is too great to allow for cooperative binding with one glucose molecule. The four other receptors have smaller intramolecular B–B distances, and therefore, each theoretically possesses the ability to bind glucose in a 1:1 complex. Indeed, comparison of the binding constants (K_b) shows that 3,3′-*o*-BBV has a binding affinity for glucose several orders of magnitude higher than those of the other BBVs. Most importantly, 3,3′-*o*-BBV displays an enhanced selectivity for glucose over fructose and galactose. For this receptor, the apparent binding constant for

Table 2 Apparent monosaccharide binding constants (K_b) determined for the BBV receptors (5×10^{-4} M) with HPTS (4×10^{-6} M in phosphate buffer)

	K_b (M^{-1})		
	Glucose (glc)	Galactose (gal)	Fructose (fru)
4,4′-*o*-BBV	27 ± 5	37 ± 5	800 ± 100
4,4′-*m*-BBV	14 ± 5	25 ± 5	300 ± 50
3,4′-*o*-BBV	260 ± 50	70 ± 10	800 ± 100
3,4′-*m*-BBV	65 ± 10	35 ± 5	400 ± 50
3,3′-*o*-BBV	1,900 ± 200	180 ± 50	1,100 ± 200
3,3′-*m*-BBV	53 ± 10	45 ± 5	500 ± 50

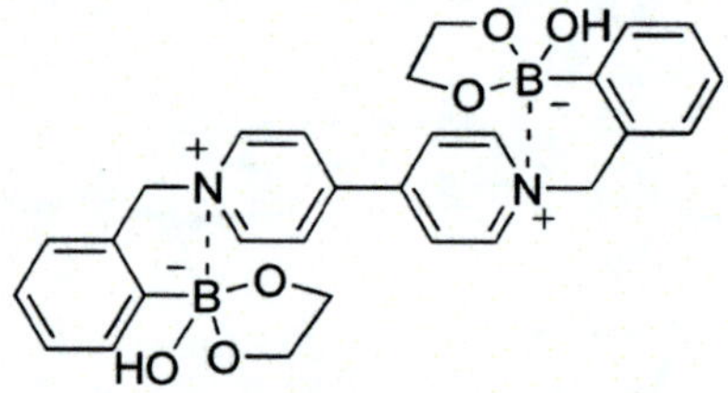

Fig. 6 Intramolecular B → N charge interaction of *ortho*-substituted benzylboronic acid bipyridinium salts bound to diols at physiological pH (ethylene glycol shown as a representative for diols)

glucose is almost twofold higher than that of fructose, and 11-fold higher than that of galactose. Although the apparent binding constant of 3,3′-*o*-BBV was found to be higher for glucose than for fructose, the fluorescence increase obtained upon addition of monosaccharide is greater for fructose. Similar behavior has been observed in other systems [83–85]. The glucose selectivities of 3,3′-*m*-BBV and the 3,4′-bipyridyl-based viologens (3,4′-*o*- and 3,4′-*m*-BBV) are also improved relative to that of the 4,4′-BBVs, though to a lesser extent: the order of selectivity for these viologens is fru >> glc > gal. Further inspection of the binding constants reveals a general trend, in which the *ortho*-substituted bipyridyls have higher affinities than their corresponding *meta*-derivatives, which was attributed to an intramolecular B → N charge interaction unique to *ortho*-substituted benzylboronic acid viologens. This interaction stabilizes boronate ester formation at physiological pH, thus resulting in higher binding constants (Fig. 6) [86, 87].

The fluorescence titration curve of HPTS with 4,4′-*o*-BBV (which cannot cooperatively bind glucose) and 3,3′-*o*-BBV upon glucose addition revealed strong differences. These two receptors share in common the *ortho*-boronic acid positioning but differ in their quenching strengths. The binding curve for 3,3′-*o*-BBV displays an early saturation and a slight sigmoidal shape, whereas 4,4′-*o*-BBV gives a nearly linear response within the same glucose concentration range (see original publication) [49].

The affinities of the BBVs for α-hydroxy carboxylic acids (at pH 7.4, α-hydroxycarboxylates), which are also known to bind to boronic acids [88–90], were also determined. The BBVs were tested for their ability to bind lactate, malate, and tartrate. Erythritol was included to facilitate comparison with tartrate. Tartaric

Table 3 Apparent analyte binding constants (K_b) determined for the BBVs (5×10^{-4} M) with HPTS (4×10^{-6} M in phosphate buffer)

	K_b (M^{-1})			
	Tartrate	Malate	Lactate	Erythritol
4,4′-*o*-BBV	160 ± 50	60 ± 10	6 ± 2	13 ± 5
4,4′-*m*-BBV	47 ± 10	30 ± 5	5 ± 2	16 ± 5
3,4′-*o*-BBV	230 ± 50	100 ± 20	15 ± 5	20 ± 5
3,4′-*m*-BBV	64 ± 10	15 ± 5	4 ± 2	20 ± 5
3,3′-*o*-BBV	1,250 ± 100	160 ± 50	26 ± 5	38 ± 10
3,3′-*m*-BBV	250 ± 50	60 ± 10	18 ± 5	34 ± 5

acid is an optically active component in wine and other grape-based drinks. Boronic acid-based sensors have previously been used to detect and quantify tartrate for possible applications in the beverage industry [26, 91–94]. Malic acid is structurally similar to tartaric acid and is also present in wine. Thus, the ability to distinguish between these two analytes is both useful and challenging. Lactate is present in the body, and levels increase during exercise or when the body is under stress. Therefore, being able to monitor lactate concentration is also important.

The calculated apparent binding constants for the α-hydroxycarboxylates are shown in Table 3. Both malate and tartrate elicited large fluorescence increases and moderate to high apparent binding constants with all the BBVs, whereas erythritol and lactate consistently produced weak fluorescence modulations and low binding constants. The results suggest that 3,3′-*o*-BBV may also cooperatively bind tartrate. All receptors displayed very low affinities for lactate – both in terms of binding constants and fluorescence signal enhancements. This is advantageous because the concentration of lactate in the blood can sometimes surpass that of glucose and could, therefore, interfere with glucose detection if the sensor were indiscriminate toward these two analytes.

The flexibility of our two-component system also allows for manipulation of the quencher-to-dye ratio (Q:D) to optimize glucose sensitivity in the physiological range. When BBV quenchers are used in a probe with HPTS, optimum performance is observed at Q:D = 125:1. The reason why a high Q:D ratio is needed has not been fully elucidated; that may be the result of weak interaction between the dye and quencher. Also, the quenching process may involve the cooperative action of multiple viologens.

The foregoing study highlights the need for receptors that can form probes with greater sensitivity. Bis-viologens have more positive reduction potentials than *mono*-viologens, indicating that they are better electron acceptors [95]. Since quenching efficacy seems to be dependent on electron affinity, bis-viologens are expected to be better quenchers than *mono*-viologens. In addition, the linking of two viologen moieties into a single quencher molecule provides four positive charges, as opposed to only two on the BBV quenchers. This should enhance Q:D interaction and the formation of a ground state complex. Based on these assumptions, a set of glucose-responsive bis-viologen quenchers was synthesized to determine the effect of this structure modification on quenching and glucose modulation.

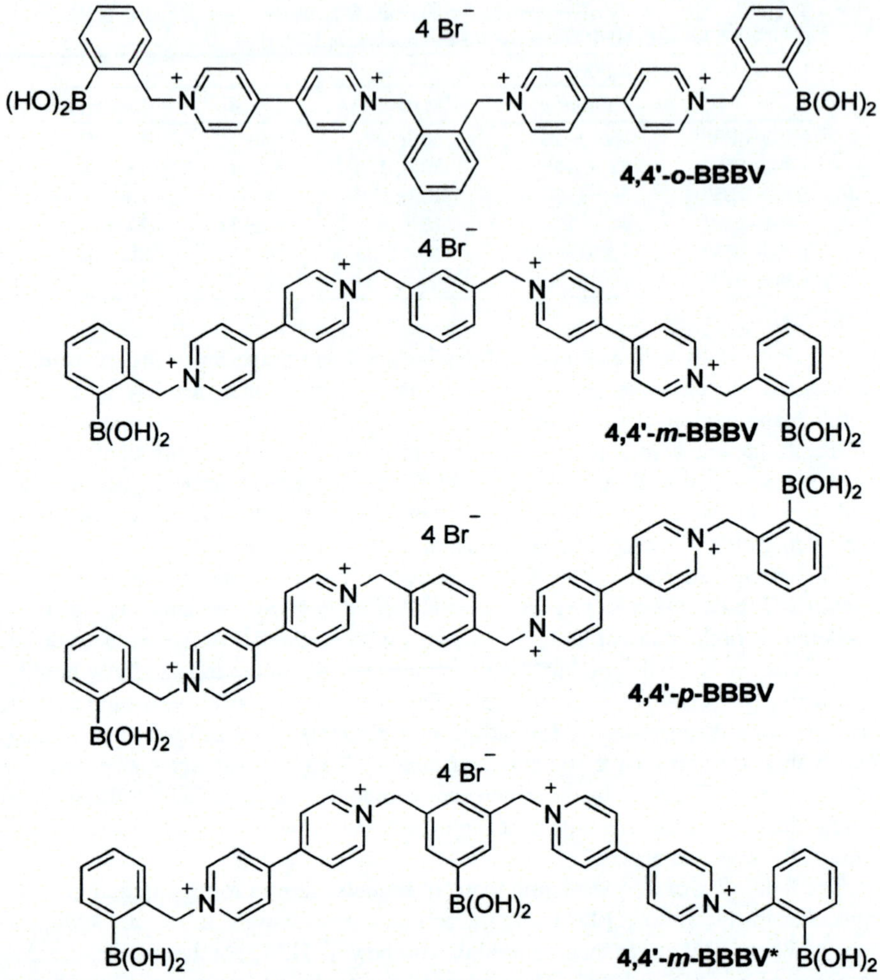

Fig. 7 Structures of boronic acid substituted bis-viologen quenchers: 4,4′-*o*-BBBV, 4,4′-*m*-BBBV, 4,4′-*p*-BBBV, and 4,4′-*m*-BBBV* with an additional boronic acid moiety at the xylylene core

Four bis-viologens (Fig. 7) were synthesized by coupling two 4-(*N*-benzylboronic acid pyridinium) pyridine moieties with the appropriate bis-bromomethylbenzene. The *mono*-viologen quencher 4,4′-*o*-BBV was used as a control compound [46].

The synthetic scheme for the BBBV compounds (Fig. 8) makes use of our standard preparative procedures for BBV compounds. First, 2-bromomethylphenylboronic acid was reacted with an excess of 4,4′-dipyridyl in acetone to afford the *mono*-substituted 4,4′-dipyridyl adduct. Combining an excess of this adduct with the appropriate bis-bromomethylbenzene in DMF yields the desired product.

Stern–Volmer plots were generated by titrating a HPTS solution with the new quenchers (Fig. 9).

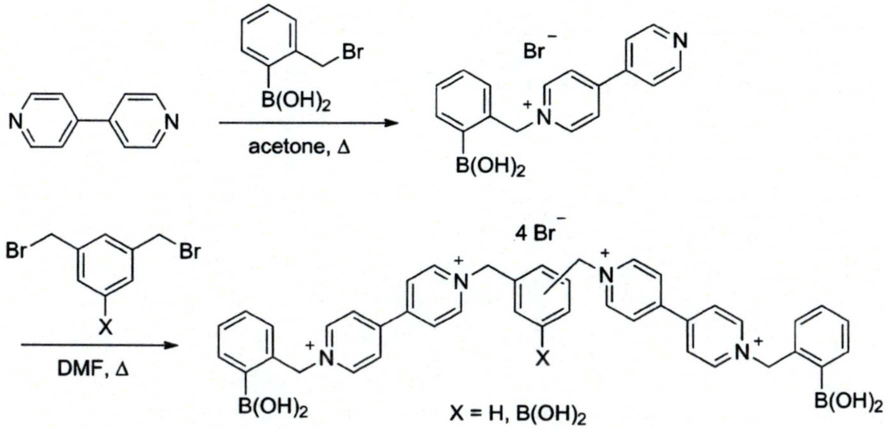

Fig. 8 Synthesis of BBBV receptors

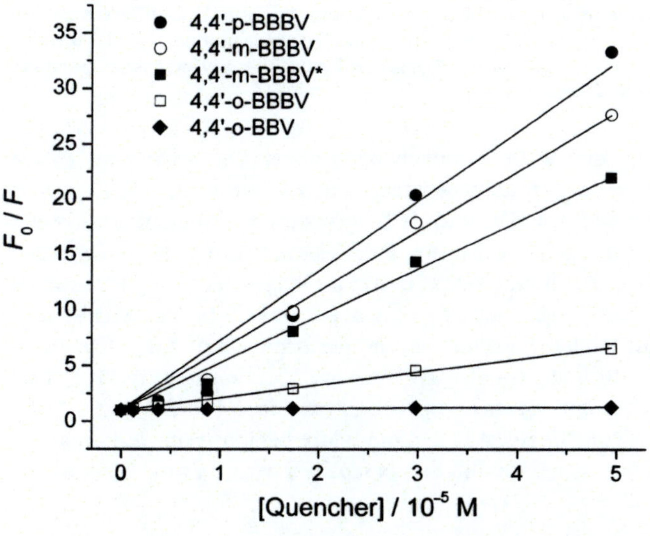

Fig. 9 Stern–Volmer plot for the quenching of HPTS (4 × 10^{-6} M) with the different quenchers. Adapted and reprinted with permission from [46]. Copyright 2008 Elsevier Ltd

As predicted, the BBBV compounds are clearly superior quenchers compared to 4,4′-*o*-BBV Among the three BBBV positional isomers, 4,4′-*o*-BBBV is the weakest quencher while, 4,4′-*p*-BBBV is the best. Assuming pimerization reduces the ability of the bis-viologen to accept electrons, the trend in quenching efficacy is understandable. Pimerization has been reported for nonfunctionalized bis-viologens linked by an *o*-xylylene group [96–98].

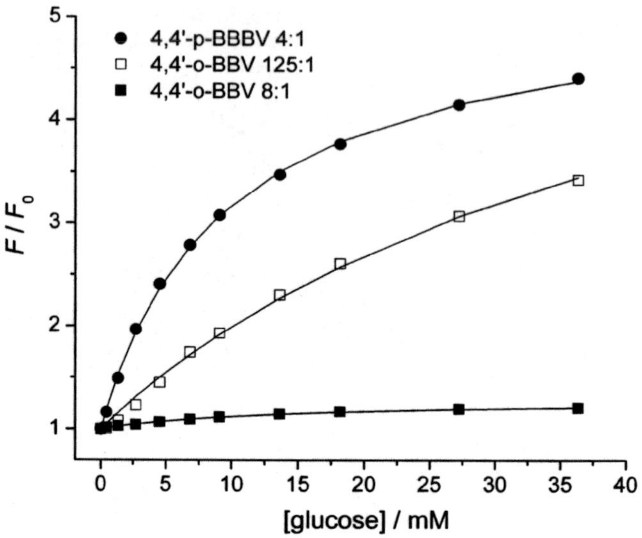

Fig. 10 Glucose sensing with 4,4′-*p*-BBBV and 4,4′-*o*-BBV at different quencher–dye ratios. The data were fit with a 1:1 binding algorithm. The concentration of HPTS was 4×10^{-6} M. The Q:D ratios used are seen in the legend. Adapted and reprinted with permission from [46]. Copyright 2008 Elsevier Ltd

The ability to adjust the quencher-to-dye ratio is a unique advantage of the two-component sensing system. As mentioned previously, a Q:D ratio of 125:1 for probes composed of BBVs and HPTS gives the optimum glucose modulation in the physiological range. Because the BBBVs are much stronger quenchers than their BBV counterparts, high Q:D ratios are no longer necessary for efficient quenching of HPTS. This should lead to enhanced sensitivity for AIDAs made with bis-viologen quenchers. To illustrate, the performance of 4,4′-*p*-BBBV at Q:D = 4:1 is compared to that of 4,4′-*o*-BBV at two Q:D ratios (Fig. 10). The appropriate ratios for making a direct comparison between 4,4′-*o*-BBV and 4,4′-*p*-BBBV are 8:1 and 4:1, respectively. As the plot vividly demonstrates, a dramatic improvement in fluorescence signal modulation is realized by use of the bis-viologen quencher.

3 Solution-Phase Sensor Arrays with Boronic Acid-Appended Bipyridinium Salts

Having demonstrated the utility of our two-component sensing ensemble for measuring glucose and other monosaccharides, we began to explore other applications for this technology. The first attempt was to employ the soluble systems in a fluorescent sensor array. Our thought was that such arrays could be used to detect and discriminate among bioanalytes that are capable of reacting with boronic acid receptors. The detection of biomolecules, including proteins and carbohydrates, is of paramount interest in medical diagnostics. Better methods are needed to expand the range of analytes and improve sensitivity and selectivity. Our sensing concept provides a new way to address these needs.

3.1 Recognition of Neutral Saccharides

There has been an ongoing interest by many researchers in the detection of small biomolecules using substrate-selective receptors [7, 29, 99–104]. The alternative to substrate-selective receptors in nature is the use of differential receptors, as in the mammalian tongue and nose, for sensing a wide range of tastes and smells [105]. This technique works by combining a large number of nonspecific receptors to create very specific recognition patterns. Fingerprints, generated from multiple differential binding interactions, provide unique diagnostic patterns for the identification of individual analytes or even mixtures thereof. Copying nature's cross-responsive strategy [106], chemists have recently reported impressive examples of differential sensor arrays [107, 108] for metal cations [109, 110], volatile organic compounds [111] and bioanalytes, such as amino acids [35, 39, 112], peptides [36, 40], proteins [113–116], phosphates [117], nucleotides [37], steroids [118] phenethylamines [119], terpenes [120], and carbohydrates [38, 48, 121, 122] using colorimetric and fluorescent indicators. Potential applications of sensor arrays in a multianalyte environment have also been described [123].

The modular sensing ensemble of an IDA with its variable indicator/receptor/analyte ratio [124] is ideally suited for implementation in solution-phase sensor arrays. Distinct fingerprints of analytes can easily be created by using receptors with slightly differing affinities for analytes and indicators. Further, the utilization of one single receptor and multiple indicators in dynamic combinatorial libraries can be advantageous [40, 125]. Strong arguments have been put forth for large arrays in the case of volatile analytes [106]. However, many analytical chemists are interested in small solution-phase sensor arrays for water-soluble analytes. Such arrays would be easier and faster to analyze.

In this section, we describe the use of designed receptor units in an AIDA-based fluorescent sensor array to differentiate neutral mono- and disaccharides in aqueous solution at neutral pH. With this array, we were able to discriminate among 12 saccharides at 2 mM concentration [48]. The array comprises the anionic fluorescent dye HPTS combined individually with six cationic bis-boronic acid-appended benzyl-viologens (BBVs, Fig. 3). A concentration ratio of 1:125:500 for the ensemble HPTS/BBV/saccharide was chosen to obtain the competitive conditions, which are necessary for an AIDA to work. Twelve saccharides were selected for the discrimination experiment: D-ribose, D-arabinose, L-rhamnose, D-xylose, D-lyxose, D-glucose, D-mannose, D-galactose, D-fructose, L-sorbose, melibiose, and lactulose. The selected saccharides all possess free hydroxy groups in the 1- and 2-position, allowing them to bind to the boronic acid [24, 126, 127]. The increase in

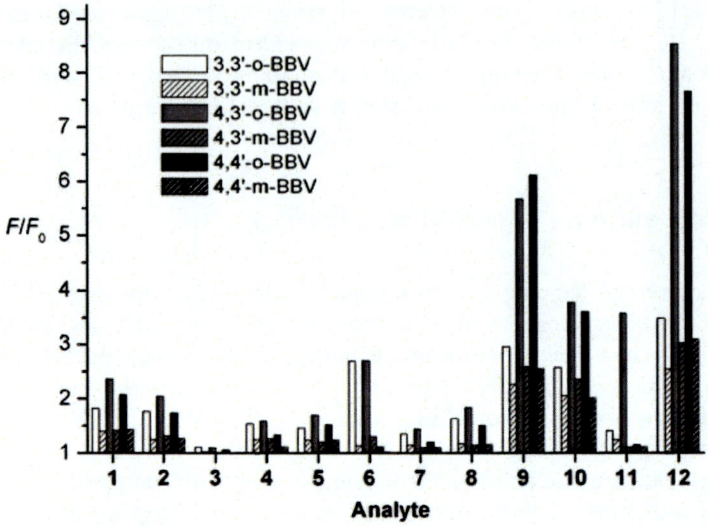

Fig. 11 Relative fluorescence increase (F/F_0) of HPTS (4.0×10^{-6} M) with all six BBV-receptors (3,3′-*o*-, 3,3′-*m*-, 4,3′-*o*-, 4,3′-*m*-, 4,4′-*o*-, and 4,4′-*m*-BBV; 5.0×10^{-4} M) after adding saccharides [(1) D-ribose, (2) D-arabinose, (3) L-rhamnose, (4) D-xylose, (5) D-lyxose, (6) D-glucose, (7) D-mannose, (8) D-galactose, (9) D-fructose, (10) L-sorbose, (11) melibiose, (12) lactulose] at a final concentration of 2.0×10^{-3} M (phosphate buffer, 39 mM). Errors, given in a 95.5% confidence interval, are lower than 2.3%

fluorescence intensity F/F_0 induced by adding the saccharide to the BBV/HPTS sensing ensemble was measured with a fluorescence plate reader.

Comparing the fingerprints of all saccharides (Fig. 11), the ketoses, D-fructose and lactulose, showed the highest response for all six BBV-receptors. All *ortho*-BBV receptors showed greater F/F_0 values than *meta*-BBVs. As illustrated earlier in Fig. 6, the superior performance of *ortho*-BBVs has been attributed to the intramolecular electrostatic interaction between the anionic boronate ester and the positively charged nitrogen of the viologen.

Multivariate analysis of the F/F_0 data was performed with the statistics program SYSTAT. We chose linear discriminant analysis (LDA), one of the most widely used classification procedures [128]. LDA is a supervised method for separating classes of objects and for assigning new components to appropriate classes. The discriminants are linear combinations of the measured sensor responses. Discriminant functions are calculated to maximize the Mahalanobis distance between classes relative to the variation within classes. It is important to note that LDA may give slightly optimistic results. Therefore, only "jackknifed" classification matrices were taken to analyze discrimination results. In "jackknifed" classification matrices, one measurement at a time was taken as an unknown; the rest of the data was used as the training set. 100% accurate discrimination was achieved for all 12 saccharides with all 6 BBV-receptors. The F value of each receptor in the LDA, which indicates the relative importance, showed that the contribution of the

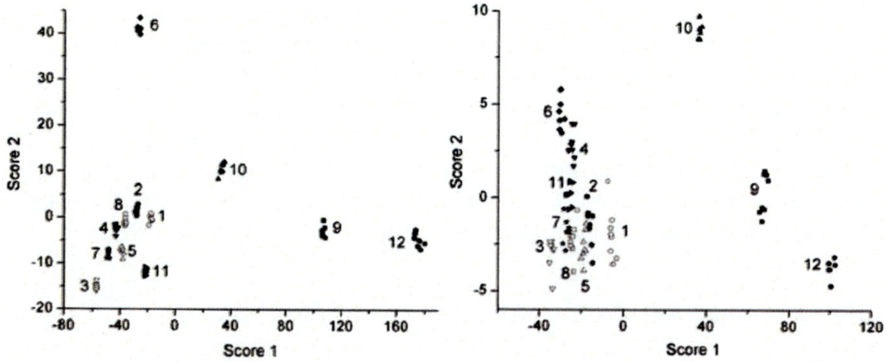

Fig. 12 Two-dimensional LDA score plots for the identification of 12 saccharides using three receptors with differing discriminating power: 3,3′-*o*-, 4,3′-*o*-, and 4,4′-*o*-BBV (left) and 3,3′-*m*-, 4,3′-*m*-, and 4,4′-*m*-BBV (right). Saccharides (2.0×10^{-3} M): (1) D-ribose, (2) D-arabinose, (3) L-rhamnose, (4) D-xylose, (5) D-lyxose, (6) D-glucose, (7) D-mannose, (8) D-galactose, (9) D-fructose, (10) L-sorbose, (11) melibiose, (12) lactulose. Adapted and reprinted with permission from [48]. Copyright 2007 Wiley-VCH Verlag GmbH & Co KGaA

meta-BBVs was minor compared with the corresponding *ortho* derivatives. This suggested that only three *ortho* receptors are needed to fully discriminate the chosen group of saccharides. This was confirmed by a 100% accurate discrimination when using only 3,3′-*o*-, 4,3′-*o*-, and 4,4′-*o*-BBV in an LDA. The results of two combinations (3,3′-*o*-, 4,3′-*o*-, 4,4′-*o*-BBV and 3,3′-*m*-, 4,3′-*m*-, 4,4′-*m*-BBV) are depicted in graphical score plots (Fig. 12). The data of all 12 saccharides appear visually distinguishable for the 3,3′-*o*-, 4,3′-*o*-, 4,4′-*o*-BBV combination (Fig. 12, left). Chemical and structural similarities of the saccharides were reflected by the relative position of their clustered data. The scores of the aldoses D-ribose, D-arabinose, D-xylose, D-lyxose, D-mannose, and D-galactose, are positioned in proximity to each other. By contrast, the data of D-glucose and the ketoses D-fructose, L-sorbose, and lactulose are well separated from the others, as ketoses generally exhibit higher binding affinities to boronic acids than aldoses. The extraordinary behavior of D-glucose can be explained by the exceptional contribution of 3,3′-*o*-BBV and 4,3′-*o*-BBV in the LDA (see Fig. 11). To verify that the discrimination between the analytes did not arise from small differences in the concentrations of the saccharide stock solutions, the concentration was reduced by 5%. Thus, F/F_0 data of 3,3′-*o*-, 4,3′-*o*-, and 4,4′-*o*-BBV with combined saccharide concentrations of 1.9 and 2.0 mM were used to perform an LDA; there was still 100% accurate discrimination. By contrast, the combination of three *meta*-BBVs leads to substantial cluster overlap of the analytes D-arabinose, D-lyxose, D-mannose, D-galactose, and melibiose (Fig. 12, right).

To evaluate the contribution of each *ortho*- and *meta*-BBV-receptor to the discriminatory power in the multivariate analysis, we combined all six BBV-receptors into three member groups. The resulting 20 distinct combinations $\left(\binom{6}{3} = \frac{6!}{3!(6-3)!} = 20\right)$ were again used in an LDA for the discrimination of the given set of saccharides. Ten of the 20 combinations containing at least two *ortho*-BBV-receptors were able to achieve 100% accurate discrimination. By contrast, not all of the saccharides were correctly differentiated by the ten other combinations, which contain at least two *meta*-BBV-receptors. The conclusion is that *ortho*-BBV-receptors have greater discriminatory power than *meta*-BBV-receptors. From the standpoint of a sensor array design, the use of *ortho*-BBV-receptors is thus the most desirable. In the next section, we demonstrate the discriminatory power of the three *ortho*-BBV-receptors in the recognition of phospho sugars and nucleotides.

3.2 Recognition of Phospho Sugars and Nucleotides

Anionic biomolecules are very important components in biological systems [129]. The majority of enzyme substrates and cofactors, such as peptides, nucleotides and nucleic acids, phospholipids, and carbohydrates, are anionic. Therefore, recognition of small anionic biomolecules using biosensors and artificial receptors is also of high interest in medical diagnostics [100]. Anion recognition in water is difficult [130]. Highly specific artificial anion receptors have been designed by mimicking nature's analyte-selective receptors, typically by using more than one recognition moiety [131, 132]. Very few studies of phospho sugar and nucleotide sensing with boronic acid-appended receptors have been reported [133–138]. Especially germane to the present review, selective nucleotide recognition with fluorescent IDAs of cyclic bipyridinium receptors and 8-hydroxypyrene-1,3,6-trisulfonic acid trisodium salt (HPTS) has recently been reported [139, 140].

This section covers our investigation of discrimination among anionic carbohydrates in aqueous solution at physiological pH using a BBV sensor array. The analytes tested included phospho sugars and nucleotides, as well as three neutral saccharides for comparison at five different concentrations in the low millimolar range. Several multivariate analysis methods were used to determine the "discrimination limit." The binding affinities of all analytes with the three BBV receptors were determined; the calculated apparent binding constants and maximum fluorescence values were also used to distinguish phospho sugars and nucleotides with the help of multivariate analyses [45].

Biologically important phospho sugars and nucleotides that can be found as intermediates, cofactors and direct allosteric controls in the pathway of glycolysis were selected for the discrimination study (Fig. 13) [141]. They include glucose-1-phosphate, glucose-6-phosphate, fructose-6-phosphate, fructose-1,6-diphosphate, and the nucleotides adenosine monophosphate (AMP), adenosine diphosphate (ADP), and adenosine triphosphate (ATP). Ribose-5-phosphate, guanosine triphosphate (GTP), and the neutral saccharides D-glucose, D-fructose, and D-ribose were included for comparison. All chosen analytes, except glucose-1-phosphate, possess at least one *cis*-diol moiety, which allows them to bind covalently to boronic acid receptors.

The modular carbohydrate sensing ensemble again comprises the anionic fluorescent dye HPTS and the array of three cationic *ortho*-BBV receptors (3,3′-*o*-, 4,3′-*o*-, and 4,4′-*o*-BBV). They were chosen for the sensor array because, as stated in

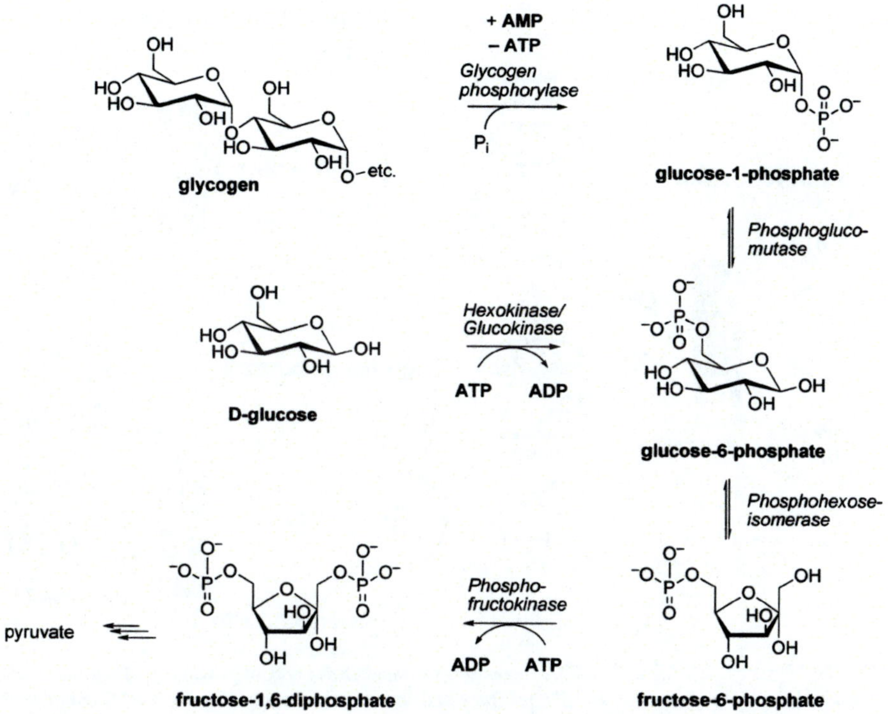

Fig. 13 Analytes used for the discrimination study: intermediates, cofactors, and direct allosteric controls (+ positive, – negative) in the pathway of glycolysis. Enzyme names are given in italics

Sect. 3.1, *ortho*-BBV-receptors have greater discriminatory power than *meta*-BBV-receptors. Diol-containing saccharides modulate fluorescence quenching in our system by affecting the boronic acid/boronate equilibrium on binding to the BBV receptors, but anionic saccharides are expected to modulate the signal by an additional electrostatic interaction mechanism. Elevated ionic strength after addition of phospho sugars or nucleotides weakens further the affinity of the electrostatic-based ground-state complex of the BBV with HPTS. As a consequence of this synergistic effect, it is expected that the anionic dye will be displaced to a greater extent, resulting in a stronger fluorescence signal (Fig. 14). Three benzyl viologens (3,3′-, 4,3′-, and 4,4′-BV) were used as controls.

The static quenching constants calculated from Stern–Volmer plots for the BBV receptors and BV quenchers are similar (see Table 1). Thus, the BV compounds can serve as non-boronic-acid-containing control quenchers to investigate the synergistic effect of the anionic substrates. The competitive AIDA conditions were obtained by choosing a ratio of 1:125:125–2,500 for the sensing ensemble HPTS/BBV/analytes. The analyte concentration was varied from 5.0×10^{-4} to 1.0×10^{-2} M to measure binding affinities and to create concentration-dependent recognition patterns in the physiological concentration range [141].

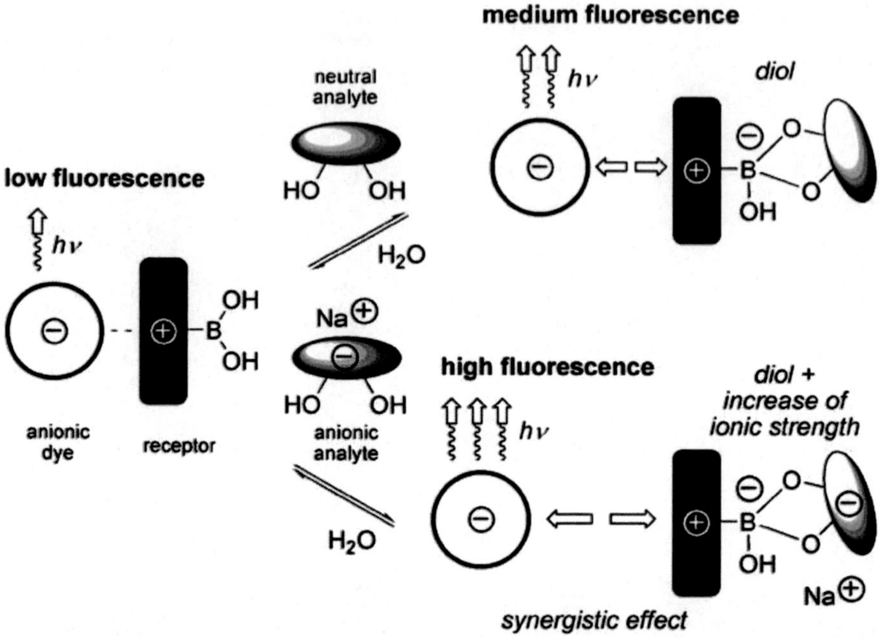

Fig. 14 Synergistic effects in AIDA: proposed mechanism for sensing neutral and anionic diol-containing analytes based on a BBV receptor and the anionic fluorescent dye HPTS. Reprinted with permission from [45]. Copyright 2008 Elsevier Ltd

To investigate the ionic strength effect of the anionic analytes on the fluorescence signals with the BBV receptors, we compared the relative fluorescence increase (F/F_0) of HPTS with the BBVs (3,3′-*o*-, 4,3′-*o*-, 4,4′-*o*-BBV) and the corresponding BVs (3,3′-, 4,3′-, and 4,4′-BV) at an analyte concentration of 10 mM (Fig. 15). Neutral saccharides, such as glucose, fructose, and ribose, exhibited no ionic strength effect. By contrast, all anionic phospho sugars showed weak ability to displace HPTS from the ground-state complex with the BVs. A stronger displacement of HPTS was observed for the nucleotides ADP, ATP, and GTP. These findings indicated a weak electrostatic contribution of the anionic analytes to the displacement assay and tentatively explain the stronger fluorescence intensities compared with that of the corresponding neutral saccharides by a synergistic effect (for comparison see Sect. 3.1).

Relative fluorescence intensities (F/F_0) of the three *ortho*-BBV receptors were used to perform concentration-dependent multivariate analyses. To evaluate at which analyte concentration the discrimination reaches its "discrimination limit" (DL is lowest analyte concentration where the discrimination is still 100%), F/F_0 data of the sensor array at five different analyte concentrations (0.5, 1, 2, 5, and 10 mM) was included. By contrast, the "discriminatory limit" for sensor arrays has been determined by the number of sensors and by the dynamic range of the detector [142]. The most important classification procedures were chosen, including linear discriminant analysis (LDA), principal component analysis (PCA), and hierarchical

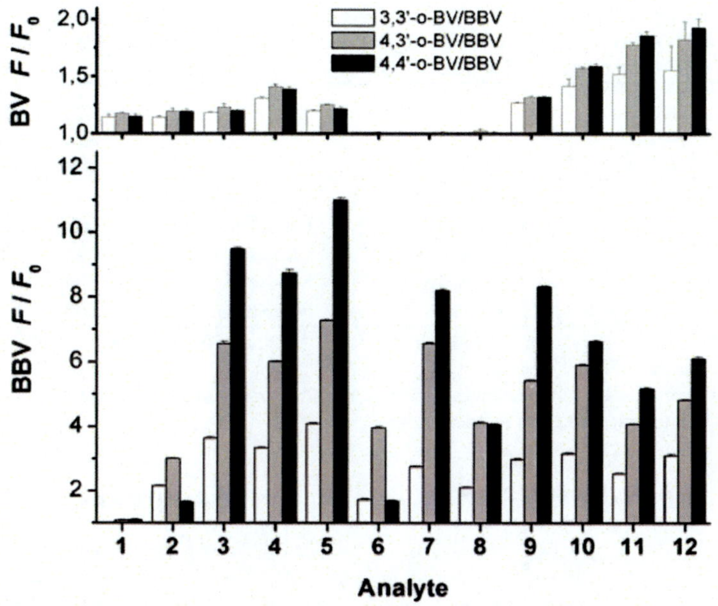

Fig. 15 Relative fluorescence increase (F/F_0) of HPTS (4.0×10^{-6} M) with viologen quenchers (3,3′- (*white bars*), 4,3′- (*gray bars*), and 4,4′-viologen (*black bars*); 5.0×10^{-4} M) after adding analytes. Bottom: *ortho*-BBV receptors. Top: BV quenchers. Analytes at a final concentration of 1.0×10^{-2} M (phosphate buffer, pH 7.4, 39 mM): (1) glucose-1-phosphate, (2) glucose-6-phosphate, (3) fructose-6-phosphate, (4) fructose-1,6-diphosphate, (5) ribose-5-phosphate, (6) glucose, (7) fructose, (8) ribose, (9) AMP, (10) ADP, (11) ATP, (12) GTP. Errors are given in a 95.5% confidence interval. Adapted and reprinted with permission from [45]. Copyright 2008 Elsevier Ltd

cluster analysis (HCA), and the calculated results were compared with each other [128]. The results of the analyte-concentration-dependent LDA, PCA, and HCA are given in Table 4.

The discrimination was strongly reduced by changing the analyte concentration from 10 to 0.5 mM. The DL concentration with all three multivariate analysis methods is at 4 mM. At lower analyte concentrations, fructose, its phosphorylated derivatives, and nucleotides could no longer be distinguished with 100% certainty. This was shown by LDA: F/F_0 finger prints were too similar in their shape and the discriminatory power of the BBV sensor array was consequently reduced. The results of the concentration-dependent LDA are visualized by two-dimensional plots. Two score plots of LDAs at concentrations of 2 and 4 mM (DL) are shown (Fig. 16). At analyte concentration of 4 mM (Fig. 16, right), the graphical plot exhibits clearly separated clusters. By contrast, the clusters of ATP and GTP at 2 mM concentration are overlapping (Fig. 16, left). At lower concentrations, the clusters of fructose-6-phosphate, fructose-1,6-diphosphate, and fructose are also no longer resolved (the ellipse in the left figure highlights cluster overlap). Since the supervised LDA models the similarities between the data of the same group of analytes, it can give better results than unsupervised methods. Therefore, two

Table 4 Results of analyte-concentration dependent multivariate analyses based on F/F_0 values from three BBV receptors[a] with phospho sugars, nucleotides and control analytes for the determination of the "discrimination limit" (DL)[b]

			LDA
Analyte concentration (M)	PCA	HCA	Accurate discrimination (%)[c]
5.0×10^{-4}	Cluster overlap	Cluster overlap	83
1.0×10^{-3}	Cluster overlap	Cluster overlap	88
2.0×10^{-3}	Cluster overlap	Cluster overlap	98
4.0×10^{-3} (DL)[b]	No overlap	No overlap	100
1.0×10^{-2}	No overlap	No overlap	100

[a]3,3′-*o*-BBV, 4,3′-*o*-BBV, and 4,4′-*o*-BBV (5.0×10^{-4} M) and HPTS (4.0×10^{-6} M) in aqueous phosphate buffer (pH 7.4, 39 mM)
[b]The "discrimination limit" is the lowest analyte concentration where the discrimination is still 100%
[c]From jackknifed classification matrices

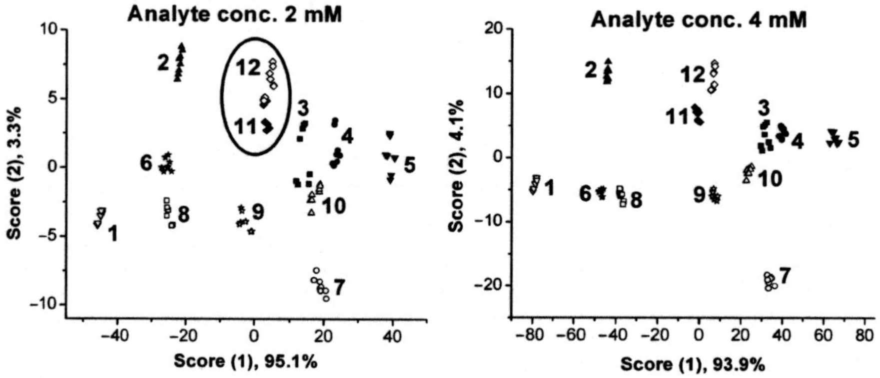

Fig. 16 LDA score plots for the discrimination among phospho sugars and nucleotides at two analyte concentrations (*left*: 2.0×10^{-3} M, *right*: 4.0×10^{-3} M DL) based on F/F_0 values from three *ortho*-BBV receptors. The proportion of the total dispersion is given in %. The ellipse in the left figure highlights cluster overlap. Phospho sugars, nucleotides, and neutral analytes: (1) glucose-1-phosphate, (2) glucose-6-phosphate, (3) fructose-6-phosphate, (4) fructose-1,6-diphosphate, (5) ribose-5-phosphate, (6) glucose, (7) fructose, (8) ribose, (9) AMP, (10) ADP, (11) ATP, (12) GTP. Adapted and reprinted with permission from [45]. Copyright 2008 Elsevier Ltd

further multivariate analyses such as PCA and HCA were performed with the given set of analytes. Concentration-dependent PCA and HCA gave very similar results to LDA with a discrimination limit at 4 mM. At 2 mM, the clusters of ATP, GTP, fructose-6-phosphate, and ADP are no longer separable (see the original publication for the LDA and PCA score plots and HCA dendrograms at other analyte concentrations) [45].

The parameters, namely, apparent binding constants (K_b), and maximum relative fluorescence intensities (F_{max}/F_0), of all analytes to the three *ortho*-BBV receptors have been determined by a 1:1 algorithm [see (1)]. These parameters contain intrinsic concentration information and can be also used to discriminate among

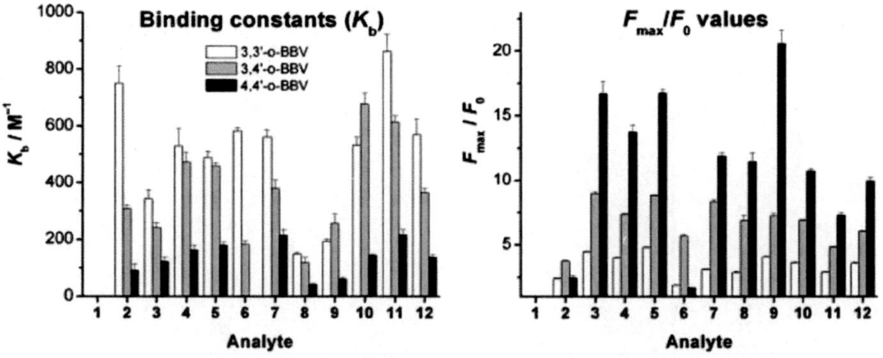

Fig. 17 Apparent binding constants K_b (*left*) and maximum fluorescence values F_{max}/F_0 (*right*) determined for 3,3′-*o*-BBV (*white bars*), 4,3′-*o*-BBV (*gray bars*), and 4,4′-*o*-BBV (*black bars*, all 5.0×10^{-4} M) and HPTS (4.0×10^{-6} M) in phosphate buffer (pH 7.4, 39 mM). Analytes: (1) glucose-1-phosphate, (2) glucose-6-phosphate, (3) fructose-6-phosphate, (4) fructose-1,6-diphosphate, (5) ribose-5-phosphate, (6) glucose, (7) fructose, (8) ribose, (9) AMP, (10) ADP, (11) ATP, (12) GTP. Errors are given in a 95.5% confidence interval. Adapted and reprinted with permission from [45]. Copyright 2008 Elsevier Ltd

the group of analytes. Averaged values of K_b and F_{max}/F_0 are shown as 3D diagnostic patterns in Fig. 17. The trend in the strength of K_b of the BBV receptors with the analytes is as follows: 3,3′-*o*-BBV > 4,3′-*o*-BBV > 4,4′-*o*-BBV (Fig. 17, left). As expected, the strength of F_{max}/F_0 followed the opposite trend: 4,4′-*o*-BBV > 4,3′-*o*-BBV > 3,3′-*o*-BBV. Analytes with high binding constants saturate at lower concentrations causing lower F_{max}/F_0 values.

Usually, optical sensor arrays use direct values of absorbance or fluorescence intensities as an input for multivariate analyses. More complex variables can be created by a concentration dependent analysis to introduce variance. However, a greater scattering of variables can lead to a reduction of the discriminatory power of an array. To evaluate whether the apparent binding constants (K_b) and maximum fluorescence values (F_{max}/F_0) of the BBV receptors can be used as fingerprints for the identification of the given set of analytes, LDA, PCA, and HCA were performed. When using six "sensors" (K_b and F_{max}/F_0 for each of the three BBV receptors), 100% accurate classification was achieved for the 12 analytes with all discrimination methods. The result of the LDA is depicted in a score plot (Fig. 18), in which the data of all analytes appear visually distinguishable. Glucose and glucose-6-phosphate are positioned in close proximity to each other because they exhibit very similar binding affinities to the BBV receptors (Fig. 17). By contrast, glucose-1-phosphate with no binding affinity is well separated from the others. Additional PCA and HCA confirmed the correct classification. Remarkably, when using only binding constants of the three BBV receptors in the multivariate analyses, only 92% accurate discrimination was achieved with the LDA and cluster overlapping was observed with PCA and HCA. We concluded that using only apparent binding constants from binding curves in a multivariate analysis does not provide enough information to differentiate between the chosen set of analytes.

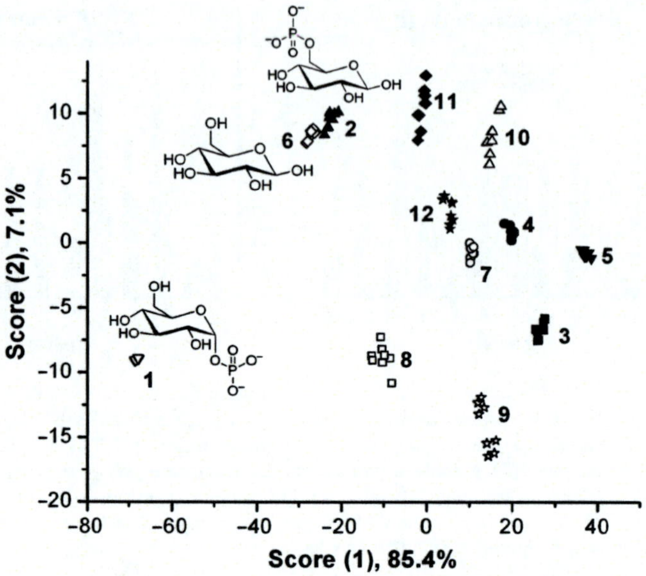

Fig. 18 Two-dimensional LDA score plot for the identification of 12 analytes based on K_b and F_{max}/F_0 values from three BBV receptors. The proportion of the total dispersion is given in %. Analytes (0.5×10^{-3} to 1.0×10^{-2} M): (1) glucose-1-phosphate, (2) glucose-6-phosphate, (3) fructose-6-phosphate, (4) fructose-1,6-diphosphate, (5) ribose-5-phosphate, (6) glucose, (7) fructose, (8) ribose, (9) AMP, (10) ADP, (11) ATP, (12) GTP. The structures of glucose and phospho derivatives are shown. Adapted and reprinted with permission from [45]. Copyright 2008 Elsevier Ltd

4 Enzyme Assays

Carbohydrate transformations are essential to biological function, and these reactions take place in a range of processes, from metabolism to cell surface modification. Enzymes that carry out these reactions include isomerase, mutase, transferase, and phosphorylase classes. Development of therapeutic drugs often arises from activation or inhibition of such enzymes [143, 144]. Further, the transfer of a glycosyl group represents one of the most important reactions in nature and has a wide range of industrial applications [145]. The attachment of a carbohydrate moiety to an acceptor substrate can drastically influence both the physicochemical and biological properties of an organic molecule [146]. Prominent examples include the increased solubility of hydrophobic compounds, the improved pharmacokinetics of drugs, the optimized activity spectrum of antibiotics, and the modulation of flavors and fragrances.

Simple, rapid methods to monitor enzymatic reactions are essential for the discovery of new enzymes for carbohydrate transformation, and agents that regulate such enzymes [147, 148]. The majority of enzyme assays are developed to test isolated enzymes or enzyme containing samples such as culture suspensions in 96-well microtiter plates or similar parallel liquid-phase systems [149].

However, carbohydrate reactions represent a challenge for high-throughput screening in that natural enzymatic substrates and products are not chromophoric [150]. Simple enzyme assays are based on synthetic substrates that release a colored or fluorescent product upon reaction: A number of enzyme substrates have been designed following the natural chromogenic glycosidase substrate indicant [151]. Cleavage of the glycosidic bond forms an unstable hydroxyindole intermediate, which dimerizes oxidatively in air to form indigo as a blue precipitate. Numerous glycosides of fluorescent or colored phenols are used to test glycosidases, e.g., nitrophenyl β-galactoside for detection of β-galactosidase activity [152]. Some carbohydrate enzyme assays are based on FRET (Fluorescence Resonance Energy Transfer) as the detection principle. For example, assays for fucosyl transferases [153] and α-amylase [154] have been presented. Glycosides labeled with BODIPY dyes were used to screen sialyl transferases in living cells, relying on the fact that this substrate but not its sialylated product is cell-permeable [147]. Enzyme reactions may also be assayed using indicators that respond indirectly to product formation or substrate consumption. The indicator may be as simple as a pH-indicator or as complex as a functionalized nanoparticle. Hydrolytic enzymes such as glycosidases have been used as secondary enzymes to follow the production of chromogenic substrates from nonreactive precursors through a primary enzyme such as a glycosidase, glycosyl transferase, and glycosynthases [155–158]. A pH-indicator assay was reported for screening glycosyl transferases based on the acidification induced by glycosyl transfer from UDP-GalNAc [159] and from glycosyl fluorides [160]. A label-free, real-time glycosyltransferase assay based on a fluorescent artificial chemosensor has recently been presented [161].

We anticipated that the AIDA fluorescent system would be capable of continuously monitoring enzymatic reactions, and we chose the enzymes sucrose phosphorylase (SPO) and phosphoglucomutase (PGM) as proof-of-concept systems. An isomerase catalyzing the conversion of glucose-1-phosphate to glucose-6-phosphate, PGM is ubiquitous in all organisms, and is important to human health because of its role in glycogenolysis and glycogenesis. The enzyme is also a target of lithium salts used to treat bipolar disorder in humans [162, 163] and affects the virulence of the pathogen *Pseudomonas aeruginosa* [164, 165]. One method for monitoring PGM activity is mass spectrometry [166]. Both an amperometric biosensor [167] and a fluorescent biosensor [168] have been developed for this purpose. The transferase SPO, found in bacteria, converts sucrose to glucose-1-phosphate and fructose. It has been also shown to glycosylate a number of noncarbohydrate acceptors, including aliphatic alcohols [169], glyceric acid, acetic acid [170], ascorbic acid [171], benzoic acid [172], and caffeic acid [173]. In contrast to information-rich techniques such as mass spectrometry, a simple optical-readout assay is the most common assay method for determining activity of both PGM [174, 175] and SPO [176]. One of the most straightforward methods to monitor an enzyme-catalyzed reaction in real-time consists in further converting the reaction product by a second enzyme to form a second product and so on until one of these follow-up reactions produces a detectable signal. The vast majority of enzyme-coupled assays involve an oxidoreductase at the end to produce reduction equivalents, such as NADH or NADPH. For example, SPO and PGM are assayed by the enzyme cascade sucrose phosphorylase, phosphoglucomutase, and glucose-6-phosphate dehydrogenase to produce NADPH, which is detected by UV/Vis spectroscopy at $\lambda = 340$ nm [176]. While this method is widely used, it has the intrinsic disadvantage of relying on multiple cascading enzyme reactions [177] and relies on absorbance in the UV region where other biomolecules can interfere.

In this section, we describe the use of our modular carbohydrate sensing system in selective enzymatic assays for SPO and PGM, which allowed the successful determination of the enzyme kinetics in 24-well plates [43]. The modular sensing ensemble comprises HPTS and an analyte-responsive quencher, either 4,4′-*o*-BBV or 3,3′-*o*-BBV. The key issue in label-free continuous enzyme assays for SPO and PGM is the selectivity of the BBVs. Using the proper BBV receptor, selective binding of the enzymatic product alters the charge state of the viologen, and hence the fluorescent signal of the probe (Fig. 19). The viologen, 4,4′-*o*-BBV, is used to assay SPO and 3,3′-*o*-BBV is used to assay the enzymatic reaction of PGM.

The binding characteristics of the BBV/HPTS combination with fructose and glucose-6-phosphate are depicted in Table 2 and Fig. 17. Binding constants are a quantitative measure of receptor affinity but do not always predict fluorescent signal modulation. It is the signal modulation (F/F_0) that determines sensitivity of an assay. The products of both enzymatic reactions give a strong signal modulation

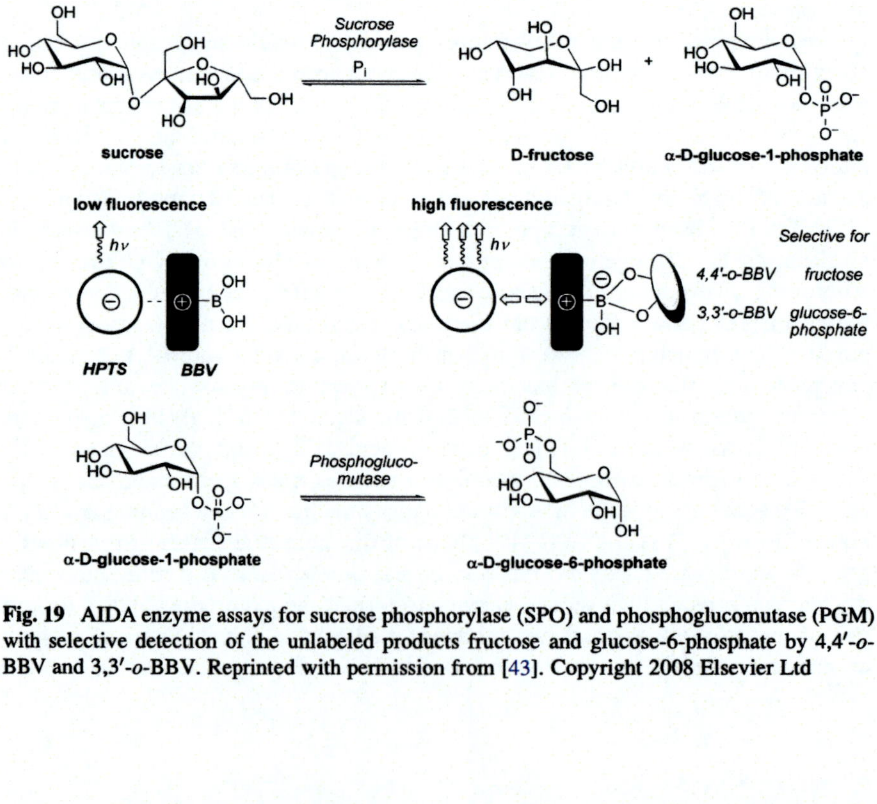

Fig. 19 AIDA enzyme assays for sucrose phosphorylase (SPO) and phosphoglucomutase (PGM) with selective detection of the unlabeled products fructose and glucose-6-phosphate by 4,4′-*o*-BBV and 3,3′-*o*-BBV. Reprinted with permission from [43]. Copyright 2008 Elsevier Ltd

even in the presence of their respective enzymatic substrates, showing that the receptors used in each assay are product-selective. The sensitivity for the two products, as measured by the slope of the calibration curve, was higher for fructose ($462 \pm 22\ M^{-1}$ using 4,4′-*o*-BBV) than for glucose-6-phosphate ($192 \pm 6\ M^{-1}$ using 3,3′-*o*-BBV) The working range for the fructose assay was 7.0×10^{-5} to 1.0×10^{-3} M, and for the glucose-6-phosphate assay was 1.9×10^{-4} to 2.0×10^{-3} M. The substrates sucrose and glucose-1-phosphate produced no observable signal modulation, even up to 10 mM concentration. In general, only reducing sugars show appreciable binding affinity with boronic acids [178].

After profiling the response of the AIDA ensemble to the two reaction products (Fig. 19), the enzymatic reactions were carried out in cuvettes in the presence of the appropriate probes and monitored by fluorimeter. A continuous fluorescence increase over 5 min was observed upon addition of the enzyme, and the modulation was roughly twice as intense for the SPO reaction. This is consistent with the higher sensitivity for fructose shown in the calibration curves for enzymatic products. The same assays were then carried out in multiwell plates and read by fluorescence plate reader, which allowed us to monitor several reactions simultaneously.

The enzymatic reactions were initiated with addition of the enzyme to a well containing the substrate, fluorescent probe ensemble, and any necessary cofactors, and the fluorescence emission was read at 30 s intervals for a period of 10 min. Both BBV receptors gave a response in the SPO assay (Fig. 20, left), with 4,4′-*o*-BBV producing a linear response whereas the signal with 3,3′-*o*-BBV saturated after several minutes. The higher sensitivity with 4,4′-*o*-BBV is explained by the greater quenching achieved with that receptor. The lower the initial (quenched) signal, the greater the range over which the signal can be increased by saccharide binding and resultant quencher dissociation. For the PGM assay, only 3,3′-o-BBV gave a signal

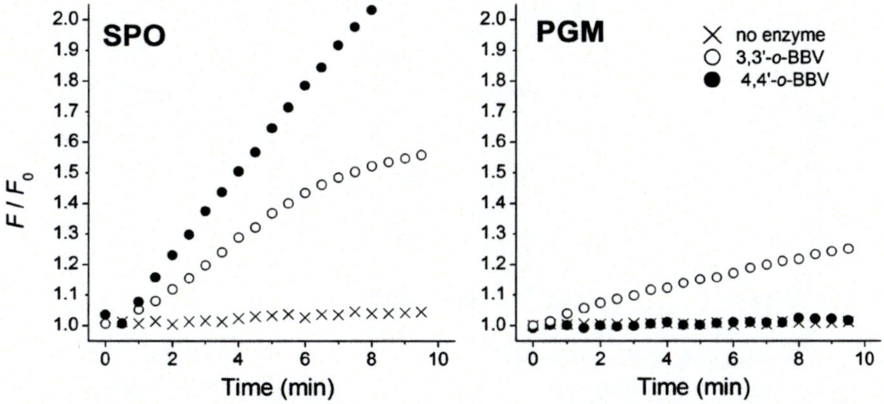

Fig. 20 Real-time fluorescence assay for SPO (*left*) and PGM (*right*) showing the different selectivity of the BBV/HPTS probes. The enzymatic reactions are both followed with 3,3′-*o*-BBV (*open circle*) and 4,4′-*o*-BBV (*filled circle*). Background with no enzyme (×) using either 4,4′-*o*-BBV (SPO, *left*) and 3,3′-o-BBV (PGM, *right*) as the receptor. Adapted and reprinted with permission from [43]. Copyright 2008 Elsevier Ltd

increase over the course of the reaction (Fig. 20, right), showing the selectivity of the probe. Enzymatic activity was determined as initial velocity (V_i) using the calibration curves to convert fluorescence increase per unit time into product evolution per unit time. The linear fluorescence increase for the first several minutes of the reactions corresponded to product concentrations falling in the linear portion of the calibration curves for fructose and glucose-6-phosphate, respectively. Good agreement with the values given by the supplier confirmed that the probes were accurately measuring the enzymatic rate. For SPO, we measured an initial velocity of 0.25 mM min^{-1} (manufacturer: 0.15 mM min^{-1}), and for PGM we measured 0.14 mM min^{-1} (manufacturer: 0.13 mM min^{-1}). It appears that the boronic acid-appended viologen, in concentrations of 125 μM, did not disrupt enzyme function [179]. We also carried out standard enzyme-coupled colorimetric assays on the same enzymes and measured comparable activity. In an additional experiment, we saw that increasing the enzyme concentration resulted in increased velocity, verifying that the kinetics of the enzymes, and not the AIDA, are the rate-limiting process in the assay.

Michaelis–Menten constants were determined for SPO by measuring V_i at varying substrate concentrations (Fig. 21). The value we obtained for K_m is 8.7 mM, while reports in the literature range from 1.7 to 5.3 mM. These parameters can vary depending on temperature, pH, and buffer medium. A comparison of kinetic parameters is provided in the original publication [43].

We took advantage of the simple optical assay to demonstrate enzyme inhibition with the PGM reaction. Lithium ions [163] were used to demonstrate the rapid screening possibilities with the new PGM assay, using sodium and potassium salts as controls. Lithium acts on PGM by displacing magnesium at the active site, and so we chose a magnesium concentration of 2 mM, close to that found in biological systems. As shown in Fig. 22, the reaction velocity is decreased in the presence of 1 mM

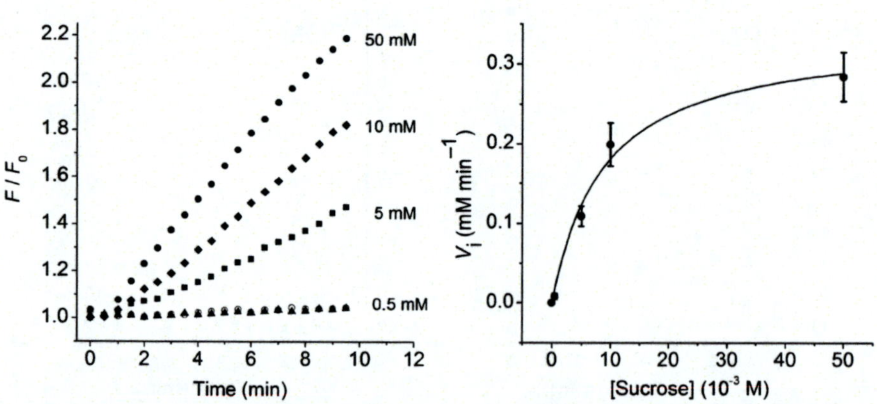

Fig. 21 *Left*: Fluorescence response of the SPO assay at different sucrose concentrations of 0.5 (*filled triangle*), 5 (*filled square*), 10 (*filled diamond*), and 50 (*filled circle*) mM with enzyme concentration of 3.3 μg mL^{-1} (0.15 U mL^{-1}). Background with no SPO (*open circle*). *Right*: plot of initial velocity V_i against sucrose concentration fitted with Michaelis–Menten equation. Adapted and reprinted with permission from [43]. Copyright 2008 Elsevier Ltd

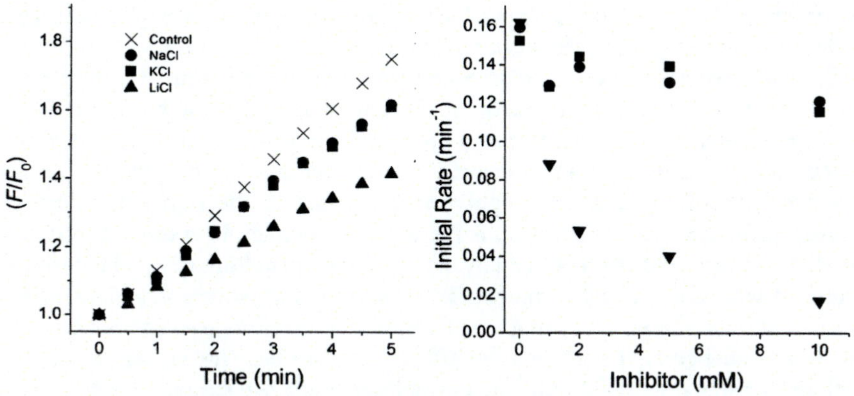

Fig. 22 Inhibition of PGM activity with 0.54 μg mL^{-1} (0.13 U mL^{-1}) enzyme and 2 mM Mg^{2+}. *Left*: Fluorescence increase measured at 30-s intervals in the presence of 1 mM chloride salts of sodium (*circle*), potassium (*square*), and lithium (*triangle*). The control (×) contains no added salts. *Right*: Initial rate (fluorescence intensity increase per minute) of enzymatic reaction in the presence of increasing salt concentrations. Adapted and reprinted with permission from [43]. Copyright 2008 Elsevier Ltd

lithium chloride with little effect from sodium chloride or potassium chloride. The inhibition is seen even more distinctly by plotting reaction rate against inhibitor concentration (Fig. 22, right). The IC_{50} for the reaction falls between 1 and 2 mM lithium, which is consistent with observations in the literature [180]. These findings are relevant to the therapeutic use of lithium, which is given at concentrations up to 1 mM and could potentially compete with magnesium ions occurring at similar concentrations [162, 163]. Given the sensitivity and simplicity of this assay, it can be used to screen for new, more potent inhibitors of PGM.

5 Conclusion

Selective recognition of glucose in the presence of other analytes in blood and interstitial fluid is essential for the successful functioning of a boronic acid-based glucose sensor. Among the many potential interfering analytes, fructose and lactate are of particular concern. Since most boronic acids display higher binding affinities for fructose, selectivity of glucose over fructose is a very challenging objective [8]. In this review, the ways that this problem has been addressed are discussed.

We have demonstrated that good analyte selectivity can be achieved by synthetically altering the bis-boronic acid-appended benzyl viologen (BBV) receptor unit to allow for cooperative binding of the two boronic acids with glucose. The nitrogen-positioning in the bipyridyl rings, as well as the boronic acid-positioning around the benzyl rings, was modified to create receptors with unique binding environments relative to one another. One particular receptor, 3,3′-*o*-BBV, displays a very large binding constant for glucose and shows greater selectivity for glucose over fructose. Other receptors showed selectivity for glucose over galactose and displayed enhanced glucose/fructose binding ratios.

The BBV receptors were also screened against three different α-hydroxycarboxylates in a neutral aqueous medium. Tartrate and malate both elicited large fluorescence increases and moderate to high apparent binding constants for all of the receptors. Lactate was found to have very low binding constants with all of the BBV receptors. To gain more insight into the nature of the signal transduction mechanism of BBVs and HPTS, a charge-transfer complex between 4,4′-*m*-BBV and HPTS in the solid phase was analyzed by X-ray crystallography. The findings correlated with UV–Vis solution data. To increase the sensitivity for glucose, we have synthesized a new family of glucose-responsive bis-viologens (BBBV). These compounds quench HPTS much more effectively than BBV quenchers and show much higher signal modulation in the physiological glucose range.

To address the problem of glucose discrimination from other saccharides under physiological conditions, powerful sensor arrays with BBVs and the fluorescent dye HPTS have been created. *Ortho-* and *meta*-BBV receptors display different discriminating power. Using multivariate techniques, a group of 12 neutral saccharides could be discriminated at 2 mM analyte concentration with only three *ortho*-BBV receptors in the sensor array. Thus, for the discrimination of phospho sugars and nucleotides, the same array was used. The binding of anionic diol-containing analytes to the BBVs resulted in stronger fluorescence signals as a result of a synergistic displacement effect. The "discrimination limit" was determined by analyte concentration-dependent multivariate analysis for the first time. The binding parameters K_b and F_{max}/F_0 with their intrinsic concentration information were also used to discriminate among the group of analytes. The combination of K_b and F_{max}/F_0, with all three BBV receptors, gave excellent discrimination results.

To demonstrate the potential of our saccharide receptors in applications other than continuous glucose monitoring and saccharide discrimination, we developed new label-free assays for carbohydrate-modifying enzymes. The saccharide receptors 4,4′-*o*-BBV and 3,3′-*o*-BBV were used to achieve product selectivity in two enzyme assays: phosphoglucomutase and sucrose phosphorylase. The method requires no substrate/dye conjugate, radio-labeling, or sophisticated equipment for detection. The inexpensive fluorescent assays are suitable for high-throughput screening of potential inhibitors or enzyme libraries.

References

1. Lorand JP, Edwards JO (1959) Polyol complexes and structure of the benzeneboronate ion. J Org Chem 24:769–774
2. Sugihara JM, Bowman CM (1958) Cyclic benzeneboronate esters. J Am Chem Soc 80:2443–2446
3. Kuivila HG, Keough AH, Soboczenski EJ (1954) Areneboronates from diols and polyols. J Org Chem 19:780–783

4. Yan J, Fang H, Wang B (2005) Boronolectins and fluorescent boronolectins: an examination of the detailed chemistry issues important for the design. Med Res Rev 25(5):490–520
5. Wang W, Gao X, Wang B (2002) Boronic acid-based sensors. Curr Org Chem 6 (14):1285–1317
6. Striegler S (2003) Selective carbohydrate recognition by synthetic receptors in aqueous solution. Curr Org Chem 7(1):81–102
7. James TD, Shinkai S (2002) Artificial receptors as chemosensors for carbohydrates. Top Curr Chem 218:159–200
8. Mader HS, Wolfbeis OS (2008) Boronic acid based probes for microdetermination of saccharides and glycosylated biomolecules. Microchim Acta 162(1–2):1–34
9. Egawa Y, Gotoh R, Niina S, Anzai J-I (2007) Ortho-azo substituted phenylboronic acids for colorimetric sugar sensors. Bioorg Med Chem Lett 17(13):3789–3792
10. Ni W, Fang H, Springsteen G, Wang B (2004) The design of boronic acid spectroscopic reporter compounds by taking advantage of the pKa-lowering effect of diol binding: nitrophenol-based color reporters for diols. J Org Chem 69(6):1999–2007
11. Ward CJ, Patel P, James TD (2002) Boronic acid appended azo dyes-colour sensors for saccharides. J Chem Soc Perkin Trans 1(4):462–470
12. Ward CJ, Patel P, James TD (2002) Molecular color sensors for monosaccharides. Org Lett 4 (4):477–479
13. DiCesare N, Lakowicz JR (2001) New color chemosensors for monosaccharides based on azo dyes. Org Lett 3(24):3891–3893
14. Wentholt Iris M, Vollebregt Marit A, Hart Augustus A, Hoekstra Joost B, DeVries JH (2005) Comparison of a needle-type and a microdialysis continuous glucose monitor in type 1 diabetic patients. Diabetes Care 28(12):2871–2876
15. Wilson Darrell M, Block J (2005) Real-time continuous glucose monitor use and patient selection: what have we learned and where are we going? Diab Technol Ther 7(5):788–791
16. Heinemann L, Schmelzeisen-Redeker G (1998) Non-invasive continuous glucose monitoring in type I diabetic patients with optical glucose sensors. Diabetologia 41(7):848–854
17. Garg SK, Hoff HK, Chase HP (2004) The role of continuous glucose sensors in diabetes care. Endocrinol Metab Clin North Am 33(1):163–173
18. Van den Berghe G, Wilmer A, Hermans G, Meersseman W, Wouters PJ, Milants I, Van Wijngaerden E, Bobbaers H, Bouillon R (2006) Intensive insulin therapy in the medical ICU. N Engl J Med 354(5):449–461
19. Van den Berghe G, Wouters P, Weekers F, Verwaest C, Bruyninckx F, Schetz M, Vlasselaers D, Ferdinande P, Lauwers P, Bouillon R (2001) Intensive insulin therapy in critically ill patients. N Engl J Med 345(19):1359–1367
20. Robertson A, Shinkai S (2000) Cooperative binding in selective sensors, catalysts and actuators. Coord Chem Rev 205(1):157–199
21. Tsukagoshi K, Shinkai S (1991) Specific complexation with mono- and disaccharides that can be detected by circular dichroism. J Org Chem 56(13):4089–4091
22. Shiomi Y, Saisho M, Tsukagoshi K, Shinkai S (1993) Specific complexation of glucose with a diphenylmethane-3,3′-diboronic acid derivative: correlation between the absolute configuration of mono- and disaccharides and the circular dichroic-activity of the complex. J Chem Soc Perkin Trans 1(17):2111–2117
23. James TD, Sandanayake KRAS, Shinkai S (1994) A glucose-specific molecular fluorescence sensor. Angew Chem Int Ed 106(21):2287–2289
24. Norrild JC, Eggert H (1995) Evidence for mono- and bisdentate boronate complexes of glucose in the furanose form. Application of 1JC-C coupling constants as a structural probe. J Am Chem Soc 117(5):1479–1484
25. Yang W, He H, Drueckhammer DG (2001) Computer-guided design in molecular recognition: design and synthesis of a glucopyranose receptor. Angew Chem Int Ed 40(9):1714–1718
26. Wiskur SL, Ait-Haddou H, Lavigne JJ, Anslyn EV (2001) Teaching old indicators new tricks. Acc Chem Res 34(12):963–972

27. Anslyn EV (2007) Supramolecular analytical chemistry. J Org Chem 72(3):687–699
28. Martinez-Manez R, Sancenon F (2003) Fluorogenic and chromogenic chemosensors and reagents for anions. Chem Rev 103(11):4419–4476
29. Lakowicz JR (2006) Principles of fluorescence spectroscopy, 3rd edn. Springer, New York
30. de Silva AP, Gunaratne HQN, Gunnlaugsson T, Huxley AJM, McCoy CP, Rademacher JT, Rice TE (1997) Signaling recognition events with fluorescent sensors and switches. Chem Rev 97(5):1515–1566
31. Ogoshi T, Harada A (2008) Chemical sensors based on cyclodextrin derivatives. Sensors 8:4961–4982
32. Nguyen BT, Anslyn EV (2006) Indicator-displacement assays. Coord Chem Rev 250 (23–24):3118–3127
33. Shabbir SH, Regan CJ, Anslyn EV (2009) A general protocol for creating high-throughput screening assays for reaction yield and enantiomeric excess applied to hydrobenzoin. Proc Natl Acad Sci 106(26):10487–10492
34. Kitamura M, Shabbir SH, Anslyn EV (2009) Guidelines for pattern recognition using differential receptors and indicator displacement assays. J Org Chem 74(12):4479–4489
35. Leung D, Anslyn EV (2008) Transitioning enantioselective indicator displacement assays for amino acids to protocols amenable to high-throughput screening. J Am Chem Soc 130 (37):12328–12333
36. Rochat S, Gao J, Qian X, Zaubitzer F, Severin K (2010) Cross-reactive sensor arrays for the detection of peptides in aqueous solution by fluorescence spectroscopy. Chem Eur J 16 (1):104–113
37. Buryak A, Pozdnoukhov A, Severin K (2007) Pattern-based sensing of nucleotides in aqueous solution with a multicomponent indicator displacement assay. Chem Commun 23:2366–2368
38. Zaubitzer F, Buryak A, Severin K (2006) Cp*Rh-based indicator-displacement assays for the identification of amino sugars and aminoglycosides. Chem Eur J 12(14):3928–3934
39. Buryak A, Severin K (2005) A chemosensor array for the colorimetric identification of 20 natural amino acids. J Am Chem Soc 127(11):3700–3701
40. Buryak A, Severin K (2005) Dynamic combinatorial libraries of dye complexes as sensors. Angew Chem Int Ed 44(48):7935–7938
41. Buryak A, Severin K (2004) An organometallic chemosensor for the sequence-selective detection of histidine- and methionine-containing peptides in water at neutral pH. Angew Chem Int Ed 43(36):4771–4774
42. Singaram B, Wessling RA (2003) Polyhydroxyl-substituted organic molecule sensing method and device. United States Patent 6653141
43. Vilozny B, Schiller A, Wessling RA, Singaram B (2009) Enzyme assays with boronic acid appended bipyridinium salts. Anal Chim Acta 649(2):246–251
44. Sharrett Z, Gamsey S, Hirayama L, Vilozny B, Suri JT, Wessling RA, Singaram B (2009) Exploring the use of APTS as a fluorescent reporter dye for continuous glucose sensing. Org Biomol Chem 7(7):1461–1470
45. Schiller A, Vilozny B, Wessling RA, Singaram B (2008) Recognition of phospho sugars and nucleotides with an array of boronic acid appended bipyridinium salts. Anal Chim Acta 627 (2):203–211
46. Sharrett Z, Gamsey S, Levine P, Cunningham-Bryant D, Vilozny B, Schiller A, Wessling RA, Singaram B (2008) Boronic acid-appended bis-viologens as a new family of viologen quenchers for glucose sensing. Tetrahedron Lett 49:300–304
47. Sharrett Z, Gamsey S, Fat J, Cunningham-Bryant D, Wessling RA, Singaram B (2007) The effect of boronic acid acidity on performance of viologen-based boronic acids in a two-component optical glucose-sensing system. Tetrahedron Lett 48(29):5125–5129
48. Schiller A, Wessling RA, Singaram B (2007) A fluorescent sensor array for saccharides based on boronic acid appended bipyridinium salts. Angew Chem Int Ed 46(34):6457–6459

49. Gamsey S, Miller A, Olmstead MM, Beavers CM, Hirayama LC, Pradhan S, Wessling RA, Singaram B (2007) Boronic acid-based bipyridinium salts as tunable receptors for monosaccharides and α-hydroxycarboxylates. J Am Chem Soc 129(5):1278–1286
50. Cordes DB, Miller A, Gamsey S, Singaram B (2007) Simultaneous use of multiple fluorescent reporter dyes for glucose sensing in aqueous solution. Anal Bioanal Chem 387:2767–2773
51. Thoniyot P, Cappucio FE, Gamsey S, Cordes DB, Wessling RA, Singaram B (2006) Continuous glucose sensing with fluorescent thin-film hydrogels. 2. Fiber optic sensor fabrication and in vitro testing. Diab Technol Ther 8(3):279–287
52. Gamsey S, Suri JT, Wessling RA, Singaram B (2006) Continuous glucose detection using boronic acid-substituted viologens in fluorescent hydrogels: linker effects and extension to fiber optics. Langmuir 22(21):9067–9074
53. Gamsey S, Baxter NA, Sharrett Z, Cordes DB, Olmstead MM, Wessling RA, Singaram B (2006) The effect of boronic acid-positioning in an optical glucose-sensing ensemble. Tetrahedron 62(26):6321–6331
54. Cordes DB, Gamsey S, Singaram B (2006) Fluorescent quantum dots with boronic acid substituted viologens to sense glucose in aqueous solution. Angew Chem Int Ed 45:3829–3832
55. Cordes DB, Miller A, Gamsey S, Sharrett Z, Thoniyot P, Wessling R, Singaram B (2005) Optical glucose detection across the visible spectrum using anionic fluorescent dyes and a viologen quencher in a two-component saccharide sensing system. Org Biomol Chem 3 (9):1708–1713
56. Cordes DB, Gamsey S, Sharrett Z, Miller A, Thoniyot P, Wessling RA, Singaram B (2005) The interaction of boronic acid-substituted viologens with pyranine: the effects of quencher charge on fluorescence quenching and glucose response. Langmuir 21(14):6540–6547
57. Cappuccio FE, Suri JT, Cordes DB, Wessling RA, Singaram B (2004) Evaluation of pyranine derivatives in boronic acid based saccharide sensing: significance of charge interaction between dye and quencher in solution and hydrogel. J Fluoresc 14(5):521–533
58. Suri JT, Cordes DB, Cappuccio FE, Wessling RA, Singaram B (2003) Monosaccharide detection with 4,7-phenanthrolinium salts: charge-induced fluorescence sensing. Langmuir 19(12):5145–5152
59. Suri JT, Cordes DB, Cappuccio FE, Wessling RA, Singaram B (2003) Continuous glucose sensing with a fluorescent thin-film hydrogel. Angew Chem Int Ed 42(47):5857–5859
60. Camara JN, Suri JT, Cappuccio FE, Wessling RA, Singaram B (2002) Boronic acid substituted viologen based optical sugar sensors: modulated quenching with viologen as a method for monosaccharide detection. Tetrahedron Lett 43(7):1139–1141
61. Zisser H (2009) Accuracy of a novel intravascular fluorescent continuous glucose sensor. American Diabetes Association Meeting, New Orleans, LA
62. De Borba EB, Amaral CLC, Politi MJ, Villalobos R, Baptista MS (2000) Photophysical and photochemical properties of pyranine/methyl viologen complexes in solution and in supramolecular aggregates: a switchable complex. Langmuir 16(14):5900–5907
63. Blondeel G, De Keukeleire D, Harriman A, Milgrom LR (1985) Fluorescence of covalently bound zinc porphyrin–viologen complexes. Chem Phys Lett 118(1):77–82
64. Yanuck MD, Schmehl RH (1985) Effects of non-electrolyte polymers on photoinduced electron transfer reactions. Influence of aqueous poly(vinyl alcohol) on quenching of tetraanionic porphyrins by methyl viologen. Chem Phys Lett 122(1–2):133–138
65. Nakashima K, Kido N (1996) Fluorescence quenching of 1-pyrenemethanol by methylviologen in polystyrene latex dispersions. Photochem Photobiol 64(2):296–302
66. Kusumoto Y, Uchikoba M (1991) The formation of ground-state nonfluorescent complex between chlorophyll a and methylviologen in methanol solution. Chem Lett 11:1985–1988
67. Kusumoto Y, Watanabe J, Kurawaki J, Satake I (1987) Selective fluorescence quenching of chlorophyll a-*N*-methylmyristamide system by methyl viologen in aqueous sodium dodecyl sulfate solution. Chem Lett 7:1417–1420

68. Chen L, McBranch DW, Wang H-L, Helgeson R, Wudl F, Whitten DG (1999) Highly sensitive biological and chemical sensors based on reversible fluorescence quenching in a conjugated polymer. Proc Natl Acad Sci 96(22):12287–12292
69. Wang D, Wang J, Moses D, Bazan GC, Heeger AJ (2001) Photoluminescence quenching of conjugated macromolecules by bipyridinium derivatives in aqueous media: charge dependence. Langmuir 17(4):1262–1266
70. Gaylord BS, Wang S, Heeger AJ, Bazan GC (2001) Water-soluble conjugated oligomers: effect of chain length and aggregation on photoluminescence-quenching efficiencies. J Am Chem Soc 123(26):6417–6418
71. Pia E, Toba R, Chas M, Peinador C, Quintela JM (2006) Synthesis of new viologen macrocycles with intramolecular charge transfer. Tetrahedron Lett 47(12):1953–1956
72. Monk PMS, Hodgkinson NM (1997) Charge-transfer complexes of the viologens: effects of complexation and the rate of electron transfer to methyl viologen. Electrochim Acta 43 (3–4):245–255
73. Tsukahara K, Kaneko J, Miyaji T, Abe K (1996) Charge-transfer interaction between bipyridinium and naphthyl groups of optically active viologens in an aqueous solution. Tetrahedron Lett 37(18):3149–3152
74. Yoon KB (1993) Electron- and charge-transfer reactions within zeolites. Chem Rev 93 (1):321–339
75. Alvaro M, Ferrer B, Fornes V, Garcia H (2003) Comparison between MCM-41 and periodic mesoporous organosilica: charge-transfer donor-viologen complexes as probes. Chem Phys Chem 4(6):612–617
76. Matos MS, Gehlen MH (1998) Spectroscopic and kinetic study of the molecular association between pyrene and benzyl viologen. Spectrochim Acta A Mol Spectrosc 54A (12):1857–1867
77. Farrugia LJ (1997) Ortep-3 for windows. J Appl Cryst 30:565
78. Yoon KB, Kochi JK (1989) Shape-selective access to zeolite supercages. Arene charge-transfer complexes with viologens as visible probes. J Am Chem Soc 111(3):1128–1130
79. Kidowaki M, Tamaoki N (2003) Unique crystal structures of donor-acceptor complexes: crossed arrangement of two charge-transfer columns. Chem Commun 290–291
80. Yoshikawa H, Nishikiori S-I (2005) Crystal structures and spectroscopic properties of polycyano-polycadmate host clathrates including a CT complex guest of methylviologen dication and aromatic donor. Dalton Trans 18:3056–3064
81. Nishikiori S-I, Yoshikawa H, Sano Y, Iwamoto T (2005) Inorganic-organic hybrid molecular architectures of cyanometalate host and organic guest systems: specific behavior of the guests. Acc Chem Res 38(4):227–234
82. Willner I, Eichen Y, Rabinovitz M, Hoffman R, Cohen S (1992) Structure and thermodynamic and kinetic properties of eosin–bipyridinium complexes. J Am Chem Soc 114 (2):637–644
83. Cao H, Diaz DI, DiCesare N, Lakowicz JR, Heagy MD (2002) Monoboronic acid sensor that displays anomalous fluorescence sensitivity to glucose. Org Lett 4(9):1503–1505
84. Gray CW Jr, Houston TA (2002) Boronic acid receptors for a-hydroxycarboxylates: high affinity of Shinkai's glucose receptor for tartrate. J Org Chem 67(15):5426–5428
85. Zhao J, Davidson MG, Mahon MF, Kociok-Koehn G, James TD (2004) An enantioselective fluorescent sensor for sugar acids. J Am Chem Soc 126(49):16179–16186
86. Collins BE, Sorey S, Hargrove AE, Shabbir SH, Lynch VM, Anslyn EV (2009) Probing intramolecular B–N interactions in ortho-aminomethyl arylboronic acids. J Org Chem 74 (11):4055–4060
87. Zhu L, Shabbir SH, Gray M, Lynch VM, Sorey S, Anslyn EV (2006) A structural investigation of the N–B interaction in an o-(*N,N*-dialkylaminomethyl)arylboronate system. J Am Chem Soc 128(4):1222–1232
88. Katzin LI, Gulyas E (1966) Optical rotatory dispersion studies on the borotartrate complexes and remarks on the aqueous chemistry of boric acid. J Am Chem Soc 88(22):5209–5212

89. Friedman S, Pace B, Pizer R (1974) Complexation of phenylboronic acid with lactic acid. Stability constant and reaction kinetics. J Am Chem Soc 96(17):5381–5384
90. Kustin K, Pizer R (1969) Temperature-jump study of the rate and mechanism of the boric acid-tartaric acid complexation. J Am Chem Soc 91(2):317–322
91. Lavigne JJ, Anslyn EV (1999) Teaching old indicators new tricks: a colorimetric chemosensing ensemble for tartrate/malate in beverages. Angew Chem Int Ed 38 (24):3666–3669
92. Wiskur SL, Floriano PN, Anslyn EV, McDevitt JT (2003) A multicomponent sensing ensemble in solution: differentiation between structurally similar analytes. Angew Chem Int Ed 42(18):2070–2072
93. Wiskur SL, Lavigne JJ, Metzger A, Tobey SL, Lynch V, Anslyn EV (2004) Thermodynamic analysis of receptors based on guanidinium/boronic acid groups for the complexation of carboxylates, a-hydroxycarboxylates, and diols: driving force for binding and cooperativity. Chem Eur J 10(15):3792–3804
94. Manimala JC, Wiskur SL, Ellington ED, Anslyn EV (2004) Tuning the specificity of a synthetic receptor using a selected nucleic acid receptor. J Am Chem Soc 126 (50):16515–16519
95. Bird CL, Kuhn AT (1981) Electrochemistry of the viologens. Chem Soc Rev 10(1):49–82
96. Geuder W, Hünig S, Suchy A (1986) Single and double bridged viologenes and intramolecular pimerization of their cation radicals. Tetrahedron 42(6):1665–1677
97. Monk PMS, Hodgkinson NM, Ramzan SA (1999) Spin pairing dimerisation of the viologen radical cation: kinetics and equilibria. Dyes Pigm 43(3):207–217
98. Neta P, Richoux MC (1985) Intramolecular association of covalently linked viologen radicals. J Chem Soc Faraday Trans 2(81):1427–1443
99. Thomas SW, Joly GD, Swager TM (2007) Chemical sensors based on amplifying fluorescent conjugated polymers. Chem Rev 107(4):1339–1386
100. Wolfbeis OS (2006) Fiber-optic chemical sensors and biosensors. Anal Chem 78 (12):3859–3874
101. Bentley R (2006) The nose as a stereochemist. Enantiomers and odor. Chem Rev 106:4099–4112
102. Jelinek R, Kolusheva S (2004) Carbohydrate biosensors. Chem Rev 104:5987–6016
103. Tsukube H, Shinoda S (2002) Lanthanide complexes in molecular recognition and chirality sensing of biological substrates. Chem Rev 102:2389–2404
104. Bell TW, Hext NM (2004) Supramolecular optical chemosensors for organic analytes. Chem Soc Rev 33:589–598
105. Mori K, Takahashi YK, Igarashi KM, Yamaguchi M (2006) Maps of odorant molecular features in the mammalian olfactory bulb. Physiol Rev 86:409–433
106. Lewis NS (2004) Comparisons between mammalian and artificial olfaction based on arrays of carbon black-polymer composite vapor detectors. Acc Chem Res 37(9):663–672
107. Wright AT, Anslyn EV (2006) Differential receptor arrays and assays for solution-based molecular recognition. Chem Soc Rev 35:14–28
108. Albert KJ, Lewis NS, Schauer CL, Sotzing GA, Stitzel SE, Vaid TP, Walt DR (2000) Cross-reactive chemical sensor arrays. Chem Rev 100(7):2595–2626
109. Palacios MA, Wang Z, Montes VA, Zyryanov GV, Anzenbacher P (2008) Rational design of a minimal size sensor array for metal ion detection. J Am Chem Soc 130(31):10307–10314
110. Hewage HS, Anslyn EV (2009) Pattern-based recognition of thiols and metals using a single squaraine indicator. J Am Chem Soc 131(36):13099–13106
111. Janzen MC, Ponder JB, Bailey DP, Ingison CK, Suslick KS (2006) Colorimetric sensor arrays for volatile organic compounds. Anal Chem 78(11):3591–3600
112. Leung D, Folmer-Andersen JF, Lynch VM, Anslyn EV (2008) Using enantioselective indicator displacement assays to determine the enantiomeric excess of α-amino acids. J Am Chem Soc 130(37):12318–12327

113. You C-C, Miranda OR, Gider B, Ghosh PS, Kim I-B, Erdogan B, Krovi SA, Bunz UHF, Rotello VM (2007) Detection and identification of proteins using nanoparticle-fluorescent polymer "chemical nose" sensors. Nat Nanotechnol 2(5):318–323
114. Sandanaraj BS, Demont R, Thayumanavan S (2007) Generating patterns for sensing using a single receptor scaffold. J Am Chem Soc 129(12):3506–3507
115. Miranda OR, You C-C, Phillips R, Kim I-B, Ghosh PS, Bunz UHF, Rotello VM (2007) Array-based sensing of proteins using conjugated polymers. J Am Chem Soc 129:9856–9857
116. Zhou H, Baldini L, Hong J, Wilson AJ, Hamilton AD (2006) Pattern recognition of proteins based on an array of functionalized porphyrins. J Am Chem Soc 128:2421–2425
117. Zyryanov GV, Palacios MA, Pavel Anzenbacher J (2007) Rational design of a fluorescence-turn-on sensor array for phosphates in blood serum. Angew Chem Int Ed 46:7849–7852
118. Green E, Olah MJ, Abramova T, Williams LR, Stefanovic D, Worgall T, Stojanovic MN (2006) A rational approach to minimal high-resolution cross-reactive arrays. J Am Chem Soc 128:15278–15282
119. Greene NT, Shimizu KD (2005) Colorimetric molecularly imprinted polymer sensor array using dye displacement. J Am Chem Soc 127(15):5695–5700
120. Adams MM, Anslyn EV (2009) Differential sensing using proteins: exploiting the cross-reactivity of serum albumin to pattern individual terpenes and terpenes in perfume. J Am Chem Soc 131(47):17068–17069
121. Edwards NY, Sager TW, McDevitt JT, Anslyn EV (2007) Boronic acid based peptidic receptors for pattern-based saccharide sensing in neutral aqueous media, an application in real-life samples. J Am Chem Soc 129:13575–13583
122. Lee JW, Lee J-S, Chang Y-T (2006) Colorimetric identification of carbohydrates by a pH indicator/pH change inducer ensemble. Angew Chem Int Ed 45(39):6485–6487
123. Zhang C, Suslick KS (2007) Colorimetric sensor array for soft drink analysis. J Agric Food Chem 55:237–242
124. Piatek AM, Bomble YJ, Wiskur SL, Anslyn EV (2004) Threshold detection using indicator-displacement assays: an application in the analysis of malate in Pinot Noir grapes. J Am Chem Soc 126(19):6072–6077
125. Buryak A, Severin K (2006) Easy to optimize: dynamic combinatorial libraries of metal-dye complexes as flexible sensors for tripeptides. J Comb Chem 8(4):540–543
126. Eggert H, Frederiksen J, Morin C, Norrild JC (1999) A new glucose-selective fluorescent bisboronic acid. First report of strong a-furanose complexation in aqueous solution at physiological pH. J Org Chem 64(11):3846–3852
127. Bielecki M, Eggert H, Norrild JC (1999) A fluorescent glucose sensor binding covalently to all five hydroxy groups of a-D-glucofuranose: a reinvestigation. J Chem Soc Perkin Trans 2 (3):449–456
128. Jurs PC, Bakken GA, McClelland HE (2000) Computational methods for the analysis of chemical sensor array data from volatile analytes. Chem Rev 100(7):2649–2678
129. Wright EM, Diamond JM (1977) Anion selectivity in biological systems. Physiol Rev 57 (1):109–156
130. Schmidtchen FP (2006) Reflections on the construction of anion receptors: is there a sign to resign from design? Coord Chem Rev 250(23–24):2918–2928
131. Gamez P, Mooibroek T, Teat S, Reedijk J (2007) Anion binding involving acidic heteroaromatic rings. Acc Chem Res 40(6):435–444
132. O'Neil EJ, Smith BD (2006) Anion recognition using dimetallic coordination complexes. Coord Chem Rev 250(23–24):3068–3080
133. Imada T, Kijima H, Takeuchi M, Shinkai S (1996) Selective binding of glucose-6-phosphate, 3,4-dihydroxyphenylalanine (DOPA) and their analogs with a boronic-acid-appended metalloporphyrin. Tetrahedron 52(8):2817–2826
134. Cabell LA, Monahan M-K, Anslyn EV (1999) A competition assay for determining glucose-6-phosphate concentration with a tris-boronic acid receptor. Tetrahedron Lett 40(44):7753–7756

135. Zhang T, Anslyn EV (2006) A colorimetric boronic acid based sensing ensemble for carboxy and phospho sugars. Org Lett 8(8):1649–1652
136. Patterson S, Smith BD, Taylor RE (1997) Fluorescence sensing of a ribonucleoside 5′-triphosphate. Tetrahedron Lett 38(36):6323–6326
137. Kanekiyo Y, Naganawa R, Tao H (2004) Fluorescence detection of ATP based on the ATP-mediated aggregation of pyrene-appended boronic acid on a polycation. Chem Commun (8):1006–1007
138. Luvino D, Smietana M, Vasseur J-J (2006) Selective fluorescence-based detection of dihydrouridine with boronic acids. Tetrahedron Lett 47(52):9253–9256
139. Atilgan S, Akkaya EU (2004) A calixpyridinium-pyranine complex as a selective anion sensing assembly via the indicator displacement strategy. Tetrahedron Lett 45 (50):9269–9271
140. Neelakandan PP, Hariharan M, Ramaiah D (2006) A supramolecular on-off-on fluorescence assay for selective recognition of GTP. J Am Chem Soc 128(35):11334–11335
141. Elliott WH, Elliott DC (2005) Biochemistry and molecular biology. Oxford University Press, New York
142. Palacios MA, Nishiyabu R, Marquez M, Anzenbacher P (2007) Supramolecular chemistry approach to the design of a high-resolution sensor array for multianion detection in water. J Am Chem Soc 129(24):7538–7544
143. Erion MD, Dang Q, Reddy MR, Kasibhatla SR, Huang J, Lipscomb WN, van Poelje PD (2007) Structure-guided design of amp mimics that inhibit fructose-1,6-bisphosphatase with high affinity and specificity. J Am Chem Soc 129(50):15480–15490
144. Dang Q, Kasibhatla SR, Reddy KR, Jiang T, Reddy MR, Potter SC, Fujitaki JM, van Poelje PD, Huang J, Lipscomb WN, Erion MD (2007) Discovery of potent and specific fructose-1,6-bisphosphatase inhibitors and a series of orally-bioavailable phosphoramidase-sensitive prodrugs for the treatment of type 2 diabetes. J Am Chem Soc 129(50):15491–15502
145. Davies G, Sinnott ML, Withers SG (1998) Glycosyl transfer. Compr Biol Catal 1:119–208
146. Kren V (2008) Glycoside vs. aglycon: the role of glycosidic residue in biological activity. In: Fraser-Reid B, Tatsuta K, Thiem J (eds) Glycoscience. Springer, Berlin, pp 2589–2644
147. Aharoni A, Thieme K, Chiu CPC, Buchini S, Lairson LL, Chen H, Strynadka NCJ, Wakarchuk WW, Withers SG (2006) High-throughput screening methodology for the directed evolution of glycosyltransferases. Nat Meth 3(8):609–614
148. Hennig A, Bakirci H, Nau WM (2007) Label-free continuous enzyme assays with macrocycle-fluorescent dye complexes. Nat Meth 4(8):629–632
149. Reymond J-L, Fluxa VS, Maillard N (2009) Enzyme assays. Chem Commun 1:34–46
150. López-Sánchez M, Ayora-Cañada MJ, Molina-Díaz A, Siam M, Huber W, Quintás G, Armenta S, Lendl B (2009) Determination of enzyme activity inhibition by FTIR spectroscopy on the example of fructose bisphosphatase. Anal Bioanal Chem 394:2137–2144
151. Oberthür C, Graf H, Hamburger M (2004) The content of indigo precursors in Isatis tinctoria leaves – a comparative study of selected accessions and post-harvest treatments. Phytochemistry 65(24):3261–3268
152. Li Y, Wang H, Lu L, Li Z, Xu X, Xiao M (2009) Purification and characterization of a novel β-galactosidase with transglycosylation activity from *Bacillus megaterium* 2-37-4-1. Appl Biochem Biotechnol 158:192–199
153. Maeda T, Nishimura S (2008) FRET-based direct and continuous monitoring of human fucosyltransferases activity: an efficient synthesis of versatile GDP-L-Fucose derivatives from abundant D-galactose. Chem Eur J 14:478–487
154. Murayama T, Tanabe T, Ikeda H, Ueno A (2006) Direct assay for [alpha]-amylase using fluorophore-modified cyclodextrins. Bioorg Med Chem 14(11):3691–3696
155. Indurugalla D, Watson JN, Bennet AJ (2006) Natural sialoside analogues for the determination of enzymatic rate constants. Org Biomol Chem 4(24):4453–4459

156. Mayer C, Jakeman DL, Mah M, Karjala G, Gal L, Warren RAJ, Withers SG (2001) Directed evolution of new glycosynthases from Agrobacterium [beta]-glucosidase: a general screen to detect enzymes for oligosaccharide synthesis. Chem Biol 8(5):437–443
157. Vankayalapati H, Singh G (1999) Synthesis of fucosidase substrates using propane-1,3-diyl phosphate as the anomeric leaving group. Tetrahedron Lett 40(20):3925–3928
158. Dicioccio RA, Piskorz C, Salamida G, Barlow JJ, Matta KL (1981) Synthesis and use of p-nitrophenyl-2-O-(α-L-fucopyranosyl)-β-D-galactopyranoside for the rapid detection of substrate-specific α-L-fucosidases. Anal Biochem 111:176–183
159. Persson M, Palcic MM (2008) A high-throughput pH indicator assay for screening glycosyltransferase saturation mutagenesis libraries. Anal Biochem 378(1):1–7
160. Ben-David A, Shoham G, Shoham Y (2008) A universal screening assay for glycosynthases: directed evolution of glycosynthase XynB2(E335G) suggests a general path to enhance activity. Chem Biol 15(6):546–551
161. Wongkongkatep J, Miyahara Y, Ojida A, Hamachi I (2006) Label-free, real-time glycosyltransferase assay based on a fluorescent artificial chemosensor. Angew Chem Int Ed 45 (4):665–668
162. Csutora P, Karsai A, Nagy T, Vas B, Kovacs GL, Rideg O, Bogner P, Miseta A (2006) Lithium induces phosphoglucomutase activity in various tissues of rats and in bipolar patients. Int J Neuropsychopharmacol 9:613–619
163. Csutora P, Strassz A, Boldizsar F, Nemeth P, Sipos K, Aiello DP, Bedwell DM, Miseta A (2005) Inhibition of phosphoglucomutase activity by lithium alters cellular calcium homeostasis and signaling in *Saccharomyces cerevisiae*. Am J Physiol Cell Physiol 289(1): C58–C67
164. Rossolini GM, Mantengoli E (2005) Treatment and control of severe infections caused by multiresistant Pseudomonas aeruginosa. Clin Microbiol Infect 11:17–32
165. Naught LE, Tipton PA (2001) Kinetic mechanism and pH dependence of the kinetic parameters of pseudomonas aeruginosa phosphomannomutase/phosphoglucomutase. Arch Biochem Biophys 396:111–118
166. Gao H, Leary JA (2004) Kinetic measurements of phosphoglucomutase by direct analysis of glucose-1-phosphate and glucose-6-phosphate using ion/molecule reactions and Fourier transform ion cyclotron resonance mass spectrometry. Anal Biochem 329(2):269–275
167. Cui Y, Barford JP, Renneberg R (2006) Amperometric determination of phosphoglucomutase activity with a bienzyme screen-printed biosensor. Anal Biochem 354:162–164
168. Nakata E, Wang H, Hamachi I (2008) Ratiometric fluorescent biosensor for real-time and label-free monitoring of fine saccharide metabolic pathways. Chem Bio Chem 9(1):25–28
169. Mieyal J, Simon M, Abeles R (1972) Mechanism of action of sucrose phosphorylase: 3. The reaction with water and other alcohols. J Biol Chem 247(2):532–542
170. Nomura K, Sugimoto K, Nishiura H, Ohdan K, Nishimura T, Hayashi H, Kuriki T (2008) Glucosylation of acetic acid by sucrose phosphorylase. Biosci Biotechnol Biochem 72 (1):82–87
171. Kwon T, Kim C, Lee J-H (2007) ransglucosylation of ascorbic acid to ascorbic acid 2-glucoside by a recombinant sucrose phosphorylase from *Bifidobacterium longum*. Biotechnol Lett 29(4):611–615
172. Sugimoto K, Nomura K, Nishiura H, Ohdan K, Hayashi H, Kuriki T (2007) Novel transglucosylating reaction of sucrose phosphorylase to carboxylic compounds such as benzoic acid. J Biosci Bioeng 104(1):22–29
173. Shin M, Cheong N-Y, Lee J-H, Kim K (2009) Transglucosylation of caffeic acid by a recombinant sucrose phosphorylase in aqueous buffer and aqueous-supercritical CO_2 media. Food Chem 115:1028–1033
174. Inoue H, Kondo S, Hinohara Y, Juni N, Yamamoto D (2003) Enhanced phosphorylation and enzymatic activity of phosphoglucomutase by the Btk29A tyrosine kinase in Drosophila. Arch Biochem Biophys 413:207–212

175. Treptau T, Kissmehl R, Wissmann JD, Plattner H (1995) A 63 kDa phosphoprotein undergoing rapid dephosphorylation during exocytosis in Paramecium cells shares biochemical characteristics with phosphoglucomutase. Biochem J 309:557–567
176. Goedl C, Schwarz A, Minani A, Nidetzky B (2007) Recombinant sucrose phosphorylase from Leuconostoc mesenteroides: Characterization, kinetic studies of transglucosylation, and application of immobilised enzyme for production of [alpha]-d-glucose 1-phosphate. J Biotechnol 129(1):77–86
177. Galbán J, Andreu Y, Sierra JF, Marcos SD, Castillo JR (2001) Intrinsic fluorescence of enzymes and fluorescence of chemically modified enzymes for analytical purposes: a review. Luminescence 16(2):199–210
178. Dowlut M, Hall DG (2006) An improved class of sugar-binding boronic acids, soluble and capable of complexing glycosides in neutral water. J Am Chem Soc 128(13):4226–4227
179. Yang W, Gao X, Wang B (2003) Boronic acid compounds as potential pharmaceutical agents. Med Res Rev 23(3):346–368
180. Ray WJ, Szymanki ES, Ng L (1978) The binding of lithium and of anionic metabolites to phosphoglucomutase. Biochim Biophys Acta Enzymol 522(2):434–442

Mapping and Immunomodulation of the Cell Surface Protein Architecture with Therapeutic Implications: Fluorescence Is a Key Tool of Solution

Péter Nagy, Andrea Balogh, János Szöllősi, and János Matkó

1 Introduction

Understanding the molecular mechanisms of cell-to-cell communication is one of the major challenges in today's biology, especially in the immune and nervous systems, where such communication leads to immediate effector functions, as well as to storage of memory [1–3].

A basic platform of cellular communication is the plasma membrane of individual cells. Here, the membrane receptors are continuously alert for sensing "physiological signals," and on the contrary, the whole cell surface is also exposed to various, sometimes unexpected external, spatiotemporally random stimulations (e.g., foreign materials, microbial attacks, drugs, stress, cell death stimuli, etc.). It is further complicated by the chemical and radiological therapies (mainly in case of tumors) where cells and tissues are exposed again to several damaging stimuli through their plasma membrane. Therefore, exploring the principles of molecular organization in the cellular membrane seems indispensable to understand the mechanisms of cellular communication, to find modulation strategies with therapeutic purposes.

Furthermore, based on several recent publications, a new concept arises that molecular compartmentation in the cell membrane (often called as "supramolecular organization" or "homo- and hetero-clustering") may sometimes become a major

J. Matkó (✉)
Department of Immunology, Eötvös Lorand University,
Pazmany Peter setany 1/C, 1117 Budapest, Hungary

Immunology Research Group of the Hungarian Academy of Sciences,
Eötvös Lorand University, Budapest, Hungary
e-mail: matko@elte.hu

C.D. Geddes (ed.), *Reviews in Fluorescence 2009*, Reviews in Fluorescence 2009,
DOI 10.1007/978-1-4419-9672-5_8, © Springer Science+Business Media, LLC 2011

factor [4–6] in determining cellular activation and fate, or disease progression, independently of the genetic background or altered expression levels of functional proteins. This can often appear as a superposition of multiple signal pathways, at the single cell level. Thus, the "being at right time at right place" principle for the key molecules can also significantly dictate the functional cellular response, in addition to the "genome-alteration" view of many frequent medical disorders.

To answer such questions, fluorescence developments provided a solid technical background (applying GFP fusion products, special fluorescent markers, Quantum dots, etc.). In addition, the continuous development of imaging technologies (FCS, live cell confocal imaging, FRET imaging, STED-microscopy, etc.) served also as excellent tools in this direction. The progressively developing imaging technology, electronics, and detector devices as well as the chemistry-based fluorescent labeling technology together largely potentiated the mapping of cell surface protein and lipid constituents with a further, largely improved spatial and temporal resolution [7, 8].

In the meantime, lipid raft microdomains were also supposed as major platforms/targets for a number of microbial agents, such as bacteria, viruses, protozoa, bacterial products, etc. [9–11]. Furthermore, many other disease-linked phenomena were coupled to membrane microdomains (rafts), such as processing of pathogenic prion and Alzheimer's amyloid-β proteins [10]. This scenario initiated a new area of research, namely, *identification of the molecular mechanisms involved in infection and searching for strategies to modulate these processes*.

In several tumors (e.g., breast cancer), cell-to-cell contacts and the molecular associations of oncogenic receptors (e.g., ErbB proteins) in the plasma membrane are also considered as critical factors [12]. Therefore, we believe that the research in the direction of identifying the molecular background of plasma membrane organization, cell surface distribution of such oncogenic molecular factors seems essential. New types of anticancer drugs enter clinical trials, revolutionizing the management and treatment of cancer patients [13]. Some of these drugs specifically interfere with the clustering of growth factor receptors. Pertuzumab, a monoclonal anti-ErbB2 antibody, inhibits the heterodimerization of ErbB2 with other members of the ErbB family, which takes place when growth factors activate them [14]. Protein clustering, which seemed to be simple a decade ago, appears to be ever more complex by the demonstration of new types of protein associations. The identification of transient receptor dimers [15, 16] and large-scale receptor clustering [17, 18] implies that transmembrane signaling involves the formation of multisubunit signaling platforms. The complexity of transmembrane signaling and the mechanism of action of novel anticancer agents can only be understood if the composition of signaling platforms is described. Fulfillment of this goal will require the application of sophisticated biophysical approaches that can analyze protein clustering in a quantitative way as opposed to classical molecular biological approaches that are restricted to a quantitative description of protein associations. Integration of this knowledge into an understanding of receptor signaling at the systems level has the potential of developing individualized treatment approaches to cancer [19]. The importance of this research area is underlined by the fact that resistance could develop against novel types of anticancer agents as well, which can only be understood and tackled if signaling mediated by receptor clusters is clearly understood [20].

Recent developments in molecular biology and in chemistry-based developments of fluorescent probes significantly promoted the studies on the molecular level organization of an essential biological platform, the plasma membranes of various cell types. These technologies include, among others, immunocytochemistry, application of single chain antibody constructs (scFv), the use of biotinylation and the connected avidin-detection technology, application of GFP constructs, fluorescent nanobeads, quantum dots, etc. [21–31].

Concerning the technical developments of fluorescence imaging, a number of variations have become available to analyze the above-mentioned questions, particularly molecular interactions, clusters, and networks in cellular systems. Among others, they include, for example, fluorescence correlation spectroscopy (FCS) for monitoring molecular mobility, Förster-type resonance energy transfer (FRET) variations for monitoring molecular proximity at the nanometer scale, various modalities of single particle tracking (SPT) techniques for detecting molecular mobility and the complexity of its constraints in biological membranes, etc. Finally, we should emphasize a recent great improvement in the spatial resolution of far-field optical microscopy initiated by Stefan Hell's group [7, 8, 32], namely, stimulation emission depletion (STED) microscopy, which allows observations with a 40–80 nm spatial resolution, far below the classical Abbe diffraction limit. These developments together with the improved time resolution [33–36] opened a new window for biologists, especially in the field of dynamic "nano-microdomain" scale compartmentation of biological membranes.

For the past 20 years, our research has been focused on investigating cell surface distribution and interaction patterns of receptor proteins in immune and tumor cells and the role of lipid microdomains in membrane compartmentation [6, 37]. The present overview focuses on several interesting and biologically important questions of immunobiology and cancer biology, addressed by our laboratories, utilizing a broad repertoire of fluorescent technologies. The most significant technological and scientific achievements of the past few years are summarized briefly. We believe that our fluorescence technology developments and their application in assessing molecular interactions may be useful to many other specific biological questions and research areas, as well.

In the following chapters, we describe a comprehensive view of plasma membrane microdomains and their biological functions, as to our current knowledge. Then, new technological (theory, method, software) developments on the cell-FRET technology, which have been made recently in our laboratories, are briefly described. Finally, we demonstrate and discuss our latest findings about our new anticholesterol monoclonal antibodies that may serve as markers of clustered cholesterol in cell-free and live cell samples and can functionally modulate raft-dependent immunological processes, such as infection of T-cells or macrophages by human immunodeficiency virus (HIV) strains.

2 Membrane Micro-compartmentation: A Current View of Lipid Raft Microdomains

Plasma membranes of cells are considered as the major interface in cell-to-cell communication. The plasma membrane has been shown to be highly heterogeneous, in spite of its overall fluid nature under physiological conditions (37°C). The majority of the membrane surface under these conditions is fluid, but a significant fraction of its constituents are able to spontaneously form ordered liquid phase (Lo) domains that may determine the functional properties of the cell membrane in various cell types. The microdomains were first defined on chemical basis, as resistant to nonionic detergents [38]. Interestingly, one of the pioneers of this chemical definition announced later that the detergent resistant membrane fractions (DRMs) are not quite equivalent to the "preexisting rafts" in live cells [39]. The raft concept of membrane microdomains [40] evolved in the past few years in a direction that these membrane domains obviously exist in all plasma membranes, but their size, stability/lifetime is still debated. One of the main reasons for this uncertainty is the lack of proper markers enabling their visualization [39]. Most of the presently used markers, although can selectively bind to raft constituents, may cause artificial perturbation in the size and composition of native rafts. The most accepted protein markers of membrane microdomains are the cell-specific GPI-anchored proteins (e.g., CD48, Thy.1, AP, HA-RP), the constitutively raft-associated CD4, or the acylated src kinase proteins at the inner membrane leaflet (such as Lyn, Fyn kinases). Concerning the lipids, there are two kinds of them accepted as general markers of such microdomains, based on their chemical/structural properties, sphingomyelin (SPM) or the variously glycosylated sphingolipids/gangliosides (GSL). They are usually labeled by fluorescent subunit B of cholera toxin, but it should be noted that this interaction may result in perturbation of the preexisting ganglioside assembly at the cell surface during the monitoring [41, 42]. A third category of markers is cholesterol itself that is considered as a major stabilizing factor of these domains [43]. New markers for clustered membrane cholesterol became also available [44] allowing extended studies on the nature/behavior of such membrane microdomains [45] (Fig. 1).

However, a basic question still remained open: is there any reasonably good (selective) lipid raft marker at all? The answer is: perhaps. A potential lipid-like marker of rafts, a synthetic compound, C-Laurdan, was reported by Gratton and coworkers, which is able to incorporate into live cell membranes without any lipid specificity. Its ability to detect microdomains is based on the unique property that its fluorescence emission spectrum is sensitive to the environment where it is located. So, its spectrum is significantly shifted between fluid or condensed (tightly packed) phases of biological membranes, and thus characterizes membrane microenvironments based on a theoretically defined term, "generalized polarization" [46, 47]. It was successfully applied in imaging of condensed membrane domains in phagocytes [48, 49]. Thus, so far it is the only probe which marks the "physical essence" of raft microdomains instead of their chemical composition.

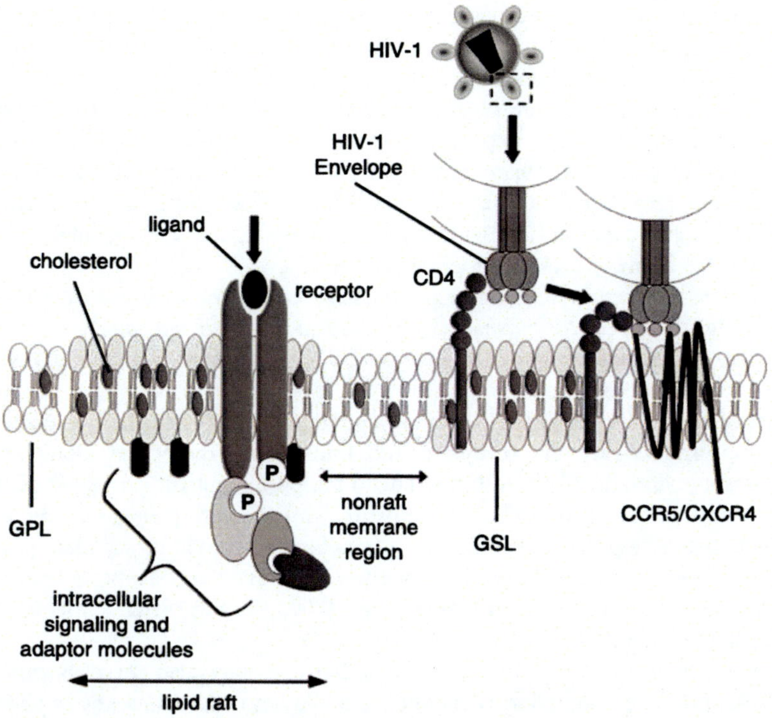

Fig. 1 Lipid rafts and their function in the cell membrane. A schematic view of plasma membrane microdomains (lipid rafts) enriched with cholesterol and glycosphingolipids (GSL). Small (20–100 nm) dynamic, unstable and large (300 nm to 1 μm) more stable rafts, like "*signaling rafts*" (*left*) or "*microbe target rafts*" (*right*) may coexist in a dynamic equilibrium, at the cell surface. Their components, localization, and behavior can be efficiently studied using various fluorescent compounds, antibodies, or GFP-analogs marking lipid or protein constituents of the domains. The signaling rafts (*left*) may spatiotemporally coordinate signal transduction by recruiting receptors, coreceptors, signal-transducing molecules, and adaptors to each other, thereby making possible formation of the so-called "signalosomes". Through recognizing and binding to their raft-associated receptor, CD4, HIV-1, viruses can associate to target cells, and then, by forming multiple contacts with the chemokine coreceptors (CXCR4/CCR5), they attach firmly to the cell surface prior to membrane fusion and internalization. Raft microdomains serve as targets for many widely known other viruses and bacteria as well (see chapter "Principal Component Global Analysis of Series of Fluorescence Spectra")

However, besides the chemical detergent-resistance definitions, several other technologies are also frequently used for detecting domain structure of plasma membranes, such as cell-FRET [50–53], single particle tracking (SPT) [54–57], single dye tracing (SDT) [58] or other techniques based on measurement of molecular mobility properties, such as fluorescence correlation spectroscopy (FCS) [59–61], or simply a quantitative confocal microscopic colocalization analysis [62, 63]. These in situ data, complemented with some specialized techniques (e.g., force measurements against motion of cell surface antigens associated to bead-coupled antibodies, using laser-tweezer technology) [64] provided ample and solid evidence for the existence of small, transient lipid raft microdomains in plasma membranes of live cells. Some basic questions about them, however, still remained unresolved, which might be solved in the near future by utilizing for example. the improved resolution of optical microscopic imaging (STED microscopy) [8]. Whatever the basic property of rafts in the plasma membrane is, they are thought to be there, and what is more important, their size and functional significance is regulated by the cell itself, and can certainly play an essential role in many cellular processes.

Thus, the membrane microdomains (rafts and caveolae) are considered today as *critical regions of cellular plasma membranes where signal transduction, microbial attacks or death/stress signals are focused.* The functional significance of these microdomains seems multiple: to positively control sensitivity of cells to low density of stimulatory ligands; to provide an efficient signal transduction machinery by recruiting/focusing the receptive and transducing molecular elements into "signalosome domains"; providing a "spatio-temporal control" on signal transduction by concentrating transducing elements while isolating negative regulatory elements (such as protein phosphatases) temporarily; providing an ideal platform for such viruses and bacteria for cellular entry that have their receptors on the cells in raft-associated form (e.g., influenza virus, HIV, measles virus, Varicella zoster virus, etc.) [11, 65–69].

Concerning the latest point, these microdomains may also serve as organized targets for immunomodulation or chemical modulation with therapeutic implications in diseases mediated by lipid rafts. The fluorescence technologies mentioned above may serve as key tools in revealing the molecular architecture/association and mobility pattern of membrane proteins (such as virus receptors, coreceptors, oncogenic receptor tyrosine kinases, etc.) and can also monitor the changes appearing under pathological situations or upon immunomodulation.

3 Förster-Type Resonance Energy Transfer as an "Intermolecular Distance Ruler"

In addition to the mere characterization of the expression profile of proteins the understanding of their molecular interactions also plays a pivotal role in current biological research. In the past few decades, Förster-type or FRET has become an established and accepted tool for the investigation of the protein interactome. As discussed in detail later, its utility stems from the dependence of its efficiency on the separation distance between the investigated, fluorescently labeled proteins. This led Stryer to call FRET a spectroscopic ruler [70]. Several excellent reviews are available, which cover the physical background and manifestations of the FRET phenomenon and also describe several available and possible future approaches for measuring FRET [50, 51, 71, 72]. Therefore, the discussion in the present paper is limited to the principles necessary to understand the techniques described herein. The FRET interaction takes place between a fluorescent donor molecule and an acceptor, the former donating its energy to the latter, i.e., the excited donor relaxes to the ground state by FRET. Since FRET competes with fluorescence, it results in the decreased fluorescence quantum yield of the donor, probably the most important manifestation of FRET called donor quenching. Mostly, but not necessarily, the acceptor is also fluorescent; therefore, the acceptor fluoresces in a FRET interaction without being directly excited. This phenomenon is called sensitized acceptor emission. In order for FRET to take place and be measurable, several conditions have to be fulfilled:

- The distance between the donor and the acceptor has to be between 2 and 10 nm.
- The emission spectrum of the donor has to overlap with the excitation spectrum of the acceptor. The extent of overlap is characterized by the overlap integral (J).
- The emission dipole of the donor and the absorption dipole of the acceptor have to be close to parallel. The relative orientation of the dipole vectors is characterized by the orientation factor (κ^2).
- The fluorescence quantum yield of the donor has to be relatively high.

The efficiency of FRET, i.e., the fraction of excited donor molecules relaxing by FRET, depends on the rate constants of FRET (k_{FRET}), fluorescent (k_{f}), and nonfluorescent decays (k_{nf}) according to the following equation:

$$E = \frac{k_{\mathrm{FRET}}}{k_{\mathrm{FRET}} + k_{\mathrm{f}} + k_{\mathrm{nf}}} \tag{1}$$

The rate constant of FRET is a function of the overlap integral (J), the index of refraction of the medium (n), the separation distance between the donor and the acceptor (R), the orientation factor (κ^2) and the rate constant of fluorescence (k_{f}):

$$k_{\mathrm{FRET}} = \mathrm{const}\ k_{\mathrm{f}} J n^{-4} R^{-6} \kappa^2 \tag{2}$$

All the parameters except for the separation distance and the orientation factor can be assumed to be constant, so changes in the FRET efficiency can be attributed to alterations in R and κ^2. In most cases, the average rate constant of FRET in an ensemble of molecules is measured; therefore, the question whether R^{-6} and κ^2 are averaged independently, i.e., $\langle R^{-6}\rangle\langle\kappa^2\rangle$, or their product is averaged, i.e., $\langle R^{-6}\kappa^2\rangle$, is an important question. Independent averaging can be assumed if the rotations of the donor and the acceptor are fast and they sample all possible relative orientations during the excited-state lifetime of the donor. In this case, called dynamic averaging, the mean of κ^2 was determined to be 2/3, and the equation describing the rate of FRET is reduced to the following form:

$$k_{\mathrm{FRET}} = CR^{-6} \tag{3}$$

where C is a pooled constant. Dynamic averaging usually takes place if the donor and acceptor are linked to the investigated targets by flexible linkers (e.g., antibodies labeled by fluorophores via flexible chemical bonds), but as the rigidity of the linkage increases (e.g., GFP-tagging, DNA intercalating dyes), the assumption of dynamic averaging introduces serious errors into FRET calculations [73, 74]. In most cases the efficiency of FRET is described by the following equation:

$$E = \frac{R_o^6}{R_o^6 + R^6} \tag{4}$$

where R_o is the distance at which the FRET efficiency is 50%. R_o is characteristic of a given donor–acceptor pair and can be calculated according to the following equation:

$$R_o = \text{const } (J\kappa^2 Q_D n^{-4})^{1/6} \tag{5}$$

Although R_o can be determined by anyone, in most cases it is taken from published tables available in the scientific literature and on the internet. These determinations were carried out with the assumption of dynamic averaging, i.e., $\kappa^2 = 2/3$; therefore (4), using published R_o values is only correct if dynamic averaging takes place.

Although FRET can indeed be used for the calculation of the distance between the donor and the acceptor, in most cases its application is limited to the determination of the FRET efficiency and changes thereof. Interpretation of these parameters is also complex. Changes in the FRET efficiency are usually interpreted as changes in the average separation distance between the donor and the acceptor, a supposition hinging upon the validity of the assumption of dynamic averaging. In addition, changes in the average separation distance between donors and acceptors can be brought about by two phenomena which can be considered as two extreme possibilities:

- The altered FRET efficiency reflects a uniform change in the donor–acceptor distance, i.e., the ensemble of donors and acceptors can be considered as a homogenous population with regard to their separation distance. This assumption usually does not hold if molecular associations are investigated.
- Usually only a fraction of the donor or the acceptor associates with the other; therefore, an increased FRET efficiency can be interpreted as an increased fraction of donors associating with acceptors.

The second point is the underlying principle of the application of FRET for the measurement of protein clustering. The extent of protein clustering may change by alterations in the fraction of clustered molecules or the size of individual clusters, i.e., the number of proteins/cluster. Conventional FRET, called hetero-FRET, when the interaction takes place between spectroscopically different molecules, is relatively insensitive for cluster size [75]. As discussed in the next chapter, another flavor of FRET, homo-FRET, is exquisitely suited for the measurement of cluster sizes. Therefore, in most FRET experiments aimed at measuring protein clustering, an increased FRET efficiency is primarily caused by an increased fraction of donors interacting with acceptors. Another factor to be considered is whether a protein cluster is the result of specific interactions or random associations. This can be decided by looking at the dependence of FRET efficiency on the acceptor density and the donor–acceptor ratio [76]. In the next chapter, we introduce some new approaches and software developed in our laboratories for the quantitative determination of protein clustering in intact cells.

4 AccPbFRET, Ri-FRET, TS-FRET, Homo-FRET, and FRET-Sensitized Acceptor Bleaching (FSAB) as New Cell-FRET Variations for Protein-Interaction Mapping: Technical and Software Developments

4.1 Calculation of FRET Based on Acceptor Photobleaching and Ratiometric Measurements Using ImageJ

Although the determination of FRET only requires solving relatively simple linear equations, the lack of user-friendly, easy-to-use image analysis programs capable of such calculations has hindered the wide-spread application of FRET in image cytometry. The simplest way to measure FRET is by comparing the intensity of the donor in two samples labeled by the donor only and double-labeled by the donor and the acceptor. As discussed elsewhere, simplicity of this approach comes at the price of unreliability [71]. If donor intensity could be measured in the same sample both in the presence and absence of the acceptor, most of the factors leading to the unreliability of the measurement of donor quenching could be eliminated. This is achieved by the method of acceptor photobleaching FRET (accPbFRET) [77]. A sample is labeled by the donor and the acceptor, and the donor intensity is measured (I_{D1}) followed by photobleaching of the acceptor by intense illumination. Photodestruction of the acceptor annuls its quenching effect on the donor; therefore, the donor intensity after acceptor photobleaching (I_{D2}) can be regarded as the unquenched intensity of the donor in the absence of FRET. The efficiency of FRET can be calculated according to the following equation:

$$E = 1 - \frac{I_{D1}}{I_{D2}} \tag{6}$$

The real-life situation turns out to be much more complex, preventing the application of the above equation in most cases. Among others, photoconversion and incomplete photobleaching of the acceptor, unwanted bleaching of the donor, and spectral cross talk have to be taken into account [78, 79]. Consideration of all these factors leads to the following equation:

$$E = 1 - \frac{(1-\alpha)(I_{D1} - \delta I_{A1})}{\gamma(I_{D2} - (\alpha\delta + (1-\alpha)\varepsilon)I_{A1}) - \alpha(I_{D1} - \delta I_{A1})} \tag{7}$$

where I_{A1} and I_{A2} are the intensities of the acceptor before and after photobleaching, respectively, α is a factor correcting for incomplete bleaching of the acceptor, γ takes unwanted photodestruction of the donor into account. δ is a correction factor compensating for the overspill of acceptor fluorescence into the donor channel. ε corrects for the generation of photodestruction products from the acceptor by the photobleaching illumination [80]. All intensities are assumed to be background-corrected in the above equations. We have written a user-friendly ImageJ (http://rsb.info.nih.gov/ij/) [81] plugin which implements (7) (http://www.biophys.dote.hu/accpbfret/) [80]. Stepensky has also written an ImageJ plugin for the calculation of accPbFRET; however, this program is based on a less comprehensive equation (http://rsb.info.nih.gov/ij/plugins/fret/fret-calc.html) [82].

The determination of FRET efficiency based on measurement of sensitized acceptor emission is complicated by spectral overspill between the donor, FRET, and acceptor channels. The principles of such a measurement have been described elsewhere [71, 83, 84]. Briefly, three independent fluorescence intensities are measured in the donor (I_1), FRET (I_2), and acceptor (I_3) channels, which are described by the following equations:

$$I_1\left(\lambda_{\mathrm{D,ex}}, \lambda_{\mathrm{D,em}}\right) = I_{\mathrm{D}}(1-E) + I_{\mathrm{A}}S_4 + \frac{S_4}{S_2}I_{\mathrm{D}}E\alpha \tag{8}$$

$$I_2\left(\lambda_{\mathrm{D,ex}}, \lambda_{\mathrm{A,em}}\right) = I_{\mathrm{D}}(1-E)S_1 + I_{\mathrm{A}}S_2 + I_{\mathrm{D}}E\alpha \tag{9}$$

$$I_3\left(\lambda_{\mathrm{A,ex}}, \lambda_{\mathrm{A,em}}\right) = I_{\mathrm{D}}(1-E)S_3 + I_{\mathrm{A}} + \frac{S_3}{S_1}I_{\mathrm{D}}E\alpha \tag{10}$$

where the first and second wavelength symbol in the parentheses specify the excitation and emission wavelengths, respectively, I_{D} and I_{A} are the unquenched donor and directly excited acceptor intensities, respectively, S_1–S_4 are spectral overspill correction factors. S_1 and S_3 compensate for the overspill of donor intensity into the FRET and acceptor channels, respectively, and S_2 and S_4 compensate for the overspill of acceptor fluorescence into the FRET and donor channels, respectively. α is the ratio of the fluorescence intensities of an equal number of excited donor and acceptor molecules measured in the donor and FRET channels, respectively. The determination of the aforementioned calibration factors has been described elsewhere [71, 83, 84]. Solving (8)–(10) for the FRET efficiency yields:

$$E = \frac{S_1S_2(I_1(S_1 - S_2S_3) + I_3(S_2 - S_1S_4) + I_2(S_3S_4 - 1))}{S_1S_2(I_1S_1 + I_3S_2 - I_1S_2S_3 - I_3S_1S_4 + I_2(S_3S_4 - 1)) - \alpha(S_1 - S_2S_3)(I_1S_2 - I_2S_4)} \tag{11}$$

The complexity of the above equation necessitates the use of custom-written computer programs. We and others have implemented this intensity-based ratiometric FRET (RiFRET) approach in ImageJ (http://www.biophys.dote.hu/rifret/, http://www.unil.ch/cig/page16989.html) [85, 86].

4.2 *Two-Sided FRET (TS-FRET) Measurements for the Characterization of the Interactions of Three Distinct Molecular Species*

Complexes of signal transducing proteins are not limited to dimers. A wise combination of two different FRET approaches allowed us to examine the correlation between the formation of two distinct species of dimers (Fig. 2) [87]. Let us consider dimers AB and BC. Molecule A is labeled by a donor-tagged antibody conjugated to fluorescein. Molecule B is labeled by a Cy3-tagged antibody, which serves as an acceptor for fluorescein, and as a donor for the Cy5-tagged antibody labeling molecule C. The BC dimer is first characterized by acceptor photobleaching FRET measurements described above [77, 80] followed by the determination of FRET efficiency between molecules A and B using the donor photobleaching FRET method (pbFRET) [71, 88, 89]. In pbFRET the FRET efficiency between the acceptor and a photolabile donor is determined by comparing the photobleaching rate of the donor in the presence and absence of the acceptor. This can be achieved by separately labeling two samples with donor only or with a mixture of donor and acceptor. Since the total number of fluorophores emitted by a fluorophore before photobleaching is characteristic of the molecule in question, processes decreases the fluorescence quantum yield will lead to decreased rate of photobleaching, i.e., longer time required for photodestruction of the molecule. Since FRET quenches the donor,

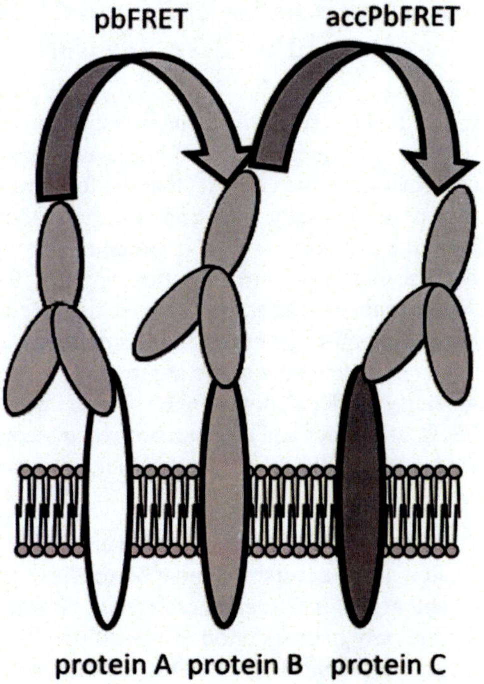

Fig. 2 Two-sided FRET measurements for the pairwise investigation of two interactions. Proteins A and B are labeled by a photolabile donor (e.g., fluorescein) and a photostable acceptor (e.g., Cy3), respectively. Protein C is labeled by a photolabile fluorophore (e.g., Cy5), which serves as an acceptor for the fluorophore labeling protein B. First, the BC interaction is characterized by acceptor photobleaching FRET (accPbFRET) followed by the investigation of the AB dimer using donor photobleaching FRET (pbFRET) measurements

the photobleaching time constant of the donor in the presence of the donor (τ_{DA}) will be larger than in its absence (τ_D). The FRET efficiency can be calculated according to the following equation:

$$E = 1 - \frac{\tau_D}{\tau_{DA}} \tag{12}$$

The photobleaching time constants can be determined by fitting the decay of fluorescence of single cells or pixels according to the following equation:

$$I = \sum_i I_{o,i} e^{-\frac{t}{\tau_i}} \tag{13}$$

The number of exponentials has to be determined experimentally. The effective photobleaching time constants, which have to be substituted into (12), can be determined as an intensity-weighted average:

$$\tau = \frac{\sum_i I_{o,i} \tau_i}{\sum_i I_{o,i}} \tag{14}$$

4.3 *Determination of Protein Cluster Size by Flow-Cytometric Homo-FRET Measurements*

In this and the following section we introduce FRET-based approaches which not only reveal the presence of molecular associations, but also characterize their composition. Given the fact that the formation of protein associations is not limited to dimers, the characterization of the size of protein associations, i.e., the number of monomers/cluster, is also important. The photon count histogram method [90], fluorescence correlation spectroscopy [91, 92] and its modifications (e.g., number and brightness analysis [93], raster image correlation spectroscopy [94]) are capable of determining cluster size. As pointed out earlier conventional hetero-FRET is relatively insensitive to the number of proteins in a cluster [75]. However, another modality of FRET, homo-FRET, has the potential to estimate cluster size. In homo-FRET, the donor and acceptor are spectroscopically identical, i.e., they are the same kind of fluorophores (e.g., two AlexaFluor488 molecules). This circumstance has several consequences:

- The conventional ways for measuring FRET (donor quenching, sensitized emission of the acceptor) cannot be used in the case of homo-FRET, since the donor and acceptor fluorescences cannot be separated from each other.
- The only manifestation of homo-FRET is decreased fluorescence anisotropy of the fluorophore [95]. This is the consequence of the almost completely

depolarized excitation of the acceptor by FRET. Since in homo-FRET the fluorescence of the acceptor cannot be discriminated from that of the donor, the phenomenon results in depolarized fluorescence emission from the ensemble of molecules. The extent of depolarization is determined by how restricted the rotations and orientations of the fluorophores are [96].

- The migration of energy in homo-FRET is not unidirectional, since the acceptor can serve as a donor and pass its energy back to the original donor or to any other fluorophore in the cluster. This is why the phenomenon has also been called energy migration FRET (emFRET) [97]. Consequently, the excitation energy is distributed all over the cluster in homo-FRET, and the extent of fluorescence depolarization depends on the number of fluorophores in a cluster.

Runnels and Scarlata derived a simple formula relating cluster size to fluorescence depolarization [95]:

$$r_k = r_1 \frac{\left(1+d^6\right)}{1+kd^6} + r_{\mathrm{FRET}} \frac{(k-1)d^6}{1+kd^6} \tag{15}$$

where r_k is the fluorescence anisotropies of a cluster of k fluorophores, r_1 and r_{FRET} are the anisotropies of the initially excited molecule and that of a molecule excited by homo-FRET, respectively, and d is the distance between the fluorophores in the cluster normalized to R_0. Although (15) directly relates cluster size (k) to anisotropy (r_k), anisotropy as a single measure cannot be used to determine the number of fluorophores in a cluster since anisotropy is influenced by factors other than FRET (e.g., molecular rotations). Therefore, either time-dependent anisotropy measurements have to be carried out [96] or steady-state anisotropy measurements have to be combined with the determination of the dependence of fluorescence anisotropy on fluorophore density [17, 97, 98]. This can be achieved by photobleaching and concomitant investigation of the observed changes in anisotropy using microscopy [97, 98]. We have developed a flow-cytometric approach for the quantitative determination of the number of proteins in a cluster (Fig. 3) [17]. It was assumed that a certain percentage of the proteins are monomeric (mon), while the rest forms clusters of N-mers. Cells are labeled with a mixture of fluorescent and unlabeled antibodies against the same epitope, and a series of samples is generated by varying the proportion of the labeled antibody (saturation, s) from 0 to 100%. The number of fluorophores determined by the proportion of labeled antibodies (s) influences fluorescence anisotropy according to (15); hence, anisotropy will be a function of both saturation (s) and the cluster size (N):

$$r_{s,k,N} = \frac{(1-\mathrm{mon})\sum_{k=0}^{N}\left[\binom{N}{k} s^k (1-s)^{N-k} \frac{k}{N}\left(r_1 \frac{(1+d^6)}{1+kd^6} + r_{\mathrm{FRET}} \frac{(k-1)d^6}{1+kd^6}\right)\right] + s\cdot \mathrm{mon}\cdot r_1}{(1-\mathrm{mon})\sum_{k=0}^{N}\binom{N}{k} s^k (1-s)^{N-k}\frac{k}{N} + s\cdot \mathrm{mon}} \tag{16}$$

d and r_1 have to be determined from independent measurements. Although the method requires a sophisticated flow cytometer capable of recording polarized fluorescence emissions, it provides a statistically reliable measure of cluster size.

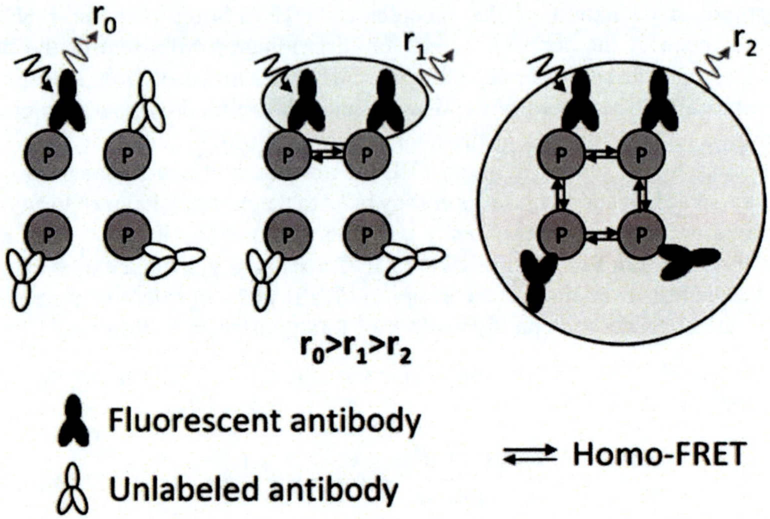

Fig. 3 Flow-cytometric homo-FRET measurements for the characterization of large-scale protein clusters. Cells are labeled with a mixture of unlabeled and fluorescently labeled antibodies. Several samples are generated in which the ratio of the labeled and unlabeled antibodies varies. If only one protein is bound to a fluorescently labeled antibody in a cluster of four (*left panel*), then the measured anisotropy (r_o) will be maximal, and only determined by the limiting anisotropy of the fluorophore and its rotational mobility. If the number of fluorescently labeled antibodies increases in the cluster (*middle* and *right panels*), the measured anisotropies (r_1, r_2) will gradually decrease since homo-FRET takes place and not only the primarily excited fluorophore will fluoresce but also those excited by homo-FRET. The size of the molecular ensemble participating in homo-FRET and fluorescence emission increases with the fraction of labeled antibodies as indicated by the growing size of the *shaded circle*

4.4 FRET-Sensitized Acceptor Bleaching (FSAB) Measurements for the Determination of the Fraction of Clustered Proteins

Since the fractions of bound donor and acceptor molecules are not normally revealed in conventional FRET measurements, we and others have developed methods for this purpose. The approach of Gadella and Jovin is based on the comparison of the photobleaching kinetics of the donor in the presence and absence of the acceptor [99]. The dual-channel photobleaching approach of Clayton et al. compares the photobleaching of the signals measured in the donor and the FRET channel for the calculation of the fraction of acceptor-bound donors [100]. In the rare case when the FRET efficiency in a single donor–acceptor pair is known and all donor–acceptor pairs can be assumed to form the same type of dimer, the fraction of acceptor-bound donors and donor-bound acceptors can be calculated by measuring fluorescence intensities in the donor, FRET, and acceptor channels [101].

The method elaborated in our laboratories [102] is based on the concept developed by Mekler [103, 104]. We assumed that two classes of acceptors exist with regard to their association with donors: acceptors bound to donors, i.e., within FRET distance from donors (A_{bound}) and free acceptors (A_{free}). A suitable pair of a photostable donor and a photolabile acceptor has to be selected. The donor is illuminated by a high-intensity beam, and nearby photolabile acceptors are excited and photobleached preferentially. Owing to the fact that acceptors are also excited directly at the wavelength of the donor, some free acceptor molecules outside the range of FRET are also bleached. The bleached fraction of free acceptors is designated by bleaching correction factor (BCF). Assuming that all acceptors within FRET distance from donors are completely bleached the fraction of donor-bound acceptors can be calculated according to the following equations:

$$F_{\mathrm{bleached}} = \frac{A_{\mathrm{free}}\mathrm{BCF} + A_{\mathrm{bound}}}{A_{\mathrm{free}} + A_{\mathrm{bound}}} \tag{17}$$

$$A_{\mathrm{bound}} + A_{\mathrm{free}} = A_0 \tag{18}$$

Solving for A_{bound} yields

$$A_{\mathrm{bound}} = A_0 \frac{F_{\mathrm{bleached}} - \mathrm{BCF}}{1 - \mathrm{BCF}} \tag{19}$$

If the bleaching of donor-bound acceptors is not complete, the unbleached fraction has to be taken into account. If the bleached fraction of the bound acceptor population is designated by $\mathrm{BCF}_{\mathrm{FRET}}$, the fraction of bound acceptors can be calculated according to the following equation:

$$F_{\mathrm{bleached}} = \frac{A_{\mathrm{free}}\mathrm{BCF} + A_{\mathrm{bound}}\mathrm{BCF}_{\mathrm{FRET}}}{A_{\mathrm{free}} + A_{\mathrm{bound}}} \Rightarrow A_{\mathrm{bound}} = A_0 \frac{F_{\mathrm{bleached}} - \mathrm{BCF}}{\mathrm{BCF}_{\mathrm{FRET}} - \mathrm{BCF}} \tag{20}$$

5 Application of the New Cell-FRET Technologies in Cancer Research

We have applied the FRET-based approaches described above for the investigation of the epidermal growth factor (EGF) receptor family of receptor tyrosine kinases. The receptor family consists of four members (ErbB1–4; ErbB1 is also known as EGF receptor). According to the widespread assumption ErbB proteins are monomeric in their unstimulated state, and an extensive network of homo- and heterodimerizations is induced upon ligand binding [19, 105]. ErbB2 is known to be an orphan receptor functioning as a coreceptor for the other, ligand-binding ErbB proteins [106]. The distinct biological behavior of ErbB2 is the manifestation of its special structure different from those of other ErbB proteins [107]. According to our expectation, the peculiar structure and function of ErbB2 has to be reflected in its distinct clustering properties. We characterized the association state of ErbB1 and ErbB2 in resting, unstimulated cells using flow-cytometric homo-FRET measurements and found that the majority of ErbB1 is monomeric, while ErbB2 displays a strong tendency to form large-scale clusters of ~100 proteins [17]. After stimulation with EGF, a ligand of ErbB1, the average size of ErbB2 clusters decreased, while that of ErbB1 clusters increased. This finding, together with the negative correlation between the tyrosine phosphorylation level and cluster size of ErbB2, implies that the large-scale clusters of ErbB2, which we described, are pools of inactive ErbB2 molecules from which ligand-activated ErbB1 (and ErbB3) can recruit ErbB2 to form heterodimers or higher-order heteroassociations.

We, then, used the FSAB approach to characterize the fraction of clustered ErbB1 and ErbB2 in quiescent and EGF-stimulated breast cancer cells [102]. We found that almost half of the population of ErbB1 proteins was heteroassociated with ErbB2 in unstimulated cells and that this fraction slightly decreased upon EGF stimulation. In accordance with our previous homo-FRET data [17], the majority of ErbB2 formed homoclusters, i.e., it did not heteroassociate with ErbB1, in quiescent cells. After challenging the cells with EGF the fraction of ErbB2 in complex with ErbB1 doubled.

Although it was known from previous investigations that ErbB molecules also interact with non-ErbB proteins [108, 109], the influence of the abundance of one type of interaction on others has not been directly assessed. The strong interaction between β1-integrins and ErbB2 inhibited the formation of ErbB2 homoassociations [87]. By influencing the amount of ErbB2 available for heterodimerization with ligand-binding ErbB proteins, such correlations can alter the efficiency of signal transduction.

According to our model based on flow-cytometric homo-FRET and FSAB data unstimulated ErbB proteins are not uniformly monomeric with ErbB1 forming heteroassociations with ErbB2 and ErbB2 being present in large-scale homoclusters. After ligand stimulation, these preformed associations are restructured and new clusters are also created. Our results also support the view that higher-order associations of proteins are also present in addition to dimers. Although FRET-based methods are certainly invaluable in studying protein associations, other methods (single particle tracking, fluorescence correlation spectroscopy, etc.) have also contributed significantly to our understanding of signal transduction events [16, 110]. These methods display different sensitivity for distinct types of clusters differing in their stability (transient vs. long-term) and size (dimers vs. higher-order associations) [111]. Some of the contradictions between results yielded by the different methods are probably explained by some clusters overlooked by some methods, while detected by others (Fig. 4). The integration of results provided by the different quantitative approaches for the characterization of the behavior of receptors are required to understand the complex network of interactions taking place between resting and stimulated ErbB molecules and other signaling proteins.

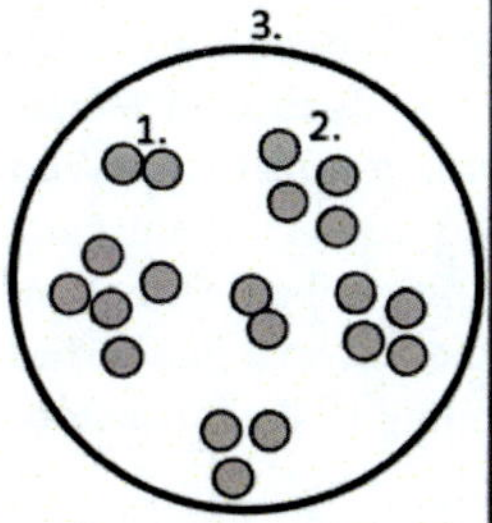

#	Type and size of cluster	Forces, interactions	Detection method
1	small-scale clusters, direct molecular associations (dimers)	protein-protein	X-ray crystallography, NMR
2	medium-scale clusters (co-confinement, 2-100 proteins)	protein-lipid, preferential distribution	hetero-FRET, N&B, FCS, SPT, Western, PLA; homo-FRET
3	large-scale clusters (e.g. caveola, immune synapse; 100-1000 proteins)		CLSM, SNOM

Fig. 4 Methods used for the investigation of protein clustering display differential sensitivity for distinct cluster sizes. Several distinct cluster types exist regarding their size and the forces holding them together. The smallest type of association (typically dimers) is created by specific molecular forces acting between the proteins (type 1). Such associations are revealed by X-ray crystallography and NMR. Preferential distribution into membrane microdomains and confinement by the cytoskeleton lead to the second level of molecular interactions (type 2). Most molecular biological and biophysical methods, used for the investigation of protein clustering, are able to detect both small- and medium-scale clusters, but they cannot discriminate between them. Hetero-FRET, number and brightness analysis (N&B), fluorescence correlation spectroscopy (FCS), single particle tracking (SPT), coprecipitation/western blotting, and proximity ligation assay (PLA) are more sensitive for small-scale clusters, whereas homo-FRET is more sensitive for medium-scale clusters. The largest protein clusters (type 3) are those which are resolvable by conventional microscopy, confocal laser scanning microscopy (CLSM) and scanning near-field optical microscopy (SNOM)

6 Novel Anticholesterol Antibodies as Markers and Diagnostic Tools of Clustered Cholesterol and Live Cell Membrane Raft Microdomains: Visualization by FRET and CLSM Imaging

Since cholesterol is a major structural constituent of most eukaryotic cell membranes and is enriched in membrane microdomains, visualizing organization and trafficking of cholesterol is an important field of lipid raft biology. One of the solutions is the use of fluorescent cholesterol reporter molecules. They can be classified into two groups: cholesterol binding molecules (e.g., filipin III antibiotics or BCθ-toxin, a recombinant derivative of perfringolysin O) and fluorescent cholesterol analogs, such as dehydroergosterol or NBD-cholesterol. An excellent review is available on the features, applicability, and disadvantages of these reporter molecules [112].

Cholesterol-specific monoclonal antibodies (mAbs) can also be suitable tools for cholesterol detection. Swartz et al. found cholesterol to be an excellent immunogen when they immunized mice with cholesterol-rich liposomes together with lipid A as an adjuvant. The generated cholesterol-specific 2C5-6 mAb bound to crystalline cholesterol and lipoproteins had complement-fixing capacity [113, 114] and was also applied to visualize cholesterol in the skeletal muscle of rats [115]. Perl-Treves and coworkers used another strategy to produce monoclonal anticholesterol antibodies (ACHAs). They have implanted cholesterol monohydrate crystals in the spleen of mice [116]. The produced 58B1 monoclonal antibody did not recognize the cholesterol molecule itself, but rather defined molecular motifs exposed on some of the crystal faces. Direct visualization of clustered plasma membrane cholesterol with the 58B1 mAb could also be achieved, but only when cells were artificially enriched with cholesterol [117]. However, these monoclonal antibodies were all IgM isotype, and none of them were capable of binding to the surface of intact cells.

Using the immunization protocol of Swartz et al., besides IgM-type antibodies, we managed to produce IgG isotype cholesterol-specific antibodies (AC1 and AC8) [44]. Both monoclonal antibodies bound efficiently to cholesterol, and other sterols containing free 3β-hydoxyl group, adsorbed on polystyrene plates. This indicates that the 3β-hydoxyl group is an important moiety of the epitope. Our monoclonal ACHAs also recognized cholesterol in human lipoproteins (VLDL, LDL, and HDL), as assessed by competitive ELISA and FRET measurements in a spectrofluorimeter (Fig. 5a). In contrast to previously generated monoclonal ACHAs, our novel IgG antibodies bound with low/intermediate avidity to the membrane of murine and human lymphocytes and macrophage cell lines (Fig. 5b). Their cell

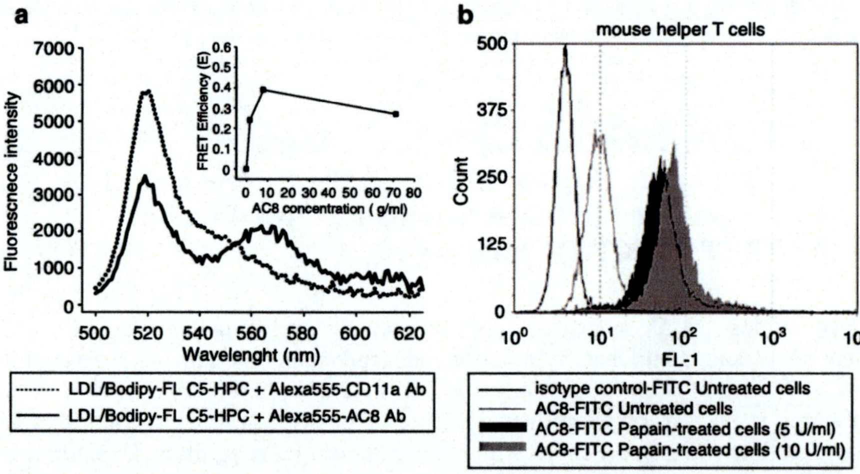

Fig. 5 Monoclonal IgG cholesterol-specific antibodies bind to membrane cholesterol of human lipoproteins and immunocytes. (**a**) In fluorescence resonance energy transfer (FRET) measurements LDL labeled with 2-(4,4,-difluoro-5,7-dimethyl-4-bora-3a,4a-diaza-s-indacene-3-pentanoyl)-1-hexadecanoyl-*sn*-glycero-3-phosphocholine (BODIPY-FL C_5-HPC) was incubated with Alexa555-AC8 or Alexa555-anti-CD11a antibodies (as a negative control). FRET efficiencies (see *inset*) with AC8 (*solid line*) or anti-CD11a (*dotted line*) acceptors were calculated from either the decrease of corrected donor emission intensities at 519 nm or the acceptor sensitization at 570 nm, upon excitation at 460 nm. (**b**) Binding of AC8 antibody to murine T-helper cells (2/13), as detected by flow cytometry using Alexa488-conjugated ACHA. Histograms of ACHA binding to cells without (*gray line*) and with 5 or 10 U/ml papain treatment (*black* and *gray areas*, respectively) are displayed, with isotype control (*black line*)

surface binding could be further enhanced by a limited papain digestion of long, protruding, extracellular protein domains (Fig. 5b), indicating that the accessibility of the supposedly small epitope is a critical factor for antibody binding. The high degree of colocalization between the IgG ACHA and cholera toxin B (CTX-B, correlation coefficient, cc: 0.69) or anticaveolin-1 antibodies (cc: 0.63) indicates that rafts and caveolae (Fig. 6) can be considered as preferential binding sites for ACHAs in the plasma membrane. In lymphocytes and macrophages, cholesterol is

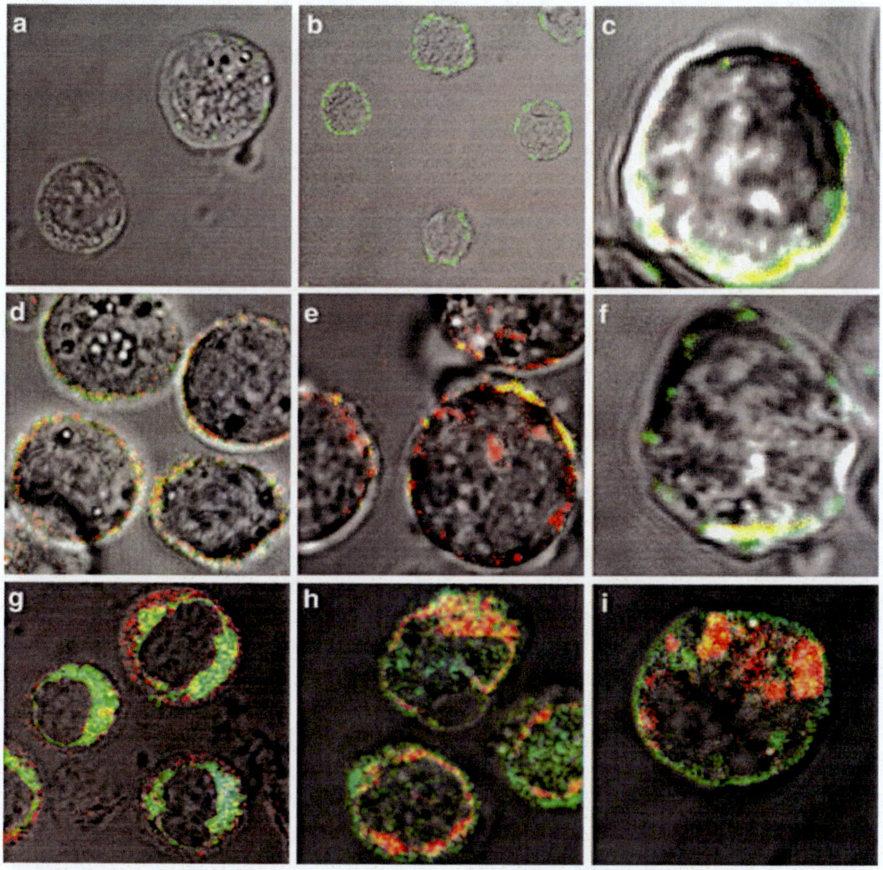

Fig. 6 AC8 antibody marks plasma- and intracellular membrane regions enriched in clustered cholesterol. Representative images of 2/13 mouse Th cells labeled with isotype control (**a**) or Alexa488-AC8 antibodies (**b**) are shown, as assessed by a confocal microscope using a ×60 oil-immersion objective (numerical aperture, 1.45). Representative images also show a high colocalization of AC8 with CTX-B (G_{M1}) on resting (**c**) and ConA-activated Th cells (in caps) (**f**). It also strongly colocalized with caveolin-1 (**d**) in macrophages, and to a smaller extent with anti-CD71/transferrin receptor (**e**) in T cells. Intracellularly bound AC8 showed partial colocalization with the Golgi complex-specific BODIPY FL C5-ceramide (**g**), with a marker of the endoplasmic reticulum and the Golgi complex, BODIPY 558/568 brefeldin A (**h**) and more weakly with LysoTracker (**i**)

locally enriched in lipid rafts or, to a lesser extent, in clathrin-coated pits [6], consistent with our findings (cc: 0.39) (Fig. 6). Furthermore, redistribution of lipid rafts upon mitogenic T lymphocyte stimulus, reflected by cap-like CTX-B staining, could also be monitored by AC8 (Fig. 6). The AC8 antibody intracellularly colocalized with markers of Golgi complex (cc: 0.3) and ER and more weakly with marker of lysosomes (Fig. 6). These organelles have relatively high cholesterol content, since transport of cholesterol occurs through them [118, 119]. Specificity of our antibodies was further confirmed using cholesterol-modulating agents. Either cholesterol oxidase or filipin III could substantially reduce binding of AC8 to pretreated cells or DRMs [44]. These results all suggest that the AC8 antibody is capable of marking cholesterol-rich lipid raft microdomains at the cell surface or in intracellular membranes, alike. Moreover, studies on lipid raft properties with polarization CLSM, FCS imaging [45, 120] clearly indicated that the cell surface proteins are distributed among distinct small raft domains in the plasma membrane, reflecting heterogeneity in the protein composition of elementary lipid raft microdomains.

7 Biological Activity of the New Anticholesterol Antibodies: The Molecular Background of Their HIV-1, Infection Inhibitory Effect Is Assessed by FRET, Confocal Microscopic Colocalization, and FCS Lateral Mobility Analysis

Exposure of cholesterol epitopes may change during apoptosis or under pathological conditions (e.g., tumors, pathogen infections, cardiovascular diseases) possibly due to the alteration of molecular composition, curvature or "surface smoothness" of the involved membrane regions or other cholesterol-containing structures. In accordance with this, elevated ACHA level in human sera has been revealed in atherosclerosis, chronic viral infections (HIV-1 and HCV) and cancer [121–124].

Binding properties of the new IgG ACHAs [44] allow further analysis of the functional properties of lipid rafts in the function of immune cells, such as antigen presentation by APCs [125, 126], receptor-mediated signal transduction in lymphocytes or myeloid cells [127], and the uptake of various pathogens and particles (e.g., influenza virus, apoptotic bodies) by phagocytes [128, 129]. The question is whether binding of ACHAs to membrane microdomains of these cells can modulate raft-dependent processes and, if so, how. Such modulatory effects, however, are most likely limited to those cells to which ACHAs can spontaneously bind to a sufficient extent.

The ability of monoclonal IgG ACHAs to modulate immune functions related to HIV-1 infection was demonstrated by our collaboration partner, who found that target cell-bound AC1 and AC8 antibodies could inhibit HIV-1 (x4 T-tropic IIIB and R5 Ada-M strains) production in vitro [130]. The inhibitory effects of AC1 and

AC8 cholesterol-specific antibodies on monocyte-derived macrophages and MT-4 T cells were comparable to those previously reported for anti-PIP antibody exhibiting neutralizing activity and blocking infection of PBMCs by two primary isolates of HIV-1 [131].

In the plasma membrane of the investigated cell types, the primary HIV-1 receptor, CD4, is constitutively and highly, whereas the chemokine receptor, CXCR4, is much more weakly raft-localized. Their localization and clustering in distinct membrane microdomains on HIV-1 permissive target cells was proposed earlier [132]. Binding of the cholesterol-specific mAb (AC8), but not of its Fab fragment or antibodies against nonraft membrane proteins (CD71), to the cells increased both CD4-CXCR4 colocalization and the raft-association of CXCR4 by approximately twofold (Fig. 7a). These observations were confirmed by microscopic FRET measurement and ganglioside staining, showing higher FRET efficiency (Fig. 7b) between CD4 and CXCR4 and increased average size of G_{M1} rafts (from 200–300 nm to 0.5–1 μm). Our results suggest that ACHAs may induce lateral clustering of cholesterol-rich lipid rafts (or caveolae) and, thus, modulate the interaction pattern between raft gangliosides, CD4 and CXCR4, which are all critical in membrane attachment/entry of the virus.

In contrast to other fluorescence techniques, the parameter of primary interest in FCS microscopy is not the emission intensity itself, but rather the spontaneous intensity fluctuations caused by minute deviations of small system from thermal equilibrium. In general, all physical parameters that give rise to fluctuations in the fluorescence signal are accessible by FCS. It is, for example, rather straightforward to determine local concentrations, mobility coefficients or characteristic rate constants of inter- or intramolecular reactions of fluorescently labeled biomolecules in nanomolar concentrations [133]. Therefore, it is a versatile tool to determine 2D or 3D diffusional rates of the labeled molecules. In the case of diffusion in the plasma membrane, data can be fitted to a single component 2D diffusion model with a triplet term using the program QuickFit (written in the laboratory of J. Langowski, DKFZ, Heidelberg):

$$G(\tau) = \frac{1}{N}\frac{\left(1 - T + Te^{\frac{-\tau}{\tau_{tr}}}\right)}{1 - T}\left[\left(1 + \frac{\tau}{\tau_d}\right)^{-1}\right] \tag{21}$$

where N is the average number of molecules in the detection volume, T is the fraction of dyes being in the triplet state within the detection volume, τ_{tr} is the phosphorescence lifetime and τ_d is the diffusion time, which is the average time spent by the dye in the detection volume.

Therefore, lateral diffusion features of the HIV-1 receptors were investigated by FCS that further supported our concept. In untreated control immune cells, the diffusional rate of CXCR4 was significantly higher than that of CD4. The lateral mobility of CXCR4 but not of CD4 decreased substantially upon AC8 binding. This suggests that these two receptors move with similar diffusion coefficients (Fig. 7c) after AC8 antibody engagement. All these data are consistent with a picture that

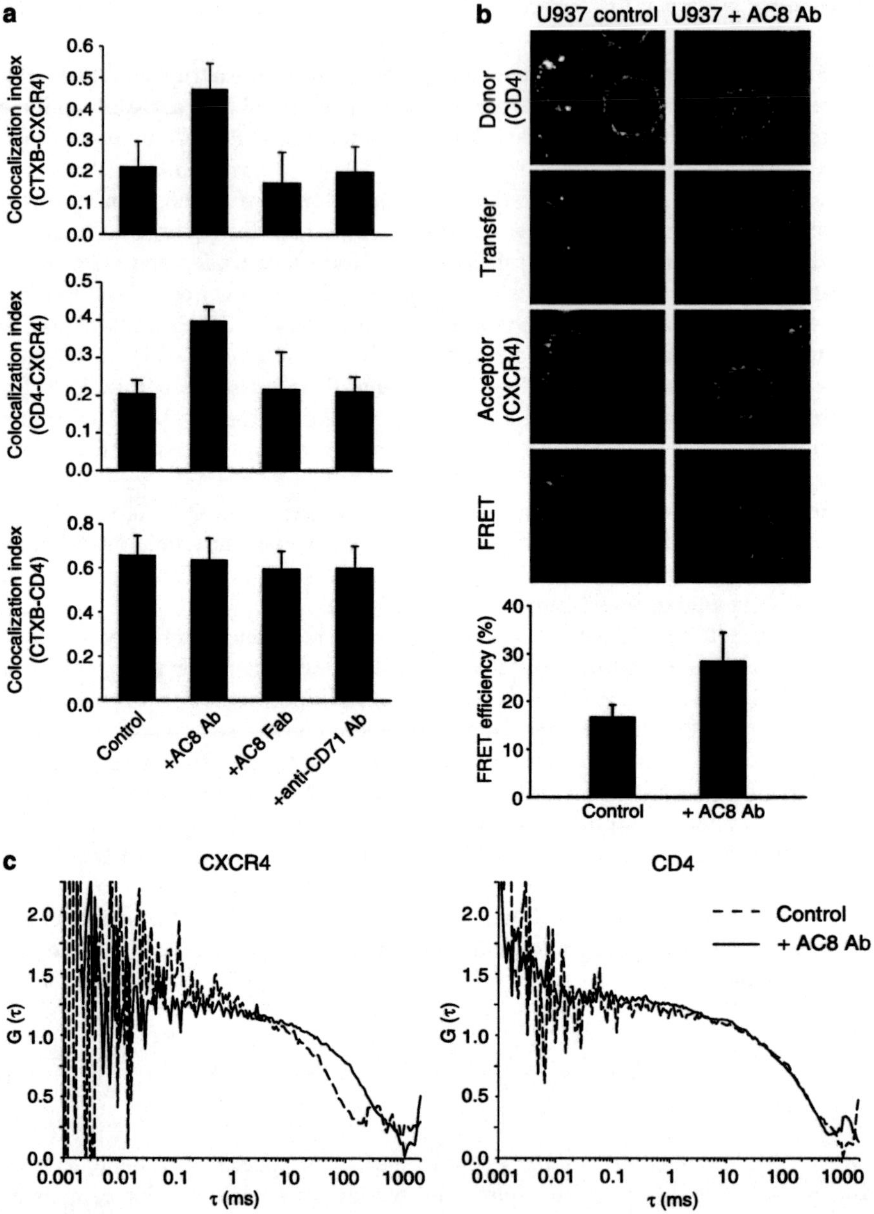

Fig. 7 HIV-1 target cell-bound AC8 antibody remodels the lateral plasma membrane interaction pattern of chemokine receptor CXCR4 with CD4 and lipid rafts. U937 monocyte-macrophage cells were treated with AC8, with different control antibodies or left untreated, prior to labeling of CD4, CXCR4 or G_{M1} gangliosides with respective fluorophore-conjugated antibodies or reagents. Colocalization (**a**), FRET efficiency (**b**) and 2D diffusion (**c**) of HIV-1 receptors were then measured. (**a**) Colocalization indices are shown as mean ± SD for CXCR4, CD4, and CTXB in different combinations. (**b**) FRET measurements were carried out by confocal microscopy. Correction factors (S1, S2, S3, and S4) and factor α were calculated from only-donor and only-acceptor labeled samples and determined by the ImageJ RiFRET plug-in. Representative images and FRET efficiency (mean ± SD) between CD4 and CXCR4 are displayed. (**c**) Lateral mobility of the molecules at 22°C was monitored by FCS imaging. Figures show representative fluorescence autocorrelation curves in control/untreated (*dashed line*) and AC8-treated cells (*solid line*)

AC8 engagement promotes colocalization of the chemokine receptor and CD4 into the same compartment. It has been revealed earlier that the diffusion of CD4 could be restricted by elevated expression of G_{M3} ganglioside in a mouse T cell line [134]. This, in turn, reduced Env-mediated fusion of HIV-1 by interfering with the lateral association of HIV-1 receptors [135].

On the basis of these substantial changes in the dynamic membrane organization of HIV-1 receptors without masking the CD4 and CXCR4 HIV-1 receptors, it is postulated that the AC8 mAb can mimic the action of viral glycoprotein gp120 in terms of coaggregating HIV-1 receptors and coreceptors. These data also concur with the findings that gp120-CD4 engagement leads to a cholesterol-sensitive lateral redistribution of membrane microdomains and a subsequent assembly of the virus internalization complex [136]. Furthermore, Nguyen et al. demonstrated a dynamic reorganization of chemokine receptors and lipid rafts to the site of CD4 engagement, using anti-CD4-beads [137].

Induction of broadly neutralizing antibodies is an essential prerequisite for an effective anti-HIV-1 vaccine. These antibodies can be specific to Env glycoproteins of the virus, such as 2F5, 4E10 or 2G12 [138–140]. However, alternative approaches for interfering with HIV-1 entry into target cells utilizing drugs or antibodies against cellular structures (CD4, chemokine receptors, HLA class I and II, adhesion molecules) involved in the entry process seem to be equally important [141]. Since cholesterol also has a major role in these processes, cholesterol-specific antibodies may also merit consideration. Therefore, the novel ACHAs, recognizing properly clustered membrane cholesterol only, seem useful to investigate the eventual role of cholesterol in the process of virus entry, and may serve as a molecular basis for the development of new, combined lipid raft-oriented approaches in HIV-1 therapy. Investigation of IgG ACHAs as modulators of other raft-dependent immune functions is currently underway in our laboratory.

8 Conclusions: Perspectives

In order to study the functional behavior of live cells, the dynamic association patterns of membrane proteins have to be investigated thoroughly using sophisticated state-of-the-art biophysical tools. The latest developments in fluorescence imaging and flow-cytometric techniques provide an excellent tool for deciphering the stoichiometry and compartmentation of signaling molecules at the cell surface. In this review, we demonstrate how flow-cytometric homo-FRET studies can be utilized to quantitate the oligomerization of ErbB proteins. In addition, using FRET-sensitized acceptor photobleaching also provides useful information about the associated fractions of molecules. If the homo-FRET studies are combined with fluorescence lifetime imaging (FLIM) and fluorescence polarization imaging, we can extract information about macromolecular associations from biological samples with high spatial resolution. The two-sided FRET and the multifluorophore FRET, where three or more fluorophores are utilized to reveal the molecular composition of molecular complexes, also provide new insights into molecular aggregation processes. The combination of FRET with fluorescent proteins opens a new field in cancer research, whereby the alterations in signaling cascades can be monitored in live cells using imaging techniques. If molecular details are important, single molecular FRET is the proper choice as an experimental tool yielding unique information about conformational changes.

The increasing number of available FRET modalities encourages us to apply FRET imaging techniques in various fields of cancer research. At the same time, however, it is also necessary to devote more effort to optimizing both the probes and the imaging systems to increase the amount of information we can retrieve from living cells. As shown by the example of immunomodulatory effects of cholesterol-specific antibodies, the proximity-analysis methods (FRET, CLSM colocalization) combined with fluorescence correlation spectroscopic mobility analysis (FCS) serve also as an excellent tool in exploring mechanism of action for different therapeutic compounds, including antibodies, acting on various cell membrane targets (either receptors or lipid microdomains). We are convinced that applying such fluorescence methods in STED microscopy with a largely improved spatial resolution will further broaden our view on cell membranes and, as a consequence, the scale of therapeutic interventions based on cell membrane targeting.

Acknowledgments This work was supported by research grants from the Hungarian Scientific Research Fund (K72677, K68763, K62648, T49696), from the European Commission (LSHC-CT-2005-018914), from the New Hungary Development Plan cofinanced by the European Social Fund and the European Regional Development Fund (TÁMOP-4.2.2-08/1-2008-0019), and from National Office of Research and Development (Pázmány Grant, RET-06/2006). The financial support of the Hungarian Academy of Sciences is also gratefully acknowledged.

References

1. Edelman GM (1993) Neural Darwinism: selection and reentrant signaling in higher brain function. Neuron 10:115–125
2. Shaw AS, Allen PM (2001) Kissing cousins: immunological and neurological synapses. Nat Immunol 2:575–576
3. Dustin ML, Colman DR (2002) Neural and immunological synaptic relations. Science 298:785–789
4. Trautmann A (2005) Microclusters initiate and sustain T cell signaling. Nat Immunol 6:1213–1214
5. Pike LJ (2006) Rafts defined: a report on the keystone symposium on lipid rafts and cell function. J Lipid Res 47:1597–1598
6. Matkó J, Szöllősi J (2005) Membrane microdomain signaling. Lipid rafts. In: Mattson MP (ed) Biology and medicine. Humana, Totowa, NJ, pp 15–46
7. Willig KI, Kellner RR, Medda R, Hein B, Jakobs S, Hell SW (2006) Nanoscale resolution in GFP-based microscopy. Nat Methods 3:721–723
8. Eggeling C, Ringemann C, Medda R, Schwarzmann G, Sandhoff K, Polyakova S, Belov VN, Hein B, von Middendorff C, Schonle A, Hell SW (2009) Direct observation of the nanoscale dynamics of membrane lipids in a living cell. Nature 457:1159–1162

9. Fantini J (2007) Interaction of proteins with lipid rafts through glycolipid-binding domains: biochemical background and potential therapeutic applications. Curr Med Chem 14:2911–2917
10. Taylor DR, Hooper NM (2007) Role of lipid rafts in the processing of the pathogenic prion and Alzheimer's amyloid-β proteins. Semin Cell Dev Biol 18:638–648
11. Jolly C, Sattentau QJ (2005) Human immunodeficiency virus type 1 virological synapse formation in T cells requires lipid raft integrity. J Virol 79:12088–12094
12. Feldner JC, Brandt BH (2002) Cancer cell motility-on the road from c-ErbB-2 receptor steered signaling to actin reorganization. Exp Cell Res 272:93–108
13. Shawver LK, Slamon D, Ullrich A (2002) Smart drugs: tyrosine kinase inhibitors in cancer therapy. Cancer Cell 1:117–123
14. Franklin MC, Carey KD, Vajdos FF, Leahy DJ, de Vos AM, Sliwkowski MX (2004) Insights into ErbB signaling from the structure of the ErbB2-pertuzumab complex. Cancer Cell 5:317–328
15. Andrews NL, Lidke KA, Pfeiffer JR, Burns AR, Wilson BS, Oliver JM, Lidke DS (2008) Actin restricts FceRI diffusion and facilitates antigen-induced receptor immobilization. Nat Cell Biol 10:955–963
16. Chung I, Akita R, Vandlen R, Toomre D, Schlessinger J, Mellman I (2010) Spatial control of EGF receptor activation by reversible dimerization on living cells. Nature 464 (7289):783–787. doi:10.1038/nature08827
17. Szabó Á, Horváth G, Szöllősi J, Nagy P (2008) Quantitative characterization of the large-scale association of ErbB1 and ErbB2 by flow cytometric homo-FRET measurements. Biophys J 95:2086–2096
18. Nagy P, Jenei A, Kirsch AK, Szöllősi J, Damjanovich S, Jovin TM (1999) Activation-dependent clustering of the ErbB2 receptor tyrosine kinase detected by scanning near-field optical microscopy. J Cell Sci 112:1733–1741
19. Citri A, Yarden Y (2006) EGF-ErbB signalling: towards the systems level. Nat Rev Mol Cell Biol 7:505–516
20. Nahta R, Esteva FJ (2006) Herceptin: mechanisms of action and resistance. Cancer Lett 232:123–138
21. Nelson AL (2010) Antibody fragments: hope and hype. MAbs 2:77–83
22. Filpula D (2007) Antibody engineering and modification technologies. Biomol Eng 24:201–215
23. Alguel Y, Leung J, Singh S, Rana R, Civiero L, Alves C, Byrne B (2010) New tools for membrane protein research. Curr Protein Pept Sci 11:156–165
24. Magliery TJ, Regan L (2006) Reassembled GFP: detecting protein-protein interactions and protein expression patterns. Methods Biochem Anal 47:391–405
25. Miyawaki A, Nagai T, Mizuno H (2005) Engineering fluorescent proteins. Adv Biochem Eng Biotechnol 95:1–15
26. Kampani K, Quann K, Ahuja J, Wigdahl B, Khan ZK, Jain P (2007) A novel high throughput quantum dot-based fluorescence assay for quantitation of virus binding and attachment. J Virol Methods 141:125–132
27. Day RN, Davidson MW (2009) The fluorescent protein palette: tools for cellular imaging. Chem Soc Rev 38:2887–2921
28. Herbst KJ, Ni Q, Zhang J (2009) Dynamic visualization of signal transduction in living cells: from second messengers to kinases. IUBMB Life 61:902–908
29. Chan FK (2004) Monitoring molecular interactions in living cells using flow cytometric analysis of fluorescence resonance energy transfer. Methods Mol Biol 261:371–382
30. Witternberg NJ, Haynes CL (2009) Using nanoparticles to push the limits of detection. Wiley Interdiscip Rev Nanomed Nanobiotechnol 1:237–254
31. Kalab P, Pralle A (2008) Chapter 21: quantitative fluorescence lifetime imaging in cells as a tool to design computational models of ran-regulated reaction networks. Methods Cell Biol 89:541–568

32. Willig KI, Rizzoli SO, Westphal V, Jahn R, Hell SW (2006) STED microscopy reveals that synaptotagmin remains clustered after synaptic vesicle exocytosis. Nature 440:935–939
33. Stockl M, Plazzo AP, Korte T, Herrmann A (2008) Detection of lipid domains in model and cell membranes by fluorescence lifetime imaging microscopy of fluorescent lipid analogues. J Biol Chem 283:30828–30837
34. Stockl MT, Herrmann A (2010) Detection of lipid domains in model and cell membranes by fluorescence lifetime imaging microscopy. Biochim Biophys Acta 1798(7):1444–1456
35. Muzzey D, van Oudenaarden A (2009) Quantitative time-lapse fluorescence microscopy in single cells. Annu Rev Cell Dev Biol 25:301–327
36. de Almeida RF, Loura LM, Prieto M (2009) Membrane lipid domains and rafts: current applications of fluorescence lifetime spectroscopy and imaging. Chem Phys Lipids 157:61–77
37. Vereb G, Szöllősi J, Damjanovich S, Matkó J (2004) Exploring membrane microdomains and functional protein clustering in live cells with flow and image cytometric methods. In: Geddes CD, Lakowicz JR (eds) Reviews in fluorescence. Kluwer/Plenum, New York, pp 99–120
38. Brown DA, Rose JK (1992) Sorting of GPI-anchored proteins to glycolipid-enriched membrane subdomains during transport to the apical cell surface. Cell 68:533–544
39. Brown DA (2006) Lipid rafts, detergent-resistant membranes, and raft targeting signals. Physiology (Bethesda) 21:430–439
40. Simons K, Ikonen E (1997) Functional rafts in cell membranes. Nature 387:569–572
41. Nagy P, Vereb G, Sebestyén Z, Horváth G, Lockett SJ, Damjanovich S, Park JW, Jovin TM, Szöllősi J (2002) Lipid rafts and the local density of ErbB proteins influence the biological role of homo- and heteroassociations of ErbB2. J Cell Sci 115:4251–4262
42. Harder T, Scheiffele P, Verkade P, Simons K (1998) Lipid domain structure of the plasma membrane revealed by patching of membrane components. J Cell Biol 141:929–942
43. Lingwood D, Simons K (2010) Lipid rafts as a membrane-organizing principle. Science 327:46–50
44. Bíró A, Cervenak L, Balogh A, Lorincz A, Uray K, Horváth A, Romics L, Matkó J, Fust G, László G (2007) Novel anti-cholesterol monoclonal immunoglobulin G antibodies as probes and potential modulators of membrane raft-dependent immune functions. J Lipid Res 48:19–29
45. Gombos I, Steinbach G, Pomozi I, Balogh A, Vámosi G, Gansen A, László G, Garab G, Matkó J (2008) Some new faces of membrane microdomains: a complex confocal fluorescence, differential polarization, and FCS imaging study on live immune cells. Cytom A 73:220–229
46. Bagatolli LA, Gratton E (1999) Two-photon fluorescence microscopy observation of shape changes at the phase transition in phospholipid giant unilamellar vesicles. Biophys J 77:2090–2101
47. Gaus K, Zech T, Harder T (2006) Visualizing membrane microdomains by Laurdan 2-photon microscopy. Mol Membr Biol 23:41–48
48. Gaus K, Gratton E, Kable EP, Jones AS, Gelissen I, Kritharides L, Jessup W (2003) Visualizing lipid structure and raft domains in living cells with two-photon microscopy. Proc Natl Acad Sci USA 100:15554–15559
49. Parasassi T, Gratton E, Yu WM, Wilson P, Levi M (1997) Two-photon fluorescence microscopy of laurdan generalized polarization domains in model and natural membranes. Biophys J 72:2413–2429
50. Jares-Erijman EA, Jovin TM (2006) Imaging molecular interactions in living cells by FRET microscopy. Curr Opin Chem Biol 10:409–416
51. Jares-Erijman EA, Jovin TM (2003) FRET imaging. Nat Biotechnol 21:1387–1395
52. Nagy P, Vámosi G, Bodnár A, Lockett SJ, Szöllősi J (1998) Intensity-based energy transfer measurements in digital imaging microscopy. Eur Biophys J 27:377–389
53. Vereb G, Matkó J, Szöllősi J (2004) Cytometry of fluorescence resonance energy transfer. Methods Cell Biol 75:105–152

54. Kusumi A, Suzuki K (2005) Toward understanding the dynamics of membrane-raft-based molecular interactions. Biochim Biophys Acta 1746:234–251
55. Dietrich C, Yang B, Fujiwara T, Kusumi A, Jacobson K (2002) Relationship of lipid rafts to transient confinement zones detected by single particle tracking. Biophys J 82:274–284
56. Suzuki K, Ritchie K, Kajikawa E, Fujiwara T, Kusumi A (2005) Rapid hop diffusion of a G-protein-coupled receptor in the plasma membrane as revealed by single-molecule techniques. Biophys J 88:3659–3680
57. Dietrich C, Volovyk ZN, Levi M, Thompson NL, Jacobson K (2001) Partitioning of Thy-1, GM1, and cross-linked phospholipid analogs into lipid rafts reconstituted in supported model membrane monolayers. Proc Natl Acad Sci USA 98:10642–10647
58. Schutz GJ, Kada G, Pastushenko VP, Schindler H (2000) Properties of lipid microdomains in a muscle cell membrane visualized by single molecule microscopy. EMBO J 19:892–901
59. Kahya N, Scherfeld D, Bacia K, Poolman B, Schwille P (2003) Probing lipid mobility of raft-exhibiting model membranes by fluorescence correlation spectroscopy. J Biol Chem 278:28109–28115
60. Scherfeld D, Kahya N, Schwille P (2003) Lipid dynamics and domain formation in model membranes composed of ternary mixtures of unsaturated and saturated phosphatidylcholines and cholesterol. Biophys J 85:3758–3768
61. Bacia K, Scherfeld D, Kahya N, Schwille P (2004) Fluorescence correlation spectroscopy relates rafts in model and native membranes. Biophys J 87:1034–1043
62. Vereb G, Matkó J, Vámosi G, Ibrahim SM, Magyar E, Varga S, Szöllősi J, Jenei A, Gáspár R Jr, Waldmann TA, Damjanovich S (2000) Cholesterol-dependent clustering of IL-2Rα and its colocalization with HLA and CD48 on T lymphoma cells suggest their functional association with lipid rafts. Proc Natl Acad Sci USA 97:6013–6018
63. Matkó J, Bodnar A, Vereb G, Bene L, Vámosi G, Szentesi G, Szöllősi J, Gáspár R, Horejsi V, Waldmann TA, Damjanovich S (2002) GPI-microdomains (membrane rafts) and signaling of the multi-chain interleukin-2 receptor in human lymphoma/leukemia T cell lines. Eur J Biochem 269:1199–1208
64. Pralle A, Keller P, Florin EL, Simons K, Horber JK (2000) Sphingolipid-cholesterol rafts diffuse as small entities in the plasma membrane of mammalian cells. J Cell Biol 148:997–1008
65. Manes S, del Real G, Martinez AC (2003) Pathogens: raft hijackers. Nat Rev Immunol 3:557–568
66. Parton RG, Richards AA (2003) Lipid rafts and caveolae as portals for endocytosis: new insights and common mechanisms. Traffic 4:724–738
67. Carrasco M, Amorim MJ, Digard P (2004) Lipid raft-dependent targeting of the influenza A virus nucleoprotein to the apical plasma membrane. Traffic 5:979–992
68. Wilflingseder D, Stoiber H (2007) Float on: lipid rafts in the lifecycle of HIV. Front Biosci 12:2124–2135
69. Hambleton S, Steinberg SP, Gershon MD, Gershon AA (2007) Cholesterol dependence of varicella-zoster virion entry into target cells. J Virol 81:7548–7558
70. Stryer L, Haugland RP (1967) Energy transfer: a spectroscopic ruler. Proc Natl Acad Sci USA 58:719–726
71. Nagy P, Vereb G, Damjanovich S, Mátyus L, Szöllősi J (2006) Measuring FRET in flow and image cytometry. In: Robinson JP (ed) Current protocols in cytometry. Wiley, New York, pp 12.18.11–12.18.13
72. Szöllősi J, Damjanovich S, Nagy P, Vereb G, Mátyus L (2006) Principles of resonance energy transfer. In: Robinson JP (ed) Current protocols in cytometry. Wiley, New York, pp 1.12.11–11.12.16
73. VanBeek DB, Zwier MC, Shorb JM, Krueger BP (2007) Fretting about FRET: correlation between kappa and R. Biophys J 92:4168–4178
74. Dale RE, Eisinger J, Blumberg WE (1979) The orientational freedom of molecular probes. The orientation factor in intramolecular energy transfer. Biophys J 26:161–193

75. Anikovsky M, Dale L, Ferguson S, Petersen N (2008) Resonance energy transfer in cells: a new look at fixation effect and receptor aggregation on cell membrane. Biophys J 95:1349–1359
76. Kenworthy AK, Edidin M (1998) Distribution of a glycosylphosphatidylinositol-anchored protein at the apical surface of MDCK cells examined at a resolution of <100 A using imaging fluorescence resonance energy transfer. J Cell Biol 142:69–84
77. Bastiaens PI, Majoul IV, Verveer PJ, Soling HD, Jovin TM (1996) Imaging the intracellular trafficking and state of the AB5 quaternary structure of cholera toxin. EMBO J 15:4246–4253
78. Karpova TS, Baumann CT, He L, Wu X, Grammer A, Lipsky P, Hager GL, McNally JG (2003) Fluorescence resonance energy transfer from cyan to yellow fluorescent protein detected by acceptor photobleaching using confocal microscopy and a single laser. J Microsc 209:56–70
79. Kirber MT, Chen K, Keaney JF Jr (2007) YFP photoconversion revisited: confirmation of the CFP-like species. Nat Methods 4:767–768
80. Roszik J, Szöllősi J, Vereb G (2008) AccPbFRET: an ImageJ plugin for semi-automatic, fully corrected analysis of acceptor photobleaching FRET images. BMC Bioinform 9:346
81. Rasband WS (1997) ImageJ, US National Institutes of Health, Bethesda, MD. http://res.info.nih.gov/ij
82. Stepensky D (2007) FRETcalc plugin for calculation of FRET in non-continuous intracellular compartments. Biochem Biophys Res Commun 359:752–758
83. Trón L, Szöllősi J, Damjanovich S, Helliwell SH, Arndt-Jovin DJ, Jovin TM (1984) Flow cytometric measurement of fluorescence resonance energy transfer on cell surfaces. Quantitative evaluation of the transfer efficiency on a cell-by-cell basis. Biophys J 45:939–946
84. Sebestyén Z, Nagy P, Horváth G, Vámosi G, Debets R, Gratama JW, Alexander DR, Szöllősi J (2002) Long wavelength fluorophores and cell-by-cell correction for autofluorescence significantly improves the accuracy of flow cytometric energy transfer measurements on a dual-laser benchtop flow cytometer. Cytometry 48:124–135
85. Roszik J, Lisboa D, Szöllősi J, Vereb G (2009) Evaluation of intensity-based ratiometric FRET in image cytometry – approaches and a software solution. Cytom A 75:761–767
86. Feige JN, Sage D, Wahli W, Desvergne B, Gelman L (2005) PixFRET, an ImageJ plug-in for FRET calculation that can accommodate variations in spectral bleed-throughs. Microsc Res Tech 68:51–58
87. Fazekas Z, Petrás M, Fábián A, Pályi-Krekk Z, Nagy P, Damjanovich S, Vereb G, Szöllősi J (2008) Two-sided fluorescence resonance energy transfer for assessing molecular interactions of up to three distinct species in confocal microscopy. Cytom A 73:209–219
88. Jovin TM, Arndt-Jovin DJ (1989) Luminescence digital imaging microscopy. Annu Rev Biophys Biophys Chem 18:271–308
89. Young RM, Arnette JK, Roess DA, Barisas BG (1994) Quantitation of fluorescence energy transfer between cell surface proteins via fluorescence donor photobleaching kinetics. Biophys J 67:881–888
90. Chen Y, Muller JD, So PT, Gratton E (1999) The photon counting histogram in fluorescence fluctuation spectroscopy. Biophys J 77:553–567
91. Brock R, Hink MA, Jovin TM (1998) Fluorescence correlation microscopy of cells in the presence of autofluorescence. Biophys J 75:2547–2557
92. Schwille P, Korlach J, Webb WW (1999) Fluorescence correlation spectroscopy with single-molecule sensitivity on cell and model membranes. Cytometry 36:176–182
93. Digman MA, Dalal R, Horwitz AF, Gratton E (2008) Mapping the number of molecules and brightness in the laser scanning microscope. Biophys J 94:2320–2332
94. Digman MA, Brown CM, Sengupta P, Wiseman PW, Horwitz AR, Gratton E (2005) Measuring fast dynamics in solutions and cells with a laser scanning microscope. Biophys J 89:1317–1327
95. Runnels LW, Scarlata SF (1995) Theory and application of fluorescence homotransfer to melittin oligomerization. Biophys J 69:1569–1583

96. Bader AN, Hofman EG, Voortman J, van Bergen En Henegouwen PM, Gerritsen HC (2009) Homo-FRET imaging enables quantification of protein cluster sizes with subcellular resolution. Biophys J 97:2613–2622
97. Lidke DS, Nagy P, Barisas BG, Heintzmann R, Post JN, Lidke KA, Clayton AH, Arndt-Jovin DJ, Jovin TM (2003) Imaging molecular interactions in cells by dynamic and static fluorescence anisotropy (rFLIM and emFRET). Biochem Soc Trans 31:1020–1027
98. Yeow EK, Clayton AH (2007) Enumeration of oligomerization states of membrane proteins in living cells by homo-FRET spectroscopy and microscopy: theory and application. Biophys J 92:3098–3104
99. Gadella TW Jr, Jovin TM (1995) Oligomerization of epidermal growth factor receptors on A431 cells studied by time-resolved fluorescence imaging microscopy. A stereochemical model for tyrosine kinase receptor activation. J Cell Biol 129:1543–1558
100. Clayton AH, Klonis N, Cody SH, Nice EC (2005) Dual-channel photobleaching FRET microscopy for improved resolution of protein association states in living cells. Eur Biophys J 34:82–90
101. Hoppe A, Christensen K, Swanson JA (2002) Fluorescence resonance energy transfer-based stoichiometry in living cells. Biophys J 83:3652–3664
102. Szabó Á, Szöllősi J, Nag P (2010) Coclustering of ErbB1 and ErbB2 revealed by FRET-sensitized acceptor bleaching. Biophys J 99(1):105–114
103. Mekler VM (1994) A photochemical technique to enhance sensitivity of detection of fluorescence resonance energy transfer. Photochem Photobiol 59:615–620
104. Mekler VM, Averbakh AZ, Sudarikov AB, Kharitonova OV (1997) Fluorescence energy transfer-sensitized photobleaching of a fluorescent label as a tool to study donor-acceptor distance distributions and dynamics in protein assemblies: studies of a complex of biotinylated IgM with streptavidin and aggregates of concanavalin A. J Photochem Photobiol B 40:278–287
105. Bublil EM, Yarden Y (2007) The EGF receptor family: spearheading a merger of signaling and therapeutics. Curr Opin Cell Biol 19:124–134
106. Klapper LN, Glathe S, Vaisman N, Hynes NE, Andrews GC, Sela M, Yarden Y (1999) The ErbB-2/HER2 oncoprotein of human carcinomas may function solely as a shared coreceptor for multiple stroma-derived growth factors. Proc Natl Acad Sci USA 96:4995–5000
107. Lemmon MA (2009) Ligand-induced ErbB receptor dimerization. Exp Cell Res 315:638–648
108. Mocanu MM, Fazekas Z, Petrás M, Nagy P, Sebestyén Z, Isola J, Timar J, Park JW, Vereb G, Szöllősi J (2005) Associations of ErbB2, β1-integrin and lipid rafts on Herceptin (Trastuzumab) resistant and sensitive tumor cell lines. Cancer Lett 227:201–212
109. Pályi-Krekk Z, Barok M, Isola J, Tammi M, Szöllősi J, Nagy P (2007) Hyaluronan-induced masking of ErbB2 and CD44-enhanced trastuzumab internalisation in trastuzumab resistant breast cancer. Eur J Cancer 43:2423–2433
110. Kawashima N, Nakayama K, Itoh K, Itoh T, Ishikawa M, Biju V (2010) Reversible dimerization of EGFR revealed by single-molecule fluorescence imaging using quantum dots. Chemistry 16:1186–1192
111. Lidke DS, Wilson BS (2009) Caught in the act: quantifying protein behaviour in living cells. Trends Cell Biol 19:566–574
112. Gimpl G, Gehrig-Burger K (2007) Cholesterol reporter molecules. Biosci Rep 27:335–358
113. Dijkstra J, Swartz GM Jr, Raney JJ, Aniagolu J, Toro L, Nacy CA, Green SJ (1996) Interaction of anti-cholesterol antibodies with human lipoproteins. J Immunol 157:2006–2013
114. Swartz GM Jr, Gentry MK, Amende LM, Blanchette-Mackie EJ, Alving CR (1988) Antibodies to cholesterol. Proc Natl Acad Sci USA 85:1902–1906
115. Clarke MS, Vanderburg CR, Bamman MM, Caldwell RW, Feeback DL (2000) In situ localization of cholesterol in skeletal muscle by use of a monoclonal antibody. J Appl Physiol 89:731–741

116. Perl-Treves D, Kessler N, Izhaky D, Addadi L (1996) Monoclonal antibody recognition of cholesterol monohydrate crystal faces. Chem Biol 3:567–577
117. Kruth HS, Ifrim I, Chang J, Addadi L, Perl-Treves D, Zhang WY (2001) Monoclonal antibody detection of plasma membrane cholesterol microdomains responsive to cholesterol trafficking. J Lipid Res 42:1492–1500
118. Smart EJ, Ying Y, Donzell WC, Anderson RG (1996) A role for caveolin in transport of cholesterol from endoplasmic reticulum to plasma membrane. J Biol Chem 271:29427–29435
119. Underwood KW, Jacobs NL, Howley A, Liscum L (1998) Evidence for a cholesterol transport pathway from lysosomes to endoplasmic reticulum that is independent of the plasma membrane. J Biol Chem 273:4266–4274
120. Steinbach G, Pomozi I, Zsiros O, Pay A, Horvath GV, Garab G (2008) Imaging fluorescence detected linear dichroism of plant cell walls in laser scanning confocal microscope. Cytom A 73:202–208
121. Bíró A, Horváth A, Varga L, Nemesanszky E, Csepregi A, David K, Tolvaj G, Ibranyi E, Telegdy L, Par A, Romics L, Karadi I, Horanyi M, Gervain J, Ribiczey P, Csondes M, Fust G (2003) Serum anti-cholesterol antibodies in chronic hepatitis-C patients during IFN-α-2b treatment. Immunobiology 207:161–168
122. Egri G, Orosz I (2006) Elevated anti-cholesterol antibody levels in the sera of non-small cell lung cancer patients. Interact Cardiovasc Thorac Surg 5:649–651
123. Horváth A, Banhegyi D, Bíró A, Ujhelyi E, Veres A, Horváth L, Prohaszka Z, Bacsi A, Tarjan V, Romics L, Horváth I, Toth FD, Fust G, Karadi I (2001) High level of anticholesterol antibodies (ACHA) in HIV patients. Normalization of serum ACHA concentration after introduction of HAART. Immunobiology 203:756–768
124. Horváth A, Fust G, Horváth I, Vallus G, Duba J, Harcos P, Prohaszka Z, Rajnavölgyi E, Janoskuti L, Kovács M, Császár A, Romics L, Karadi I (2001) Anti-cholesterol antibodies (ACHA) in patients with different atherosclerotic vascular diseases and healthy individuals. Characterization of human ACHA. Atherosclerosis 156:185–192
125. Gombos I, Detre C, Vámosi G, Matkó J (2004) Rafting MHC-II domains in the APC (presynaptic) plasma membrane and the thresholds for T-cell activation and immunological synapse formation. Immunol Lett 92:117–124
126. Poloso NJ, Roche PA (2004) Association of MHC class II-peptide complexes with plasma membrane lipid microdomains. Curr Opin Immunol 16:103–107
127. Pizzo P, Viola A (2004) Lipid rafts in lymphocyte activation. Microbes Infect 6:686–692
128. Rosenberger CM, Brumell JH, Finlay BB (2000) Microbial pathogenesis: lipid rafts as pathogen portals. Curr Biol 10:R823–R825
129. Yoshizaki F, Nakayama H, Iwahara C, Takamori K, Ogawa H, Iwabuchi K (2008) Role of glycosphingolipid-enriched microdomains in innate immunity: microdomain-dependent phagocytic cell functions. Biochim Biophys Acta 1780:383–392
130. Beck Z, Balogh A, Kis A, Izsepi E, Cervenak L, László G, Bíró A, Liliom K, Mocsár G, Vámosi G, Fust G, Matkó J (2010) New cholesterol-specific antibodies remodel HIV-1 target cells' surface and inhibit their in vitro virus production. J Lipid Res 51:286–296
131. Brown BK, Karasavvas N, Beck Z, Matyas GR, Birx DL, Polonis VR, Alving CR (2007) Monoclonal antibodies to phosphatidylinositol phosphate neutralize human immunodeficiency virus type 1: role of phosphate-binding subsites. J Virol 81:2087–2091
132. Kozak SL, Heard JM, Kabat D (2002) Segregation of CD4 and CXCR4 into distinct lipid microdomains in T lymphocytes suggests a mechanism for membrane destabilization by human immunodeficiency virus. J Virol 76:1802–1815
133. Schwille P, Haupts U, Maiti S, Webb WW (1999) Molecular dynamics in living cells observed by fluorescence correlation spectroscopy with one- and two-photon excitation. Biophys J 77:2251–2265

134. Rawat SS, Zimmerman C, Johnson BT, Cho E, Lockett SJ, Blumenthal R, Puri A (2008) Restricted lateral mobility of plasma membrane CD4 impairs HIV-1 envelope glycoprotein mediated fusion. Mol Membr Biol 25:83–94
135. Rawat SS, Gallo SA, Eaton J, Martin TD, Ablan S, KewalRamani VN, Wang JM, Blumenthal R, Puri A (2004) Elevated expression of GM3 in receptor-bearing targets confers resistance to human immunodeficiency virus type 1 fusion. J Virol 78:7360–7368
136. Manes S, del Real G, Lacalle RA, Lucas P, Gomez-Mouton C, Sanchez-Palomino S, Delgado R, Alcami J, Mira E, Martinez AC (2000) Membrane raft microdomains mediate lateral assemblies required for HIV-1 infection. EMBO Rep 1:190–196
137. Nguyen DH, Giri B, Collins G, Taub DD (2005) Dynamic reorganization of chemokine receptors, cholesterol, lipid rafts, and adhesion molecules to sites of CD4 engagement. Exp Cell Res 304:559–569
138. Calarese DA, Scanlan CN, Zwick MB, Deechongkit S, Mimura Y, Kunert R, Zhu P, Wormald MR, Stanfield RL, Roux KH, Kelly JW, Rudd PM, Dwek RA, Katinger H, Burton DR, Wilson IA (2003) Antibody domain exchange is an immunological solution to carbohydrate cluster recognition. Science 300:2065–2071
139. Kunert R, Wolbank S, Stiegler G, Weik R, Katinger H (2004) Characterization of molecular features, antigen-binding, and in vitro properties of IgG and IgM variants of 4E10, an anti-HIV type 1 neutralizing monoclonal antibody. AIDS Res Hum Retroviruses 20:755–762
140. Zwick MB, Komori HK, Stanfield RL, Church S, Wang M, Parren PW, Kunert R, Katinger H, Wilson IA, Burton DR (2004) The long third complementarity-determining region of the heavy chain is important in the activity of the broadly neutralizing anti-human immunodeficiency virus type 1 antibody 2F5. J Virol 78:3155–3161
141. Phogat S, Wyatt RT, Karlsson Hedestam GB (2007) Inhibition of HIV-1 entry by antibodies: potential viral and cellular targets. J Intern Med 262:26–43

Origin of Tryptophan Fluorescence

J.R. Albani

1 Introduction

In February 1985, I attended the meeting of the Biophysical Society in Baltimore. During the meeting, Professor Ludwig Brand invited researchers working in fluorescence field to his laboratory to a soirée he entitled "cheese and wine around fluorescence." In fact, during this 3 h short but interesting nonformal session, scientists main subject was "origin of tryptophan fluorescence in proteins." I still remember that this evening I have asked the following question: "why do we not consider tryptophan as a person who is going to act or react depending on his environment?" I did not get any answer. I notice now, after long years of work, that I suggested something logical although incomplete. In fact, tryptophan, like human being, has specific properties, intrinsic ones. Thus, these properties could be measured and characterized. Also, behavior of tryptophan is modified depending on its environment. However, do all fluorescence observables give the same information? Are they modified identically with the tryptophan environment perturbation?

On my arrival to the University of Lille 1 in 1988, I studied structure and dynamics properties of a small acute-phase protein, α_1-acid glycoprotein, by means of fluorescence spectroscopy. Crystallization of α_1-acid glycoprotein proved to be somewhat difficult, because of the high solubility and the large carbohydrate moiety of the protein (40% carbohydrate by weight). Up to now, no crystallographic data have been reported concerning the spatial structure of the protein, probably because of the instability of the crystals themselves. Therefore, we had an important reason to focus on fluorescence studies since this method can help obtaining information on structure and dynamics of macromolecules.

J.R. Albani (✉)
Laboratoire de Biophysique Moléculaire, Université des Sciences et Technologies de Lille, Bâtiment C6. 59655, Villeneuve d'Ascq Cédex, France
e-mail: Jihad-Rene.Albani@univ-lille1.fr

C.D. Geddes (ed.), *Reviews in Fluorescence 2009*, Reviews in Fluorescence 2009,
DOI 10.1007/978-1-4419-9672-5_9,

While performing our studies on α_1-acid glycoprotein and by comparing our data to those obtained on other proteins and on free tryptophan in solution, we were able to build a model that gives a logical explanation of tryptophan fluorescence origin. Data published by other researchers on tryptophan fluorescence were in good agreement with those obtained in our laboratory. However, the main difference was in the results interpretation. For example, fluorescence lifetimes of Trp residues are considered to originate from rotamers or conformers resulting from the rotation of the indole ring within the peptide bonds. However, how can we explain the fact that in most of the proteins, the two lifetimes 0.5 and 3 ns, attributed to the conformers, are also observed for free tryptophan in solution? Tryptophan has two overlapping $S_0 \rightarrow S_1$ electronic transitions (^1L_a and ^1L_b) which are perpendicular to each other. Both $S_0 \rightarrow {}^1L_a$ and $S_0 \rightarrow {}^1L_b$ transitions occur in the 260–300 nm range. What is the role played by each of the two transitions in generating the fluorescence lifetimes?

The best way to answer to these questions is by reading and analyzing the data. Well, let us do it.

2 Protein Structure and Tryptophan Fluorescence Lifetime

Important number of papers in the literature correlates fluorescence origin proteins primary and tertiary structures. In peptides and proteins, rotation of the polypeptide chain is observed at the level of the α-carbon atom of each amino acid. This rotation is characterized by two well-defined angles of torsion or dihedral ϕ and ψ (Fig. 1). The values of the two angles are illustrated in the Ramachandran diagram, which defines the limits of conformational freedom for each peptide bond unit and hence for the entire polypeptide chain. However, even if limited conformations are allowed for each amino acid, the number of possible conformations for the polypeptide remains high. For example, a 100-residue protein with three conformations for each amino acid would have 3^{100} possible conformations. Therefore, number of conformations that a protein can possess is very important.

A folded protein could have a set of different conformations, thus we have here a first definition of a protein structure: the global structure is a combination of substructures or conformations. The interconversion between them is not too fast. Each conformation is rigid and has a definite specific structure. This model is known as the rotamers model. However, one should not forget that a protein displays motions. Therefore, each conformation or substructure possesses local motions, i.e., a folded protein would display very complex dynamics. Therefore, when we perform spectroscopic measurements, we could monitor the mean dynamics of the different conformations.

A fast interconversion between different subspecies induces a more complex system. In this case, fluorescence measurements would be a mean value of all the different possible subspecies. Interactions that occur between different amino acid residues of a protein generate local dynamics that should be analyzed differently

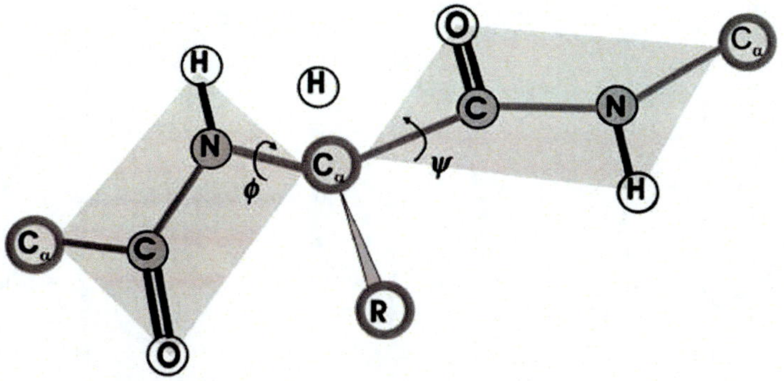

Fig. 1 Illustration of the peptide plane (*gray area*) and ϕ–Ψ angles. *The line* formed by the repeating –Cα–C–N–Cα– is the backbone of the peptide chain. Mathematically, phi (ϕ) and psi (Ψ) are the dihedral angle (also known as tensional angle) which is defined as the angle between the point (e.g., Cα) at the end of a four-point sequence and the plane (e.g., peptide plane) occupied by the other three points. In a peptide, phi–psi angles are restricted to certain ranges. A plot of their distribution is called the Ramachandran plot. *Source*: Web Book publications

from the interconversions between the different conformations. Therefore, proteins and macromolecules in general possess local flexibilities allowing diffusion of small molecules such as solvent or oxygen through the macromolecules. This flexibility varies from a domain to another and in general protein surface is more flexible than its interior. This local flexibility can be put into evidence by monitoring, e.g., oxygen diffusion as a function of temperature.

Although scientists agree with the fact that when a tryptophan is buried inside the hydrophobic core of a protein, its fluorescence is blue shifted compared to the fluorescence observed from a tryptophan present at the protein surface, origin of the fluorescence lifetime is still in great debate.

Tryptophan in peptides and proteins exhibits a bi- or multiexponential fluorescence decay. One explanation of this decay is the presence of conformers in equilibrium in the folded state. Each conformer exhibits one specific fluorescence lifetime. Conformers origin arises from indole ring rotation within the C_α–C_β bond and/or the C_β–C_γ bond. Interconversion between the rotamers is slow relative to the fluorescence time scale. Rotamers are considered also as rigid entities (Fig. 2).

Rotamers model was used to explain origin of the biexponential decay of tryptophan free in solution. In polypeptides, lifetime of each rotamer is explained as the result of the quenching interactions between the indole and quenching groups in the fluorophore. Charge transfer from the excited indole moiety to the carbonyl group of the peptide bond, excited state electron and proton transfer, and solvent-quenching are described to play a role in the deexcitation process of tryptophan fluorescence [1–4]. In rotamer A, the quenching group near the indole is the amino-group. Fluorescence quenching is weak and thus fluorescence lifetime is long (around 3–4 ns). In rotamer C, the carbonyl and amino groups surround indole group. Thus, fluorescence quenching is important and lifetime is weak (around 0.5 ns).

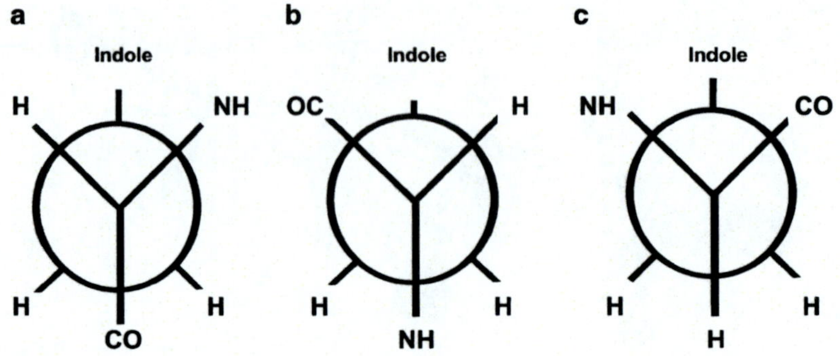

Fig. 2 The three possible rotamers of tryptophan in proteins

However, the rotamer model does not explain why rotamer B does not have any fluorescence!!! If we have to apply this model to tryptophan in polypeptides, what will be the contribution of protein structure and dynamics to the fluorescence lifetime? Also, we do not see how and why the rotamers model, if it is correct, can be applied to free tryptophan in solution.

3 Description of Some Proteins Studied in This Review

3.1 α_1-Acid Glycoprotein

α_1-Acid glycoprotein (orosomucoid) is a small acute-phase glycoprotein (Mr = 41,000) that is negatively charged at physiological pH. It consists of a chain of 183 amino acids [5], contains 40% carbohydrate by weight, and has up to 16 sialic acid residues (10–14% by weight) [6]. Five heteropolysaccharide groups are linked via an *N*-glycosidic bond to the asparaginyl residues of the protein [7] (Fig. 3). The protein contains tetra-antennary as well as di- and tri-antennary glycans.

Although biological function of α_1-acid glycoprotein is still obscure, a number of activities of possible significance have been described such as, ability to bind the β-drug adrenergic blocker, propranolol [8], and certain steroid hormones such as progesterone [6]. Many of these activities have been shown to be dependent on the glycoform of α_1-acid glycoprotein [9]. As the serum concentration of specific glycoforms of α_1-acid glycoprotein changes markedly under acute or chronic inflammatory conditions, as well as in pregnancy and tumor growth, a pathophysiological dependence change in the carbohydrate-dependent activities of the protein may occur. Therefore, relation between function of α_1-acid glycoprotein and pathophysiological changes in glycosylation was extensively studied [10–12].

α_1-Acid glycoprotein contains three Trp residues, one residue, Trp-160, is at the surface of the protein and two, Trp-122 and Trp-25, are located in the protein matrix [6, 7]. We studied structure and dynamics of α_1-acid glycoprotein in solution by

```
                                                                                 20
1                                                              *                 Arg
Glu-Ile-Pro-Leu-Cys-Ala-Asn-Leu-Val-Pro-Val-Pro-Ile-Thr-Asn-Ala-Thr-Leu-Asp-------Ile-Thr-Gly- Lys -Trp-Phe-
                                                                                 Gln

                                                 *      40
Tyr-Ile-Ala-Ser-Ala-Phe-Arg-Asn-Glu-Glu-Tyr-Asn-Lys-Ser-Val-Gln-Glu-Ile-Gln-Ala-Thr-Phe-Phe-Tyr-Phe-Thr-

  *                         60                                    Asp          Ile       * Thr-Thr
Pro-Asn-Lys-Thr-Glu-Asp-Thr-Ile-Phe-Leu-Arg-Glu-Tyr-Gln-Thr-Arg-Gln--------Gln-Cys--------Tyr-Asn-----------
                                                                  Asn          Phe         Ser-Ser

   80                               *         Ile            Val          Gln          Val     100
Tyr-Leu-Asn-Val-Gln-Arg-Glu-Asn-Gly-Thr---------Ser-Arg-Tyr----------Gly-Gly--------Glu-His------- Ala-His-Leu-
                                              Val            Glu          Arg          Phe

                                   Leu       Phe-Gly-Ser-Tyr-Leu-Asp              120
Leu-Ile-Leu-Arg-Asp-Thr-Lys-Thr--------Met----------------------------------Asp-Glu-Lys-Asn-Trp-Gly- Leu-Ser-
                                   Tyr       Leu-Ala-Phe-Asp-Val-Asn

 Phe                                                       140
---------Tyr-Ala-Asp-Lys-Pro-Glu-Thr-Thr-Lys-Glu-Gln-Leu-Gly-Glu-Phe-Tyr-Glu-Ala-Leu-Asp-Cys-Leu-
 Val

 Cys          Arg                Met              160
---------Ile-Pro---------Ser-Asp-Val---------Tyr-Thr-Asp-Trp-Lys-Lys-Asp-Lys-Cys-Glu-Pro-Leu-Glu-Lys-Gln-His-
 Arg          Lys                Val

                             180
Glu- Lys-Glu-Arg-Lys-Gln-Glu-Glu-Gly-Glu-Ser.
```

Fig. 3 Primary structure of α_1-acid glycoprotein. The five heteropolysaccharide units are linked *N*-glycosidacilly to the asparagine residues that are marked with a star. *Sources*: Schmid K, Kaufmann H, Isemura S, Bauer F, Emura J, Motoyama T, Ishiguro M, Nanno S (1973) Biochemistry 12:2711–2724 and Dente L, Pizza MG, Metspalu A, Cortese R (1987) Structure and expression of the genes coding for human α_1-acid glycoprotein. EMBO J 6:2289–2296

investigating fluorescence of Trp residues and of extrinsic probes such as TNS and calcofluor white. Our results showed:

1. The three Trp residues contribute to the fluorescence of α_1-acid glycoprotein [13].
2. The N-terminal fragment of α_1-acid glycoprotein adopts a spatial conformation so that a pocket in contact with the buffer is induced [14–16].
3. One of the supposedly buried Trp residues (Trp-25) lies in fact in a hydrophobic environment in the pocket [17].
4. The progesterone-binding site is within the pocket near one Trp residue (Trp-25 residue). The important motions of the pockets facilitate binding of hormone to the protein [17].
5. Fluorescence of the extrinsic probe, calcofluor, allowed us to put into evidence that carbohydrates of α_1-acid glycoprotein possess a secondary structure which is absent when they are not bound to the protein [18–23].

3.2 *Cyclophilin B*

Cyclophilins are highly conserved proteins that are present in many types of cells with a considerably divergent phylogenetic distribution [24]. Several cyclophilin isoforms with distinct subcellular localizations have been described, cyclophilin A, B, and C.

Cyclophilin A (CyPA) (18 kDa) is an abundant cytosolic protein [25]. The 22-kDa cyclophilin B (CyPB) is located in the endoplasmic reticulum and shares 64% sequence identity to CyPA [26] (Fig. 4). Cyclophilin C (CyPC) (23 kDa) shares 55.8% sequence identity to CyPA and was found to be particularly abundant in kidneys [27]. Finally, there is cyclophilin D (CyPD) that is a mitochondrial protein [28, 29].

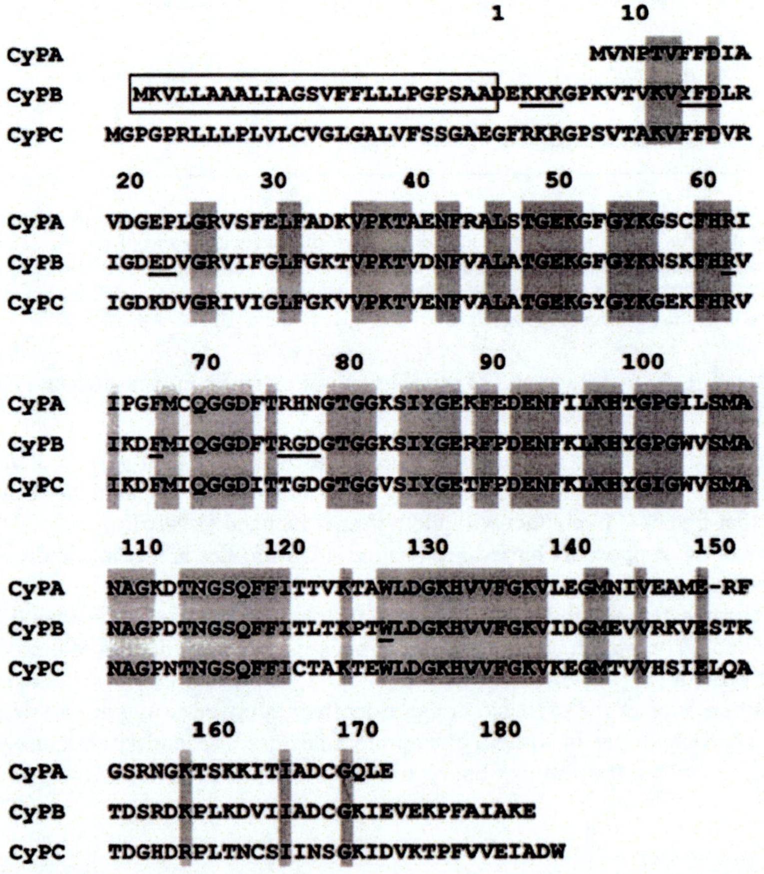

Fig. 4 Amino acids sequence alignment of CyPB with CyPA and CyPC. The MKVLLAAA-LIAGSVFFLLLPGPSAA peptide is boxed and corresponds to a signal peptide that does not exist in mature CyPB. *Gray areas* indicate amino acid residues conserved among CyPA, CyPB, and CyPC. *Source*: Carpentier M, Allain F, Haendler B, Denys A, Mariller C, Benaissa M, Spik G (1999) J Biol Chem 274:10990–10998

Cyclophilins A, B, and C could be involved in degradation of the genome during apoptosis [30] and cyclophilin D plays a role in the mitochondrial permeability transition [31]. Human cyclophilins belong to the family of peptidyl–prolyl cis–trans isomerases (PPIases). PPIases catalyze the interconversion of the cis and trans isomers of the peptidyl–prolyl bonds in peptide and protein substrates [32, 33].

The potent and clinically useful immunosuppressant cyclosporin A (CsA) binds to human cyclophilins, inducing in this way an important inhibition of their enzymatic activity [25, 34]. Recent work has shown that binding site of CsA on cyclophilin B and catalytic site of the protein are distinct [35].

Human cyclophilin B is a monomeric protein consisting of 183 amino acids and containing two tryptophan residues, Trp104 and 128. Trp128-residue belongs to the binding site of cyclosporin A and is the homologous of Trp 121 in CyPA, while Trp-104 residue belongs to the hydrophobic pocket. The two sites are localized in a large hydrophobic pocket [36]. Fluorescence studies with Trp residues and the extrinsic fluorophore TNS indicate that the pocket is rigid [37]. $\text{CyPB}_{\text{W128A}}$ mutant is unable to interact with CsA, demonstrating that the requirement of this amino acid residue in the interactions with the drug is also a common feature of CyPA and CyPB. However, $\text{CyPB}_{\text{W128A}}$ mutant retains 60% of the capacity to accelerate the cis–trans isomerization of a Pro-containing substrate, indicating that Trp-128 residue in CyPB was not essential for PPIase activity. This is not the case for CyPA where mutation of the corresponding Trp-121 residue was reported to strongly reduce the enzymatic activity [35].

Comparison of fluorescence spectra of Trp residues in wild-type cyclophilin B and $\text{CyPB}_{\text{W128A}}$ indicates that both 128 and 104-Trp residues participate to the protein emission, although this fluorescence seems to be dominated by that of Trp-104 residue. In fact, quantum yields of Trp-104 residue (0.061) is five times that of Trp-128 residue (0.012) [37]. The low quantum yield of Trp-128 can be explained by the possibility of an electron transfer from the Trp residue to both Phe-67 and Asp-79 both present within a distance of 8.5 Å. Electron transfer toward the two neighboring amino acids decreases enormously the quantum yield of Trp-128 residue. Fact that quantum yield of Trp-104 residue is very close in wild type (0.054) and in the mutant (0.049) indicates the absence of energy transfer between Trp-128 and 104 residues in the wild type. This may be explained by the fact that the phenol rings of the two Trp residues are perpendicular one to each other (Fig. 5). Therefore, electron transfer from Trp-128 to Phe-67 and Asp-79 could be a cause of the low value of the quantum yield of Trp-128 residue.

However, another interpretation of the weak fluorescence of Trp-128 may also exist. Fluorescence quantum yield compares emitted photons to absorbed ones. However, there is no indication up to now that all absorbed photons participate in the excitation process. Therefore, the low quantum yield of Trp-128 in the wild-type protein can also be explained by the fact that this tryptophan residue was not excited by the light source [37]. The lack of excitation would be the result of structural rearrangement within the fluorophore microenvironment [38].

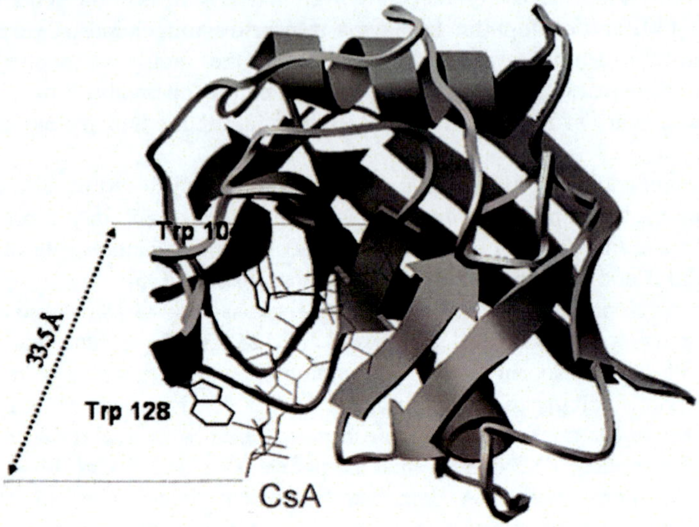

Fig. 5 Tertiary structure of cyclophilin B

3.3 Human Serum Albumin and Bovine Serum Albumin

Bovine serum albumin (BSA), a large globular protein (66,000 Da), consists of a single chain of 583 amino acids residues and forms subdomains by paired 17 disulfide bonds [39]. BSA binds free fatty acids, other lipids, and flavor compounds that can alter heat denaturation of the protein [40]. Also, BSA may play a role in lipid oxidation [41] and in the maintenance of blood pH [42]. BSA is formed from three domains I, II, and III, divided into two subdomains (A and B), and contains two tryptophan residues, Trp-134 and Trp-212, located, respectively, in domains I and II. Trp-212 residue is surrounded by a hydrophobic environment within a protein pocket while Trp-134 residue is located in a hydrophilic environment, close to the protein surface [39].

Human serum albumin (HSA) is the most abundant protein in blood plasma. It is produced in the liver. Albumin comprises about half of the blood serum protein. HSA is soluble and monomeric [43]. It has been known for a long time that human blood proteins like hemoglobin and serum albumin may undergo a slow nonenzymatic glycation, mainly by the formation of a Schiff base between ε-amino groups of lysine (and sometimes arginine) residues and glucose molecules in blood (Maillard reaction). This reaction can be inhibited in the presence of antioxidant agents [43].

The primary sequence of HSA shows that the protein is a single polypeptide with 585 residues containing 17 pairs of disulfide bridges and one free cysteine [44] (Fig. 6). HSA and many serum albumins from other species have been found to consist of three homologous domains probably derived through

*MKWVTFISLL FLFSSAYSRG VFRR*DAHKSE VAHRFKDLGE ENFKALVLIA FAQYLQQCPF EDHVKLVNEV
TEFAKTCVAD ESAENCDKSL HTLFGDKLCT VATLRETYGE MADCCAKQEP ERNECFLQHK DDNPNLPRLV
RPEVDVMCTA FHDNEETFLK KYLYEIARRH PYFYAPELLF FAKRYKAAFT ECCQAADKAA CLLPKLDELR
DEGKASSAKQ RLKCASLQKF GERAFKAWAV ARLSQRFPKA EFAEVSKLVT DLTKVHTECC HGDLLECADD
RADLAKYICE NQDSISSKLK ECCEKPLLEK SHCIAEVEND EMPADLPSLA ADFVESKDVC KNYAEAKDVF
LGMFLYEYAR RHPDYSVVLL LRLAKTYETT LEKCCAAADP HECYAKVFDE FKPLVEEPQN LIKQNCELFE
QLGEYKFQNA LLVRYTKKVP QVSTPTLVEV SRNLGKVGSK CCKHPEAKRM PCAEDYLSVV LNQLCVLHEK
TPVSDRVTKC CTESLVNRRP CFSALEVDET YVPKEFNAET FTFHADICTL SEKERQIKKQ TALVELVKHK
PKATKEQLKA VMDDFAAFVE KCCKADDKET CFAEEGKKLV AASQAALGL

Fig. 6 Amino acid sequence of human serum albumin

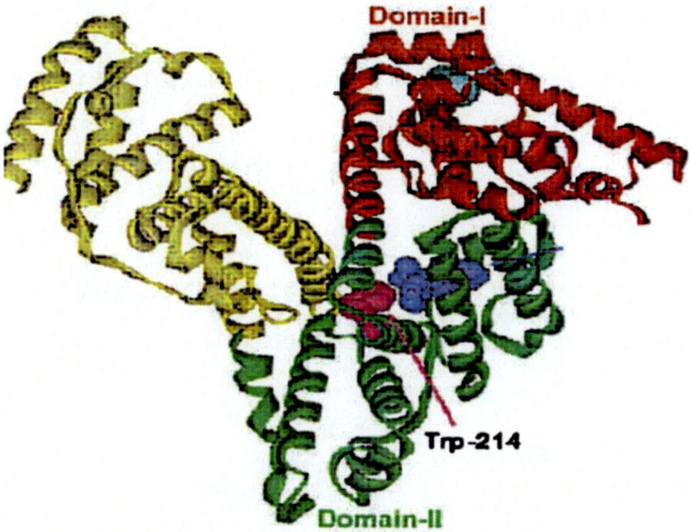

Fig. 7 Three-dimensional structure of human serum albumin. *Source*: Data Bank *(ID code 1ha2)*

gene multiplication [45]. HSA contains only one tryptophan residue (Trp-214) that participates in additional hydrophobic packing interactions at the IIA–IIIA interface [39, 46]. Trp-212 in BSA and Trp-214 in HSA are located in a similar hydrophobic microenvironment (subdomain IIA) [39].

The italicized first 24 amino acids are signal and propeptide portions not observed in the transcribed, translated, and transported protein but present in the gene. There are 609 amino acids in this sequence with only 585 amino acids in the final product observed in the blood. HSA has a molecular mass of 67 kDa [43].

Three-dimensional structure of HSA (Fig. 7) has been determined crystallographically to a resolution of 2.8 Å. It comprises three homologous domains that assemble to form a heart-shaped molecule. Each domain is a product of two subdomains that possess common structural motifs. The principal regions of ligand binding to HSA are located in hydrophobic cavities in subdomains IIA and IIIA, which exhibit close properties [46].

3.4 β-Lactoglobulin

β-Lactoglobulin is a small protein of 162 amino acid residues [47] (Mr = 18,400) (Fig. 8) which tertiary structure possesses a pocket (calyx) where hydrophobic ligands can easily bind [48] (Fig. 9). It normally exists as a dimer, each monomer is formed by 162 amino acids, with one free cysteine and two disulphide bridges [49].

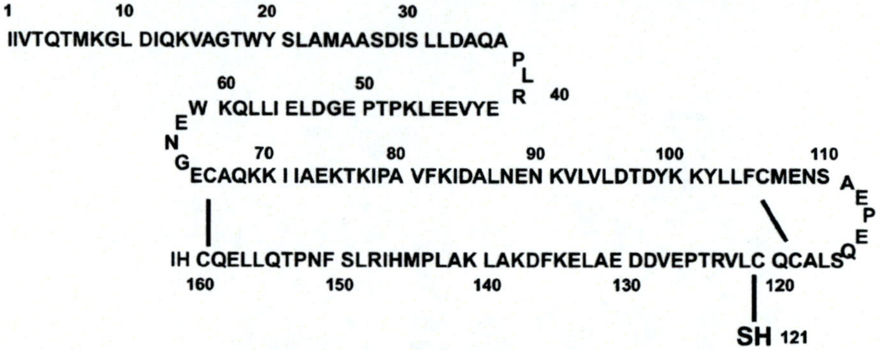

Fig. 8 Primary structure of β-LG. β-Lactoglobulin comprises 162 AA, including five Cys residues. Two disulfide linkages are located at residues Cys-106 to Cys-119 and Cys-66 to Cys-160. One free Cys is at position 121. *Source*: Liu HC, Chen WL, Mao SJT (2007) J Dairy Sci 90:547–555

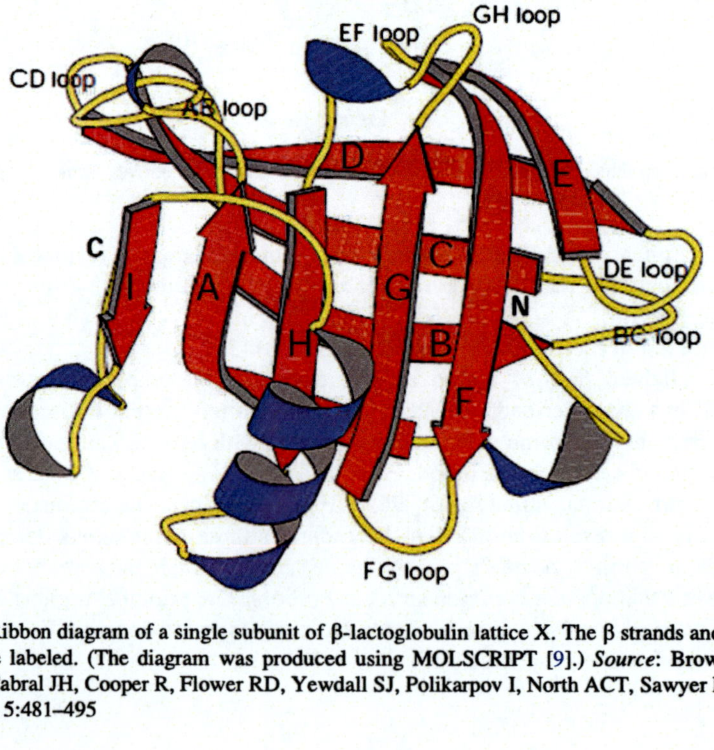

Fig. 9 Ribbon diagram of a single subunit of β-lactoglobulin lattice X. The β strands and joining loops are labeled. (The diagram was produced using MOLSCRIPT [9].) *Source*: Brownlow S, Morais Cabral JH, Cooper R, Flower RD, Yewdall SJ, Polikarpov I, North ACT, Sawyer L (1995) Structure 5:481–495

Tertiary structure of the protein varies with the pH. For example, at pH 2, β-lactoglobulin is in a native monomeric state and at pH 12, it is denatured. At some pH, mixture of both monomeric and dimeric forms could be find [50]. β-lactoglobulin contains two tryptophan residues, one (Trp-19) is surrounded by a hydrophobic environment and the second (Trp-61) in a hydrophilic one [51]. There are no evidence yet whether both tryptophan residues contribute to the protein fluorescence. β-Lactoglobulin can bind fatty acids [52] and different hydrophobic molecules [53, 54]. Nevertheless, up to now, its true function remains unclear.

β-Lactoglobulin, such as odorant binding proteins (OBPs) and α_1-acid glycoprotein, belongs to the lipocalin family proteins. All these proteins contain a β barrel composed of eight antiparallel β strands with $(+1)_8$ topology (i.e., each successive β strand is adjacent to the preceding one). Calyx of the proteins is formed by the β strands.

3.5 *Odorant Binding Protein*

Odorant binding proteins (OBPs) are small soluble proteins secreted by exocrine glands in various fluids and mucus of vertebrates (urine, nasal mucus, and vaginal secretions). They belong to the lipocalin family [55]. Many functions such as detoxification from lipid peroxidation [56], protection against insects carrying parasites [57], binding of anesthetic molecules such as halothane and isoflurane [58], acting passive transporters of hydrophobic and volatile ligands, particularly odors, through the hydrophilic mucus layer [59] have been attributed to OBPs expressed in the nasal area. pOBP purified from pig nasal mucus is secreted in the nasal mucus of vertebrates, which convey odorants to their neuronal receptors [60]. It is a monomer of 157 amino acids [61, 62] containing one disulfide bridge between cysteines at positions 63 and 155. Burova et al. [63] showed that pOBP, such as bovine β-lactoglobuline, is dimeric at physiological pH (7.2) and monomeric at acidic pH. Porcine OBP structure has been resolved by X-ray crystallography and one molecule of pOBP was found in the crystal (Fig. 9) [64]. The protein contains one Trp residue at position 16. Both analysis of crystal structure and binding experiments have shown that the Trp residue does not belong to the internal cavity forming the binding site for ligands [64, 65].

Figures 10 and 11 display, respectively, primary and tertiary structures of porcine OBP.

1	**16**	**31**
EEPQPEQDPFELSGK	**WITSYIGSSDLEKIG**	**ENAPFQVFMRSIEFD**
46	**61**	**76**
DKESKVYLNFFSKEN	**GICEEFSLIGTKQEG**	**NTYDVNYAGNNKFVV**
91	**106**	**121**
SYASETALIISNINV	**DEEGDKTIMTGLLGK**	**GTDIEDQDLEKFKEV**
136	**151**	
TRENGIPEENIVNII	**ERDDCPA**	

Fig. 10 Primary sequence of porcine odorant binding protein. *Source*: Paolini S, Scaloni A, Amoresano A, Marchese S, Napolitano E, Pelosi P (1998) Chem Senses 23:689–698

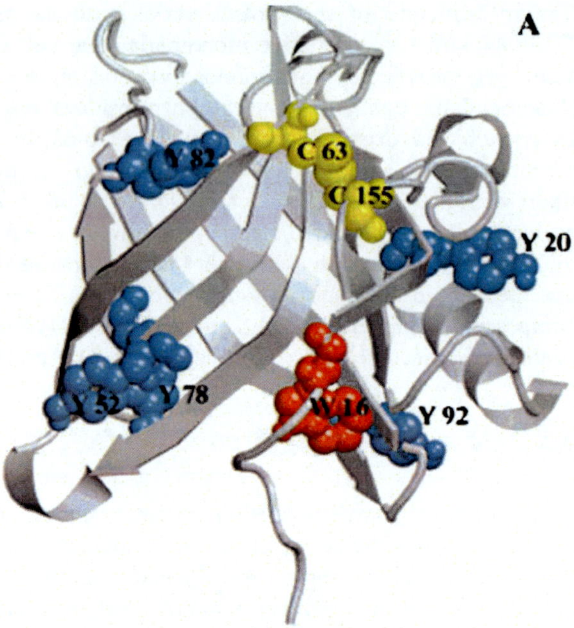

Fig. 11 Cartoon diagram of porcine odorant binding protein. Localization of Trp 16 residue (*red*), Tyr residues (*blue*), and the disulfide bridge made between Cys 63 and Cys 155 (*yellow*) are shown as spheres. *Source*: Spinelli S, Ramoni R, Grolli S, Bonicel J, Cambillau C, Tegoni M (1998) Biochemistry 37:7913–7918

4 Results

4.1 Fluorescence Lifetimes of Tryptophan Residues in Proteins

Table 1 displays fluorescence lifetimes of tryptophan free in water and of proteins of different structures and containing one, two, three, or eight tryptophan residues. Interpretation of the data does not need the presence of a protein or a peptide around the tryptophan. Our data characterize an internal reorganization of the tryptophan structure independently of its environment. These structures obtained in the excited state yield the two fluorescence lifetimes (0.4 and 3 ns) found for tryptophan whether free in solution or present within a protein. In most of the proteins, a third lifetime around 1 ns is in general recorded. This lifetime could be the result of the interaction between the Trp residue (s) and the surrounding amino acids. In human and bovine serum albumins and in OBP, this third lifetime goes from 6 to 9 ns. Values of the relative amplitudes of the fluorescence lifetimes can be dependent on the number of emitting Trp residues or/and on the type of interaction that is occurring between Trp residues and the surrounding environment. In fact, our results (Table 1) indicate that lifetimes relative amplitudes are not the same for all the proteins.

It is clear from the data displayed in Table 1 that proteins having one tryptophan residue or more emit with three fluorescence lifetimes. Thus, number of fluorescence lifetime is independent of that of tryptophan residues present within a protein,

Table 1 Fluorescence lifetimes and corresponding relative amplitudes in % of tryptophan free in solution (L-Trp) and present in different proteins

Sample	λ_{em} (nm)	τ_1	α_1	τ_2	α_2	τ_3	α_3	χ^2	Trp
L-Trp	350	0.43	4.71	3.06	95.29	–	–	1.008	
Porcine odorant binding protein	340	0.86	35.7	2.76	53.4	9.2	10.9	1.167	1
Human serum albumin	325	0.53	12.35	3.57	31.04	7.46	56.6	1.066	1
Mutant cyclophilin B (CyPBw128A)	330	0.413	35	2.096	56	5.45	9	1.086	1
Cyclophilin B	330	0.684	21.4	1.784	60.3	4.376	18.3	1.067	2
Bovine serum albumin	355	0.265	4.73	3.055	25.7	6.58	69.6	1.006	2
Bovine β-lactoglobulin	330	0.56	14.40	1.306	50.2	4.475	35.4	1.156	2
Helicase	330	0.47	17.83	1.735	37.2	4.373	44.96	1.128	3
α_1-Acid glycoprotein	330	0.24	6.67	1.4	34.88	3.410	58.45	1.046	3
Chymotrypsin	330	0.69	27.26	1.472	45.03	3.82	27.71	1.154	8
Serotransferrin	330	0.31	25.65	1.25	37.69	3.784	36.66	1.137	8

Sources: Albani JR (2007) J Fluoresc 17:406–417; Albani JR, Carpentier M, Lansiaux C (2008) J Fluoresc 18:75–85; Tayeh N, Rungassamy T, Albani JR (2009) J Pharm Biomed Anal 50:109–116; Kmiecik D, Albani JR (2010) J Fluoresc 20:973–983

although exceptions could exist. Let us discuss these results in terms of rotamers model. The purpose of this discussion is to find out how far we can go in the application of this model. Three lifetimes are observed in most of the studied proteins. Thus, presence of these lifetimes is independent of the structure surrounding the tryptophan residue(s), of the proteins tertiary structure, and of proteins dynamic properties. In other terms, presence of protein matrix around Trp residues does not play any fundamental role in the fluorescence lifetimes of Trp residues. In this case, protein backbone has no or nonsignificant effect on the fluorescence lifetimes of proteins. In other terms, rotamers model can in no way explain the presence of three lifetimes in proteins.

Also, attributing fluorescence lifetimes to conformers, means that, in proteins with more than one tryptophan residue, we should have a combination of conformers that yields three fluorescence lifetimes.

Rotamers model can in no way explain the absence of significant differences in fluorescence lifetimes between cyclophilin B wild type and the mutant $CyPBw_{128A}$. One can consider fluorescence lifetimes as the result of Trp residues interaction with their microenvironments. However, since fluorescence lifetimes do not vary significantly between wild-type protein and the mutant, one should consider the possibility of having, around the Trp residues, a common identical protein structure responsible of the three measured fluorescence lifetimes.

Emission maximum of tryptophan residue in HSA, mutant cyclophilin B, and OBP is located at 335, 325, and 340 nm, respectively. Thus, modification of the environment of a tryptophan affects fluorescence emission peak, however as we see from Table 1, in the three proteins, tryptophan residue emits with three lifetimes which two are almost identical to those found for tryptophan free in water. Thus, two fluorescence lifetimes of tryptophan residue in proteins are independent of the fluorophore environment. This implies that spatial structure of a protein does not affect these two fluorescence lifetimes.

Table 1 indicates also that tryptophan residues in both BSA and HSA fluorescence with three identical lifetimes [66]. Thus, one Trp residue emission can be described with three fluorescence lifetimes and it is not possible, in BSA, to assign a specific lifetime to a specific Trp residue. In order to check whether the three lifetimes are dependent on the protein structure or not, we measured fluorescence lifetimes of HSA Trp residue in two protein states, native and denatured by 6 M guanidine pH 7.8. Figure 12 displays the normalized emission spectra of the protein in both native and denatured state. Emission spectrum of native HSA Trp 214 residue displays a peak at 340 nm. When dissolved in 6 M guanidine solution, emission peak is shifted to 352 nm accompanied with an increase of the fluorescence intensity, indicating protein denaturation and thus an increase of the protein tryptophan residue exposure to the solvent.

Fluorescence intensity decay of native HSA, recorded at 350 nm, provides lifetime data represented by a sum of three preexponentials:

$$I(\lambda, t) = 0.0277\mathrm{e}^{-t/0.492} + 0.2928\mathrm{e}^{-t/3.961} + 0.6796\mathrm{e}^{-t/7.777},$$

where 0.277, 0.2928, and 0.6796 are the preexponential factors (α), while 0.492, 3.961, and 7.777 ns are fluorescence lifetimes. The mean fluorescence lifetime τ_0, calculated from two experiments, is equal to 7.085 ± 0.094 ns.

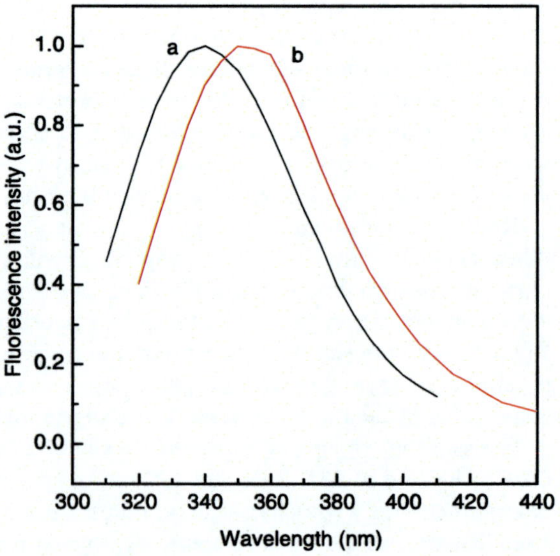

Fig. 12 Normalized fluorescence emission spectra of HSA in its native state (*a*) and denatured states (*b*)

When HSA is dissolved in 6 M guanidine, its fluorescence decay intensity can still be characterized by three preexponentials:

$$I(\lambda, t) = 0.1222e^{-t/0.380} + 0.6399e^{-t/2.350} + 0.2379e^{-t/4.725},$$

where 0.1222, 0.6399, and 0.2379 are the preexponential factors (α), while 0.380, 2.350, and 4.725 ns are fluorescence lifetimes. The mean fluorescence lifetime τ_0, calculated from two experiments, is equal to 3.314 $\pm$ 0.05 ns. Thus, denaturation of HSA caused by guanidine decreases the three fluorescence lifetimes values and does not yield a single fluorescence lifetime. This fact means that presence of three lifetimes for tryptophan residue in HSA does not depend on the protein structure only but also on the tryptophan itself.

In order to study effect of protein dynamics on fluorescence lifetimes of Trp residues, we compared fluorescence parameters of α_1-acid glycoprotein prepared in two different ways: (1) by a successive combination of ion displacement chromatography, gel filtration, and ion exchange chromatography (α_1-acid glycoprotein^c) and (2) by ammonium sulfate precipitation (α_1-acid glycoprotein^s). Perrin plots reveal that Trp residues of α_1-acid glycoprotein^c display free motions while those of α_1-acid glycoprotein^s follow protein global rotation. Excitation at 295 nm yields a fluorescence spectrum with a peak equal to 332 nm (α_1-acid glycoprotein^c) and to 337 nm (α_1-acid glycoprotein^s). Positions of the peaks are typical for proteins containing tryptophan residues in a hydrophobic environment and result from the contribution of both solvent-exposed and buried Trp residues. Fluorescence emission maximum of α_1-acid glycoprotein^c is blue shifted compared to that of α_1-acid glycoprotein^s. Thus, the two Trp residues embedded in the protein core of α_1-acid glycoprotein^c are less exposed to the solvent than those embedded in α_1-acid glycoprotein^s [67].

Time decay of fluorescence intensity measurements of α_1-acid glycoprotein^s was performed with the time correlated single photon counting. The time decay of α_1-acid glycoprotein^c was measured by the phase method. Table 2 gives the fluorescence lifetime values for α_1-acid glycoprotein measured by the two methods. One would expect to obtain different fluorescence lifetimes for the two types of preparation if protein dynamics has an influence on the lifetime data. Instead, we notice that the two preparations give very close fluorescence lifetimes. Thus, structure and dynamics of α_1-acid glycoprotein have no effect on fluorescence lifetimes of the tryptophan residues.

Table 2 Fluorescence lifetimes data of α_1-acid glycoprotein^c and α_1-acid glycoprotein^s

τ_1	f_1	τ_2	f_2	τ_3	f_3	$<\tau>$
α_1-Acid glycoprotein^c						
0.354 ± 0.034	0.101 ± 0.05	1.664 ± 0.072	0.66 ± 0.03	4.638 ± 0.342	0.238 ± 0.01	2.285
α_1-Acid glycoprotein^s						
0.26 ± 0.07	0.0955 ± 0.02	1.75 ± 0.03	0.7 ± 0.05	5.15 ± 0.09	0.203 ± 0.06	2.29

Sources: Albani JR, Vos R, Willaert K, Engelborghs Y (1995) Photochem Photobiol 62:30–34 and Albani JR (1998) Spectrochimica Acta Part A 54:175–183

4.2 Origin of Tryptophan Fluorescence Lifetimes: Substructures Generated in the Excited State

It has been suggested that fluorescence lifetimes of Trp residue(s) within random coil polypeptides (denatured proteins) are very close, the differences would come from the position of the tryptophan within the protein primary structure (end of the chain or in its middle) and/or presence of quencher group near the Trp residue(s) [68–70]. Nevertheless, the data described in this review indicate that fluorescence lifetimes characterize intrinsic properties of the fluorophore. In fact, the two shortest lifetimes around 0.4–0.5 and 2–4 ns were measured for tryptophan free in solution and present within proteins. Thus, presence of these two lifetimes is independent of any structure around tryptophan and characterizes an internal property or/and organization of the tryptophan structure in the excited state, independently of its environment. Even when proteins are denatured, three lifetimes are still observed. These structures or substructures obtained in the excited state yield the two fluorescence lifetimes found for tryptophan whether free in solution or present within a protein. The third lifetime recorded in proteins could be attributed to interaction between the Trp residue (s) and the surrounding amino acids and to possible specific properties of the protein. Also, values of relative amplitudes of the three fluorescence lifetimes can be dependent on the number of emitting Trp residues or/and on the type of interaction that is occurring between Trp residues and the surrounding environment. In fact, our results (Table 1) indicate that fluorescence lifetime relative amplitudes are not the same for the measured proteins. Lifetime amplitudes will characterize the populations of emitting species or tryptophan residue substructures. Fluorescence lifetimes and their populations (preexponential values) can be modified with the global structure of the protein.

The 7.7 ns observed in the native state of HSA is the result of the interaction of Trp-214 residue with its surrounding environment, the amino acids of the hydrophobic part of the ligand binding pocket. This long lifetime indicates that corresponding radiative (k_r) and nonradiative (k_i) constants are low which means that substructure yielding this long lifetime displays important interaction within the neighboring amino acids. The preexponential value (0.68) of this lifetime is high, which means that the population of the substructure emitting with lifetime equal to 7.7 ns is the most important between the three populations. This is in a complete opposition with the shorter lifetime (0.492 ns) which preexponential value is equal to 0.0277. Despite the presence of Trp-214 residue within a highly hydrophobic domain of HSA, this does not exclude contacts with the solvent molecules that are diffusing within the protein matrix. This diffusion is the result of local motions within HSA matrix. Protein denaturation decreases the contact between Trp-214 residue and its microenvironment and thus induces an increase of the fluorescence intensity and of the radiative constant (k_r) value. Populations of substructures obtained in the denatured states are not necessarily identical to those observed for the native protein.

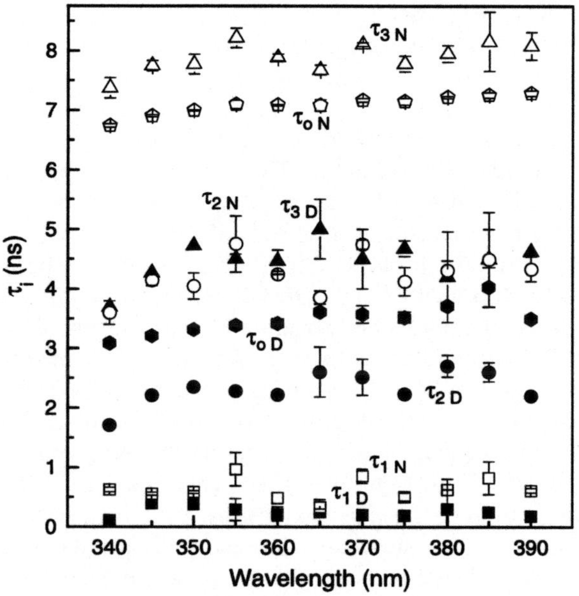

Fig. 13 Lifetimes values of HSA Trp residue with emission wavelength of both denatured (*closed symbols*) and native (*open symbols*) proteins. τ_1, τ_2, τ_3, and τ_0 are symbolized by *squares*, *circles*, *triangles*, and *pentagons*, respectively. *Source*: M. Amiri, K. Jankeje and J.R. Albani (2010) J Fluoresc 20:651–656

Figure 13 displays fluorescence lifetime variation of native and denatured HSA along emission wavelengths [71]. While the shape of this lifetimes variation is identical for both proteins states, lifetime values of denatured protein are almost half those of the native one. This decrease in the fluorescence lifetime can be explained by the fact that in the native state, interaction between HSA Trp residue and neighboring amino acids is important compared to that observed when the protein is totally denatured. Thus, in the native state, radiative rate constant decreases as the result of the important interaction between fluorophore and neighboring amino acids. In the unfolded state, interaction between Trp residue and amino acids decreases, inducing by that an increase in the radiate rate constant and thus a decrease of the measured fluorescence lifetimes compared to that recorded in the native state. Fluorescence lifetime decrease can also be explained by the difference in the local motions of tryptophan residue between native and denatured states. In fact, in the native state, HSA tryptophan-214 residue displays local motions [72–74], which are more important in the denatured state inducing an increase in the nonradiative rate constant k_i. This will decrease the value of fluorescence lifetime. Therefore, HSA structure plays a role of gap, retaining and organizing photon emission energy. Quenching efficiency characterized by the protein unfolding (passage from the native to the denatured state) is equal to

$$E = 1 - \frac{\tau_{0(D)}}{\tau_{0(N)}}, \tag{1}$$

where $\tau_{0(D)}$ and $\tau_{0(N)}$ are the mean fluorescence lifetimes in the denatured and native states, respectively. Quenching efficiency (E) was found equal to 0.5 ± 0.05 along the emission spectrum from 340 to 390 nm. This value reveals importance of the interactions between tryptophan residue and surrounding amino acids in the native state compared to the unfolded one. Denaturation induces structural modifications within the IIA subdomain where tryptophan residue is located, affecting different interactions that exist between IIA and IIIA subdomains and thus protein conformation and stability.

Figure 14 displays lifetime preexponentials variation along the emission spectra of both denatured and native HSA. First of all, the data show that we have three different fluorophore populations independent of the surrounding environment. HSA denaturation induces an important modification in the preexponential values recorded in the native state. This means that denaturation has modified the population of each of the three tryptophan substructures. This result is clear evidence that preexponential values characterize the type and nature of interaction existing between Trp-214 residue and its environment, while the number of lifetimes is an indication of the substructures that exist for the fluorophores in the excited state. Figure 14 indicates that in the denatured state, the population characterizing the two shortest lifetimes are much more important than those observed in the native state. This is not the case for the population with the longest fluorescence lifetime. Mean fluorescence lifetime value is calculated from those of the three lifetimes and of their preexponentials. Upon HSA denaturation, important decrease in the three fluorescence lifetimes (Fig. 13), and of the longest lifetime preexponential (Fig. 14) occurs, inducing a decrease of the mean fluorescence lifetime.

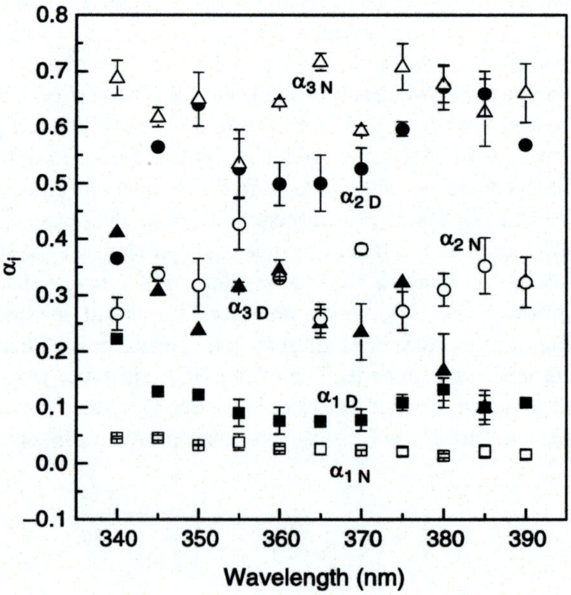

Fig. 14 Preexponential variation of fluorescence lifetimes of native (*open symbols*) and denatured (*closed symbols*) HSA with emission wavelengths. α_1, α_2, and α_3 are symbolized by *squares*, *circles*, and *triangles*, respectively. *Source*: M. Amiri, K. Jankeje and J.R. Albani (2010) J Fluoresc 20:651–656

Since emission occurs from the excited state, then it is normal that fluorescence lifetimes characterize species or substructures existing in the excited state and not in the ground one. Tryptophan molecule cannot have identical structure in both ground and excited states. In fact, energy absorbed by a fluorophore is not entirely used to excite the fluorophore. Some of the absorbed energy helps to reorganize the fluorophore molecules in the excited state inducing a structure different from that of the ground state. In other terms, light excitation is not necessary equal to photon absorption. This can be put into evidence by comparing fluorescence excitation and absorption spectra of a fluorophore. Although, the two spectra could have a similar global shape, their intensities at many wavelengths are not identical. In other terms, emission spectra intensity is not proportional to the optical density of the excitation wavelength but to the intensity of the excitation spectrum at the excitation wavelength.

Figure 15 displays fluorescence emission spectra of BSA obtained at four excitation wavelengths, 260 (a), 280 (b), 295 (c), and 300 nm (d). The spectra obtained at the four excitation wavelengths display different intensities at the emission peaks. The weakest intensity corresponds to the excitation wavelength of 300 nm and the highest intensity to 280 nm excitation wavelength. Since the spectra displayed are corrected for the inner filter effect, they are interpreted by the classical fluorescence theory as the result of the absorption at the excitation wavelength. Therefore, on the basis of this theory, one should conclude that the highest optical density is that at 280 nm followed by 295, 260, and 300 nm.

Figure 16 displays absorption spectrum of BSA. Optical density values are equal to 0.097 (280 nm), 0.020 (295 nm), 0.059 (260 nm), and 0.007 (300 nm). Thus, optical density at 260 nm is three times higher than that recorded at 295 nm. Therefore, if emission intensity is proportional to optical density at the excitation

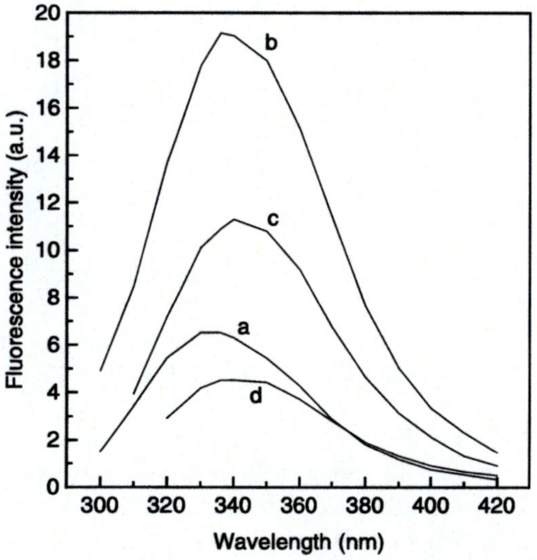

Fig. 15 Fluorescence emission spectra of bovine serum albumin at four excitation wavelengths. *a*: λ_{ex} = 260 nm; *b*: λ_{ex} = 280 nm; *c*: λ_{ex} = 295 nm; *d*: λ_{ex} = 300 nm. Source of Figs. 15–20: Albani JR (2007) J Fluoresc 17:406–417

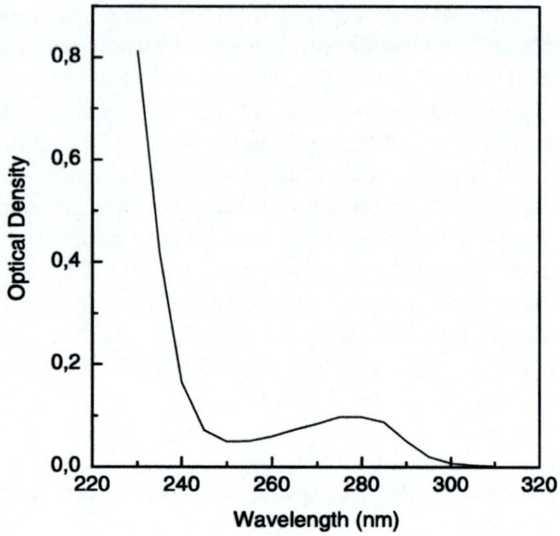

Fig. 16 Absorption spectrum of bovine serum albumin

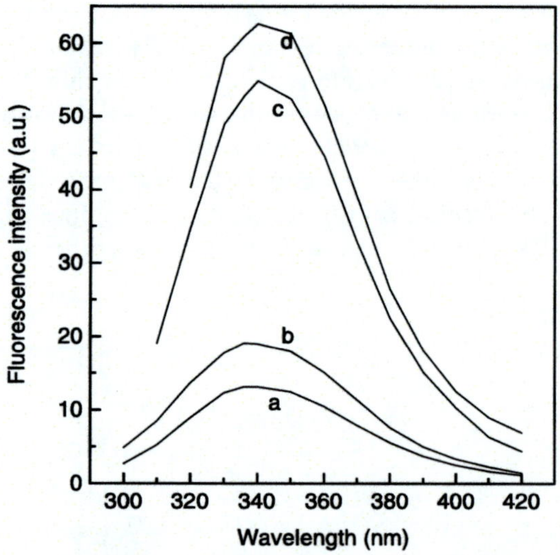

Fig. 17 Fluorescence emission spectra of bovine serum albumin normalized at the optical density at 280 nm. *a*: $\lambda_{ex} = 260$ nm; *b*: $\lambda_{ex} = 280$ nm; *c*: $\lambda_{ex} = 295$ nm; *d*: $\lambda_{ex} = 300$ nm

wavelength, it should be higher when excitation was performed at 260 nm than at 295 nm.

Normalizing emission spectra obtained in Fig. 15 for the optical density at 280 nm yields spectra displayed in Fig. 17. We notice that the four spectra do not have identical intensities and that there is no direct correlation between intensity values at the emission peak with the optical density of the excitation wavelength.

Figure 18 displays fluorescence excitation spectrum of BSA ($\lambda_{em} = 360$ nm). Although the general feature of this spectrum looks like the absorption spectrum (Fig. 16), there is an important difference between them: intensity of the excitation

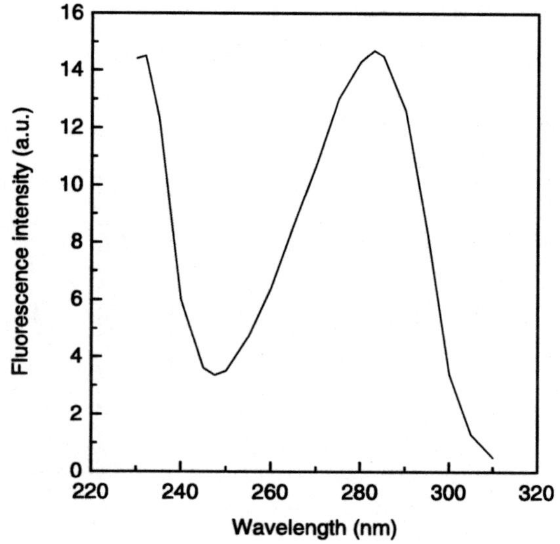

Fig. 18 Fluorescence excitation spectrum of bovine serum albumin. $\lambda_{em} = 360$ nm

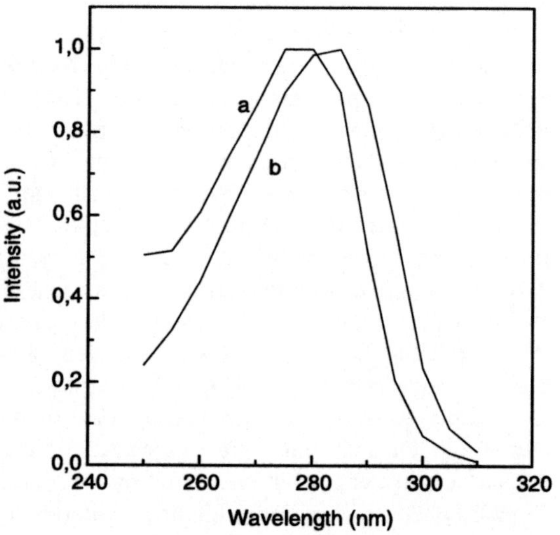

Fig. 19 Normalized absorption (*a*) and excitation (*b*) spectra of bovine serum albumin

spectrum does not vary with the wavelengths in the same way than the optical density along the absorption spectrum. An example, optical density at 260 nm is higher at 295 nm than at 260 nm while excitation intensity is higher at 295 nm than at 260 nm.

Normalizing absorption and excitation spectra at 280 nm allows us to see the differences between them (Fig. 19). Excitation peak is shifted to the red compared to the absorption peak. It is clear that the optical density at 260 nm is higher than

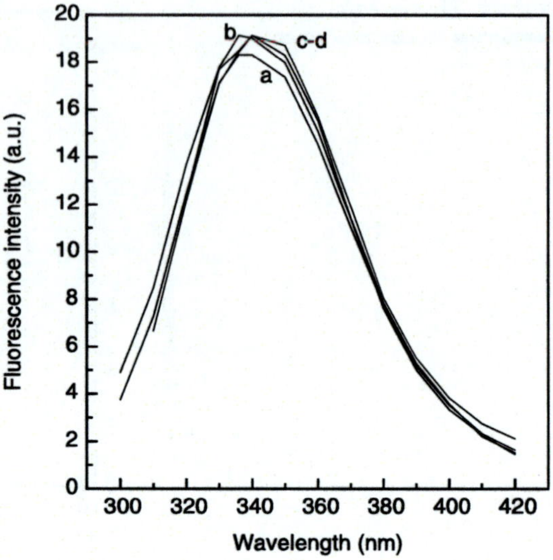

Fig. 20 Fluorescence emission spectra of bovine serum albumin normalized for the excitation intensity at 280 nm. *a*: $\lambda_{ex} = 260$ nm; *b*: $\lambda_{ex} = 280$ nm; *c*: $\lambda_{ex} = 295$ nm; *d*: $\lambda_{ex} = 300$ nm

that at 295 nm, while in the excitation spectrum, we have the opposite; the intensity at 295 nm is higher than that at 260 nm.

Normalizing emission spectra obtained in Fig. 15 for the excitation intensity at 280 nm yields the spectra displayed in Fig. 20. The four normalized spectra display very close or equal emission intensities. Thus, fluorescence emission intensities obtained at different excitation wavelengths (Fig. 15) are directly proportional to the intensity of the excitation wavelength of the excitation spectrum. The same results were obtained for HSA, serotransferrin, β-lactoglobulin, human α_1-acid glycoprotein, creatine phosphokinase, deoxyribonuclease I, rat albumin, helicase, trypsin, free tryptophan in solution, and free tyrosine in solution [38].

It is common to read that absorption is equal to excitation. However, the results described in this work show that absorption is not necessarily equivalent to excitation. Also, excitation spectrum yields information and details that can be hidden in the absorption spectrum especially at wavelengths where optical densities are very small. Data displayed show that fluorescence emission intensities of aromatic amino acids free in solution or present in proteins are proportional to the excitation spectrum at the excitation wavelength but not to the optical density at the excitation wavelength. Similarity between the results obtained with free tryptophan in solution and proteins clearly indicates that within proteins, these results are not the effects of cysteine or peptide bonds absorption. Therefore, energy absorbed by the fluorophore does not reflect necessarily the effective energy used to excite the fluorophore. Absorbed energy helps to reorient fluorophore structure within the excited state, is dissipated in the medium (solvent and/or protein matrix), and is used to excite the fluorophore molecules. Therefore, excitation is not equal to absorption, i.e., excitation and absorption spectra are not necessarily identical. Also, emission spectra recorded at different excitation wavelengths can be normalized for the intensities of the excitation spectrum and not for the optical densities.

5 Fluorescence Lifetimes of Tryptophan and Relation with $S_0 \rightarrow {}^1L_b$ and $S_0 \rightarrow {}^1L_a$ Transitions

5.1 *Fluorescence Excitation Spectra of the* $S_0 \rightarrow {}^1L_a$ *and* $S_0 \rightarrow {}^1L_b$ **Transitions** *in Water*

Tryptophan absorbs with two transitions $S_0 \rightarrow {}^1L_a$ and $S_0 \rightarrow {}^1L_b$ (Fig. 21). We measured fluorescence lifetimes and we recorded fluorescence spectra (emission and excitation) of L-tryptophan dissolved in water and in ethanol in absence and presence of high progesterone concentrations. Hormone absorbs between 220 and 280 with a peak around 250 nm, while its absorption is equal to zero beyond 280 nm. Thus, progesterone absorbs at the same wavelengths as $S_0 \rightarrow {}^1L_a$ transition while its absorption is negligible in the $S_0 \rightarrow {}^1L_b$ transition domain.

Figure 22 displays fluorescence emission spectrum of L-tryptophan in water in absence (a) and presence of 650 μM progesterone (b) ($\lambda_{ex} = 295$ nm). Presence of progesterone at high concentration does not induce any significant decrease of tryptophan fluorescence intensity.

It is important here to remind that although progesterone concentrations are important in the fluorescence cuvettes, we added only small volumes (11 μl to the 1 ml tryptophan solution in water and 4 μl to the 1 ml tryptophan solution in ethanol) of progesterone stock solution so that to avoid interference, if any, of added volumes.

Progesterone does not bind to tryptophan free in solution, as shown in Fig. 22, although its presence at high concentrations abolishes completely the ${}^1L_a$ state. Thus, upon excitation at 295 nm and in the presence of high progesterone concentrations, ${}^1L_b$ state is the primary contribution to tryptophan emission.

Figure 23 displays fluorescence excitation spectrum of L-tryptophan in water in absence (a) and presence of 650 μM progesterone (b) ($\lambda_{em} = 350$ nm). In the presence of progesterone, we observe a decrease in the intensity of the

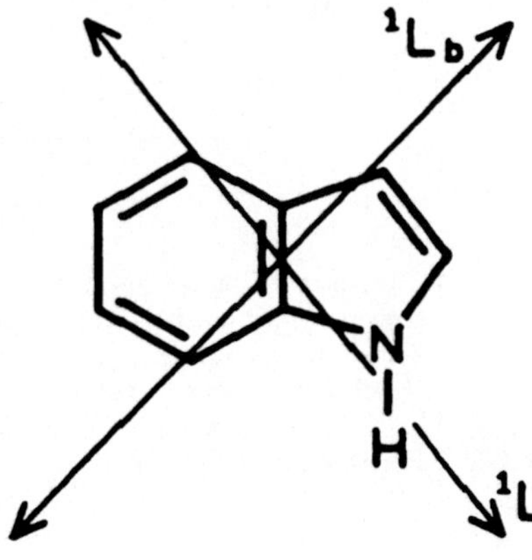

Fig. 21 The two dipole transitions of tryptophan

Fig. 22 Fluorescence emission spectra of L-tryptophan in deionized water in the absence (*a*) and in the presence (*b*) of 650 μM progesterone. $\lambda_{ex} = 295$ nm

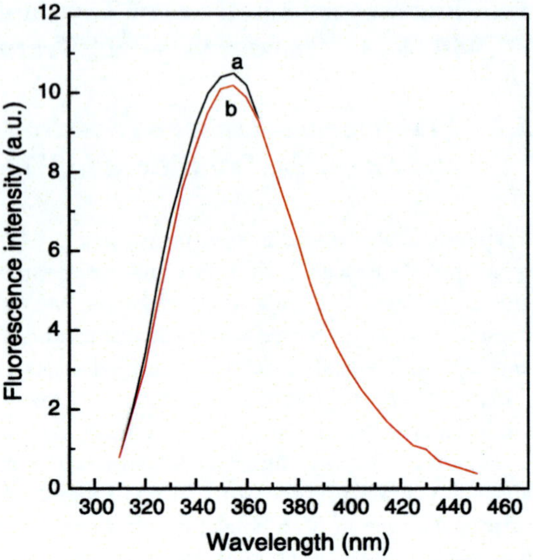

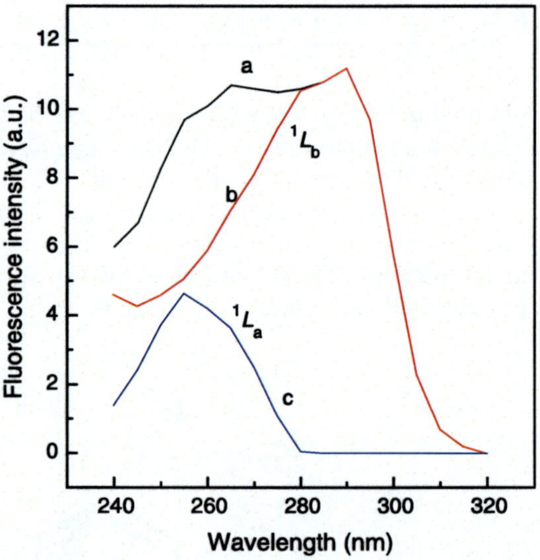

Fig. 23 Fluorescence excitation spectra of L-tryptophan dissolved in deionized water in the absence (*spectrum a*) and in the presence of 650 μM progesterone (*spectrum b*). Excitation spectrum obtained in the presence of progesterone corresponds to the $S_0 \rightarrow {}^1L_b$ transition. *Spectrum c* which corresponds to the $S_0 \rightarrow {}^1L_a$ transition is obtained by subtracting (*b*) from (*a*). λ_{max} of ^1L_a and of ^1L_b is equal to 255 and 290 nm, respectively. Source of Figs. 22–32: Albani JR (2009) J Fluoresc 19:1061–1071

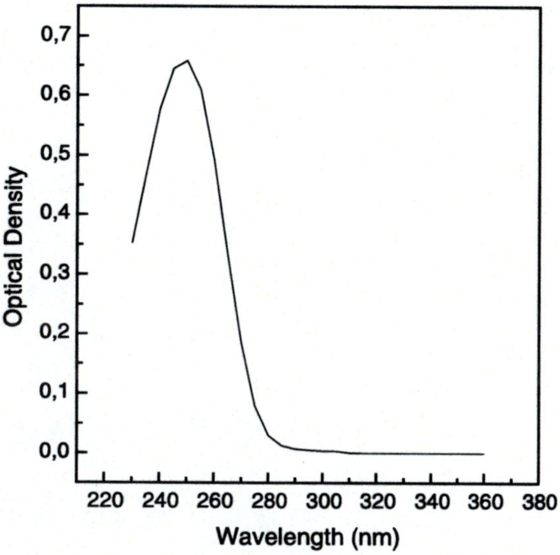

Fig. 24 Absorption spectrum of 240 μM progesterone in twice distilled ethanol measured with optical pathlength of 0.4 cm

excitation spectrum, the peak of the recorded spectrum being at 290 nm. Subtracting spectrum (b) from (a) yields spectrum (c) with a peak around 255 nm. Spectra (b) and (c) are similar to the ^1L_b and ^1L_a transitions that characterize the two singlet states of Tryptophan [75]. Emission of tryptophan in solution and in most proteins is supposed to be unstructured and thus would take place from the ^1L_a state which is considered to be the first excited state S_1 [76, 77]. Addition of progesterone to L-tryptophan in water inhibits completely the ^1L_a state and thus ^1L_b state would be the main responsible of tryptophan emission. At 295 nm, the ^1L_b state is preferentially selected since at this wavelength, the $S_0 \rightarrow {}^1L_a$ transition is completely abolished.

Figure 24 shows absorption spectrum of 240 μM progesterone dissolved in ethanol. Progesterone absorbs mainly between 220 and 280 nm. Beyond this wavelength, hormone absorption is close to zero. Thus, high progesterone concentrations act as a filter of the ^1L_a state, excitation at wavelength equal to 295 nm would select ^1L_b state.

5.2 *Does Complete Inhibition of $S_0 \rightarrow {}^1L_a$ Transition Modifies Fluorescence Lifetimes of Tryptophan in Water?*

In the absence of 650 μM progesterone, recorded fluorescence lifetimes were equal to 0.50386 ± 0.05425 and 2.775 ± 0.01158 ns ($\chi^2 = 1.0056$) and in the presence of progesterone, recorded lifetimes were equal to 0.46024 ± 0.0330 and 2.752 ± 0.0111 ns ($\chi^2 = 1.0107$). We can notice that both short and long lifetimes do not vary significantly in the presence of progesterone.

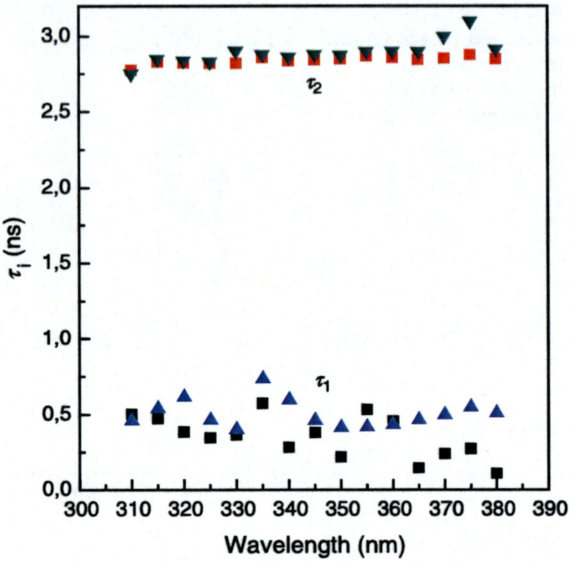

Fig. 25 Fluorescence lifetimes variation of L-tryptophan in deionized water with emission wavelength in the absence (*squares*) and the presence (*triangles*) of 650 μM progesterone. $\lambda_{ex} = 296$ nm

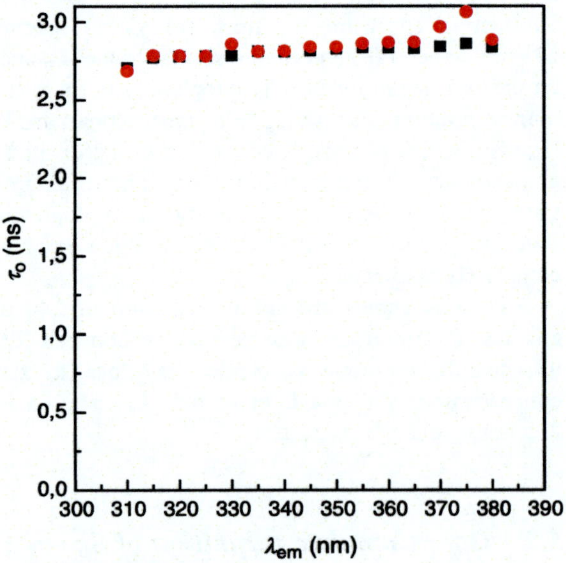

Fig. 26 Mean fluorescence lifetime variation of L-tryptophan in deionized water with emission wavelength in the absence (*squares*) and in the presence (*circles*) of 650 μM progesterone. $\lambda_{ex} = 296$ nm

In the absence and the presence of progesterone, analyzing fluorescence decays with two lifetimes increases the value of χ^2 at all emission wavelengths (results not shown).

Figure 25 displays fluorescence lifetimes variation in water in the absence and the presence of progesterone along the emission wavelengths. We can notice that the presence of progesterone does not modify any of the two lifetimes. Also, mean fluorescence lifetime is not affected by the presence of progesterone (Fig. 26).

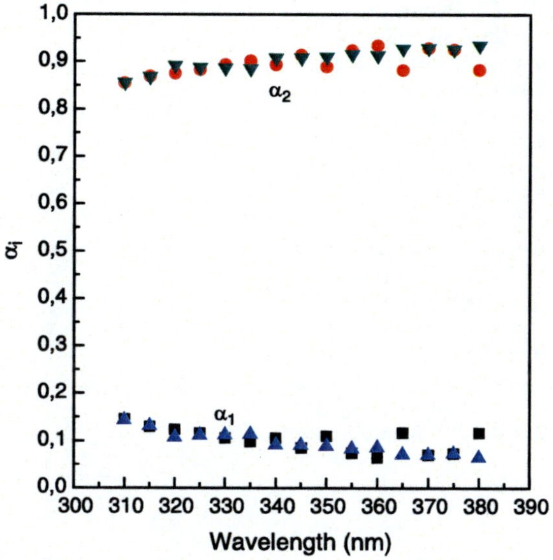

Fig. 27 Preexponential variation of L-tryptophan fluorescence lifetimes in deionized water in the absence (*squares* and *circles*) and in the presence of 650 μM progesterone (*triangles*). $\lambda_{ex} = 296$ nm

Figure 27 shows preexponential lifetimes variation of tryptophan in water in the absence and the presence of progesterone. The results clearly indicate the absence of any effect of progesterone presence in solution on the preexponential values.

Time-resolved emission spectra of tryptophan in water indicate that the two lifetimes contribute to the global emission spectrum of the fluorophore, contribution of the long fluorescence lifetime to the global emission spectrum is much more important than that of the short lifetime (results not shown).

5.3 *Fluorescence Spectral Properties of Tryptophan in Ethanol*

Figure 28 displays fluorescence excitation spectra of L-tryptophan in ethanol in the absence and the presence of 240 μM progesterone. λ_{max} of ^1L_a and of ^1L_b are equal to 265 and 282 nm, respectively. Thus, excitation at 295 nm in the presence of high progesterone concentrations allows selecting the $S_0 \rightarrow {}^1L_b$ transition only, the $S_0 \rightarrow {}^1L_a$ transition being completely abolished by high progesterone absorption.

In the absence of progesterone, fluorescence intensity decay $I(\lambda, t)$, of L-tryptophan can be adequately represented by a sum of three exponentials:

$$I(\lambda, t) = 0.12385e^{-t/0.3475} + 0.79295e^{-t/1.768} + 0.0832e^{-t/4.77},$$

where 0.12385, 0.70295, and 0.0832 are the preexponential factors, 0.3475 ± 0.0305, 1.768 ± 0.042, and 4.77 ± 0.026 ns are the decay times and λ is the emission wavelength (315 nm) ($\chi^2 = 1.073$).

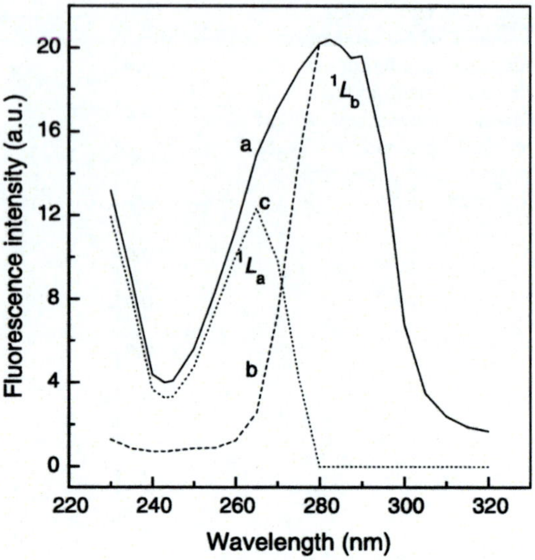

Fig. 28 Fluorescence excitation spectra of L-tryptophan dissolved in distilled ethanol in the absence (*spectrum a*) and in the presence of 240 μM progesterone (*spectrum b*). Excitation spectrum obtained in the presence of progesterone corresponds to the $S_0 \rightarrow {}^1L_b$ transition. *Spectrum c* which corresponds to the $S_0 \rightarrow {}^1L_a$ transition is obtained by subtracting (*b*) from (*a*). λ_{max} of ^1L_a and of ^1L_b is equal to 265 and 282 nm, respectively

In the presence of 240 μM progesterone, fluorescence intensity decay $I(\lambda, t)$ of L-tryptophan can be adequately represented also by a sum of three exponentials:

$$I(\lambda, t) = 0.1398\mathrm{e}^{-t/0.378} + 0.7719\mathrm{e}^{-t/1.81} + 0.0883\mathrm{e}^{-t/4.796},$$

where 0.1398, 0.7719, and 0.0883 are the preexponential factors, 0.378 ± 0.0402, 1.81 ± 0.065, and 4.796 ± 0.035 ns are the decay times and λ is the emission wavelength (315 nm) ($\chi^2 = 1.014$).

Lifetime measurements indicate that addition of progesterone to a solution of L-tryptophan-ethanol does not affect any of the fluorescence decay parameters (lifetimes and preexponentials). In the absence and the presence of 240 μM progesterone, analyzing fluorescence decays in ethanol with two lifetimes increases the value of χ^2 at all the emission wavelengths (Tables 3 and 4).

Figure 29 displays lifetime variation along emission wavelengths of L-tryptophan dissolved in ethanol in the absence and the presence of 240 μM progesterone. We notice that the three lifetimes and the mean one (τ_0) are not affected by the presence of progesterone. Also, preexponential values of the three lifetimes are not affected by the presence of progesterone as it is clearly indicated in Figs. 30 and 31. The fact that progesterone presence abolishes completely the $S_0 \rightarrow {}^1L_a$ transition without modifying any of the fluorescence lifetimes, and their preexponentials values indicate that the three lifetimes result from excited state induced by the $S_0 \rightarrow {}^1L_b$ transition.

Figure 32 displays time-resolved emission spectra of the three lifetimes in the absence and the presence of 250 μM progesterone. The two longest lifetimes are the main contributors to tryptophan emission spectrum in the absence and the presence of progesterone. Since emission peaks are identical or very close, the three emitting species are substructures of the tryptophan in ethanol solution. These substructures are generated in the excited state.

Table 3 Value of χ^2 obtained for L-tryptophan decay in ethanol and analyzed with two and three fluorescence lifetimes

Wavelength	χ^2 (3 τ)	χ^2 (2 τ)
310	1.008	2.638
315	1.129	2
320	1.096	1.645
325	1.036	1.509
330	1.070	1.303
335	0.952	1.421
340	1.166	1.484
345	0.94	1.317
350	0.896	1.448
355	0.91	1.324
360	1.29	1.785
365	0.972	1.32
370	0.968	1.45
375	1.02	1.29
380	1.112	1.65
385	1.152	1.55
390	1.076	1.612
395	1.12	1.54

Table 4 Value of χ^2 obtained for L-tryptophan decay in ethanol in the presence of 240 μM progesterone and analyzed with two and three fluorescence lifetimes

Wavelength	χ^2 (3 τ)	χ^2 (2 τ)
310	1.065	2.54
315	1.073	2.02
320	1.184	1.88
325	1.058	1.703
345	1.173	1.621
350	0.886	1.33
355	1.009	1.356
360	0.86	1.338
365	1.124	1.581
370	1.019	1.398
375	1.161	1.427
380	1.083	1.619
385	1.159	1.58
390	1.009	1.457
395	0.945	1.313
400	1.06	1.254

5.4 Analysis of the Experiments (Figs. 22–32) Obtained in Absence of the $S_0 \rightarrow {}^1L_a$ Transition

Tryptophan free in water emits with two lifetimes equal to 0.5 and 2.78 ns. Tryptophan absorbs with two transitions $S_0 \rightarrow {}^1L_a$ and $S_0 \rightarrow {}^1L_b$. Our data clearly rule out the axiom which considers that in polar solvents, emission occurs

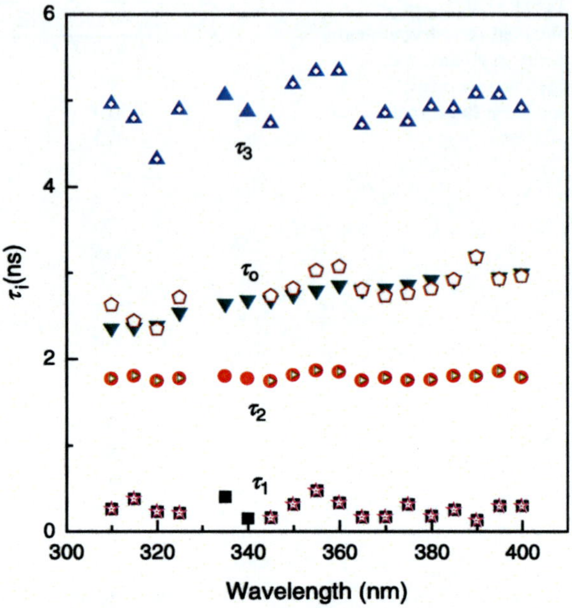

Fig. 29 Lifetimes variation of L-Trp in ethanol with emission wavelengths in the absence (*closed symbols*) and in the presence of 240 μM progesterone (*open symbols*) $\lambda_{ex} = 296$ nm

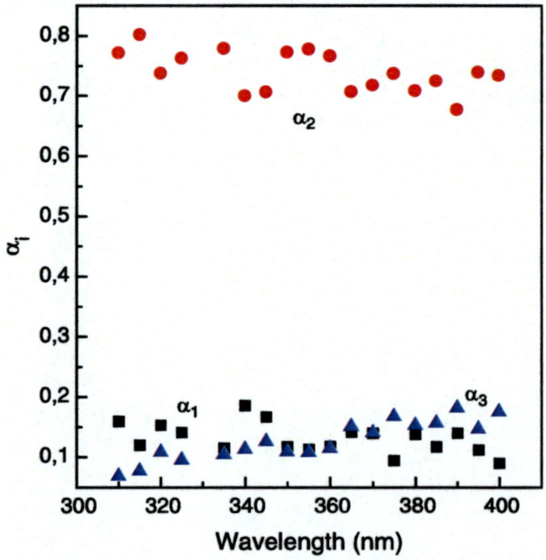

Fig. 30 Preexponential variation of Trp lifetimes in ethanol with emission wavelengths. $\lambda_{ex} = 296$ nm

principally from ^1L_a state. Also, our results rule out the fact that the 0.5 ns would be the result of the $S_0 \rightarrow {}^1L_b$ transition while the 2.78 ns would be induced by the $S_0 \rightarrow {}^1L_a$ transition. Upon excitation at 295 nm and in the presence of high progesterone concentrations, the ^1L_b state is the primary contribution to tryptophan emission.

Fig. 31 Preexponential variation of Trp lifetimes in ethanol with emission wavelengths, in the presence of progesterone. $\lambda_{ex} = 296$ nm

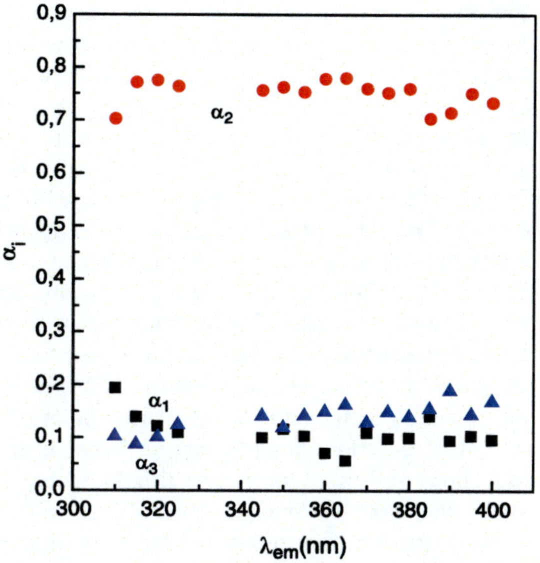

Fig. 32 Emission spectra of lifetimes of L-Trp in ethanol in the absence (*lines*) and the presence (*dashed lines*) of progesterone. $\lambda_{ex} = 296$ nm. Addition of ethanol does not affect the steady-state spectrum

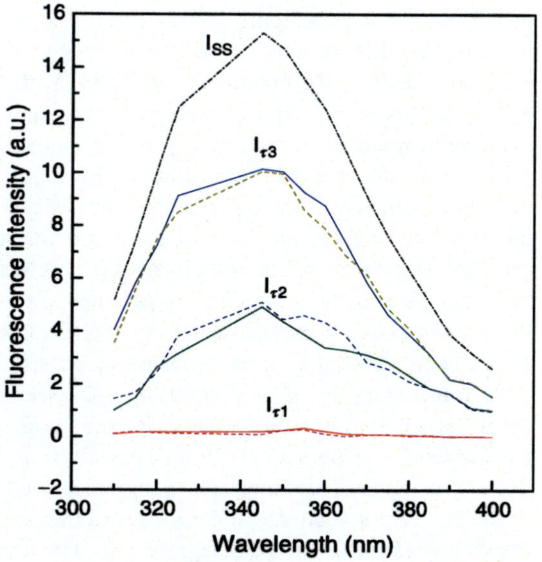

In the presence of high progesterone concentrations and thus in the absence of $S_0 \rightarrow {}^1L_a$ transition, two fluorescence lifetimes are still observed, 0.46 and 2.75 ns, values equal to those measured in the absence of progesterone. Also, presence of the hormone in solution does not affect lifetimes preexponential values which characterize the contribution of each lifetime to the global emission decay. Therefore, in a first approximation, one could conclude that $S_0 \rightarrow {}^1L_b$ transition is the main transition for tryptophan excitation. However, in the absence of experiments on the $S_0 \rightarrow {}^1L_a$ transition alone, it is more serious to say that combination of both transition states induce the excited state of tryptophan and the reorganization of the fluorophore into two interrelated structures. The 0.5 and 2.75 ns lifetimes are associated to these two structures. Finally, it is clear from our data that we cannot assign a specific fluorescence lifetime to each of ${}^1L_a$ and ${}^1L_b$ states.

Dissolved in ethanol, tryptophan emits with three lifetimes 0.3475, 1.768, and 4.77 ns. Thus, tryptophan in ethanol would adopt in the excited state substructures different from those observed in water. Three lifetimes have already been observed for tryptophan dissolved in nonaqueous solvents [78]. Since, high progesterone concentrations abolish $S_0 \rightarrow {}^1L_a$ transition, without affecting fluorescence lifetimes or their preexponential values, then the three lifetimes observed in ethanol would be generated by the $S_0 \rightarrow {}^1L_b$ transition. If this is the case, what would be the role of the $S_0 \rightarrow {}^1L_a$ transition in the excitation process? Also, on the light of our data, an important question could be asked: What is the real relation between the two transitions and the fluorescence lifetimes observed in solution (two lifetimes in water and three lifetimes in ethanol)? In other words, is there any correlation between the two transitions and the fluorescence lifetimes? Excitation spectrum characterizes electrons distribution within a molecule at the ground state while fluorescence lifetimes occur from the excited state and thus reflect electron distribution within the molecule in this state. Therefore, possibility that fluorescence lifetimes are not directly correlated with any of the two transitions is not to be excluded. In fact, we notice that while high progesterone concentrations abolish $S_0 \rightarrow {}^1L_a$ transition, fluorescence lifetimes are not modified. Also, this could simply mean that excitation of tryptophan molecules in solution induces an excited state where fluorophore molecules retain at least two conformations which are different from those observed in the ground state. In ethanol, the three lifetimes are generated by the presence of three substructures induced by an environment (ethanol) whose chemical and physical properties differ from those of water. The same analysis can be drawn for Trp residues in proteins where the fluorophore is, in the excited state, within substructures identical or very close to those observed in ethanol and affected by their surrounding environment.

In a first step, one should correlate fluorescence lifetime of tryptophan with its main structure or substructures in the excited state, independently of the surrounding environment. Then, in a second step, modifications of the fluorescence decay parameters (lifetimes and preexponentials) should be analyzed as the effect of the environment on the substructures of the fluorophore in the excited state. This latest suggestion is in good agreement with the fact that although *N*-acetyl-L-tryptophanamide (NATA) and nonterminal tryptophan residues in proteins display the same structures in the ground state, NATA emits with a single fluorescence lifetime while tryptophan residues emission is bi- or multiexponential. This clearly means that the structures generating fluorescence lifetimes are in the excited state and differ from a fluorophore to another.

Therefore, lifetimes characterize substructures formed in the excited state and not in the ground state, otherwise both molecules, NATA and tryptophan residues, should yield same lifetime values and numbers. Therefore, NATA cannot be used as a reference to understand the origin of tryptophan fluorescence in proteins. Comparison should be made only with free tryptophan in solution (water, ethanol, etc.).

The conclusion of the work described in this part of the review is both $S_0 \rightarrow {}^1L_b$ and $S_0 \rightarrow {}^1L_a$ transitions can induce tryptophan excitation and that recorded fluorescence lifetimes are assigned to substructures generated in the excited state [79].

6 Applications of Tryptophan Fluorescence Lifetimes in Studying Protein Structural Modification

6.1 Characterization of Different Conformations of HSA

HSA undergoes reversible conformational transformations with change in pH. At pH 7, HSA assumes the normal form (N) which abruptly changes to highly charged fast migrating form (F) at pH values less than 4.3, as this form moves "fast" upon gel electrophoresis. At pH less than 2.7, the F-form changes to the fully extended form (E). On the basic side of the normal pH above pH 8, the N-form changes to basic form (B) and above pH 10, the structure changes to the aged form (A) (Scheme 1) [80].

We studied these conformational changes by measuring fluorescence emission spectra and fluorescence lifetimes. Fluorescence emission spectra of HSA recorded from pH 2 to 12, indicate that the peak position at values of pH in a range 4–12 is the same (340 nm) and thus, structure surrounding Trp residue in HSA is almost stable. On the contrary, at pH 2 and 3, tryptophan is more surrounded by hydrophobic area since the observed peak is at 330 nm (Fig. 33). Thus, at pH 2 and 3, no denaturation of HSA occurs but a structural modification. Tryptophan residue at these two pH_s is more embedded in the protein core than at higher pH. Complete protein denaturation leads to an emission peak for the tryptophan around 350–355 nm.

Also, at all studied pH, three lifetimes were obtained, indicating that the presence of these lifetimes is independent of the protein structure and/or conformation. Figure 34 displays the variation of the three fluorescence lifetimes at pH 2, 8, and 12 along the emission spectrum. The shape of the three lifetimes is identical at the three pH. Although it is difficult to observe a real difference between the three

$$E \longrightarrow F \longrightarrow N \longrightarrow B \longrightarrow A$$

Scheme 1 Conformational changes of HSA induced by pH modification

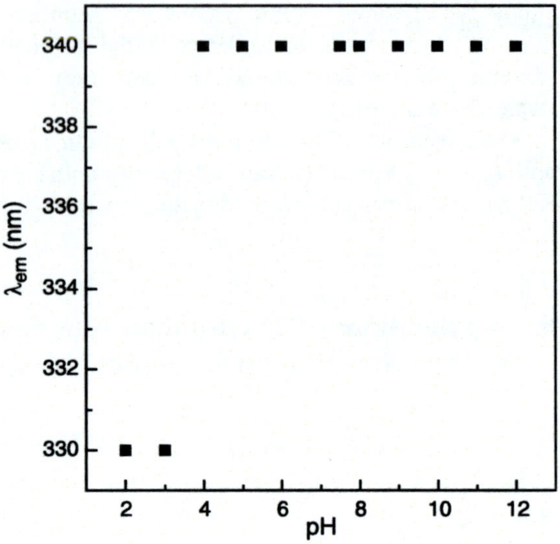

Fig. 33 Fluorescence emission peak of HSA at different pH

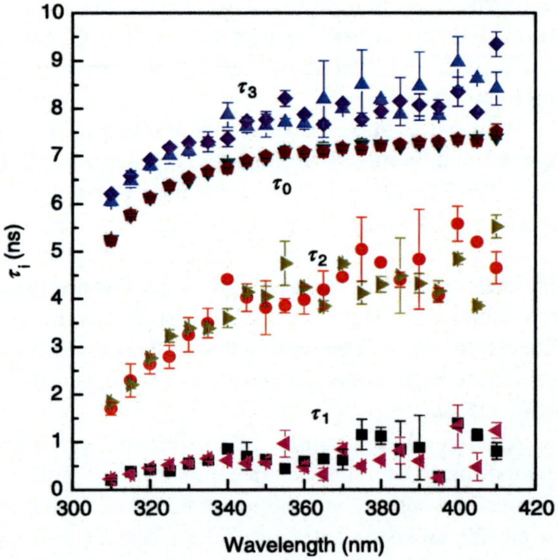

Fig. 34 Variation of the three fluorescence emission lifetimes of HSA Trp-214 residue with emission wavelength measured at pH 2 (*squares*), 8 (*triangles*), and 12 (*circles*). Source of Figs. 34–37: Amiri M, Jankeje K, Albani JR (2010) J Pharm Biomed Anal 51:1097–1102

shortest lifetimes τ_1 measured at the three pH$_s$, it is very clear that values of τ_2 and τ_3 measured at pH 2 are lower than those obtained at pH 8 and 12. This allows us to conclude that the extended form (E) of HSA at pH 2 has a structure that differs significantly from those of basic (B) or aged forms (A). This conclusion is confirmed by the variation of the mean fluorescence lifetime τ_0 with emission wavelength at different pH (Fig. 35). In fact, two different populations can be separated

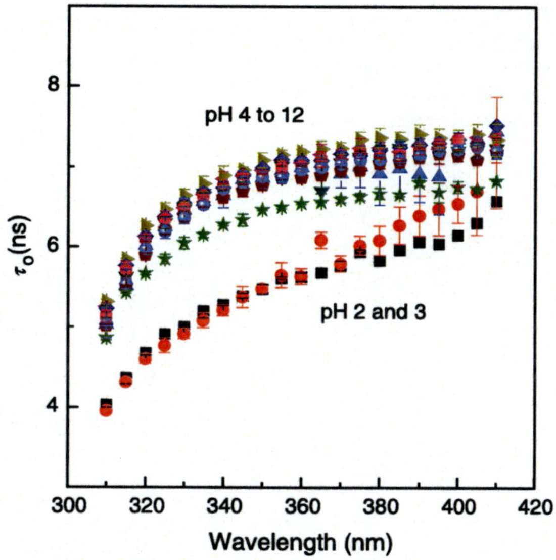

Fig. 35 Mean lifetime variation with emission wavelength from pH 2 to 12 (pK of HSA = 4.7)

from the plot obtained. A population containing protein form at pH 2 and 3 (extended form E) and a second population containing the other protein forms (from pH 4 to 12). Also, at all pH, value of τ_0 increases with the emission wavelengths (Fig. 35). This is the result of the increase of the longest fluorescence lifetime τ_3 value at high emission wavelengths (Fig. 34).

Experiments performed on HSA at different pH allow us to conclude again that number of fluorescence lifetime of tryptophan residue (s) within a protein is not dependent on the structure or the global conformation of the protein.

In order to find out whether it is possible to separate the different forms present at pH higher than 4, we plotted τ_0 as a function of the pH. Figure 36 displays the results obtained at three wavelengths, 320 (squares), 360 (circles), and 400 nm (triangles). We notice that the data obtained can be displayed in three different populations: One of pH 2 and 3 (extended form), the second from pH 4 to 8 containing HSA migrating (F) and normal (N) forms. Another type of population is obvious for pH higher than 8, characterizing the basic (B) and aged (A) forms of HSA.

The structure observed at pH 2 and 3 (extended form) corresponds to a molten globule form which is an intermediate state of the protein before reaching denaturation. In this state, the protein could be partly unfolded; importance of this partial unfolding depends on the protein studied.

By measuring the fluorescence lifetime values τ_0 at different pH, we are able to separate at least three different populations containing two to three forms of the protein. Since we were not able to differentiate all the forms one from each other, this means that fluorescence method we are using is not sensitive to the structural differences between the forms within each population or because the structural differences between the forms present within each population are not important, even absent.

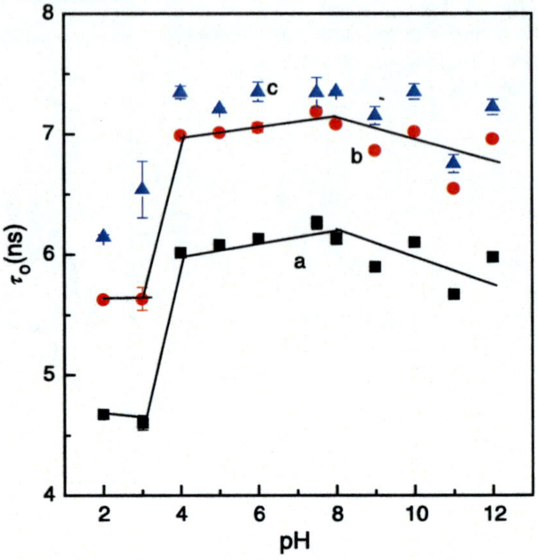

Fig. 36 Mean lifetime variation of HSA Trp residue with pH measured at 320 (*squares*), 360 (*circles*), and 400 nm (*triangles*)

Another information can be obtained from Fig. 36. The mean fluorescence lifetime measured at all the pH is lower at 320 nm than at 360 or 400 nm. At 320 nm, we are observing fluorescence mainly from an environment that is more hydrophobic than at 360 or 400 nm. Since τ_0 value depends on those of the three lifetimes and of their populations characterized by the preexponential values, Fig. 36 clearly indicates that the values of τ_2 and τ_3 are much more important at long wavelengths than lower ones, a conclusion in good agreement with the results obtained in Fig. 34. Also, the populations, characterized by the preexponential terms, of these two lifetimes may also be more important at high than at low wavelengths.

Figure 37 compares lifetimes preexponentials with emission wavelengths obtained at pH 6 and 8. The data show that from 320 nm, the first preexponential α_1 decreases along the emission spectrum, α_2 does not change while α_3 increases with the emission wavelengths. These variations can explain why τ_0 is lower at 320 than at 360 or 400 nm (Fig. 36). Also, results obtained in Fig. 37 mean that tryptophan population characterized by the first preexponential α_1 is more sensitive to hydrophobic than hydrophilic area, whereas the population characterized by α_3 is much more in a contact with hydrophilic area. The second preexponential α_2 is equally sensitive to both areas.

Thus, α_1 which characterizes the population of the shorter lifetime (τ_1), is characteristic of a hydrophobic area since α_1 decreases with the emission wavelength. At high wavelength, its value approaches even zero.

α_2 does not vary with the emission wavelength, it is stable at all wavelengths.

α_3, which characterizes the population of the longest fluorescence lifetime (τ_3), is characteristic of both hydrophilic and hydrophobic areas although its sensitivity

Fig. 37 HSA lifetimes preexponentials with emission wavelengths at pH 6 (*squares*) and pH 8 (*triangles*)

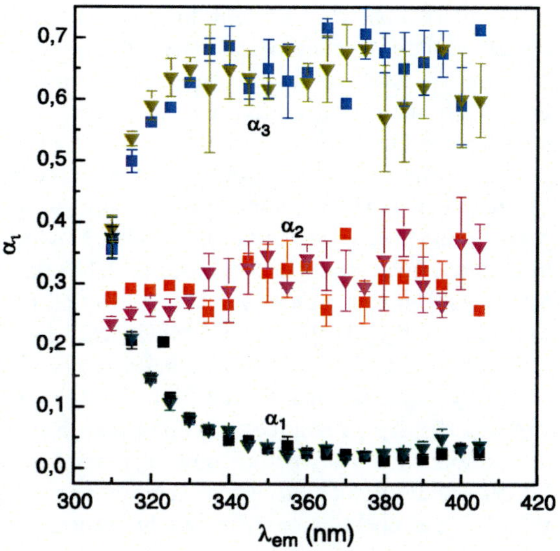

to hydrophilic area is clearly established since it increases with emission wavelengths.

The three substructures of tryptophan 214 residue, each with a specific lifetime, should have different orientations in space within HSA and thus should be surrounded by environments of different polarities.

One should be careful here in the interpretation of meaning of "different polarities." By this we do not intend to pretend that one substructure is located in a hydrophobic area while the other is in polar one. In fact, the three substructures are located in the same hydrophobic location of HSA domain IIA, but they could have different slight contact with the solvent molecules which are in the vicinity of the fluorophore, rendering one of the substructures more "hydrophilic" than the others [81].

6.2 Structural Modifications of Dimeric β Lactoglobulin and Dimeric OBP Upon 1-Aminoanthracene Binding: Fluorescence Excitation Spectra and Lifetimes Studies

Fluorophores have been defined by Edelman and McClure [82] as "small molecules which undergo changes in one or more of their fluorescence properties as a result of noncovalent interaction with a protein or other molecules." Such fluorophores, when added to macromolecules such as proteins, should not modify their local or global conformation. Only in this case, a fluorophore is considered as a good probe to study, for example, allosteric transitions, protein–protein interactions, protein–ligands interactions and structural alterations which accompany protein denaturation.

In this section, we are going to show how fluorophore binding to proteins can modify tertiary structure of the macromolecules. These structural modifications are detected by recording fluorescence excitation spectra and measuring fluorescence lifetimes of the proteins tryptophan residue(s). We choose for this study interaction between the fluorophore 1-aminoanthracene (1-AMA) and three lipocalin family proteins, dimeric β lactoglobulin, dimeric porcine OBP, and monomeric α_1-acid glycoprotein. 1-AMA is widely used to study interaction between lipocalin family proteins and their ligands [58, 60, 83–88]. However, aminoanthracene is known to be a cytotoxic compound [89] inducing DNA damaging and thus altering proteins structures. For this reason, we decided to find out whether binding of 1-AMA to lipocalin family proteins modifies their structures or not.

Let us remind the readers that the three lipocalin proteins posses a pocket (calyx) where hydrophobic ligands can bind. β-Lactoglobulin contains one hydrophobic Trp residue (Trp-19 residue) and one hydrophilic one (Trp-61 residue) [51]. There are no evidence yet whether both tryptophan residues contribute to the protein fluorescence. Porcine odorant binding protein (pOBP) is dimeric at physiological pH (7.2) and contains one Trp residue at position 16 [64]. α_1-Acid glycoprotein (orosomucoid) consists of a chain of 183 amino acids [5], contains 40% carbohydrate by weight [6], and has three Trp residues which all contribute to the protein fluorescence [13, 90]. Carbohydrate units are linked to the pocket of α_1-acid glycoprotein giving it both hydrophobic and hydrophilic properties [16]. Thus, α_1-acid glycoprotein pocket contains a hydrophobic domain where ligands such as progesterone and TNS can bind and a hydrophilic one formed mainly by the glycosylation site of the protein [14, 15]. Therefore, chemical and physical properties of the pocket in α_1-acid glycoprotein are different from those of other lipocalin family as the result of the carbohydrate residues presence. For more details on the structures of the three proteins, the readers can refer to Sect. 3.

7 Interaction Between 1-Aminoanthracene and β-Lactoglobulin

7.1 *Fluorescence Emission Spectrum of β-Lactoglobulin with 1-Aminoanthracene*

Binding of 1-AMA to β-lactoglobulin dimer solution induces a decrease in the fluorescence emission spectrum of tryptophan residues ($\lambda_{ex} = 295$ nm) accompanied by an increase of the emission intensity of 1-AMA ($\lambda_{max} = 520$ nm) (Fig. 38). Position of tryptophan emission peak in the absence of 1-AMA (spectrum 1) is equal to 335 nm, while in the presence of 20 μM AMA, it shifts to 328 nm (spectrum 8). This blue shift is the result of the increasing hydrophobicity around

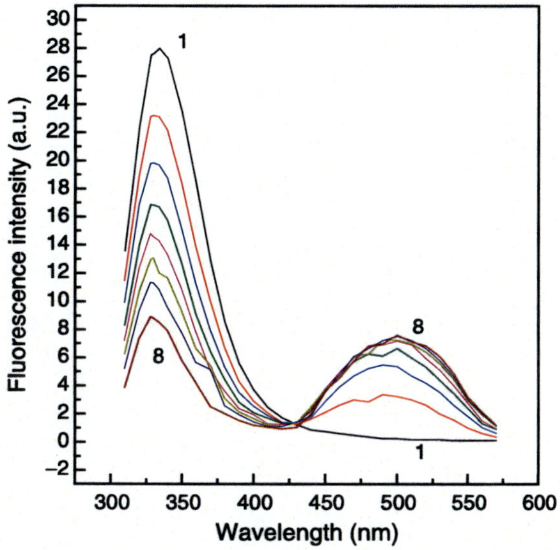

Fig. 38 Titration of 30 μM β-lactoglobulin dimer with 1-aminoanthracene. λ_{ex} = 295 nm. Position of the tryptophan maximum in the absence of 1-AMA (spectrum 1) is equal to 335 nm, while in the presence of 20 μM AMA, it shifts to 328 nm (spectrum 8). Corresponding [1-AMA] concentrations from spectra 1 to 8 are: 0, 3.1, 7.27, 9.87, 12.55, 14, 16.16, and 20 μM, respectively. Source of Figs. 38–48: Kmiecik D, Albani JR (2010) J Fluoresc 20:973–983

tryptophan residues. Intensity increase of bound 1-AMA is the result of the extrinsic fluorophore binding to the protein and to energy transfer from tryptophan residues to the extrinsic fluorophore. The following experiments have been described in ref. [91].

7.2 *Fluorescence Excitation Spectrum of β-Lactoglobulin with 1-Aminoanthracene*

Structural modification within a protein can be monitored by recording the fluorescence excitation spectrum of Trp residues in the absence and the presence of different 1-AMA concentrations. The fluorescence excitation spectrum characterizes electron distribution of the molecule in the ground state. Since 1-AMA does not emit at 340 nm, thus, at this wavelength, only excitation spectrum of the Trp residues will be recorded. Therefore, any modification of the fluorescence excitation spectrum in the presence of 1-AMA would be the result of a structural modification of the protein in the ground state.

Figure 39 displays the fluorescence excitation spectrum of Trp residues of β-lactoglobulin in the absence and the presence of increasing concentrations of 1-AMA. We observe a significant decrease in the intensity of the excitation spectrum accompanied by a shift of the recorded spectrum, from 280 nm in the absence of 1-AMA to 290 nm in the presence of 20 μM of 1-AMA. This red shift clearly indicates that binding of 1-AMA to β-lactoglobulin induces conformational modifications within the tryptophan environment and mainly within the binding site of the protein. Therefore, β-lactoglobulin conformation is not the same in the absence and the presence of 1-AMA. Also, the shift observed occurs at a ratio of 0.3 (1-AMA to β-lactoglobulin), much before the stoichiometry of the complex (2 AMA for 1 β-lactoglobulin dimer) is reached.

Fig. 39 Fluorescence excitation spectra of 30 μM β-lactoglobulin dimer with 1-AMA. $\lambda_{em} = 340$ nm. Spectrum 1: peak = 280 nm; spectrum 8: peak = 290 nm. Corresponding [1-AMA] concentrations from spectra 1 to 8 are: 0, 3.1, 7.27, 9.87, 12.55, 14, 16.16, and 20 μM, respectively. The position of the peak shifts to 285 nm in the presence of 0.3 μM of 1-AMA

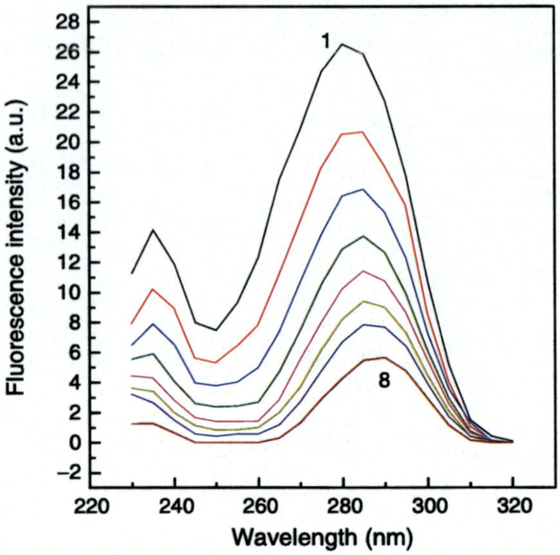

Emission spectra of β-lactoglobulin dimer in the presence of 1-AMA (Fig. 38) do not show any shift to the longest wavelengths (350–355 nm). This means that protein denaturation in the presence of 1-AMA is not occurring. However, excitation spectra clearly indicate that addition of 1-AMA to β-lactoglobulin induces conformational changes in the protein structure. Excitation spectrum characterizes tertiary structure of a molecule in the ground state. Although structural modification does not mean here protein denaturation, still β-lactoglobulin structure in the absence of 1-AMA is not the same in its presence. In this case, any dissociation constant measured from titration experiments has no significant meaning.

At pH 7.5, β-lactoglobulin is a dimer, the pocket of each monomer makes a barrel (calyx) where 1-AMA binds. However, this calyx is narrow and thus 1-AMA, although attracted by the hydrophobic binding site of β-lactoglobulin, modifies the binding site and in consequence the protein structure. Therefore, 1-AMA would be able to induce a specific binding site that fits to its size. In other terms, hydrophobic binding site adopts a structure that is imposed by the bound ligand (in this case, 1-AMA).

7.3 *Fluorescence Lifetime Measurements of β-Lactoglobulin in the Presence of 1-Aminoanthracene*

In the absence and the presence of 1-AMA, fluorescence intensity, $I(\lambda, t)$, of Trp residues in β-lactoglobulin can be adequately represented by a sum of three exponentials. In the absence of 1-AMA,

$$I(\lambda, t) = 0.412e^{-t/0.747} + 0.425e^{-t/2.1} + 0.164e^{-t/5.678},$$

where 0.412, 0.425, and 0.164 are the preexponential factors, 0.747 ± 0.05, 2.1 ± 0.07, and 5.678 ± 0.27 ns are the decay times and λ is the emission wavelength (340 nm) ($\chi^2 = 0.922$, $\lambda_{ex} = 296$ nm). The two shortest lifetimes correspond to what we can usually observe for Trp residues in Protein or for free tryptophan in solution, while the longest lifetime is the result of the interaction of the Trp residue within its environment. The mean fluorescence lifetime calculated from four experiments was found equal to 3.466 ns. Analyzing the decay curves with two lifetimes yields a χ^2 equal to 2.1. And fitting the data with four exponentials did not yield better χ^2 values than the fitting with three lifetimes.

Increasing 1-AMA concentrations in the β-lactoglobulin solution induces a dramatic increase in the longest fluorescence lifetime of the tryptophan residues. The two shortest fluorescence lifetimes are not affected by the presence of 1-AMA. The mean fluorescence lifetime increases as the result of the important increase of the longest fluorescence lifetime (Fig. 40). When the stoichiometry 2:1 of the 1-AMA-β-lactoglobulin is reached, i.e., 2 mol of 1-AMA for one dimer of β-lactoglobulin, we notice that the longest lifetime increase stops. Modification of the longest fluorescence lifetime in the presence of 1-AMA is in good agreement with the results obtained when the fluorescence excitation spectrum of β-lactoglobulin was recorded in the presence of increasing concentrations of 1-AMA. In both experiments, we are following structural modifications within β-lactoglobulin as the result of 1-AMA binding to β-lactoglobulin.

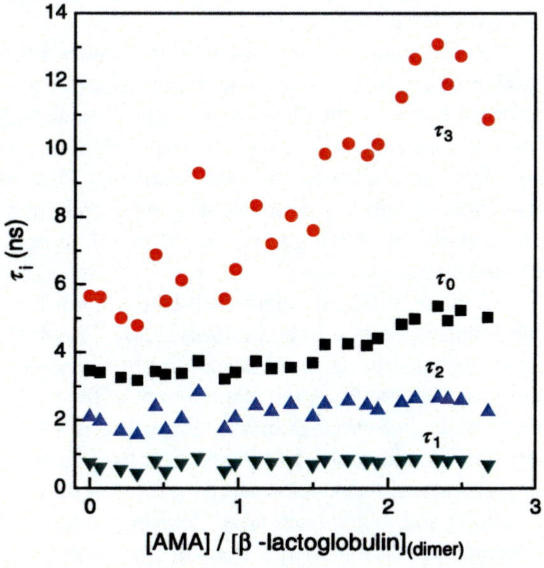

Fig. 40 Fluorescence lifetimes variation of β-lactoglobulin tryptophan residues as a function of [1-AMA]/[β-lactoglobulin] ratio. $\lambda_{ex} = 296$ nm and $\lambda_{em} = 340$ nm

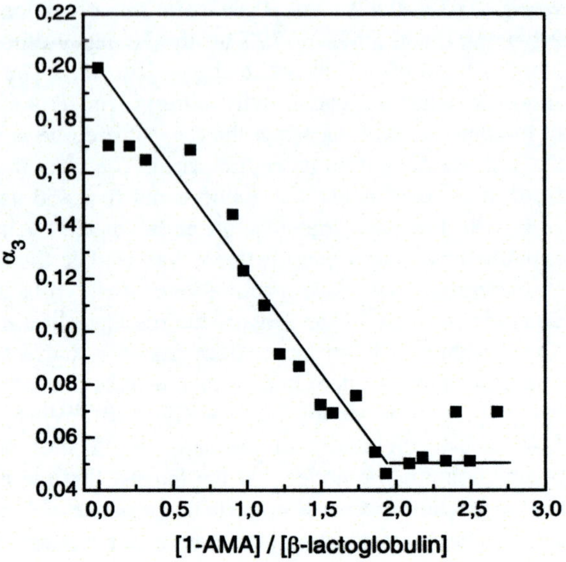

Fig. 41 Variation of longest fluorescence lifetime preexponential of β-lactoglobulin tryptophan residues as a function of [1-AMA]/[β-lactoglobulin] ratio

The increase in the longest fluorescence lifetime in the presence of 1-AMA is accompanied by a decrease in its preexponential value (α_3). This decrease reaches a plateau at a stoichiometry of 2 mol of 1-AMA for 1 β-lactoglobulin dimer (Fig. 41). The preexponential values (α_1 and α_2) of the two shortest lifetimes are affected by 1-AMA binding to β-lactoglobulin but their variations do not correspond to that observed for α_3 (Fig. 42).

The fact that only the longest fluorescence lifetime is modified upon the structural modification confirms that this lifetime would characterize the interaction existing between the fluorophore and surroundings amino acids. The other two lifetimes are close to those measured for free tryptophan in water ($\approx$0.5 and 2.5 ns) and found in almost all proteins (Table 1). Thus, these two values are in a certain way independent of the surrounding environment of tryptophan, revealing that they characterize intrinsic properties of the fluorophore, i.e., specific substructures formed upon excitation.

1-Aminoanthracene disrupts β-lactoglobulin tertiary structure even at very low fluorophore/protein concentrations ratios. Tryptophan 19 residue in β-lactoglobulin is facing into the base of the hydrophobic pocket while tryptophan 61 is within the extremely mobile loop at the mouth of the pocket [48]. Thus, the two tryptophan residues in β-lactoglobulin are in positions which render them very sensitive to structural modification within and near the pocket. This high sensitivity would affect tryptophan residues excitation spectrum and fluorescence intensity decay parameters (lifetimes and preexponentials). Conformational modifications are observed from the fluorescence excitation spectrum of the protein recorded at different concentrations of 1-AMA and from the modification of the longest fluorescence lifetime of the

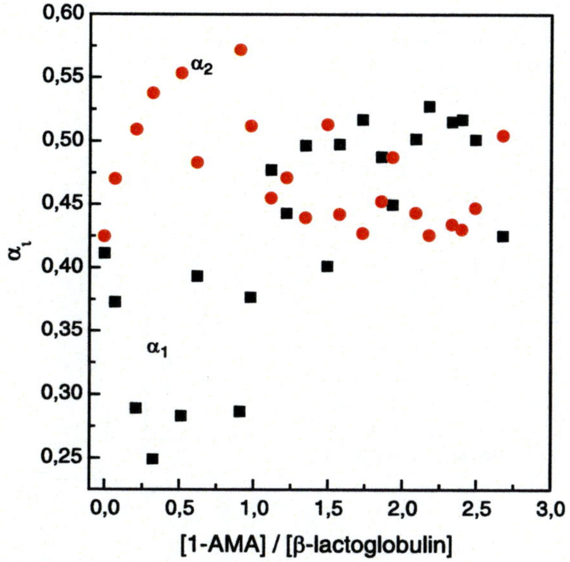

Fig. 42 Variation of α_1 and α_2 as a function of [1-AMA]/[β-lactoglobulin] ratio

tryptophan residues and of its preexponential component. In the absence of any conformational modification within the protein, the fluorescence excitation spectrum should not display any shift of its maximum and the lifetime should not vary significantly.

8 Interaction Between 1-AMA and α_1-Acid Glycoprotein

8.1 Fluorescence Excitation Spectrum of α_1-Acid Glycoprotein with 1-Aminoanthracene

Figure 43 displays the fluorescence excitation spectrum of Trp residues of α_1-glycoprotein in the absence and the presence of increasing concentrations of 1-AMA.

We observe a significant decrease in the intensity of the excitation spectrum and the absence of a shift up to [AMA]/[α_1-glycoprotein] ratio equal to 1.30. Above this ratio, a shift occurs to 285 then to 287 nm. Thus, to the difference of β-lactoglobulin, binding of 1-AMA to α_1-glycoprotein does not induce, within the stoichiometric range, any structural modification within the glycoprotein. Presence of 40% carbohydrate residues on α_1-glycoprotein would be the principal reason for that. It is already known that the presence of important amount of carbohydrate residues in α_1-acid glycoprotein (40% of carbohydrate by weight) explains the resistance and the weak antigenicity of the protein, the carbohydrates acting as protective shield.

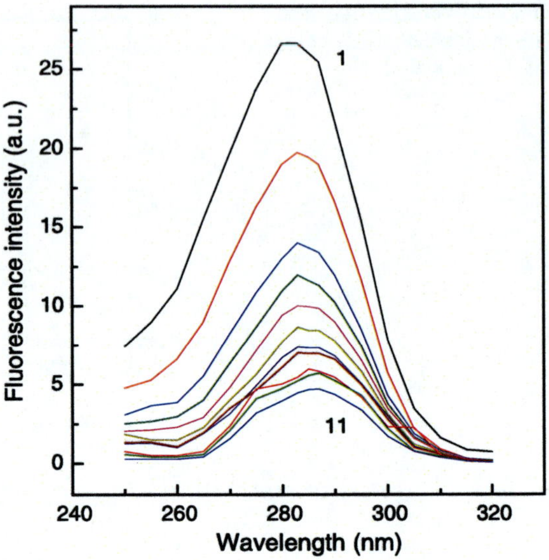

Fig. 43 Titration of 9.5 μM α_1-acid glycoprotein with 1-AMA following excitation spectra of the tryptophan residues. $\lambda_{em} = 340$ nm. Spectrum 1: $\lambda_{max} = 283$ nm; spectrum 11: $\lambda_{max} = 287$ nm. Corresponding [1-AMA] concentrations from spectra 1 to 11 are: 0, 2, 3.05, 5.1, 6.68, 8.52, 9.76, 12.35, 13.29, 16.07, and 17 μM, respectively. The stoichiometry of the complex is reached at plot 6. The *slight red shift* is observed when the ratio [1-AMA]/[α_1-acid glycoprotein] is equal to or higher than 1.3 (plots 8–11)

Carbohydrate residues stabilize the conformation of the protein matrix and prevent the unfolded or partially folded protein molecules from aggregation [92].

8.2 *Fluorescence Lifetimes Measurements of α_1-Acid Glycoprotein in the Presence of 1-Aminoanthracene*

In the absence and the presence of 1-AMA, fluorescence intensity, $I(\lambda, t)$, of Trp residues in α_1-acid glycoprotein can be adequately represented by a sum of three exponentials. In the absence of 1-AMA,

$$I(\lambda, t) = 0.230\mathrm{e}^{-t/0.795} + 0.625\mathrm{e}^{-t/2.445} + 0.145\mathrm{e}^{-t/6.18},$$

where 0.230, 0.625, and 0.145 are the preexponential factors, 0.795 ± 0.03, 2.445 ± 0.06, and 6.18 ± 0.77 ns are the decay times and λ is the emission wavelength (335 nm) ($\chi^2 = 1.03$, $\lambda_{ex} = 296$ nm). The mean fluorescence lifetime calculated from three experiments was found equal to 3.61 ns. Analyzing the decay curve with two lifetimes yields a χ^2 equal to 2.5. Also, it was not possible to fit the data with four exponentials.

Increasing 1-AMA concentrations up to 0.6 [1-AMA]/[α_1-acid glycoprotein] ratio does not affect any of the three fluorescence lifetimes. However, a slight increase of the third lifetime (from 6 to 8 ns) occurs up to 1.2 [1-AMA]/[α_1-acid glycoprotein] ratio. This increase is simply a variation correlated to statistical measurements since the data shown are from three different experiments. Beyond the 1.2 ratio, the value of the longest lifetime increases faster (Fig. 44).

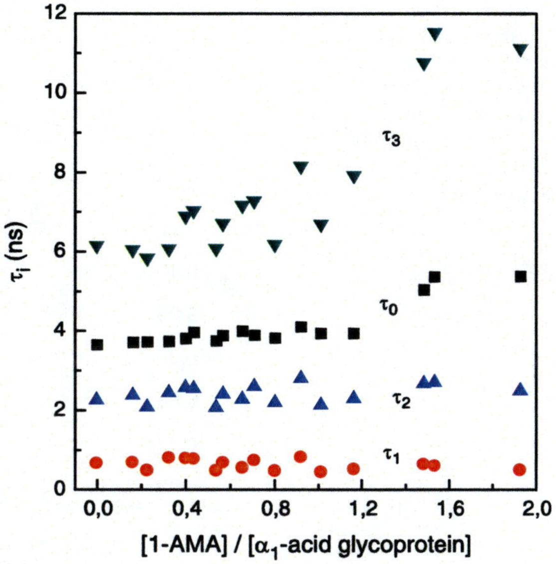

Fig. 44 Fluorescence lifetimes variation of α_1-glycoprotein tryptophan residues as a function of [1-AMA]/[α_1-acid glycoprotein] ratio. $\lambda_{ex} = 296$ nm and $\lambda_{em} = 335$ nm. Data are from three experiments

Absence of a significant increase in the longest fluorescence lifetime of α_1-acid glycoprotein in the presence of 1-AMA up to the stoichiometry of the complex (1:1) is in good agreement with the results obtained when the fluorescence excitation spectrum of α_1-acid glycoprotein was recorded in the presence of different 1-AMA concentrations. In both experiments and within the stoichiometric range, structural modifications are not observed within α_1-acid glycoprotein as the result of 1-AMA binding.

The slight increase in the longest fluorescence lifetime of α_1-acid glycoprotein, tryptophan residues in the presence of 1-AMA is accompanied by a decrease in its preexponential value (α_3). However, to the difference of the results obtained with β-lactoglobulin, this decrease does not reach a plateau at the stoichiometry (1:1) of the complex (Fig. 45). Variation of α_3 affects the values of α_1 and α_2 (Fig. 46) differently from that was observed for β-lactoglobulin (Fig. 40). The global similar variation (a decrease) of α3 in both α_1-acid glycoprotein and β-lactoglobulin in the presence of 1-AMA is a clear indication to the fact that the longest lifetime along with its preexponential value characterize the interaction between tryptophan residues with the protein matrix. Any modification of the conformation within the protein will affect both long lifetime and its preexponential factor.

It is important here to remind that stoichiometry of 1-AMA to β-lactoglobulin is 2:1 while that of 1-AMA to α_1-acid glycoprotein is 1:1. Therefore, increase of fluorescence lifetime in β-lactoglobulin occurs when [1-AMA]/[β-lactoglobulin] ratio is already equal to 0.3 while this increase is not observed for α_1-acid glycoprotein until this ratio reaches 1.2. Thus, it is possible to use 1-AMA as a fluorophore to study binding of ligands to α_1-acid glycoprotein. However, in the

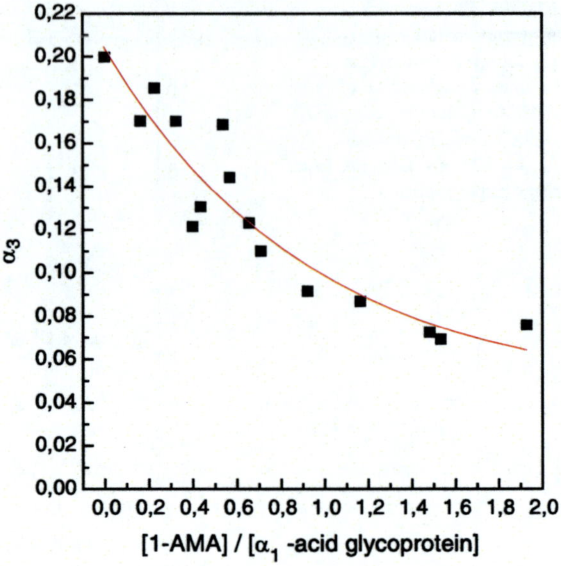

Fig. 45 Variation of longest fluorescence lifetime preexponential of α_1-glycoprotein tryptophan residues as a function of [1-AMA]/[α_1-acid glycoprotein] ratio. $\lambda_{ex} = 296$ nm and $\lambda_{em} = 335$ nm

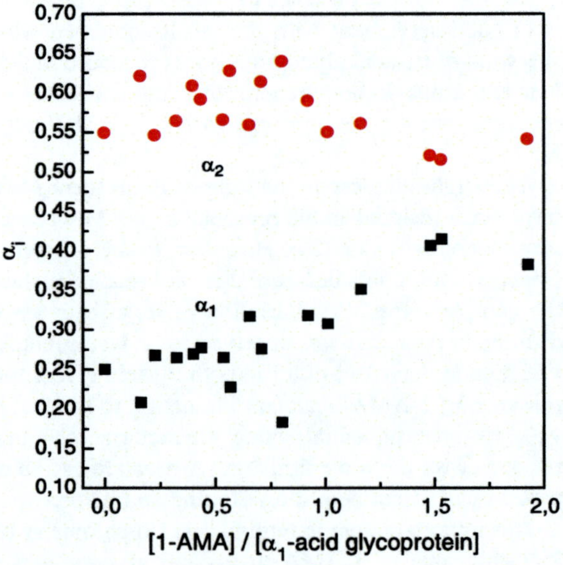

Fig. 46 Preexponential variations of the two shortest fluorescence lifetimes of α_1-acid glycoprotein tryptophan residues as a function of 1-AMA/glycoprotein ratio

presence of β-lactoglobulin, it should be used at low concentrations to avoid any modification in the protein conformation.

9 Interaction Between 1-AMA and OBP

9.1 *Fluorescence Emission Spectrum of OBP with 1-Aminoanthracene*

Addition of increasing concentrations of 1-AMA to 2.7 μM OBP dimer induces a decrease in the fluorescence intensity at 340 nm (Trp residue) accompanied with an increase of the fluorescence intensity at 490 nm (1-AMA) (data not shown). Binding of 1-AMA to OBP does not modify the emission peak position of the tryptophan residue indicating that binding site of 1-AMA is not necessarily at proximity of OBP Trp residue. In fact, analysis of OBP structure has shown that the Trp residue does not belong to the internal cavity forming the binding site for ligands [63, 64]. Intensity increase of bound 1-AMA is the result of the extrinsic fluorophore binding to the protein and to energy transfer from tryptophan residue to the extrinsic fluorophore.

It is important here to indicate that emission peak of 1-AMA bound to α_1-acid glycoprotein is located at 520 nm compared to 500 and 490 nm for β-lactoglobulin and OBP, respectively. This means that binding site of 1-AMA on α_1-acid glycoprotein is more hydrophilic than that of the other two proteins. 1-AMA binding site on OBP is the most hydrophobic which facilitates binding of the extrinsic fluorophore.

9.2 *Fluorescence Lifetimes Measurements of OBP with 1-Aminoanthracene*

In the absence of 1-AMA, fluorescence intensity, $I(\lambda, t)$, of Trp residue in OBP can be adequately represented by a sum of three exponentials:

$$I(\lambda, t) = 0.357\mathrm{e}^{-t/0.858} + 0.534\mathrm{e}^{-t/2.761} + 0.109\mathrm{e}^{-t/9.201},$$

where 0.357, 0.534, and 0.109 are the preexponential factors, 0.858 ± 0.05, 2.761 ± 0.07, and 9.201 ± 0.27 ns are the decay times and λ is the emission wavelength (340 nm) ($\chi^2 = 1.167$, $\lambda_{ex} = 296$ nm).

Analyzing the decay curve with two lifetimes yields a χ^2 equal to 2.5. However, in the presence of 1-AMA and to the difference of β-lactoglobulin and α_1-glycoprotein, fitting the data with four exponentials yield better χ^2 values than the fitting with three lifetimes indicating that 1-AMA modifies the global conformation around the Trp residue of OBP. For example, when the ratio [1-AMA]/[OBP] reaches 0.33, fluorescence intensity decay of OBP can be described as

$$I(\lambda, t) = 0.153e^{-t/0.617} + 0.6e^{-t/1.862} + 0.188e^{-t/3.98} + 0.059e^{-t/16},$$

($\chi^2 = 1.152$). Analyzing the data with three lifetimes yields a χ^2 value equal to 1.214. At high [1-AMA]/[OBP] ratios, for example at 1.7, χ^2 values obtained were 1.180 and 1.466 with four and three lifetimes, respectively.

Also, increasing 1-AMA concentrations in the OBP solution induces an important increase in both the longest and the mean fluorescence lifetimes of the tryptophan residue. When the stoichiometry 2:1 of the 1-AMA–OBP is reached, i.e., 2 mol of 1-AMA for one dimer of OBP, we notice that the longest and mean lifetimes increase stops (Figs. 47 and 48).

The increase in the longest fluorescence lifetime in the presence of 1-AMA is accompanied by an increase in its preexponential value (data not shown).

β-Lactoglobulin belongs to the lipocalin family and thus, the effect observed on β-lactoglobulin structure upon 1-AMA binding has been detected also on proteins such as OBPs. In fact, our results show that 1-AMA binding to OBP modifies fluorescence lifetimes of the only Trp residue of the protein. In the presence of 1-AMA, fluorescence intensity decrease with time can be best described with four lifetimes instead of three in the absence of 1-AMA. This means clearly that binding of 1-AMA to OBP modifies its local and global conformation. In the same conditions, when 1-AMA was added to β-lactoglobulin, it did not induce any appearance of a fourth fluorescence lifetime. This may be explained by the fact that binding sites and

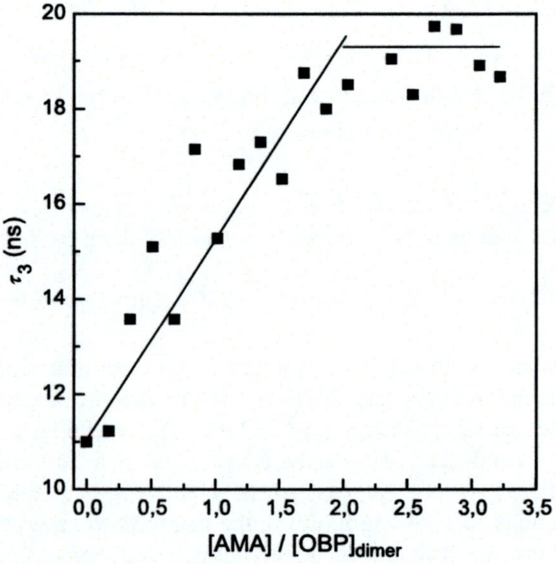

Fig. 47 Variation of the third fluorescence lifetime of OBP tryptophan residues with the [1-AMA]/[OBP] ratio. $n = 2$ monomer of AMA for 1 dimer of OBP

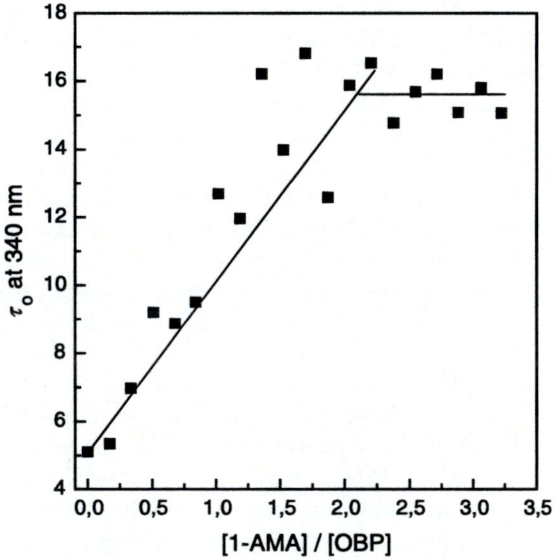

Fig. 48 Mean fluorescence lifetime of Trp residue of OBP at different [1-AMA]/[OBP] ratios. $\lambda_{ex} = 296$ nm. $n = 2$ monomer of AMA for 1 dimer of OBP

interactions of 1-AMA with the two proteins are not the same. Nevertheless, in both cases, the two protein conformations are disrupted in the presence of 1-AMA.

9.3 1-AMA is Not a Good Probe to Study Binding of Ligands to OBP

1-AMA is widely used as a probe to study interaction between OBPs and hydrophobic ligands such as 2-iso-butyl-3-metoxypyrazine (IBMP), dihydromyrcenol (DHM), or benzophenone (BZP). In general, hydrophobic ligands are added to OBP_s–1-AMA complex and binding of these ligands to the proteins is followed by analyzing the fluorescence intensity decrease of bound 1-AMA. Looking to the literature very carefully, one can notice that, in order to perform their experiments, the different authors measure the dissociation constant of the 1-AMA–OBP complex by titrating 1 μM of protein with concentrations of 1-AMA that go from 0.076 to 5 μM (see, e.g., ref. [84]). Based on the data we described in the present review, OBP tertiary structure, in the presence of the concentrations used of 1-AMA, is different from the native tertiary structure of the protein. However, some authors considered that their proteins structures are identical either in the presence or in the absence of 1-AMA. Their conclusion is based on the fact that emission peak does not shift to high wavelengths (350 or 355 nm) upon 1-AMA binding (see, e.g., ref. [88]). One should remind here that emission spectrum is not so sensitive to structural modifications as it is the case for the excitation spectrum or fluorescence lifetimes. Therefore, upon slight or important structural modifications within the protein, recording the emission spectrum will not yield any information on these modifications.

Another problem occurs when 1-AMA is used as a probe. In fact, displacing 1-AMA from its binding site on OBP is usually performed in the following conditions: 0.5 μM or 1 μM OBP samples are incubated with a fixed amount of 1-AMA (3 μM) and increasing concentrations of odorants (0.39–50 μM) are then added. First of all, the [1-AMA]/[OBP] ratios used would in principle modify protein structure. Thus, the OBP studied in the presence of 1-AMA should have different structure than the nonmodified protein. Second, displacement of 3 μM of 1-AMA with 50 μM of ligand clearly means that the ligand, in order to displace the 1-AMA, needs to modify the structure of the OBP of the 1-AMA–OBP complex. If the interaction between 1-AMA and the other hydrophobic ligands are competitive, this means that small concentrations of the ligands are sufficient to displace 1-AMA from its binding site. Also, in the case of a real competitive interaction, affinity of the ligands to OBP should be around 100 times that of 1-AMA to OBP. See, for example, displacement by hemin of the fluorophore 2,*p*-toluidinylnaphthalene-6-sulfonate (TNS) from its binding site on apocytochrome b_2 core [93] and on α_1-acid glycoprotein [16]. In both cases, TNS fluorescence decreases to zero when the stoichiometry of the proteins–hemin complex reaches 1:1. In the studies on OBP, it is clear from the concentrations of hydrophobic ligands used do not fit with competitive interactions. The large concentrations of ligands modify OBP structure so that 1-AMA can thus be released from the binding site. This fact is in good agreement with our results on both β-lactoglobulin and OBP that show that binding of 1-AMA to the lipocalin proteins induces important structural modifications within the proteins. Since important concentrations of ligands are needed to displace 1-AMA from its binding site, this means that hydrophobic binding site follows a conformation imposed by 1-AMA presence. Added ligand would exert a pressure on the protein so that new structural rearrangements within the protein and within the binding site occur in order to release 1-AMA to the solution. Our conclusion is in good agreement with the fact that hydrophobic binding site of OBP is highly mobile and can adopt the structure of the bound ligand as it is shown by the electron density map of OBP pocket in the presence of different ligands [86]. Also, the close dissociation constant values of 1-AMA–OBP complex and ligands–OBP complex are in good agreement with the fact that conformational modifications are occurring within the protein upon 1-AMA displacement with the "supposed" competitive ligands.

Thus, one should question the ability of using 1-AMA as a probe to study interaction of lipocalin proteins family with their ligands and the meaning of the measured K_d of the 1-AMA-protein and of the ligands–protein complexes.

10 Conclusions

Emission occurs from the excited state. Thus, fluorescence lifetimes characterize species or substructures existing in the excited state and not in the ground one.

Energy absorbed by a fluorophore is not entirely used to excite the fluorophore. Some of the absorbed energy helps to reorganize the fluorophore molecules in the excited state inducing a structure different from that of the ground state.

In the excited state, electronic distribution within tryptophan differs from that in the ground state. Thus, in the excited state, tryptophan sub-structures differ from those present in the ground state.

Number of fluorescence lifetimes is in general independent of that of tryptophan residues present within a protein.

Two lifetimes around 0.4–0.5 ns and 2–4 ns are measured for tryptophan free in solution and within proteins. Thus, presence of these two lifetimes is independent of any structure around tryptophan and characterizes an internal property or/and organization of the tryptophan structure in the excited state, independently of its environment.

The shortest fluorescence lifetime is not the result of the $S_0 \rightarrow {}^1L_b$ transition and the second fluorescence lifetime is not induced by the $S_0 \rightarrow {}^1L_a$ transition.

Fluorescence lifetimes are not necessarily directly related to both $S_0 \rightarrow {}^1L_b$ and $S_0 \rightarrow {}^1L_a$ transition.

Combination of both transition states induce the excited state of tryptophan and the reorganization of the fluorophore into two interrelated structures.

The third lifetime recorded in proteins could be attributed to interaction between the Trp residue(s) and the surrounding amino acids and to possible specific properties of the protein.

The values of the relative amplitudes of the three fluorescence lifetimes can be dependent on the number of emitting Trp residues or/and on the type of interaction that is occurring between Trp residues and the surrounding environment.

Lifetimes amplitudes will characterize the populations of emitting species or tryptophan residues substructures.

Fluorescence lifetimes and their populations (preexponential values) can be modified with the global structure of the protein.

Protein denaturation does not yield a single fluorescence lifetime.

Emission spectra intensity is not proportional to the optical density of the excitation wavelength but to the intensity of the excitation spectrum at the excitation wavelength.

Fluorophores used to follow protein–ligands interaction should not modify the local or/and the global structures of proteins.

Acknowledgments The author wishes to thank Dr. Daniel Kmiecik for careful reading of the manuscript.

References

1. Petrich JW, Chang MC, McDonald DB, Fleming GR (1983) On the origin of nonexponential fluorescence decay in tryptophan and its derivatives. J Am Chem Soc 105:3824–3832

2. Antonini PS, Hillen W, Ettner N, Hinrichs W, Fantucci P, Doglia SM, Bousquet JA, Chabbert M (1997) Role of Trp-187 in the annexin V–membrane interaction: a molecular mechanics analysis. Biophys J 72:1800–1811
3. Kuszaj S, Kaszycki P, Wasylewski Z (1996) Fluorescence and phosphorescence study of Tet Repressor–Operator interaction. J Protein Chem 15:607–619
4. Martinho JMG, Santos AM, Fedorov A, Baptista RP, Taipa MA, Cabral JMS (2003) Fluorescence of the single tryptophan of cutinase: temperature and pH effect on protein conformation and dynamics. Photochem Photobiol 78:15–22
5. Dente L, Pizza MG, Metspalu A, Cortese R (1987) Structure and expression of the genes coding for human alpha 1-acid glycoprotein. EMBO J 6:2289–2296
6. Kute T, Westphal U (1976) Steroid-protein interactions. XXXIV. Chemical modification of alpha1-acid glycoprotein for characterization of the progesterone binding site. Biochim Biophys Acta 420:195–213
7. Schmid K, Kaufmann H, Isemura S, Bauer F, Emura J, Motoyama T, Ishiguro M, Nanno S (1973) Structure of α_1 -acid glycoprotein. The complete amino acid sequence, multiple amino acid substitutions, and homology with the immunoglobulins. Biochemistry 12:2711–2724
8. Sager G, Nilsen OG, Jackobsen S (1979) Variable binding of propranolol in human serum. Biochem Pharmacol 28:905–911
9. Chiu KM, Mortensen RF, Osmand AP, Gewurz H (1977) Interactions of alpha1-acid glycoprotein with the immune system. I. Purification and effects upon lymphocyte responsiveness. Immunology 32:997–1005
10. Van Dijk W, Havenaar EC, Brinkman-van der Linden EC (1995) α_1-acid glycoprotein (orosomucoid): pathophysiological changes in glycosylation in relation to its function. Glycoconj J 12:227–233
11. Mackiewicz A, Mackiewicz K (1995) Glycoforms of serum α_1-acid glycoprotein as markers of inflammation and cancer. Glycoconj J 12:241–247
12. Brinkman-van der Linden EC, van Ommen EC, van Dijk W (1996) Glycosylation of alpha 1-acid glycoprotein in septic shock: changes in degree of branching and in expression of sialyl Lewis(x) groups. Glycoconj J 13:27–31
13. Albani JR (2001) Effect of binding of calcofluor white on the carbohydrate residues of α_1-acid glycoprotein (orosomucoid) on the structure and dynamics of the protein moiety. A fluorescence study. Carbohydr Res 334:141–151
14. Albani JR (1999) New insights in the conformation of α_1-acid glycoprotein (orosomucoid). Quenching resolved emission anisotropy studies. Spectrochim Acta A 55:2353–2360
15. De Ceukeleire M, Albani JR (2002) Interaction between carbohydrate residues of α_1-acid glycoprotein (orosomucoid) and progesterone. A fluorescence study. Carbohydr Res 337:1405–1410
16. Albani JR (2004) Tertiary structure of human α_1-acid glycoprotein (orosomucoid). Straightforward fluorescence experiments revealing the presence of a binding pocket. Carbohydr Res 339:607–612
17. Albani JR (2006) Progesterone binding to the tryptophan residues of human α_1-acid glycoprotein. Carbohydr Res 341:2557–2564
18. Albani JR, Plancke YD (1998/1999) Interaction between calcofluor white and carbohydrates of α_1–acid glycoprotein. Carbohydr Res 314:169–175, 318:194–200
19. Albani JR, Sillen A, Coddeville B, Plancke YD, Engelborghs Y (1999) Dynamics of carbohydrate residues of α_1-acid glycoprotein (orosomucoid) followed by red-edge excitation spectra and emission anisotropy studies of Calcofluor White. Carbohydr Res 322:87–94
20. Albani JR, Sillen A, Plancke YD, Coddeville B, Engelborghs Y (2000) Interaction between carbohydrate residues of α_1-acid glycoprotein (orosomucoid) and saturating concentrations of Calcofluor White. A fluorescence study. Carbohydr Res 327:333–340
21. Albani JR (2003) Relation between the secondary structure of carbohydrate residues of α_1-acid glycoprotein (orosomucoid) and the fluorescence of the protein. Carbohydr Res 338:1097–1101

22. Albani JR (2003) Förster energy-transfer studies between Trp residues of α_1-acid glycoprotein (orosomucoid) and the glycosylation site of the protein. Carbohydr Res 338:2233–2236
23. Albani JR (2004) Effect of the secondary structure of carbohydrate residues of α_1-acid glycoprotein (orosomucoid) on the local dynamics of Trp residues. Chem Biodivers 1:152–160
24. Galat A (1993) Peptidylproline cis–trans-isomerases: immunophilins. Eur J Biochem 216:689–707
25. Fischer G, Wittmann-Liebold B, Lang K, Kiefhaber T, Schmid FX (1989) Cyclophilin and peptidyl-prolyl cis–trans isomerase are probably identical proteins. Nature 337:476–478
26. Price ER, Zydowsky LD, Jin M, Baker CH, McKeon FD, Walsh CT (1991) Human cyclophilin B: A second cyclophilin gene encodes a peptidyl–prolyl isomerase with a signal sequence. Proc Natl Acad Sci U S A 88:1903–1907
27. Friedman J, Weissman I (1991) Two cytoplasmic candidates for immunophilin action are revealed by affinity for a new cyclophilin: one in the presence and one in the absence of CsA. Cell 66:799–806
28. Connern CP, Halestrap AP (1992) Purification and N-terminal sequencing of peptidyl–prolyl cis–trans-isomerase from rat liver mitochondrial matrix reveals the existence of a distinct mitochondrial cyclophilin. Biochem J 284:381–385
29. Kay JE (1992) Mitochondrial cyclophilins. Biochem J 288:1074–1075
30. Montague JW, Hughes FM Jr, Cidlowski JA (1997) Native recombinant cyclophilins A, B, and C degrade DNA independently of peptidylprolyl cis–trans isomerase activity: potential roles of cyclophilins in apoptosis. J Biol Chem 272:6677–6684
31. Woodfield K, Ruck A, Brdiczka D, Halestrap AP (1998) Direct demonstration of a specific interaction between cyclophilin-D and the adenine nucleotide translocase confirms their role in the mitochondrial permeability transition. Biochem J 336:287–290
32. Fisher G, Bang H (1985) The refolding of urea-denatured ribonuclease A is catalyzed by peptidyl–prolyl cis–trans isomerase. Biochim Biophys Acta 828:39–42
33. Schmid FX, Mayr LM, Mücke M, Schönbrunner ER (1993) Prolyl isomerases: role in protein folding. Adv Protein Chem 44:25–66
34. Handschumacher RE, Harding MW, Rice J, Drugge RJ (1984) Cyclophilin: a specific cytosolic binding protein for cyclosporine A. Science 226:544–547
35. Carpentier M, Allain F, Haendler B, Denys A, Mariller C, Benaissa M, Spik G (1999) Two distinct regions of cyclophilin B are involved in the recognition of a functional receptor and of glycosaminoglycans on T lymphocytes. J Biol Chem 274:10990–10998
36. Mikol V, Kallen J, Walkinshaw MD (1994) X-ray structure of a cyclophilin B/cyclosporin complex: comparison with cyclophilin A and delineation of its calcineurin-binding domain. Proc Natl Acad Sci USA 91:5183–5186
37. Albani JR, Carpentier M, Lansiaux C (2008) Fluorescence characterization of the hydrophobic pocket of Cyclophilin B. J Fluoresc 18:75–85
38. Albani JR (2007) New insight in the interpretation of tryptophan fluorescence. J Fluoresc 17:406–417
39. Peters T Jr (1985) Serum albumin. Adv Protein Chem 37:161–245
40. Kinsella JE, Whitehead DM (1989) Proteins in whey: chemical, physical, and functional properties. Adv Food Nutr Res 33:343–438
41. Smith RG, Gauthier DA, Dennis DT, Turpin DH (1992) Malate- and pyruvate-dependent fatty acid synthesis in leucoplasts from developing castor endosperm. Plant Physiol 98:1233–1238
42. Figge J, Rossing TH, Fencl V (1991) The role of serum proteins in acid-base equilibria. J Lab Clin Med 117:453–467
43. Human Serum Albumin – from Wikipedia, the free encyclopedia [online]. http://en.wikipedia.org/wiki/Human_serum_albumin
44. Dugaiczyk A, Law SW, Dennison OE (1982) Nucleotide sequence and the encoded amino acids of human serum albumin mRNA. Proc Natl Acad Sci USA 79:71–75
45. Brown JR (1976) Structural origins of mammalian albumin. Fed Proc 35:2141–2144

46. He XM, Carter DC (1992) Atomic structure and chemistry of human serum albumin. Nature (London) 358:209–215
47. Liu HC, Chen WL, Mao SJT (2007) Antioxidant nature of bovine milk *β*-lactoglobulin. J Dairy Sci 90:547–555
48. Brownlow S, Morais Cabral JH, Cooper R, Flower RD, Yewdall SJ, Polikarpov I, North ACT, Sawyer L (1995) Bovine β-lactoglobulin at 1.8 Å resolution – still an enigmatic lipocalin. Structure 5:481–495
49. Grosclaude J, Mahe MF, Mercier JC, Bonnemaire J, Tessier JH (1976) Polymorphisme des lactoproteines de bovine nepalais. Ann Genet Sel Anim 8:461–479
50. Invernizzi G, Amalikova MS, Brocca S, Lotti M, Molinari H, Grandori R (2006) J Mass Spectrom 41:717–727
51. Ya M, Sakurai K, Kalidas C, Batt C, Goto Y (2003) Reversible unfolding of bovine β-lactoglobulin mutants without a free thiol group. J Biol Chem 278:47009–47015
52. Perez MD, Diaz de Villegas MC, Sanchez L, Aranda P, Ena JM, Calvo M (1989) Interaction of fatty acids with β-lactoglobulin and albumin from ruminant milk. Biochem J 106:1094–1097
53. Spector AA, Fletcher JE (1970) Binding of long chain fatty acids to β-lactoglobulin. Lipids 5:403–411
54. Farrell HM, Behe MJ, Enyeart JA (1987) Binding of *p*-nitrophenol phosphate and other aromatic-compounds to beta-lactoglobulin. J Dairy Sci 70:252–258
55. Tegoni M, Pelosi P, Vincent F, Spinelli S, Campanacci V, Grolli S, Ramoni R, Cambillau C (2000) Mammalian odorant binding proteins. Biochim Biophys Acta 1482:229–240
56. Vincent F, Spinelli S, Ramoni R, Grolli S, Pelosi P, Cambillau C, Tegoni M (2000) Complexes of porcine odorant binding protein with odorant molecules belonging to different chemical classes. J Mol Biol 300:127–139
57. Ramoni R, Vincent F, Grolli S, Conti V, Malosse C, Boyer FD, Nagnan Le Meillour P, Spinelli S, Cambillau C, Tegoni M (2001) The insect attractant 1-octen-3-ol is the natural ligand of bovine odorant-binding protein. J Biol Chem 276:7150–7155
58. Johansson JS, Manderson GA, Ramoni R, Grolli S, Eckenhoff RG (2005) Binding of the volatile general anesthetics halothane and isoflurane to a mammalian β-barrel protein. FEBS J 272:573–581
59. Pevsner J, Snyder SH (1990) Odorant-binding protein: odorant transport function in the vertebrate nasal epithelium. Chem Senses 15:217–222
60. Nespoulous C, Briand L, Delage M-M, Tran V, Pernollet J-C (2004) Odorant binding and conformational changes of a rat odorant-binding protein. Chem Senses 29:189–198
61. Del Monte M, Andreini L, Revoltella R, Pelosi P (1991) Purification and characterization of two odorant-binding proteins from nasal tissue of rabbit and pig. Comp Biochem Physiol 99B:445–451
62. Paolini S, Scaloni A, Amoresano A, Marchese S, Napolitano E, Pelosi P (1998) Amino acid sequence, post-translational modifications, binding and labelling of porcine odorant-binding protein. Chem Senses 23:689–698
63. Burova TV, Choiset Y, Jankowski CK, Haertlé T (1999) Conformational stability and binding properties of porcine odorant binding protein. Biochemistry 38:15043–15051
64. Spinelli S, Ramoni R, Grolli S, Bonicel J, Cambillau C, Tegoni M (1998) The structure of the monomeric porcine odorant binding protein sheds light on the domain swapping mechanism. Biochemistry 37:7913–7918
65. Paolini S, Tanfani S, Fini C, Bertoli F, Pelosi P (1999) Porcine odorant-binding protein: structural stability and ligand affinities measured by Fourier-transform infrared spectroscopy and fluorescence spectroscopy. Biochim Biophys Acta 1431:179–188
66. Tayeh N, Rungassamy T, Albani JR (2009) Fluorescence spectral resolution of tryptophan residues in bovine and human serum albumins. J Pharm Biomed Anal 50:109–116
67. Albani, J (1998) Correlation between dynamics, structure and spectral properties of human α_1-acid glycoprotein (orosomucoid): A fluorescence approach. Spectrochim Acta Part A 54:175–178

68. Swaminathan R, Krishnamoorthy G, Periasamy N (1994) Similarity of fluorescence lifetime distributions for single tryptophan proteins in the random coil state. Biophys J 67:2013–2023
69. Chen LX-Q, Petrich JW, Fleming GR, Perico A (1987) Picosecond fluorescence studies of polypeptide dynamics: fluorescence anisotropies and lifetimes. Chem Phys Lett 139:55–61
70. Demchenko AP (1992) Fluorescence and dynamics in proteins. In: Lakowicz JR (ed) Topics in fluorescence spectroscopy, vol 3, Biochemical applications. Plenum, New York 65–111
71. Amiri M, Jankeje K, Albani JR (2010) Origin of fluorescence lifetimes in human serum albumin. Studies on native and denatured protein. J Fluoresc 20:651–656
72. Albani JR (2007) Principles and applications of fluorescence spectroscopy. Blackwell, Oxford
73. Munro I, Pecht I, Stryer L (1979) Subnanosecond motions of tryptophan residues in proteins. Proc Natl Acad Sci USA 76:56–60
74. Zolese G, Falcioni G, Bertoli E, Galeazzi R, Wozniak R, Wypych Z, Gratton E, Ambrosini A (2000) Steady-state and time resolved fluorescence of albumins interacting with N oleylethanolamine, a component of the endogenous N-acylethanolamines. Proteins 40:39–48
75. Strickland EH, Billups C (1973) Oscillator strengths of the ^1L_a and ^1L_b absorption bands of tryptophan and several other indoles. Biopolymers 12:1989–1995
76. Lakowicz JR (1999) Principles of fluorescence spectroscopy. Kluwer Academic/Plenum, New York
77. Valeur B, Weber G (1977) Resolution of the fluorescence excitation spectrum of indole into the ^1L_a and ^1L_b excitation bands. Photochem Photobiol 25:441–444
78. Gudgin E, Lopez-Delgado R, Ware WR (1983) Photophysics of tryptophan in H_2O, D_2O and in non aqueous solvents. J Phys Chem 87:1559–1585
79. Albani JR (2009) Fluorescence lifetimes of tryptophan: structural origin and relation with $S_0 \rightarrow {}^1L_b$ and $So \rightarrow {}^1L_a$ transitions. J Fluoresc 19:1061–1071
80. Shaw AK, Pal SK (2008) Spectroscopic studies on the effect of temperature on pH-induced folded states of human serum albumin. J Photochem Photobiol B 90:69–97
81. Amiri M, Jankeje K, Albani JR (2010) Characterization of human serum albumin forms with pH. Fluorescence lifetime studies. J Pharm Biomed Anal 51:1097–1102
82. Edelman GM, McClure WO (1968) Fluorescent probes and the conformation of proteins. Acc Chem Res 1:65–70
83. Campanacci V, Krieger J, Bette S, Sturgis JN, Lartigue A, Cambillau C, Breer H, Tegoni M (2001) Revisiting the specificity of Mamestra brassicae and Antheraea polyphemus pheromone-binding proteins with a fluorescence binding assay. J Biol Chem 276:20078–20084
84. Löbel D, Strotmann J, Jacob M, Breer H (2001) Identification of a third rat odorant binding protein (OBP3). Chem Senses 26:673–680
85. Ramoni R, Vincent F, Ashcroft AE, Accornero P, Grolli S, Valencia C, Tegoni M, Cambillau C (2002) Control of domain swapping in bovine odorant-binding protein. Biochem J 365:739–748
86. Vincent F, Ramoni R, Spinelli S, Grolli S, Tegoni M, Cambillau C (2004) Crystal structures of bovine odorant-binding protein in complex with odorant molecules. Eur J Biochem 27: 3832–3842
87. Wei Y, Brandazza A, Pelosi P (2008) Binding of polycyclic aromatic hydrocarbons to mutants of odorant-binding protein: a first step towards biosensors for environmental monitoring. Biochim Biophys Acta 1784:666–671
88. Nagnan-LeMeillour P, Lagant P, Cornard J-P, Brimau F, Le Danvic C, Vergoten G, Michalski J-C (2009) Phenylalanine 35 and tyrosine 82 are involved in the uptake and release of ligand by porcine odorant-binding protein. Biochim Biophys Acta 1794:1142–1150
89. Boudreau MD, Taylor HW, Baker DG, Means JC (2006) Dietary exposure to 2-aminoanthracene induces morphological and immunocytochemical changes in pancreatic tissues of Fisher-344 rats. Toxicol Sci 93:50–61
90. Hof M, Vajda S, Fidler V, Karpenko V (1996) Picosecond tryptophan fluorescence of human blood serum orosomucoid. Collect Czech Chem Commun 61:808–818

91. Kmiecik D, Albani JR (2010) Effect of 1-aminoanthracene (1-AMA) binding on the structure of three lipocalin proteins, the dimeric β-lactoglobulin, the dimeric odorant binding protein and the monomeric α_1-acid glycoprotein. Fluorescence spectra and lifetimes studies. J Fluoresc 20:973–983
92. Lee YC (1997) Fluorescence spectroscopy in studies of carbohydrate-protein interactions. J Biochem 121:818–825
93. Albani JR (1985) Fluorescence studies of the interaction between two cytochromes extracted from the yeast Hansenula anomala. Arch Biochem Biophys 243:292–297

Protein Folding, Unfolding, and Aggregation Processes Revealed by Rapid Sampling of Time-Domain Fluorescence

Saswata Sankar Sarkar, Anoop Saxena, Nihav Dhawale, Jayant B. Udgaonkar, and G. Krishnamoorthy

1 Introduction

Temporal information on the structural transformations occurring during the folding and unfolding reactions of proteins is of great importance for elucidating the mechanism of protein folding [1–6]. The rapidity of protein folding reactions precludes the use of standard structural tools such as NMR and X-ray crystallography. Furthermore, the conformational heterogeneity inherent in these reactions poses a further problem in using these standard tools [7–11]. Fluorescence resonance energy transfer (FRET) between two positions in a protein has the capability of providing the intramolecular distance separating the two locations [12]. Thus, in spite of the limited structural information it provides (a single intramolecular distance is revealed from a single sample), FRET has the ability to provide information on structural parameters with a time-resolution covering even the most rapid early events of protein folding [13–19]. Although, FRET can be measured either by steady-state spectroscopy or by time-resolved methods, the latter has several advantages. This chapter describes the power of time-resolved FRET measurements especially when coupled with rapid sampling (in "double-kinetics" experiments) [20, 21] and the Maximum Entropy Method (MEM) of data analysis [22, 23]. MEM is a model-free method of obtaining the distribution of fluorescence lifetimes (and, hence, the probability distribution of intramolecular distances) from fluorescence decay kinetics [20, 21]. The use of MEM obliterates the need of any a priori assumption of functional forms such as single or multiple Gaussians for intramolecular distance distributions.

G. Krishnamoorthy (✉)
Department of Chemical Sciences, Tata Institute of Fundamental Research,
Mumbai 400 005, India
e-mail: gk@tifr.res.in

C.D. Geddes (ed.), *Reviews in Fluorescence 2009*, Reviews in Fluorescence 2009,
DOI 10.1007/978-1-4419-9672-5_10,

2 Tr-FRET and Population Heterogeneity During Protein Folding/Unfolding

The main advantage of time-resolved FRET (tr-FRET) over the steady-state version (ss-FRET) is the ability to discern population heterogeneity encountered during folding/unfolding processes. To substantiate this assertion, let us consider the following two models of folding process, viz. (1) A two-state model with unfolded (U) and native (N) states with the U state transforming into the N state in a first order process without populating, substantially, any intermediate states and (2) A continuous model where the U state transforms through a continuum of progressively more folded forms, I, into the N state. In the ss-FRET measurement, the observed fluorescence intensity at time t ($F(t)$) of the folding process is given by

$$F(t) = \alpha_U(t)F_U + \alpha_N(t)F_N \quad \text{2 - State Model,} \tag{1}$$

$$F(t) = F_I(t) \quad \text{Continuous Model,} \tag{2}$$

where $\alpha_U(t)$ and $\alpha_N(t)$ are the mole fractions of U and N states, respectively and F_U and F_N are the mole fraction-normalized fluorescence intensities of U and N, respectively. In the continuous model, $F_I = F_U$ at $t = 0$, and $F_I = F_N$ at $t = \infty$. It is important to note that differentiation between the two models based on ss-FRET measurements is not possible, as discriminating between (1) and (2) requires prior knowledge on the process.

Alternatively, in the tr-FRET method, the fluorescence lifetimes (τ) and their amplitudes (β) recovered from intensity decays recorded at time t of the folding process can be represented as

$$\beta_U(t), \tau_U, \beta_N(t), \tau_N \quad \text{2 - State Model,}$$

$$\tau_I(t) \quad \text{Continuous Model.}$$

In the 2-state model, the values of β are time-dependent, as they represent the time-evolving population. By contrast, the values of τ_U and τ_N are time-independent as they are associated with the U and N states. In the continuous model, τ_I would be time-dependent due to the continuous change in the structural characteristics of the folding protein. More complex mechanisms could have aspects of both the models. It should be emphasized that τ_U, τ_N and τ_I themselves could be the mean-lifetimes of multiexponential fluorescence intensity decay processes.

The advantage of the tr-FRET method as opposed to the ss-FRET method is that it has the ability to differentiate between the 2-state and continuous models, as the set of parameters recovered from the analysis of fluorescence intensity decays would correspond to one of the two sets mentioned above. Thus, the picosecond time resolution of the fluorescence intensity measurement offers us the handle in identifying the model. The ps–ns timescale of the excited state of fluorophores is faster, by several orders of magnitude, than the rates of exchange between various conformational states and structural forms of proteins in general [24–27]. This ensures that time-domain fluorescence captures a "snap-shot" picture of the conformational and structural heterogeneity present during folding/unfolding processes. The snap-shot picture provides (1) the relative populations of various discrete structural forms (such as U and N) and (2) an estimation of the structural heterogeneity of U and any intermediate structures from an analysis of intensity decay kinetics as a distribution of lifetimes, especially by MEM [28]. A distribution of decay rates is obtained from such analyses, and as the decay rate depends on the distance between the fluorescence donor and the acceptor, this can be used to obtain the probability distribution of distance between donor and acceptor in the ensemble.

3 Instrumentation for Time-Domain Fluorescence with ms Sampling

The integrated fluorescence intensity measured in ss-FRET can be collected with sufficient S/N with a sampling rate down to microseconds and below, thus enabling monitoring of the early events of folding such as the collapse of unfolded proteins [29] and the formation of β-bends [15]. The typical sampling rate of picosecond time-resolved fluorescence setups is usually in several tens of seconds time domain, although rapid sampling of ~20 ms has been achieved in specialized setups [20, 30–32]. This has opened the possibility of performing tr-FRET measurements during folding/unfolding processes. Two such setups being used in our laboratory, one based on time-correlated-single-photon-counting (TCSPC), and the other based on Streak camera detection, are shown in Fig. 1a, b. Both the setups consist of three major components: (1) a picosecond high repetition rate laser, (2) a stopped-flow mixer, and (3) a TCSPC or Streak Camera detector. Other details are given in the figure legends.

4 Continuous Dissolution of Structure During Unfolding of a Protein

Many protein folding and unfolding processes can be described by a simple "two-state" model (U and N) with the kinetics describable by transition state theory. Transition state theory describes the process in terms of a single reaction coordinate with an energy barrier separating the two states. Thus, during folding and unfolding reactions, the protein molecules are not expected to spend appreciable time in conformations intermediate between U and N. In recent years, there has been, however, growing realization that this is not an acceptable model even for small proteins, as the landscape between U and N can be quite complex [28, 33–36].

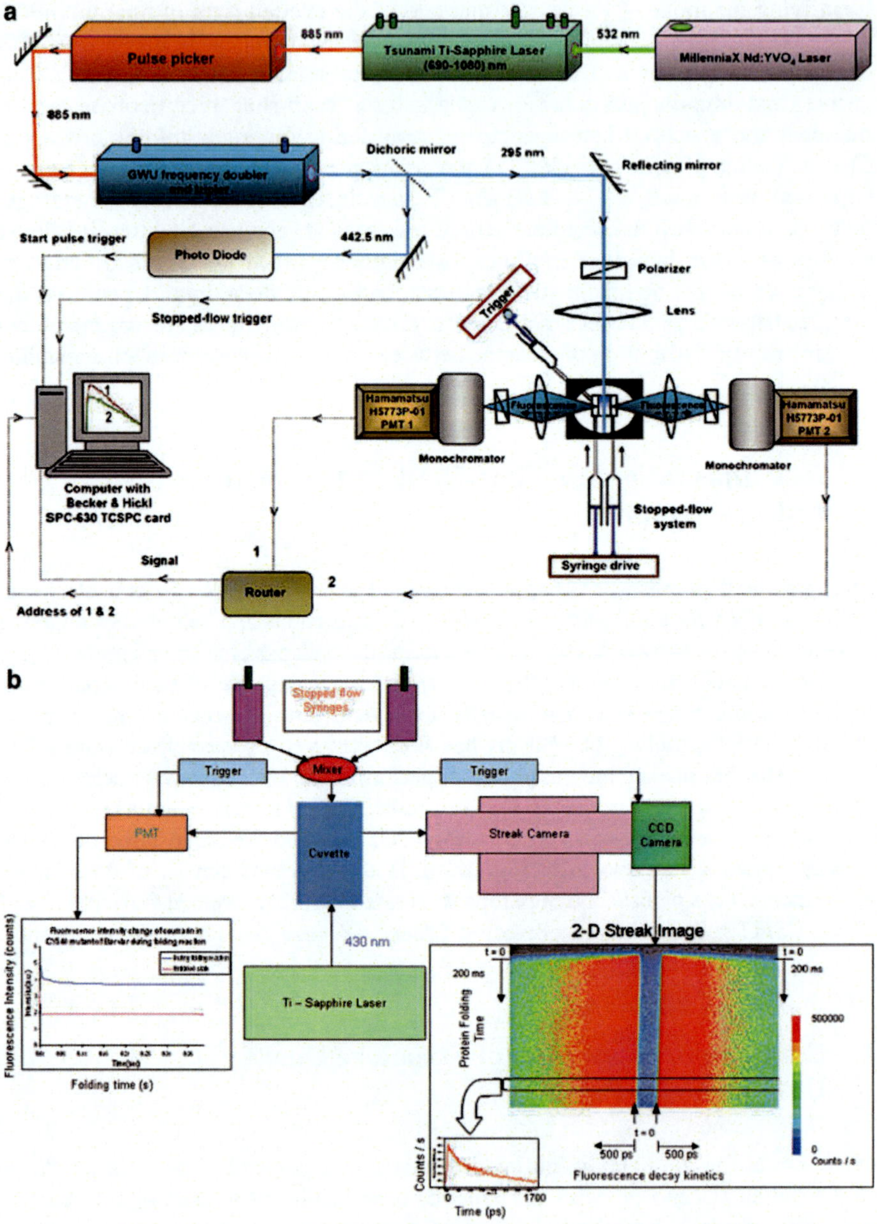
a
Pulse picker
885 nm
Tsunami Ti-Sapphire Laser
(690-1080) nm
532 nm
MillenniaX Nd:YVO4 Laser
885 nm
GWU frequency doubler and tripler
Dichoric mirror
295 nm
Reflecting mirror
Start pulse trigger
Photo Diode
442.5 nm
Polarizer
Trigger
Lens
Stopped-flow trigger
Hamamatsu H5773P-01 PMT 1
Hamamatsu H5773P-01 PMT 2
Monochromator
Monochromator
Computer with Becker & Hickl SPC-630 TCSPC card
Stopped-flow system
Syringe drive
Signal
1
2
Router
Address of 1 & 2
b
Stopped flow Syringes
Trigger
Mixer
Trigger
PMT
Cuvette
Streak Camera
CCD Camera
430 nm
Ti – Sapphire Laser
Fluorescence Intensity (counts)
Folding time (s)
2-D Streak Image
t = 0
200 ms
Protein Folding Time
500000
0
Counts / s
500 ps
500 ps
Fluorescence decay kinetics
Counts / s
Time (ps)
0
1700

In fact, the transition between the U and N states can be expected to be a continuous process. The unfolding of a small protein, monellin, was studied by the tr-FRET methodology coupled to MEM analysis with the aim of describing the energy landscape defining the unfolding process.

Single chain monellin (MNEI) is a sweet plant protein, whose folding and unfolding reactions have been studied extensively [37–39]. Time evolution of the probability distributions of four intramolecular distances in this protein was monitored, as it unfolds in 4 M guanidine hydrochloride (GdnHCl), starting from the native state [21]. Four single cysteine, single tryptophan-containing mutant forms of MNEI were used (Fig. 2a). In each of the mutant proteins, a single tryptophan residue at position four served as the donor fluorophore (D) and a thionitrobenzoate (TNB) adduct attached at one of the four cysteine sidechains (Fig. 2a) served as the FRET acceptor (A). TNB had been used as a nonfluorescent FRET acceptor of tryptophan fluorescence ($R_0 \sim 23$ Å, an ideal value for small proteins) in several earlier studies [28, 36, 53].

Fig. 1 (**a**) Experimental setup to monitor the time evolution of time-resolved fluorescence decay and time-resolved anisotropy decay during protein folding, unfolding and aggregation processes. The setup consists of (1) a picosecond high repetition rate laser system, (2) a stopped-flow module, and (3) dual channel fast time-correlated single photon counting system (SPC-630 from Becker and Hickl). A Ti-sapphire laser pumped by a Nd:YVO_4 laser (Milenia X, Spectra Physics) generated one picosecond pulses of 885 nm at 80 MHz repetition rate, which were frequency tripled to 295 nm by a frequency doubler/tripler unit (GWU, Spectra Physics) to excite tryptophan in folding-unfolding reactions. For studying aggregation processes, the laser output was at 740 nm at an 8 MHz repetition rate that was doubled to 370 nm by the same doubler/tripler unit to excite the fluorophore IAEDANS. The folding-unfolding reaction or the aggregation process was initiated by stopped-flow mixing (RX2000 from Applied Photophysics). The dead time of mixing was estimated to be ~60 ms. The fluorescence decays were collected through two polarizers, one vertical and the other horizontal with respect to the polarization of excitation, using two photomultiplier tubes (Model H5773P-01;Hamamatsu) coupled to the TCSPC card. A dilute colloidal suspension of dried nondairy coffee whitener was used to collect the instrument response function. The typical sampling time of this setup is ~150 ms (figure reproduced from ref. [21]). (**b**) Streak Camera-based "Double-Kinetic" setup: this system is essentially similar to that shown in (**a**) except that a Streak Camera is used in place of TCSPC detection. The laser system used is the same as that described in (**a**). The stopped-flow system used here (SFM-4 from Biologic, France) has a dead time of ~2 ms. Fluorescence was collected from two windows of the stopped-flow cuvette: a PMT at one window was used to obtain the ms dependence of the integrated fluorescence intensity (shown by the plot on the *left side*), and a Streak Camera unit (SC10/SSU11-10, Optronis, Germany) with a time resolution of 2 ps was used as the detector at the other window. Rapid sampling of the Streak Camera traces in ms was done by moving the streak images in the vertical direction of the streak tube by slow (ms) scanning, using a second deflection unit (TSU23-10). Thus, the horizontal direction provided the ps-resolved fluorescence transients, and the vertical direction provided the sampling of these traces at ms intervals. The dynamic range of the CCD camera used to capture the images is 16 bits. The 2D Streak Image shown on the right side was obtained in a folding reaction. The *x*-axis shows the streak image corresponding to fluorescence decay kinetics. The streak image moves from *top* to *bottom* following the initiation of the folding reaction. Thus, the intensity decay kinetics at various times during the folding reaction can be obtained by taking horizontal cuts at successive ms time intervals. The intensity vs. time plot shown as an inset in the 2D Streak Image panel corresponds to the area in between the *two horizontal lines* marked in the 2D Streak Image

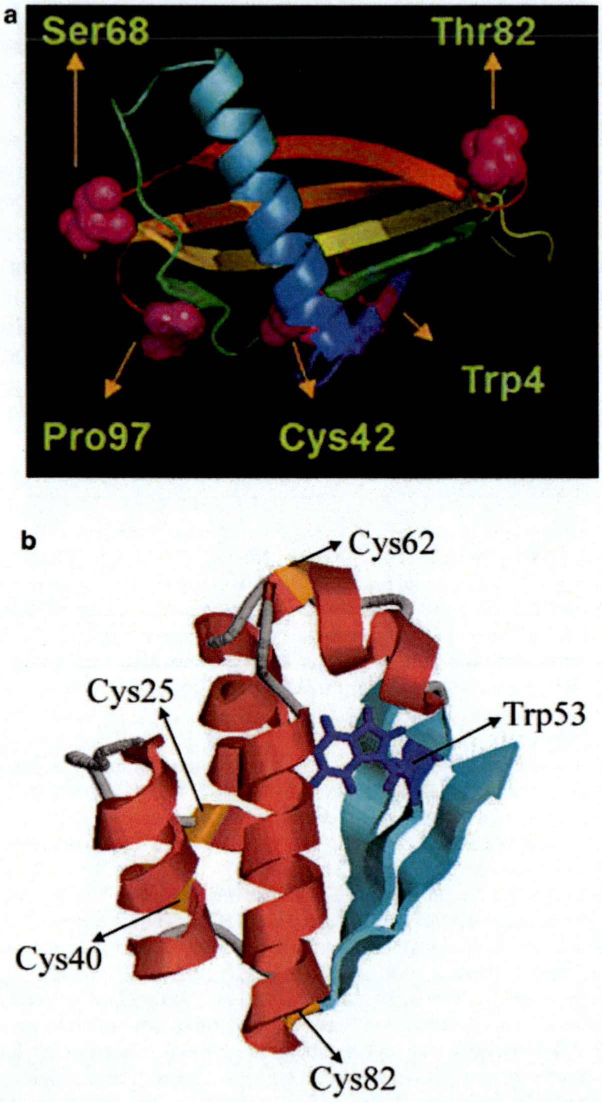

Fig. 2 (**a**) Structure of single chain monellin (PDB entry 1IV7) showing the positions of the trp residue and the four cys residues used to monitor the evolution of four different intramolecular distances of the protein during unfolding. It was drawn using the program PyMOL (figure reproduced from ref. [21]). (**b**) Location of the trp and the four cys residues in four single trp, single cys-containing mutant forms of barstar (PDB entry 1BTA). RASMOL was used to draw the structure

When the mean fluorescence lifetime (τ_m) of the donor, Trp 4, was measured during the unfolding of MNEI, by using the setup shown in Fig. 1a, the kinetics of the change of τ_m was nearly exponential for all the four mutant proteins, consistent with an apparently 2-state unfolding reaction [21]. However, as mentioned earlier, an ensemble-averaged parameter such as τ_m (which is proportional to the integrated fluorescence intensity) is not expected to reveal any heterogeneity hidden in these observations.

The heterogeneity of population got revealed once the decay kinetics monitored at various times of unfolding, was analyzed using MEM (Fig. 3). The fluorescence lifetime distributions (Fig. 3) were then transformed into probability distributions of D-A distances using Forster's equation [12], as shown in Fig. 4. This transformation was feasible as the lifetime of Trp 4, was nearly a single exponential in both the native and unfolded states of MNEI [21], a situation rarely found in proteins. Incorporation of tryptophan analogs having single lifetimes would enable a similar methodology to be adapted for other proteins as well [40].

Upon commencement of the unfolding reaction, the probability distributions of D-A distances jump from the narrow unimodal distribution of the native state to bimodal distributions for three distances, while the fourth distance remains unimodal (Fig. 4). It is interesting to note that at the first time point of observation (150 ms), at which the drastic jump from a sharp unimodal distribution to a broad bimodal distribution was observed (Fig. 4), the extent of secondary structure, as monitored by circular dichroism spectroscopy, has not decreased [21]. This indicated that the protein has swollen rapidly in size while keeping the secondary structural elements intact. As the unfolding proceeds to completion, the distributions of D–A distances evolve continuously through increasing values. One of the broad peaks observed at the first time point grows at the expense of the other, and finally, the broad distribution characteristic of random coil [41] is observed.

The physics of swelling of the polypeptide chain, upon unfolding, can be assumed to be similar to that of a freely jointed self-avoiding chain, with a few noncovalent interactions. A simple phenomenological model, which is a variation of the well-known Rouse model of polymer physics [42, 43], was used to explain the observations [21]. Time evolution of the two populations of the initially unfolded forms (Scheme 1) can be described by Gaussian probability distribution function

$$P(R,t) = \frac{1}{\sqrt{2\pi}\sigma_i(t)} \mathrm{e}\frac{(R - Ri(t))^2}{2\sigma_i{}^2(t)} \qquad i = 1, 2.,$$

where $R_i(t)$ and $\sigma_i^2(t)$ are the time-dependent mean and variance, respectively, of the distance distribution in the ensemble i. $R_i(t)$ and $\sigma_i^2(t)$ increase exponentially with time t. Thus, the protein, starting from a native conformation, swells with time to reach its finite limiting value.

Figure 4 shows that this model fits the data reasonably well. The kinetic parameters, τ_i associated with the unfolding process (Scheme 1) are found to be

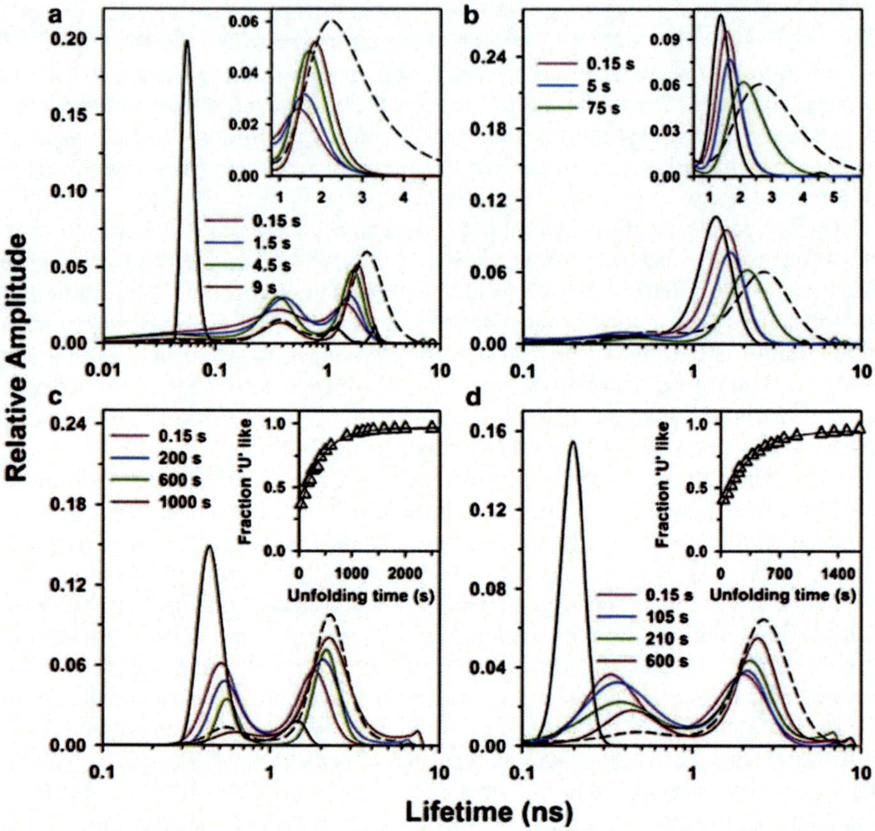

Fig. 3 Fluorescence lifetime distributions of four different TNB-labeled mutant forms of MNEI at different time points during unfolding. (**a**) Cys42-TNB, (**b**) Cys68-TNB, (**c**) Cys82TNB, (**d**) Cys97-TNB. The *black solid* and *dashed lines* represent the lifetime distributions of native protein in 0 M GdnHCl and unfolded protein in 4 M GdnHCl, respectively. A jump in GdnHCl concentration from 0 to 4 M was given to start the unfolding reaction and the subsequent evolution of the lifetime distributions during the process is shown in other colored lines as representative. The time axis is in log scale. The insets of (**a**) and (**b**) show the gradual evolution of lifetime as the protein unfolds in a linear scale. Insets of (**c**) and (**d**) represent the time course of the fraction of protein with unfolded-like distributions being referred to as the fraction of "U"-like form during the unfolding of the protein. The fraction was calculated by dividing the area under the U-like peak of the bimodal lifetime distributions, at various unfolding times, by the total area under the U distribution obtained at the end of the unfolding reaction in 4 M GdnHCl (figure reproduced from ref. [21])

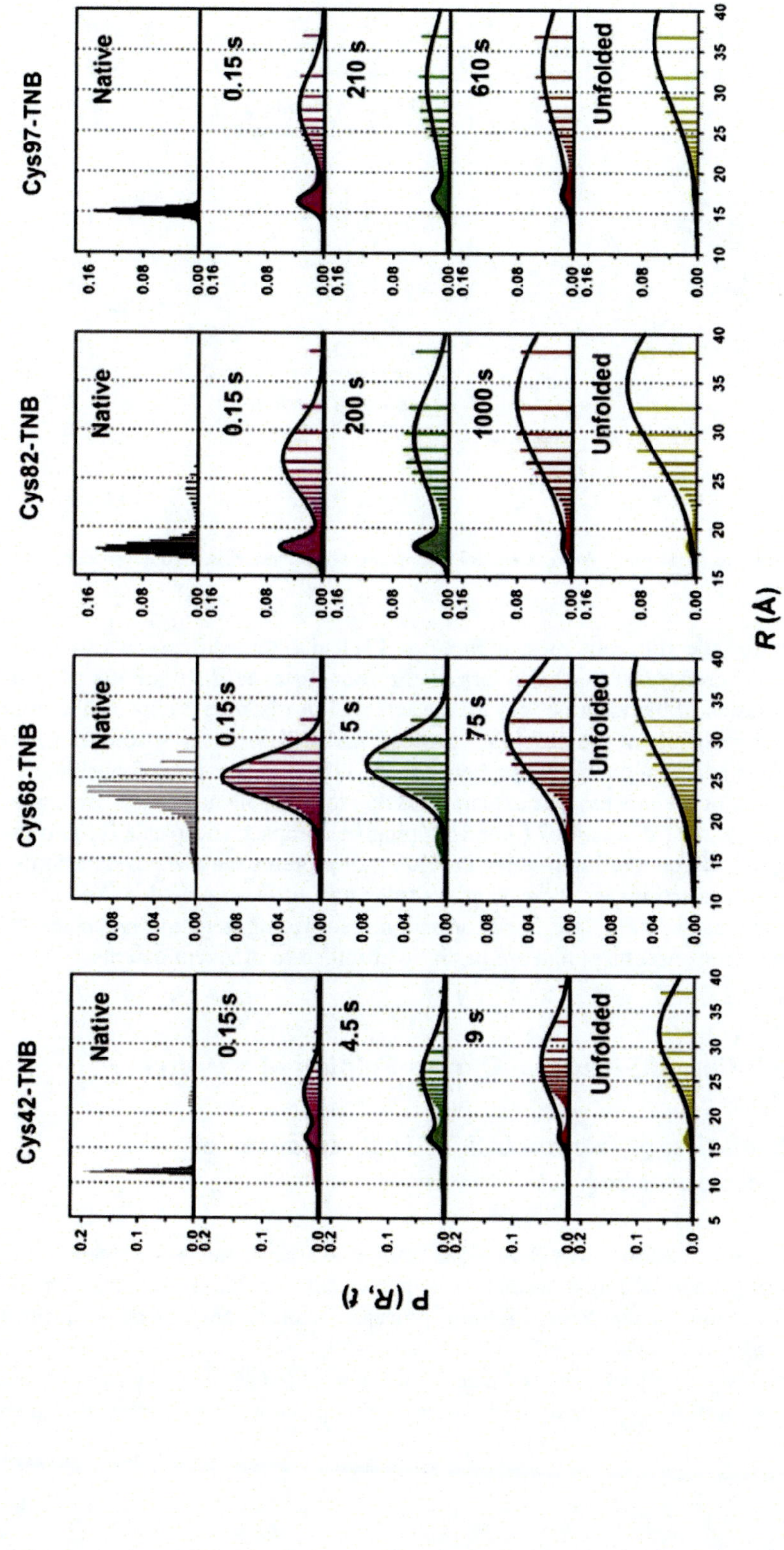

Fig. 4 Time evolution of the probability distributions of four intramolecular distances during the unfolding of monellin. The distance distributions were converted from the lifetime distributions, as described in the text (figure reproduced from ref. [21])

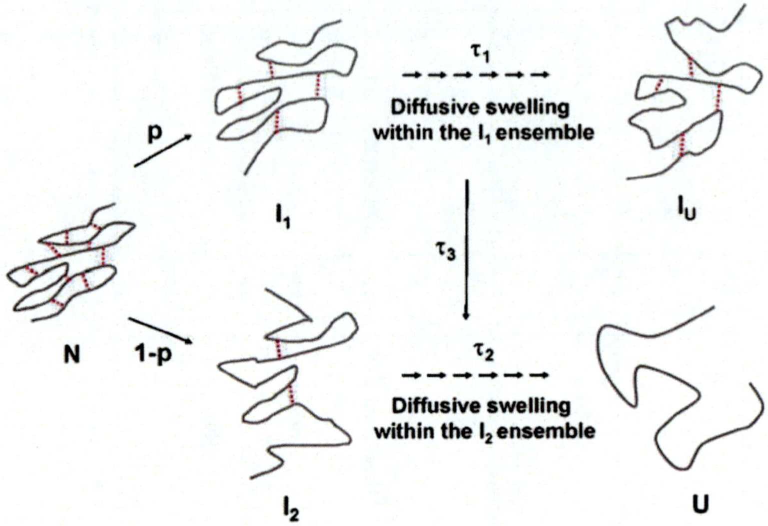

Scheme 1 Diffusive swelling during the unfolding of single chain monellin (figure reproduced from ref. [21])

different for the different mutant proteins [21] and this difference causes the apparent differences in the behaviors of the four distance distributions (Fig. 4). Rationalization of the values of τ_i with the actual D–A distance is expected to bring out atomistic details of the unfolding process, and will form the subject of future studies. The apparent contradiction between the continuous nature of swelling and the slowness of the process was explained as due to probable activation barriers for the basic motional dynamics of bond flipping in compact and partially unfolded forms. The main conclusion that the unfolding of proteins may occur via a continuum of conformations of increasing disorder can have implications in protein aggregation, such as amyloid fibril formation, because aggregation processes can commence from structures that are partially unfolded to different extents.

5 Evolution of Structure During Folding of a Protein

5.1 Evolution of Intramolecular Distances During Protein Folding

In the previous section it was shown that intramolecular distances change continuously during the unfolding of a small protein, monellin. In this section, we address a similar question for the protein folding process, by using another model protein, barstar which has been studied extensively [44–53].

Barstar is a small 89 residue bacterial protein which inhibits the ribonuclease activity of barnase. Equilibrium unfolding of this protein, when observed through steady state fluorescence measurement of the single tryptophan (Trp 53) in the core of the protein, showed a sigmoidal transition [50], suggesting a 2-state model for this transition. But when an intramolecular distance was estimated by the combined use of tr-FRET and MEM analysis, the equilibrium unfolding transition revealed aspects of continuous change, and indicated that the protein could swell toward the unfolded state in a gradual manner [36]. This observation of continuous structural change rules out a 2-state model to describe the equilibrium unfolding transition of barstar. Ensemble-averaged properties such as steady-state fluorescence and circular dichroism could not differentiate between the 2-state and a continuous model. It is worth noting that observation of continuous structural change during an equilibrium unfolding transition need not imply the presence of such continuous structural change happening during the process of folding. The kinetic process of folding was investigated as detailed below.

The process of folding of barstar occurs in three kinetic phases: (1) a very rapid burst phase occurring within a few ms, (2) a fast phase with a time constant of ~1 s, and (3) a slow process of ~20 min at 10°C, when protein that had been unfolded in high concentration of urea is diluted to ~1 M of urea [28]. Tr-FRET measurements were used to investigate the kinetics of the folding using single trp (Trp 53) and single cys-containing mutants of the protein with Trp 53 as the donor and the TNB group attached to the cys as the acceptor of the FRET pair, as before. The fast phase of folding was investigated by using the setup shown in Fig. 1a by sampling the tr-FRET data every 200 ms. Such rapidly sampled data had a good value of S/N (peak counts were >10 K), which is essential to obtain reliable D–A distances. Four mutant forms of the protein, Cys25, Cys40, Cys62, and Cys82, were used to monitor the time courses of change in the D–A distances following the initiation of folding (Fig. 2b).

In the N state, Trp 53 has nearly a single lifetime of 4.8 ns in the absence of the FRET acceptor, which gets shortened to between 0.1 and 0.7 ns, when the acceptor TNB group is attached to the cys side chain [28] in the four mutant proteins. This variation (0.1–0.7 ns) reflects the variation in the D–A distances in the four mutant proteins. Immediately after the initiation of folding (within ~100 ms), the presence of this short lifetime is seen, indicating compaction of the U state to an N-like form at the end of the burst phase. For two of the four mutant proteins (Cys62 and Cys82) (Fig. 5), the value of this lifetime is, however, seen to be significantly longer than that observed in the N state. This higher value changes in a continuous manner to that corresponding to the N state during the fast phase of folding. The other two mutant proteins (Cys25 and Cys40) show a constant value of the short lifetime during the course of folding. D–A distances calculated from the Forster equation (Fig. 6) show that Trp53-Cys62 and Trp53-Cys82 distances decrease continuously from a higher value corresponding to the initially collapsed form to those of the N state. By contrast, the Trp53-Cys25 and Trp53-Cys40 distances of the initially collapsed form have already become the same as those in the N state, and hence, show no further change in the D-A distance. Thus, these results point out that the nature of folding could vary within the same protein, viz. some regions may show 2 state-like behavior, and some other regions may change continuously. A note of caution is relevant in this interpretation. Any observed continuous variation of τ may also arise as a weighted average of 2 or more close-lying lifetimes, especially

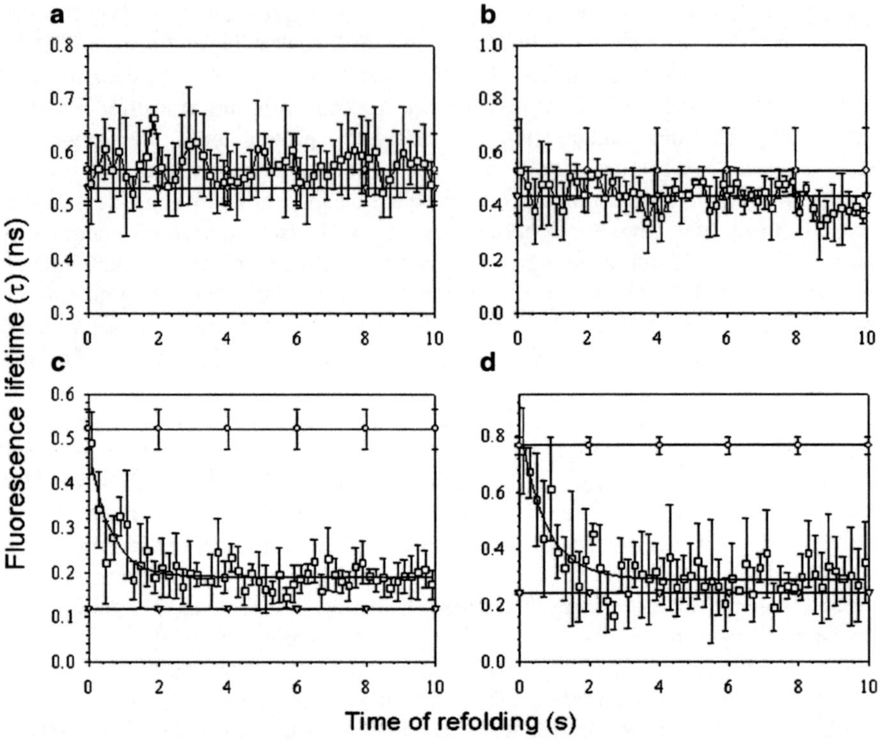

Fig. 5 Time evolution of the shortest fluorescence lifetime component (τ) corresponding to the compact N-like state during the refolding reaction of TNB labeled mutant forms of barstar upon changing the solvent from 7 to 1.25 M urea at 10°C. The different panels correspond to the different TNB labeled mutant proteins, (**a**) Cys25TNB, (**b**) Cys40TNB, (**c**) Cys62TNB, and (**d**) Cys82TNB. τ values corresponding to the fast refolding kinetics, protein unfolded in 7 M urea, and protein refolded in 1.25 M urea are shown by the *open squares*, *open circles*, and *open inverted-triangles*, respectively. The *error bars* represent the standard deviations from four different measurements. The *solid lines* through the refolding kinetic data points (*open squares*) represents fits to the exponential decay function, $y = y_0 + a(e^{-bt})$.

when the S/N of the decay traces are insufficient to resolve them as discrete lifetimes. Thus, the reliability of any conclusion based on observed continuous variation of τ will, in general, depend on the S/N of the measurement.

5.2 Compaction of the Protein Core During Folding

An important question in protein folding is to determine when the core of the protein consolidates and attains its compact rigid structure. The rotational dynamics of a tryptophan side chain is generally taken as a sensitive assay of flexibility of

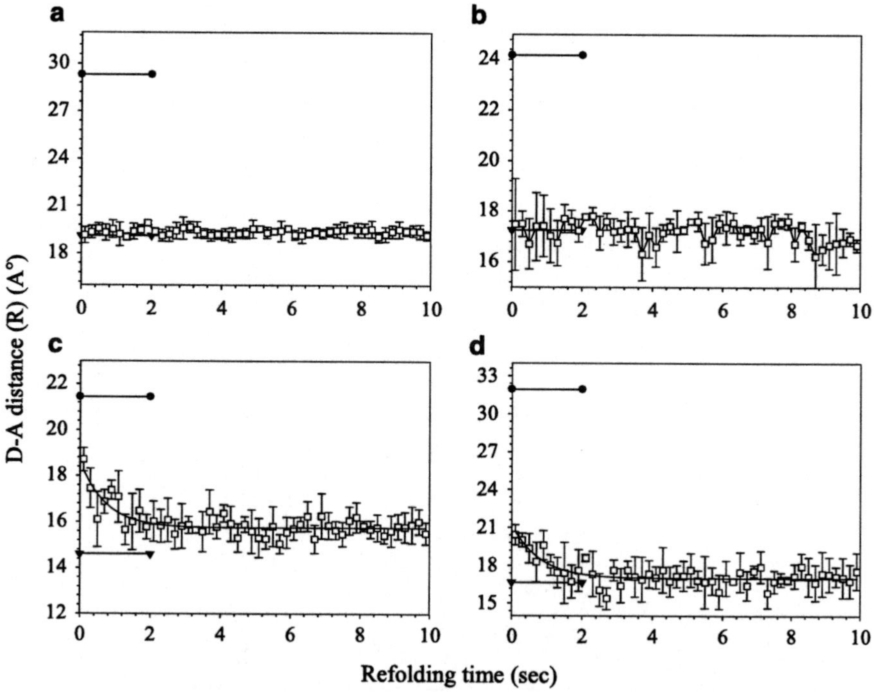

Fig. 6 Time evolution of D-A distances (R) of the compact N-like state during the refolding reactions of TNB labeled mutant variants of barstar upon changing the solvent from 7 to 1.25 M urea at 10°C. The different panels correspond to the different TNB-labeled mutant proteins, (**a**) Cys25TNB, (**b**) Cys40TNB, (**c**) Cys62TNB, and (**d**) Cys82TNB. R values corresponding to the fast refolding kinetics, protein unfolded in 7 M urea, and protein refolded in 1.25 M urea are shown by the *open squares*, *filled circles*, and *filled inverted-triangles*, respectively. The *error bars* represent the standard deviations from four different measurements. The *solid lines* through the refolding kinetic data points (*open squares*) represent fits to the exponential decay function, $y = y_0 + a(e^{-bt})$. The Cys62TNB-Trp53 and Cys82TNB-Trp53 distances decrease with a rate constant of ~1.5 s^{-1} after an initial burst phase decrease

proteins, and Trp 53, which is present in the core of barstar, is ideally suited to address this question. The rotational correlation time (Φ) of Trp53 is 1.4 and 5.5 ns, for the U and N states, respectively (Table 1). The shorter value of Φ in the U state is a reflection of the segmental flexibility of the unfolded chain. The higher value in the N state is due to global tumbling of the N state in which Trp 53 lacks any relative motion with respect to the rest of the protein. Initiation of the folding process, as described above, results in a loosely compact N-like state which is characterized by (1) the presence of local motion of Trp 53 and (2) a larger hydrodynamic size. This results in two values of Φ, viz. Φ_{int} ~ 1 ns corresponding to the local motion of Trp 53, and Φ_{glo}, corresponding to the global tumbling of the compact protein (Table 1). As the folding proceeds to completion, the amplitude associated with Φ_{int} decreases gradually to zero, signifying a disappearance of the local motion of Trp 53 as the core rigidifies. Figure 7 shows the time evolution of

Table 1 Time evolution of rotational correlation times associated with the global and local motions of Trp 53 in barstar during the fast phase of the refolding reaction upon changing the solvent from 7 M urea to 1.25 M urea at 10°C

	Rotational correlation times (ns)[b] (amplitude)	
Time[a] (s)	Φ_{glo} (β_1)	Φ_{int} (β_2)
0.0	–	1.4 (1.00)
0.2	8.2 (0.62)	0.9 (0.38)
0.6	8.1 (0.65)	1.0 (0.35)
1.0	8.1 (0.67)	1.0 (0.33)
1.4	8.1 (0.71)	0.9 (0.29)
1.8	8.1 (0.69)	1.1 (0.31)
2.2	8.3 (0.71)	1.1 (0.29)
2.6	7.9 (0.71)	0.9 (0.29)
3.0	8.0 (0.71)	0.9 (0.29)
3.4	8.2 (0.75)	1.1 (0.25)
3.8	7.9 (0.76)	0.9 (0.24)
4.2	8.1 (0.76)	0.9 (0.24)
4.6	8.0 (0.79)	0.9 (0.21)
5.0	8.0 (0.80)	0.9 (0.20)
5.4	8.1 (0.81)	1.0 (0.19)
5.8	7.9 (0.80)	1.0 (0.20)
60.0	7.0 (0.80)	0.7 (0.20)
∞[c]	5.5 (1.00)	–

[a] The first time (t = 0.0 s) corresponds to the protein in 7 M urea at 10°C. Each kinetic time point is the mean value of the time window (400 ms) spanning the measurement of fluorescence decay during refolding

[b] Errors associated with Φ_{glo}, Φ_{int}, β_1, and β_2 were ~0.3 ns, ~0.2 ns, ~0.05, and ~0.05 respectively

[c] The values of the parameters obtained after a long time (∞) equilibration of refolded barstar in 1.25 M urea are similar to the values of the parameters in the absence of any urea

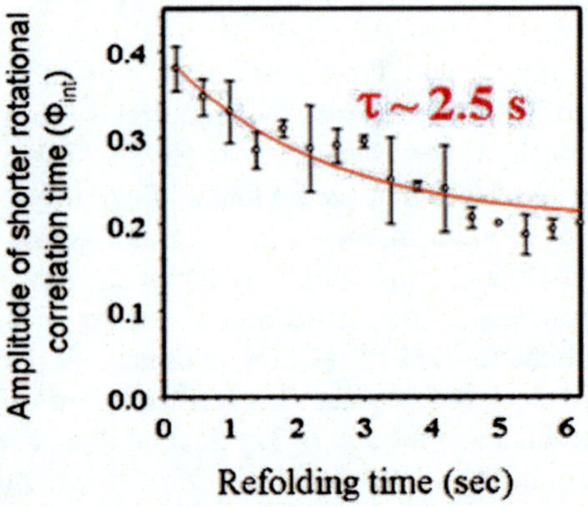

Fig. 7 Time evolution of the amplitude of the rotational correlation time (Φ_{int}) representing the local dynamics of Trp 53 in time-resolved fluorescence anisotropy experiments during the fast phase of the folding reaction of barstar at 10°C. Exponential fitting of the data yields the time constant (τ ~ 2.5 s) of compact core formation during this phase

this amplitude. It is interesting to note that the entire process of core rigidification occurs in two distinct phases: one with a time constant of ~2.5 s seen in this work (Fig. 7) and another slower process (~270 s) seen in earlier studies [28]. Rationalization of such observations in many proteins could provide deep insight into this important aspect of protein folding.

6 Rotational Diffusion and the Kinetics of Protein Aggregation

It is well known that many if not all proteins under some critical physical conditions of concentration, pH, and temperature [54–60] (or a combination of these conditions) misfold and then form aggregates. These aggregates can be either amorphous or ordered and be varying in size from small oligomers (consisting of a few tens of monomers) to much larger structures (consisting of thousands and more of monomers) [61–65]. Ordered aggregates generally give rise to structured fibrils, which have been the subject of intense study [66–69]. Recent work from our laboratory has shown that the internal structure of protein fibrils is formed by self-complimentary interactions in a process that may be akin to the process of crystallization [70]. Protein aggregation not only is an intriguing question in the field of polymer science as to why and how these aggregates form but also has a great relevance in medical science too, as such aggregates formed by a few specific proteins are believed to be the cause of many neurodegenerative diseases such as Alzheimer's and Parkinson's disease [71–74]. Furthermore, recent studies point toward the smaller aggregates as the toxic agent leading to disease [75–77].

From the viewpoint of thermodynamics, the formation of protein aggregates reduces the total free energy available to the system, but the kinetics of the process is as yet not fully understood [75, 78–80]. Of the several models that have been proposed to explain aggregate formation, the isodesmic model and the nucleation–elongation model could be considered as representing two extremes. In the isodesmic process, the rate constant for monomers binding onto monomers is the same as that of monomers binding onto oligomers, and the aggregate forms at a steady pace with a single elementary rate constant. By contrast, in the nucleation–elongation model, the rate-determining step is the formation of an oligomer that acts as a nucleus for monomers to further bind onto it. Thus, the rate constant for the addition of a monomer to either another monomer or an aggregate smaller than the size of the nucleus is smaller than the rate constant associated with the addition to an aggregate of size equal to or larger than the nucleus. The first model assumes a largely featureless energy landscape that favors large aggregate formation, while the second implies that there is a local energy barrier (corresponding to the size of the nucleus) that has to be traversed before elongation is energetically favored. Crossing of the nucleation stage is usually protein concentration dependent. It has been argued that differentiating between the two mechanisms is nontrivial, and that the actual mechanism may lie in between the two models. The kinetics of aggregation is expected to provide information on the mechanism of aggregate formation.

The rotational correlation time estimated from time-resolved fluorescence anisotropy is a sensitive indicator of the overall size of macromolecular complexes in solution [12]. Although a variety of physical techniques such as dynamic light scattering (DLS) and fluorescence correlation spectroscopy (FCS) are very popular in estimating macromolecular size in solution, they suffer from low sensitivity on the molecular size, especially when the sizes are very large. DLS and FCS rely on measurement of the translational diffusion coefficient, $D_t = kT/(6\pi\eta r)$, where η and r are the solvent viscosity and the molecular radius, respectively. By contrast, the rotational diffusion constant, $D_r = kT/(8\pi\eta r^3)$, scales as the third power of r, thus making it more sensitive to molecular size. Rapid sampling of rotational diffusion is a powerful and sensitive way of real-time monitoring of the aggregation process.

Barstar, a small protein used extensively in protein folding studies is also an excellent model protein for aggregation studies. It undergoes aggregation at low pH (<3) values, and further proceeds to form fibrils when either heated [81–83] or treated with trifluoroethanol [84]. Low pH-induced aggregation is a relatively fast process and hence the time evolution of the aggregation process, following a jump in pH from 8.0 to 2.7, was monitored by rapid sampling of fluorescence anisotropy decay kinetics by using the setup shown in Fig. 1a. A relatively long lifetime

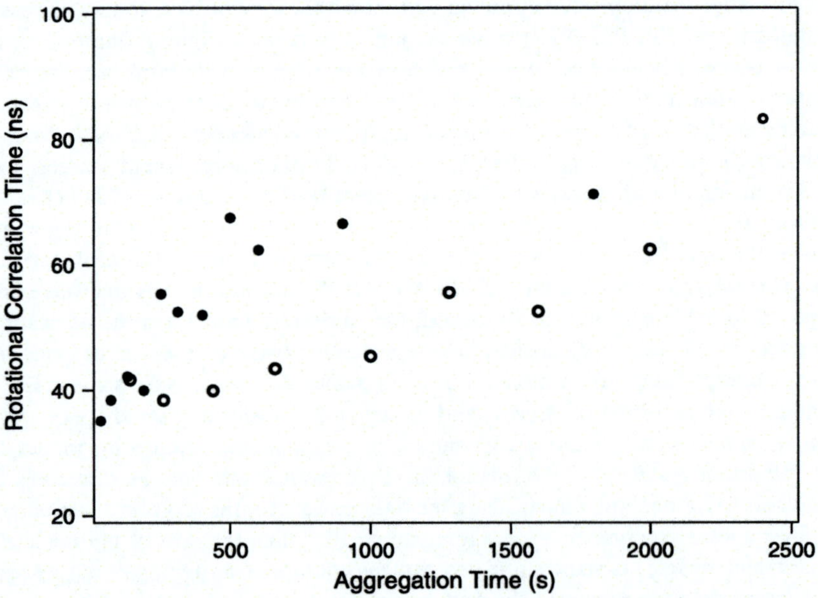

Fig. 8 Change in the rotational correlation time associated with global tumbling during the aggregation process for 50 μM (*solid circles*) and 14 μM (*open circles*) of the IAEDANS-labeled Cys82 mutant form of barstar. A jump of pH from 8.0 to 2.7 triggered the aggregation process. Each anisotropy trace, collected for 10 s during the process, was fitted as a sum of two exponentials. The increase in the longer rotation correlation time, representing the global tumbling motion of the molecule, indicates the increase in hydrodynamic radius, which in turn indicates progressive formation of larger oligomeric forms during aggregation

fluorophore, IAEDANS (~20 ns in the bound form) enables estimation of rotational correlation times up to ~100 ns, at a reasonable level of accuracy. Figure 8 shows the time evolution of the rotational correlation time (Φ) associated with the global tumbling dynamics of the aggregate growing in size. The following inferences are worth noting: (1) there is a significant dependence of the aggregation process on the concentration of protein; and (2) the saturation of the value of Φ around ~70 ns would indicate an aggregate size of ~17 monomers ($\Phi = \eta r^3/kT$), assuming a spherical geometry for the aggregate. This value agrees with the estimate from NMR spectroscopy [85]. More detailed work to understand the mechanism of low pH-induced aggregation is in progress.

7 Summary and Future Directions

In this chapter we have demonstrated the power of time-domain fluorescence spectroscopy in revealing the time evolution of structural features such as intramolecular distances, the flexibility of side chains and the overall hydrodynamic size, during protein folding, unfolding and aggregation. We believe that such information is vital in elucidating the mechanisms of these processes. For example, rationalization of the site-specific differences in the kinetic parameters observed during the folding and unfolding of the same protein is likely to shed light on the forces that stabilize the structure. As mentioned earlier, reliable estimation of intramolecular distance distributions during folding, unfolding, and misfolding/aggregation processes requires (1) a single lifetime of the fluorescence donor in the absence of the acceptor (for unambiguous use of MEM) and (2) the capability to acquire time-domain fluorescence data with good S/N every ms of the process. Future developments in both these fronts are awaited.

Acknowledgments We thank Prof. Deepak Dhar and Dr. Santosh Kumar Jha for their collaboration in some of the research work described here. We are grateful to Prof. N. Periasamy for providing the software used in the analysis of time-domain fluorescence data, and for his advice in using it effectively.

References

1. Kim PS, Baldwin RL (1982) Specific intermediates in the folding reactions of small proteins and the mechanism of protein folding. Annu Rev Biochem 51(1):459–489
2. Matthews CR (1993) Pathways of protein folding. Annu Rev Biochem 62(1):653–683
3. Myers JK, Oas TG (2002) Mechanisms of fast protein folding. Annu Rev Biochem 71 (1):783–815
4. Baldwin RL (2008) The search for folding intermediates and the mechanism of protein folding. Annu Rev Biophys 37(1):1–21
5. Udgaonkar JB (2008) Multiple routes and structural heterogeneity in protein folding. Annu Rev Biophys 37(1):489–510

6. Pande VS, Grosberg AY, Tanaka T, Rokhsar DS (1998) Pathways for protein folding: is a new view needed? Curr Opin Struct Biol 8(1):68–79
7. Neri D, Wider G, Wuthrich K (1992) Complete 15 N and 1 H NMR assignments for the amino-terminal domain of the phage 434 repressor in the urea-unfolded form. Proc Natl Acad Sci USA 89(10):4397–4401
8. Alexandrescu AT, Abeygunawardana C, Shortle D (1994) Structure and dynamics of a denatured 131-residue fragment of staphylococcal nuclease: a heteronuclear NMR study. Biochemistry 33(5):1063–1072
9. Zhang O, Forman-Kay JD (1995) Structural characterization of folded and unfolded states of an SH3 domain in equilibrium in aqueous buffer. Biochemistry 34(20):6784–6794
10. Pan H, Barbar E, Barany G, Woodward C (1995) Extensive nonrandom structure in reduced and unfolded bovine pancreatic trypsin inhibitor. Biochemistry 34 (43):13974–13981
11. Dyson HJ, Wright PE (2004) Unfolded proteins and protein folding studied by NMR. Chem Rev 104(8):3607–3622
12. Lakowicz JR (2006) Principle of fluorescence spectroscopy, 3rd edn. Springer, New York
13. Ballew RM, Sabelko J, Gruebele M (1996) Observation of distinct nanosecond and microsecond protein folding events. Nat Struct Biol 3(11):923–926
14. Ballew RM, Sabelko J, Gruebele M (1996) Direct observation of fast protein folding: the initial collapse of apomyoglobin. Proc Natl Acad Sci USA 93(12):5759–5764
15. Munoz V, Thompson PA, Hofrichter J, Eaton WA (1997) Folding dynamics and mechanism of β-hairpin formation. Nature 390(6656):196–199
16. Qiu L, Pabit SA, Roitberg AE, Hagen SJ (2002) Smaller and faster: The 20-residue Trp-cage protein folds in 4 μs. J Am Chem Soc 124(44):12952–12953
17. Kubelka J, Eaton WA, Hofrichter J (2003) Experimental tests of villin subdomain folding simulations. J Mol Biol 329(4):625–630
18. Yang WY, Gruebele M (2004) Folding λ-repressor at its speed limit. Biophys J 87(1):596–608
19. Sinha KK, Udgaonkar JB (2008) Barrierless evolution of structure during the submillisecond refolding reaction of a small protein. Proc Natl Acad Sci USA 105(23):7998–8003
20. Lyubovitsky JG, Gray HB, Winkler JR (2002) Mapping the cytochrome c folding landscape. J Am Chem Soc 124(19):5481–5485
21. Jha SK, Dhar D, Krishnamoorthy G, Udgaonkar JB (2009) Continuous dissolution of structure during the unfolding of a small protein. Proc Natl Acad Sci USA 106(27):11113–11118
22. Brochon JC (1994) Maximum entropy method of data analysis in time-resolved spectroscopy. Methods Enzymol 240:262–311
23. Swaminathan R, Periasamy N (1996) Analysis of fluorescence decay by maximum entropy method: Influence of noise and analysis parameters on the width of the distribution of lifetimes. Proc Indian Acad Sci (Chem Sci) 108:39–49
24. Rhoades E, Gussakovsky E, Haran G (2003) Watching proteins fold one molecule at a time. Proc Natl Acad Sci USA 100(6):3197–3202
25. Kuzmenkina EV, Heyes CD, Nienhaus GU (2005) Single-molecule Forster resonance energy transfer study of protein dynamics under denaturing conditions. Proc Natl Acad Sci USA 102 (43):15471–15476
26. Mukhopadhyay S, Krishnan R, Lemke EA, Lindquist S, Deniz AA (2007) A natively unfolded yeast prion monomer adopts an ensemble of collapsed and rapidly fluctuating structures. Proc Natl Acad Sci USA 104(8):2649–2654
27. Munoz V (2007) Conformational dynamics and ensembles in protein folding. Annu Rev Biophys Biomol Struct 36:395–412
28. Sridevi K, Lakshmikanth GS, Krishnamoorthy G, Udgaonkar JB (2004) Increasing stability reduces conformational heterogeneity in a protein folding intermediate ensemble. J Mol Biol 337(3):699–711
29. Jha SK, Udgaonkar JB (2009) Direct evidence for a dry molten globule intermediate during the unfolding of a small protein. Proc Natl Acad Sci USA 106(30):12289–12294

30. Beechem JM (1997) Picosecond fluorescence decay curves collected on millisecond time scale: Direct measurement of hydrodynamic radii, local/global mobility, and intramolecular distances during protein-folding reactions. Methods Enzymol 278:24–49
31. Pletneva EV, Gray HB, Winkler JR (2005) Snapshots of cytochrome c folding. Proc Natl Acad Sci USA 102(51):18397–18402
32. Kimura T, Lee JC, Gray HB, Winkler JR (2009) Folding energy landscape of cytochrome cb562. Proc Natl Acad Sci USA 106(19):7834–7839
33. Bryngelson JD, Onuchic JN, Socci ND, Wolynes PG (1995) Funnels, pathways, and the energy landscape of protein folding: a synthesis. Protein 21:167–195
34. Dill KA, Chan HS (1997) From Levinthal to pathways to funnels. Nat Struct Biol 4(1):10–19
35. Chavez LL, Onuchic JN, Clementi C (2004) Quantifying the roughness on the free energy landscape: entropic bottlenecks and protein folding rates. J Am Chem Soc 126(27):8426–8432
36. Lakshmikanth GS, Sridevi K, Krishnamoorthy G, Udgaonkar JB (2001) Structure is lost incrementally during the unfolding of barstar. Nat Struct Biol 8(9):799–804
37. Kimura T, Uzawa T, Ishimori K, Morishima I, Takahashi S, Konno T, Akiyama S, Fujisawa T (2005) Specific collapse followed by slow hydrogen-bond formation of β-sheet in the folding of single-chain monellin. Proc Natl Acad Sci USA 102(8):2748–2753
38. Patra AK, Udgaonkar JB (2007) Characterization of the folding and unfolding reactions of single-chain monellin: evidence for multiple intermediates and competing pathways. Biochemistry 46(42):11727–11743
39. Kimura T, Maeda A, Nishiguchi S, Ishimori K, Morishima I, Konno T, Goto Y, Takahashi S (2008) Dehydration of main-chain amides in the final folding step of single-chain monellin revealed by time-resolved infrared spectroscopy. Proc Natl Acad Sci USA 105 (36):13391–13396
40. Broos J, Maddalena F, Hesp BH (2003) In vivo synthesized proteins with monoexponential fluorescence decay kinetics. J Am Chem Soc 126(1):22–23
41. Swaminathan R, Krishnamoorthy G, Periasamy N (1994) Similarity of fluorescence lifetime distributions for single tryptophan proteins in the random coil state. Biophys J 67 (5):2013–2023
42. Doi M (2001) Introduction to polymer physics. Oxford University Press, Oxford, pp 65–88
43. Dua A, Adhikari R (2008) Non-Markovian fluctuations in Markovian models of protein dynamics. http://arxiv.org/abs/0810.2232
44. Khurana R, Udgaonkar JB (1994) Equilibrium unfolding studies of barstar: evidence for an alternative conformation which resembles a molten globule. Biochemistry 33(1):106–115
45. Shastry MCR, Agashe VR, Udgaonkar JB (1994) Quantitative analysis of the kinetics of denaturation and renaturation of barstar in the folding transition zone. Protein Sci 3:1409–1417
46. Agashe VR, Udgaonkar JB (1995) Thermodynamics of denaturation of barstar: evidence for cold denaturation and evaluation of the interaction with guanidine hydrochloride. Biochemistry 34(10):3286–3299
47. Shastry MCR, Udgaonkar JB (1995) The folding mechanism of barstar: evidence for multiple pathways and multiple intermediates. J Mol Biol 247(5):1013–1027
48. Agashe VR, Shastry MCR, Udgaonkar JB (1995) Initial hydrophobic collapse in the folding of barstar. Nature 377(6551):754–757
49. Swaminathan R, Nath U, Udgaonkar JB, Periasamy N, Krishnamoorthy G (1996) Motional dynamics of a buried tryptophan reveals the presence of partially structured forms during denaturation of barstar. Biochemistry 35(28):9150–9157
50. Sridevi K, Juneja J, Bhuyan AK, Krishnamoorthy G, Udgaonkar JB (2000) The slow folding reaction of barstar: the core tryptophan region attains tight packing before substantial secondary and tertiary structure formation and final compaction of the polypeptide chain. J Mol Biol 302(2):479–495
51. Sridevi K, Udgaonkar JB (2002) Unfolding rates of barstar determined in native and low denaturant conditions indicate the presence of intermediates. Biochemistry 41(5):1568–1578

52. Sinha KK, Udgaonkar JB (2005) Dependence of the size of the initially collapsed form during the refolding of barstar on denaturant concentration: evidence for a continuous transition. J Mol Biol 353(3):704–718
53. Saxena AM, Udgaonkar JB, Krishnamoorthy G (2006) Characterization of intra-molecular distances and site-specific dynamics in chemically unfolded barstar: evidence for denaturant-dependent non-random structure. J Mol Biol 359(1):174–189
54. Wood SJ, Maleeff B, Hart T, Wetzel R (1996) Physical, morphological and functional differences between pH 5.8 and 7.4 aggregates of the Alzheimer's amyloid peptide Aβ. J Mol Biol 256(5):870–877
55. Krebs MRH, Wilkins DK, Chung EW, Pitkeathly MC, Chamberlain AK, Zurdo J, Robinson CV, Dobson CM (2000) Formation and seeding of amyloid fibrils from wild-type hen lysozyme and a peptide fragment from the β-domain. J Mol Biol 300(3):541–549
56. Zurdo J, Guijarro JI, Jiménez JL, Saibil HR, Dobson CM (2001) Dependence on solution conditions of aggregation and amyloid formation by an SH3 domain. J Mol Biol 311 (2):325–340
57. Uversky VN, Li J, Fink AL (2001) Evidence for a partially folded intermediate in alpha-synuclein fibril formation. J Biol Chem 276:10737–10744
58. Hoyer W, Antony T, Cherny D, Heim G, Jovin TM, Subramaniam V (2002) Dependence of α-synuclein aggregate morphology on solution conditions. J Mol Biol 322(2):383–393
59. Fung SY, Keyes C, Duhamel J, Chen P (2003) Concentration effect on the aggregation of a self-assembling oligopeptide. Biophys J 85(1):537–548
60. Baldwin AJ, Bader R, Christodoulou J, MacPhee CE, Dobson CM, Barker PD (2006) Cytochrome display on amyloid fibrils. J Am Chem Soc 128(7):2162–2163
61. Yong W, Lomakin A, Kirkitadze MD, Teplow DB, Chen S-H, Benedek GB (2002) Structure determination of micelle-like intermediates in amyloid β-protein fibril assembly by using small angle neutron scattering. Proc Natl Acad Sci USA 99(1):150–154
62. Lashuel HA, Petre BM, Wall J, Simon M, Nowak RJ, Walz T, Lansbury PT Jr (2002) α-Synuclein, especially the Parkinson's disease-associated mutants, forms pore-like annular and tubular protofibrils. J Mol Biol 322(5):1089–1102
63. Bitan G, Kirkitadze MD, Lomakin A, Vollers SS, Benedek GB, Teplow DB (2003) Amyloid β-protein (Aβ) assembly: Aβ40 and Aβ42 oligomerize through distinct pathways. Proc Natl Acad Sci USA 100(1):330–335
64. Chen Y-R, Glabe CG (2006) Distinct early folding and aggregation properties of Alzheimer amyloid-β peptides Aβ40 and Aβ42. J Biol Chem 281(34):24414–24422
65. Ono K, Condron MM, Teplow DB (2009) Structure-neurotoxicity relationships of amyloid β-protein oligomers. Proc Natl Acad Sci USA 106(35):14745–14750
66. Sunde M, Serpell LC, Bartlam M, Fraser PE, Pepys MB, Blake CCF (1997) Common core structure of amyloid fibrils by synchrotron X-ray diffraction. J Mol Biol 273(3):729–739
67. Serpell LC, Berriman J, Jakes R, Goedert M, Crowther RA (2000) Fiber diffraction of synthetic α-synuclein filaments shows amyloid-like cross-β conformation. Proc Natl Acad Sci USA 97(9):4897–4902
68. Jimenez JL, Nettleton EJ, Bouchard M, Robinson CV, Dobson CM, Saibil HR (2002) The protofilament structure of insulin amyloid fibrils. Proc Natl Acad Sci USA 99(14):9196–9201
69. Jaroniec CP, MacPhee CE, Bajaj VS, McMahon MT, Dobson CM, Griffin RG (2004) High-resolution molecular structure of a peptide in an amyloid fibril determined by magic angle spinning NMR spectroscopy. Proc Natl Acad Sci USA 101(3):711–716
70. Jha A, Udgaonkar JB, Krishnamoorthy G (2009) Characterization of the heterogeneity and specificity of interpolypeptide interactions in amyloid protofibrils by measurement of site-specific fluorescence anisotropy decay kinetics. J Mol Biol 393(3):735–752
71. Cookson MR (2005) The biochemistry of Parkinson's disease. Annu Rev Biochem 74 (1):29–52
72. Robinson PA (2008) Protein stability and aggregation in Parkinson's disease. Biochem J 413:1–13

73. Irvine GB, El-Agnaf OM, Shankar GM, Walsh DM (2008) Protein aggregation in the brain: the molecular basis for Alzheimer's and Parkinson's diseases. Mol Med 14(7–8):451–464
74. Roychaudhuri R, Yang M, Hoshi MM, Teplow DB (2009) Amyloid beta-protein assembly and Alzheimer disease. J Biol Chem 284(8):4749–4753
75. Sokolowski F, Modler AJ, Masuch R, Zirwer D, Baier M, Lutsch G, Moss DA, Gast K, Naumann D (2003) Formation of critical oligomers is a key event during conformational transition of recombinant Syrian hamster prion protein. J Biol Chem 278(42):40481–40492
76. Chiti F, Dobson CM (2006) Protein misfolding, functional amyloid, and human disease. Annu Rev Biochem 75(1):333–366
77. Lansbury PT, Lashuel HA (2006) A century-old debate on protein aggregation and neurodegeneration enters the clinic. Nature 443(7113):774–779
78. Lomakin A, Chung DS, Benedek GB, Kirschner DA, Teplow DB (1996) On the nucleation and growth of amyloid beta-protein fibrils: detection of nuclei and quantitation of rate constants. Proc Natl Acad Sci USA 93(3):1125–1129
79. Spires TL, Hannan AJ (2007) Molecular mechanisms mediating pathological plasticity in Huntington's disease and Alzheimer's disease. J Neurochem 100(4):874–882
80. Frieden C (2007) Protein aggregation processes: in search of the mechanism. Protein Sci 16 (11):2334–2344
81. Gast K, Modler AJ, Damaschun H, Kröber R, Lutsch G, Zirwer D, Golbik R, Damaschun G (2003) Effect of environmental conditions on aggregation and fibril formation of barstar. Eur Biophys J 32(8):710–723
82. Kumar S, Mohanty SK, Udgaonkar JB (2007) Mechanism of formation of amyloid protofibrils of barstar from soluble oligomers: evidence for multiple steps and lateral association coupled to conformational conversion. J Mol Biol 367(4):1186–1204
83. Kumar S, Udgaonkar JB (2009) Conformational conversion may precede or follow aggregate elongation on alternative pathways of amyloid protofibril formation. J Mol Biol 385 (4):1266–1276
84. Kumar S, Udgaonkar JB (2009) Structurally distinct amyloid protofibrils form on separate pathways of aggregation of a small protein. Biochemistry 48(27):6441–6449
85. Juneja J, Bhavesh NS, Udgaonkar JB, Hosur RV (2002) NMR identification and characterization of the flexible regions in the 160 kDa molten globule-like aggregate of barstar at Low pH. Biochemistry 41(31):9885–9899

Theme and Variation on *N*-Aryl-1, 8-Napthalimides: Minimal Modification to Red-Shifted Fluorescence and Applications in Fluorescent Chemosensors

Premchendar Nandhikonda, Zhi Cao, and Michael D. Heagy

1 Introduction and Background

Our continuing efforts into the development of *N*-aryl-1,8-naphthalic dicarboximides (NI) as dual fluorescent (DF) dyes for biomedical applications has led to new insights into the photophysical features that these simple dyes can be designed to display. Consequently, the development of new DF dyes with improved fluorescent properties represents a major focus of our research. The first section of this review presents results involving a "minimal modification approach" to red-shifted absorption and fluorescence in NIs and affords some key design concepts for improved DF dyes. In this section, we demonstrate the significant effect of appropriately placed charges can have on the emission properties of these unique dyes. In the next section, dual fluorescent probes for the ions of potassium and sodium are introduced with the NI framework and crown ether receptors. The ratiometric features of these dyes from absorption as well as fluorescence spectroscopy are highlighted. Finally, in the third section, we demonstrate dual fluorescence detection of saccharides with the same DF dye component as our ion probe, in this case, however, a simple phenylboronic acid is utilized as a saccharide binding component.

2 Minimal Modification Approach to Red-Shifted Fluorescence in 1,8-Naphthalimides

A key design aim in the development of functional dyes involves the synthesis of structural elements that extend the absorption to longer visible wavelengths. Such properties are sought in both dye-sensitized solar cells (DSSCs) for panchromatic

M.D. Heagy (✉)
Department of Chemistry, New Mexico Institute of Mining and Technology, Socorro, NM 87801, USA
e-mail: mheagy@nmt.edu

C.D. Geddes (ed.), *Reviews in Fluorescence 2009*, Reviews in Fluorescence 2009,
DOI 10.1007/978-1-4419-9672-5_11,

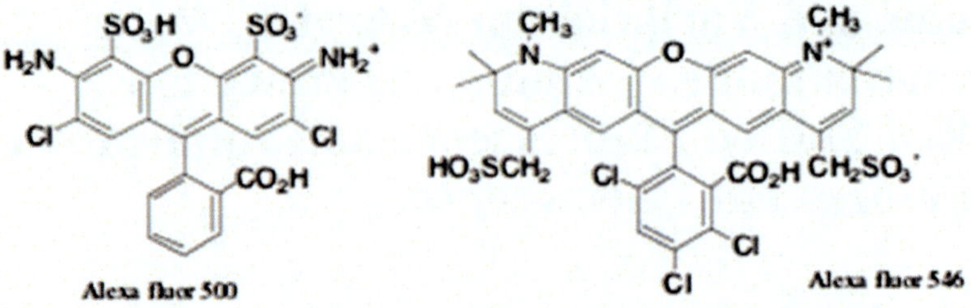

Fig. 1 Comparison of structural features required for 1,685 cm^{-1} shift in absorption

absorption [1, 2] and in biological settings by extending the emission beyond the blue autofluorescence of cellular media [3]. Extending the conjugation and thereby increasing the size of the fluorophore remains a conventional approach toward increasing absorption and emission wavelengths. The Alexa fluor class of fluorescent dyes provides a typical example with a 46-nm bathochromic shift difference in absorption (Fig. 1).

An alternative approach requiring far fewer synthetic steps involves placement of positive charge along with a functional groups that strongly polarize the molecule in both the ground and excited states. This synthetic design is an application of internal charge transfer (ICT) whereby introduction of donor/acceptor groups has been shown to extend the absorption and emission wavelengths without incorporating additional aromatic rings into the chromophore structure [4]. Recently, Klymchenko et al. demonstrated the effect of proximal charge on the spectroscopic behavior of 3-hydroxyflavone (3 HF) derivatives [5]. Placement of a positively charged ammonium group within close proximity to 4-carbonyl stabilizes the Frank-Condon state and the emissive excited states due to electrostatic interaction, and results in greater fluorescence quantum yield and relatively large red-shifts of absorption and emission (typically 980–1,100 cm^{-1}) in low-polarity solvents [4].

In the case of *N*-arylnaphthalimides (NI), the dicarboximide functionality features a molecular orbital arrangement that allows us to explore the effects of donor/acceptor groups as substituents on both the NI ring and *N*-aryl component. Recently, there has been an upsurge in the number of reports on photophysics and applications of NI-based dyes [6–9]. Beginning with Bérces, Kossanyi, and coworkers, 1,8- and 2,3-naphthalimides were shown to display long wavelength emission due to large ICT character from donor group substitution on the naphthalimide ring [10]. This research group also found that protonation of the pyridyl arene component to form *N*-pyridinium-1,8-naphthalimides resulted in a 17,500 cm^{-1} Stokes shift. Takahashi et al. utilized the electron-deficient properties of the NI scaffold and created molecular dyads by appending electron-rich dianisidyl group as the imide arene [11]. This design resulted in a 12,500 cm^{-1} Stokes shift. In both reports, however, the absorption wavelength maxima are well below 400 nm. Molecular orbital calculations for both systems show that the LUMO is localized on the acceptor NI component. This observation prompted us to alter the π-densities and their locations as an approach to longer wavelength absorption. Herein, we report on the implementation of appropriately placed substituent groups and the successful outcome of this minimum modification strategy.

3 Results

Our strategy for manipulation of the NI π-system involves reversing the electron densities between HOMO and LUMO in comparison to these earlier reports. By attaching a donor diethylamino group on the NI ring and coupling electron-deficient pyridine in a manner similar to Bérces et al. [5], increased polarization through placement of positive charge and generating the *N*-pyridinium salt is expected to further enhance this polarization. Figure 2 shows the HOMO and to a greater extent LUMO localized away from the NI ring and centered at the pyridinium component. Using the three possible isomers of aminopyridine for the imide formation allowed us to study their photophysical differences with respect to position of the proximal pyridinium cation. The *ortho*-, *meta*-, and *para*-isomers also provided a systematic study on charge separation, and compared to the 3-hydroxyflavone system, place the coulombic interactions at closer distances.

Scheme 1 outlines the synthetic route used to prepare compounds **4–6a**. Nucleophilic aromatic substitution proceeded smoothly using diethylamine with the bromo-derivatives (**1–3**) in dry DMF [12]. Next, *N*-alkylation using 1,3-propanesultone has been shown to be selective for the pyridine nitrogen relative to similar heterocyclic systems bearing a diethylamino moiety [13]. Yields obtained from this reaction were excellent despite the need of chromatography for separation from starting materials.

Absorption spectroscopy of these dyes was carried out in solvents that allowed both the neutral dyes (**4–6**) to be compared with the sultone salts (**4a–6a**). As shown in Table 1, all of the dyes displayed red-shifted absorption spectra when comparing sultonated species with their respective neutral precursor. The largest bathochromic shifts occurred in the low-polarity solvent, ethyl acetate. In particular, the *para*-isomer **6a** displayed bathochromic behavior on par with the Alexa Fluor series mentioned above. Quaternization of the nitrogen heteroatom should increase the

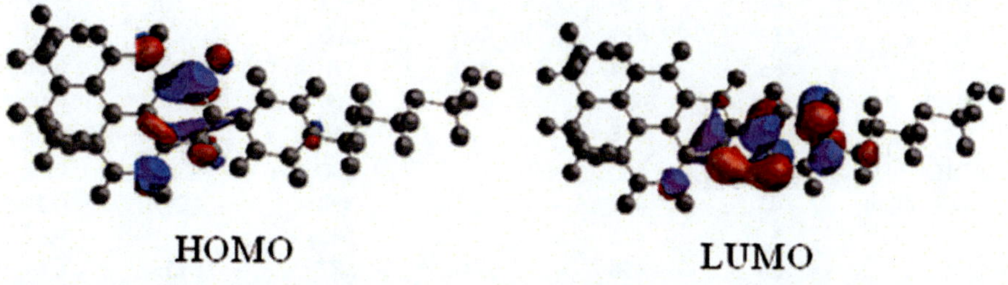

Fig. 2 Calculated distribution of excited state charges predicted for NI-pyridinium salts shows how electrostatic stabilization of the excited state is responsible for shifting the absorption and emission bands to the red [47]. Graphically depicted using Chem3D

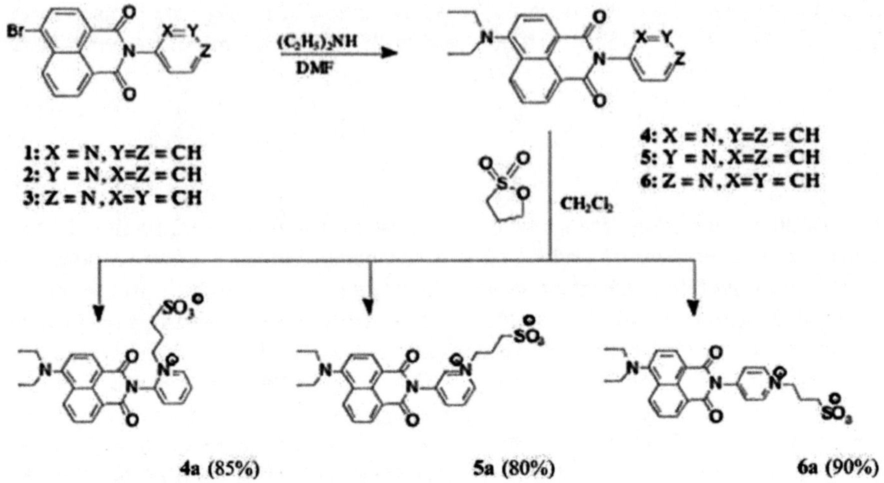

Scheme 1 Synthetic pathway for generation of N-sultonated pyridinium salts 4a-6a

Table 1 Spectral properties of the studied naphthalimides

Solvent	Entry	Abs $\lambda_{\mu\alpha\xi}$ (nm)	Shift (cm^{-1})	Em $\lambda_{\mu\alpha\xi}$ (nm)	Shift (cm^{-1})	ϕ_F
$CHCl_3$	**4**	424		505		0.04
($\varepsilon = 4.8$)	**4a**	430	329	508	117	0.07
	5	426		511		0.08
	5a	432	326	512	65	0.01
	6	428		508		0.05
	6a	433	269	514	230	0.02
EtOAc	**4**	411		561		0.09
($\varepsilon = 6.1$)	**4a**	439	1,552	572	342	0.26
	5	412		560		0.06
	5a	433	1,177	565	158	0.01
	6	414		565		0.13
	6a	445	1,682	517	−1,643	0.02
Acetone	**4**	428		564		0.07
($\varepsilon = 20.4$)	**4a**	439	585	573	278	0.38
	5	417		560		0.04
	5a	434	940	570	292	0.08
	6	421		568		0.03
	6a	438	922	565	−94	0.05
MeOH	**4**	430		580		0.02
($\varepsilon = 32.6$)	**4a**	442	632	590	292	0.11
	5	425		582		0.02
	5a	438	698	587	158	0.02
	6	425		580		0.02
	6a	438	698	585	121	0.03

(continued)

Table 1 (continued)

Solvent	Entry	Abs $\lambda_{\mu\alpha\xi}$ (nm)	Shift (cm^{-1})	Em $\lambda_{\mu\alpha\xi}$ (nm)	Shift (cm^{-1})	ϕ_F
CAN	**4**	425		570		0.03
($\varepsilon = 36.0$)	**4a**	442	688	584	420	0.17
	5	421		566		0.08
	5a	437	870	577	342	0.02
	6	421		572		0.02
	6a	439	974	576	121	0.001
DMSO	**4**	426		527		0.01
($\varepsilon = 46.5$)	**4a**	436	538	538	388	0.01
	5	428		525		0.03
	5a	440	637	531	215	0.01
	6	427		529		0.04
	6a	439	640	531	70	0.01
Water	**4**	–	–	–	–	–
($\varepsilon = 78.4$)	**4a**	440	–	604	–	0.01
	5	–	–	–	–	–
	5a	440	–	604	–	0.01
	6	–	–	–	–	–
	6a	440	–	600	–	0.01

electron acceptor properties of the *N*-aryl group. This proximal charge appears to be responsible for the red-shift in absorption as it strongly withdrawals electron density from the imide nitrogen of the fluorophore compared to the neutral system [14]. A general trend for solvents more polar than chloroform is observed within each isomeric group such that *para*-sultone **6a** displayed the largest degree of red-shifted absorption relative to *ortho*- and *para*-isomers **4a** and **5a**. For the least polar solvent chloroform, the largest change in red-shifted absorption was observed with *ortho*-isomer **4**. Given the low dielectric constant for this solvent, it is likely that the proximal charges are less stabilized by solvent but have a greater electrostatic stabilization via their proximity in an *ortho*-ammonium arrangement. For the entire series, however, dyes **5**, **5a** and **6**, **6a** generally gave larger bathochromic shifts in absorption. In the case of *ortho*-isomer **4a**, where steric hindrance is the strongest, the *N*-aryl group should be twisted from the plane of the fluorophore to the greatest extent. This larger twist angle is expected to decrease the electron-withdrawing effects of the proximal charge and result in less red-shifted absorption. Solvents less polar than chloroform such as hexane could not be included in the study due to insolubility of the sultone salts **4a**–**6a**. Conversely, neutral species **4**–**6** were insoluble in water, whereas salts **4a**–**6a** readily dissolved and their spectral data are given in Table 1.

To examine the excited state features of these dyes, steady-state fluorescence spectroscopy also revealed bathochromic behavior for each set of sultonated dyes relative to their neutral precursors. For both *ortho* sultone systems (**4** and **4a**) and *meta* versions (**5** and **5a**), bathochromic shifts were observed although not to the extent found in the optical absorption. Earlier reports on *N*-aryl-1,8-NI have pointed out that these dyes undergo adiabatic ring rotation in the excited state toward a co-planar configuration; the so-called *excited state with extended conjugation* mechanism [15]. This photophysical effect may provide the necessary energy to overcome some of the barrier to rotation imposed by *ortho*-substitution and lead to longer wavelength emission for isomer **4a** because of the decreased charge separation. Compared to the optical data, a wide range of emission was observed in the case of dye **6**, where a large hypsochromic shift was found in ethyl acetate and to a lesser extent in acetone. In solvents such as acetonitrile and methanol, bathrochromic shifts of magnitudes similar to *ortho* and *meta* systems were evident. Comparisons between isomers show that *ortho*-isomer with the quaternized nitrogen has the largest fluorescence quantum yield. This observation is attributed to the reduced rotational freedom of the *ortho*-pyridinium salt relative to *meta* and *para*-isomers. Because of the insolubility of dyes **4–6** in water, no comparison between shifts in absorption can be made, nonetheless, for dyes **4a–6a**, the longest emission wavelengths were recorded in this solvent ranging between 600 and 604 nm. Such long wavelength emission properties in combination with solubility in aqueous solution could be useful for biological applications.

To summarize, a minimal modification approach on *N*-pyridyl-1,8-naphthalimides was shown to be effective in shifting the wavelengths of both absorption and emission toward the red for all three isomers of the pyridyl moiety. In some cases, such as **6a** in ethyl acetate, a 1,682 cm^{-1} shift in absorbance was observed. Such large increases in wavelength compared favorably to the 1,685 cm^{-1} red-shift of Alexafluor series, but with fewer synthetic steps. Moreover, an increase in fluorescence quantum yield was observed for each of the charged dyes relative to their neutral precursors. This bathochromic behavior is attributed to the polarization induced on the NI scaffold via positive charge of the cationic nitrogen in compounds **4a–6a**. Our initial studies involving simple 1,8-NI dyes establish that large red-shifts in absorption can be accomplished by altering the π-molecular orbital system of 1,8-NI from their usual electron-deficient framework. These findings bode well for designing panchromatic dyes used in DSSC applications as well as fluorescent platforms such as Lucifer yellow anhydride, a commonly used NI-based biological probe.

4 Dual Fluorescent Probes for Sodium and Potassium Ions

4.1 Introduction

The development of wavelength ratiometric probes for Na^+ and K^+ has important implications in clinical research such as hyponatremia and hypertensive screening [16, 17]. In blood, the concentrations of K^+ and Na^+ are near 4.5 and 120 mM, whereas in resting cells the concentrations are effectively reversed at 140 and 4 mM, respectively [18, 19]. Presently, wavelength ratiometric probes are available but problems regarding solubility, adequate spectral changes, and selectivity persist [20]. Therefore, designs for new probes of sodium and potassium that function in water with high selectivity remain highly desirable for biological studies. Because of their ability to selectively bind to alkali and alkaline earth metal cations, crown ether derivatives have attracted considerable attention [21]. Crown ethers have been widely used as the receptor component in conjunction with the fluorophore make them attractive targets as fluorescent probes. To date, sodium and potassium chemosensors based on such fluorescent platforms as benzofuranisophthalate [19], 4-(dimethylamino)benzonitrile [22], coumarin groups [23], phenolichydroxyl [24], anthracenylmethyl [25], spirobenzopyran [26], and 4-amino-1,8-napthalimide [27] have all been explored.

A relatively unexplored class of fluorescent platforms for ion sensing are 1,8-naphthalimides (NI). Recently, there has been an upsurge in the number of reports on photophysics and applications of NI-based dyes [6]. In the case of *N*-arylnaphthalimide, the dicarboximide functionality features a molecular orbital arrangement that augments the effects of donor/acceptor groups on both the NI ring and *N*-aryl component. *N*-Phenylnaphthalimides have been found to exhibit dual fluorescence when appropriately substituted at both the *N*-phenyl aromatic ring and naphthalene system [28]. These compounds display dual luminescence with two clearly resolved emission bands in the visible region from the locally excited (LE) state and a strongly red shifted long wave length emitted by an ICT state. Dual fluorescence (DF) compounds feature the advantage of ratiometric detection as they reduce or eliminate distortions caused by photobleaching, indicator concentrations, and illumination stability [29]. In this report, we introduce two NI-based ion sensors that provide ratiometric absorption spectra as well as DF emission for internally calibrated ion analysis of sodium and potassium. Therefore, the crown ether probes **7** and **8** were synthesized based on DF systems previously reported by our group [30].

5 Results

In Scheme 2, compounds **7** and **8** were conveniently synthesized from 4-sulfo-1, 8-naphthalic anhydride, potassium salt, and 4′-aminobenzo-15-crown-5 and 4′-aminobenzo-18-crown-6, respectively. No chromatography was required and isolated yields for both **7** and **8** were nearly quantitative. Figure 5 shows ratiometric absorption spectra of **7** with increasing concentrations of Na^+ in buffered ($[CH_3(CH_2)_3]_4NCl$)

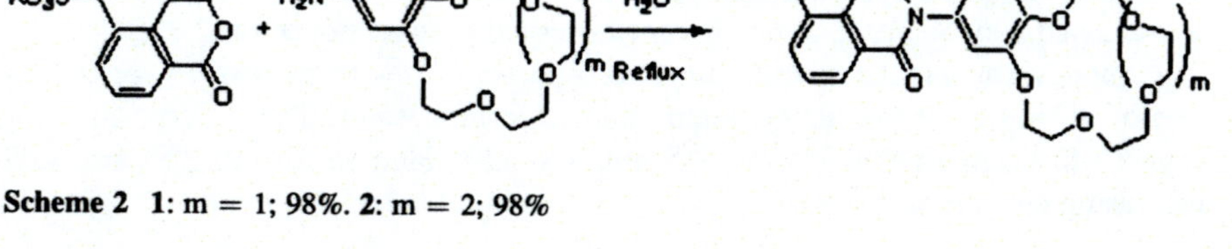

Scheme 2 **1**: m = 1; 98%. **2**: m = 2; 98%

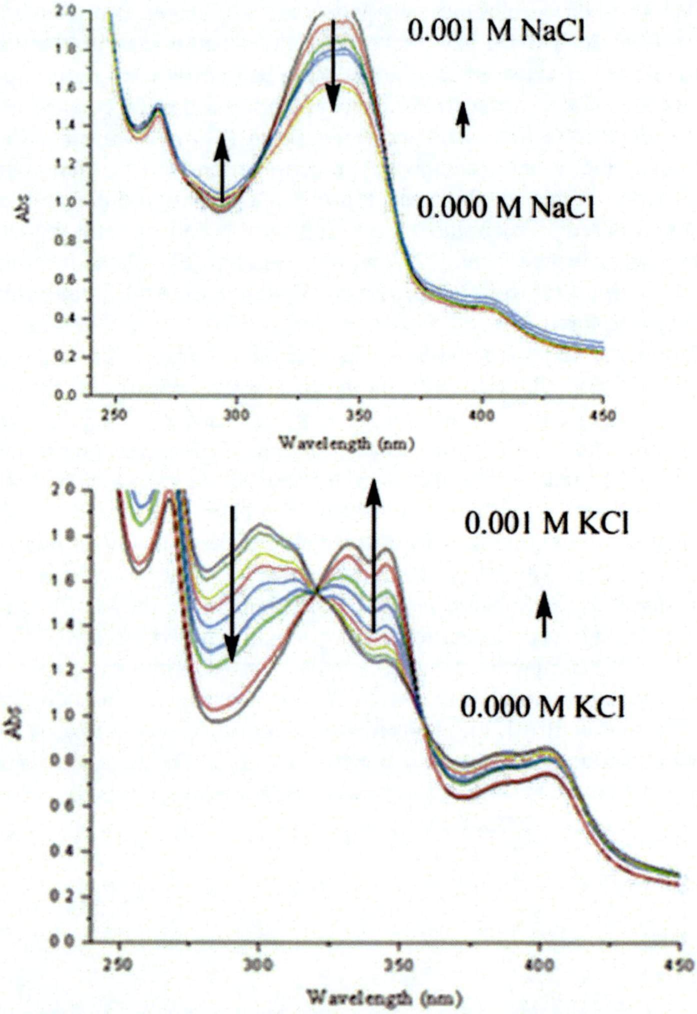

Fig. 3 Absorption spectra for **7** (1×10^{-5} M) with NaCl (1×10^{-3} M) and **8** (1×10^{-5} M) with KCl (1×10^{-3} M) aqueous solutions

water. The absorption spectrum of **7**, which originally peaked at 345 nm, was blue shifted to 293 nm with increasing concentration of Na^+. This spectrum was accompanied by the appearance of one isosbestic point at 311 nm. When probe **7** was assayed with K^+, little or no changes observed (spectra not shown).

Figure 3a shows ratiometric absorption spectra of **8** with increasing concentration of K^+ in buffer (tetrabutyl ammonium chloride) water. The absorption spectrum of **8**, originally peaked at 345 nm, was also blue shifted to 300 nm with increasing concentration of K^+.

In the presence of K^+, the absorption spectrum was accompanied by the appearance of two isosbestic points at 357 and 320 nm. In the case of Na^+, the absorption spectrum was also blue shifted to 310 nm. This titration resulted in relatively little change in optically density and the appearance of one isosbestic point at 320 nm in Fig. 4b. Therefore, **7** selectively binds Na^+ over K^+, whereas **8** displays higher sensitivity to K^+ relative to Na^+.

In Fig. 5, sensor **7** shows a substantial increase in its dual emission with Na^+ in aqueous media. Fluorescence intensity of **7** was enhanced from 0 to 0.5 mM of NaCl

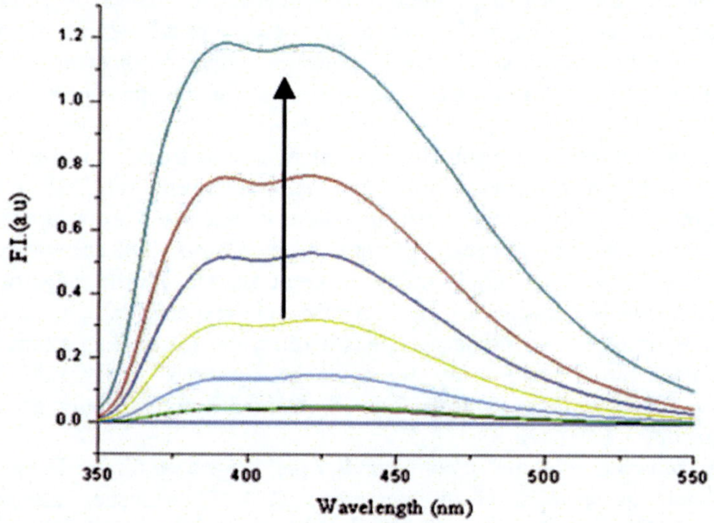

Fig. 4 Fluorescence response for **7** (1×10^{-5} M) with NaCl (1×10^{-3} M) aqueous buffered system

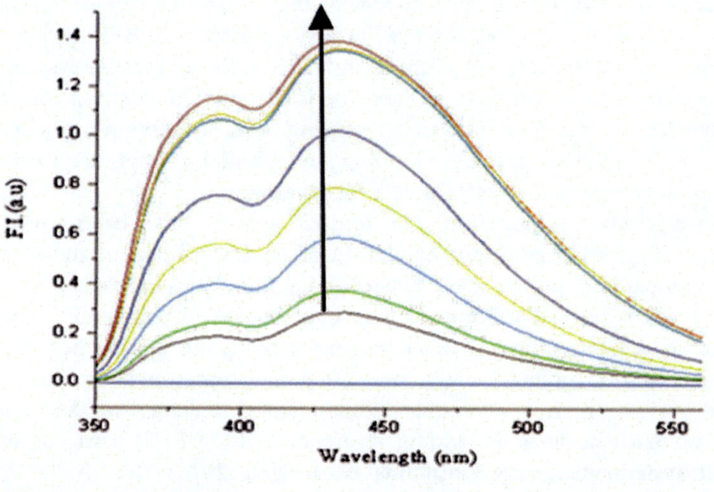

Fig. 5 Fluorescence response for **8** (1×10^{-5} M) with KCl (1×10^{-3} M) aqueous buffered system

Table 2 Spectroscopic data of compounds **7** and **8**

Sensors	$K_{D(absorb)}$ (mM)	$K_{D(fluoresce)}$ (mM)	λ_{ex} (nm)	λ_1 (nm)	λ_2 (nm)
7	1.5 ± 0.002	1.12 ± 0.002	345	374	435
8	0.6 ± 0.002	0.4 ± 0.002	345	380	440

in an aqueous solution containing tetrabutylammonium chloride (pH 7.2). The fluorescence intensity of both emission peaks (382 and 424 nm) increased concomitantly up to 0.5 mM and then leveled off at higher concentration. However, the addition of K^+ did not change fluorescence intensity of sensor **7**. Therefore, it shows that **7** selectively binds Na^+ even in the presence of K^+ because of potassium rejection from the smaller size of the 15-crown-5 cavity. Job plots confirm the 1:1 binding stoichiometry. Based on the fluorescence assay, the apparent K_D for Na^+ is 1.12 mM.

In Fig. 5, fluorescence intensity of **8** was enhanced from 0 to 6 mM of KCl in aqueous medium containing tetrabutylammonium chloride (pH 7.2). In the case of compound **8**, fluorescence intensity was not enhanced proportionally with sodium. The fluorescence intensity of both emission bands (382 and 441 nm) also showed an OFF-ON signaling in the presence of analyte. As the range of K^+ ion in human blood is between 3.5 and 5.3 mM, this response falls exactly within the dynamic range of our probe. Job plots confirm the 1:1 binding stoichiometry. Based on the fluorescence assay, the apparent K_D for K^+ is 0.4 mM. Table 2 shows the dissociation constants for **7** and **8** obtained from absorption and fluorescence spectra.

The crown ether moieties of compounds **7** and **8** were synthesized based on DF systems previously reported by our group [13]. Fluorescence intensities of 15-crown-5 (**7**) and 18-crown-6 (**8**) were substantially enhanced upon addition of Na^+ and K^+ cations. This OFF-ON signal is attributed to the electron accepting properties of NI. Thus, alkali metal cations disrupt the PET quenching process that occurs between oxygen lone pairs of crown ether and NI fluorophore. Most of the earliest and best known sodium (SBFI) and potassium (PBFI) probes contain azacrown groups exhibit lower affinities relative to the oxycrown systems. While the azacrown systems provide an ideal optical range for sodium ion, they are unsuitable for extracellular potassium sensing. This observation occurs mainly because PBFI exhibits insufficient binding strength and interference from sodium, as well as showing small spectral shifts for potassium.

In closing, whereas numerous Na^+ and K^+ sensors have been developed for water solubility, ratiometric probes, and ideal binding constants, they are sizable systems compared to probes **7** and **8**. Such large molecules can be an issue in their ability to enter cells. Furthermore, our one step synthesis holds considerable synthetic advantage compared to the elaborate multistep approaches required by previously reported probes. Finally, these DF probes represent a rare example of the "OR-OR switching in a reversible excited state" mechanism. Such signaling devices obviate the need of adding reference dyes, FRET pairs, or other dye ensemble systems necessary for internal calibration [29].

6 Dual Fluorescent Probes for Glucose

6.1 Introduction

With cases of diabetes reaching epidemic proportions, there continues to be a strong demand for methods of detecting saccharide concentration in blood for those patients who are suffering from this chronic disease [31]. Since the ability of recognition of saccharides, boronic acids with diol moiety are widely investigated for their huge potential biomedical applications [20–22, 32–34]. Unfortunately, most of carbohydrate sensors based on boronic acid moiety continue to have water solubility and those based on enzymes exhibit poor stability or consumption of substrate during the detection procedure which also limit further biosensing applications [35, 36]. Because of their ability to bind to the diols of sugars, phenylboronic acid and its derivatives have been developed for saccharide sensing based on different measurements, such as fluorescence [37, 38], UV-vis absorption [39, 40], and other methods [41, 42]. Some key challenges in this field that continue to limit the number of useful probes are: (1) the design of synthesis of reporters that have excitation and emission wavelength above 500 nm, (2) low molecular weight, (3) photostability, and (4) perhaps most important water solubility.

As a highly photostable and fluorescent probe, naphthalic anhydrides and their derivatives have been widely used for fluorescent tags and receptor antagonists [43, 44]. Of particular interest, when appropriately substituted at both the naphthalic and phenyl rings of *N*-aryl-1,8-naphthalimide, a clear dual fluorescence was observed [45, 46]. For instance, by introducing a nitro group into the naphthalic anhydride ring, two emission bands(430 nm/550 nm, respectively) of the dye molecule were reported by our group.

A bis-boronic acid-based probe was first synthesized by Shinkai et al. [47] when 3-aminophenylboronic acid was added into a protoporphyrin system. At pH 10.5, in the presence of fructose, the fluorescence signal of this probe could be increased 100-fold. In addition, other studies [48, 49] have shown that bis-boronic acid probe designs exhibit higher binding affinity specific for glucose than monoboronic acid probes by comparing K_d values of both sensors, 10^{-5} to 10^{-4} M for bis-boronic acid probe and 10^{-3} to 10^{-2} M for monoboronic acid probe, respectively. Unlike changing distances between the boronic acids through synthetic modifications of bis-boronic acid probes, recent investigations on simple monoboronic acid sensors showed that fluorophores and substituents on fluorophores also could contribute to saccharide selectivity [49]. Therefore, more complicated synthetic schemes of bis-boronic acid probes could be avoided by employing appropriate substituents of fluorophore, without losing saccharide binding efficiency.

7 Results

In this work, we synthesized two different *N*-phenylnaphthalimide-based monoboronic acid sensors. Upon consideration of the structurally related Lucifer yellow dye and its high water solubility [50], we utilized the 4-sulfo potassium salt group of 1,8- naphthalic anhydride. By changing substituted –$B(OH)_2$ positions on the phenyl ring, we investigated steric effect on saccharide binding of different probes and structural configuration during this procedure.

To explore fluorescent properties through sugar binding, we synthesized two sensors **9–10**, by using a common fluorophore 4-sulfo-1,8-naphthalic anhydride as a starting material. 3-Aminophenylboronic acid and 2-aminophenylboronic pinacol ester were selected for investigation of isomeric effects on different boronic acid binding pathways with sugar. Sensor **9** was prepared in single-step through reaction of commercial potassium 4-sulfo-1,8-naphthalic anhydride and 3-aminophenylboronic acid hemisulfate. The remaining sensor was prepared by two steps. After reaction between naphthalic anhydride and aminophenylboronic acid, the product was treated with H_2O_2 for 1 h and sensors **10** was obtained. A synthetic scheme for construction of those isomers is given in Scheme 3. High water solubility and significant emission signal change are observed in our experiments.

Beyond our expectations, both saccharide sensors displayed a greater optical sensitivity to glucose than fructose. Numerous papers indicate that most of the monoboronic saccharide chemsensors favor fructose more than glucose. Some reports [31–33, 39] indicated that the affinity of probe to fructose is approximately 100 times greater than glucose for monoboronic acid sensors. Probes **9–19** show

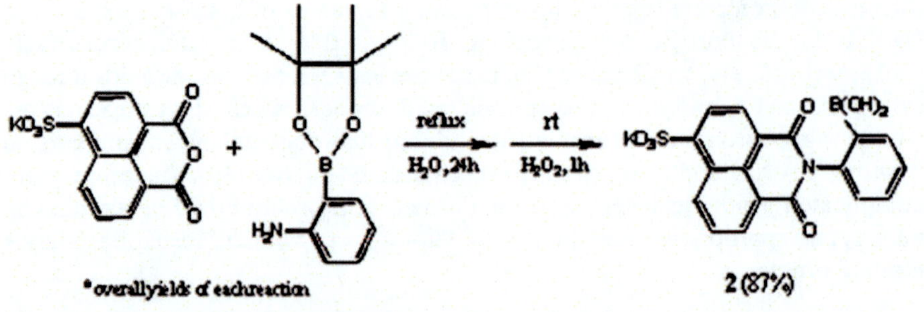

Scheme 3 Synthetic route of *ortho*- and *meta*-phenyl monoboronic acid sensors for saccharides based on 4-sulfo-1,8-naphthalic anhydride

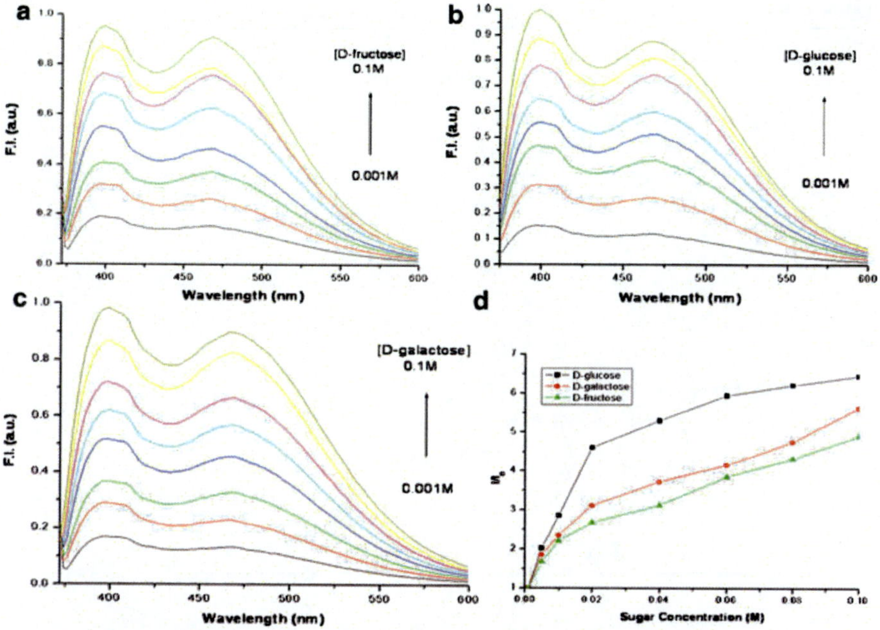

Fig. 6 Fluorescent spectral changes of boronic acid sensor **9** upon addition of different saccharides in phosphate buffer (0.1 M) at pH 8: $\lambda_{ex} = 360$ nm, $\lambda_{1em} = 400$ nm, $\lambda_{2em} = 474$ nm. **(a)** Fructose; **(b)** glucose; **(c)** galactose; **(d)** plots of fluorescent intensity changes of **9** as a function of sugar concentration

large fluorescence intensity changes through chelation-enhanced fluorescence (CHEF) via three most abundant monosaccharides in human blood and receptors interactions, as shown in Fig. 6a–c. Next, we examined the selectivity of these sensors to common monosaccharides at pH = 8 condition. Figure 6d shows the relative fluorescence of **9** at 400 and 474 nm as a function of carbohydrate concentration. Photophysical properties of monoboronic acid sensors **9–10** are listed in Table 3. The increase in fluorescence intensity ratios (*I* in the presence of saccharide, I_0 in the absence of saccharide) for this series shows an increase of about 2–5 times, respectively. Sensor **9** displays an increase in the ratio of intensities for three common monosaccharides, showing the largest increase in fluorescence for glucose (Fig. 7). Similar observations were made by our group previously in which the largest quenching effect took place on glucose [21]. The dissociation constant K_d for fructose was found to be 6.5 mM, while a higher K_d of 28.5 mM was obtained by calculation for glucose at pH 8. Though several bis-boronic acid sensors have been synthesized to favor glucose over fructose among saccharide bindings, their limitations to complex glucose make them less efficient as glucose probes at high blood glucose levels. Consequently, sensor **9** shows an advantage at pH 8 by displaying the largest fluorescence increase to glucose while maintaining an affinity within physiological limits, similar to our sensor

Table 3 Photophysical properties of **9–10** monoboronic acid sensors

Entry	Sensor	λ_{ex} (nm)	λ_{1em} (nm)	λ_{2em} (nm)	ϕ_F
9	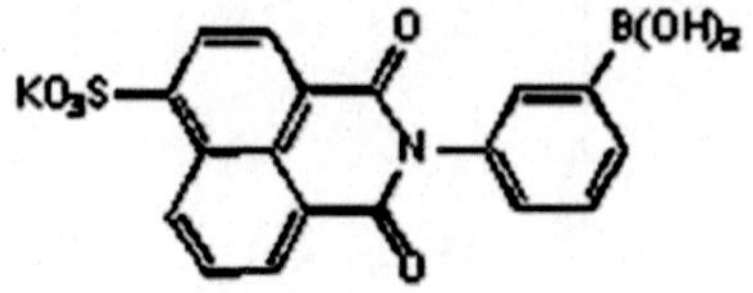 **9**	360	400	474	0.092
10	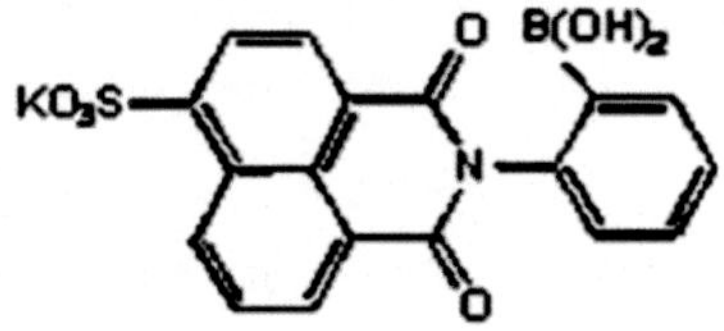 **10**	340	391		0.112

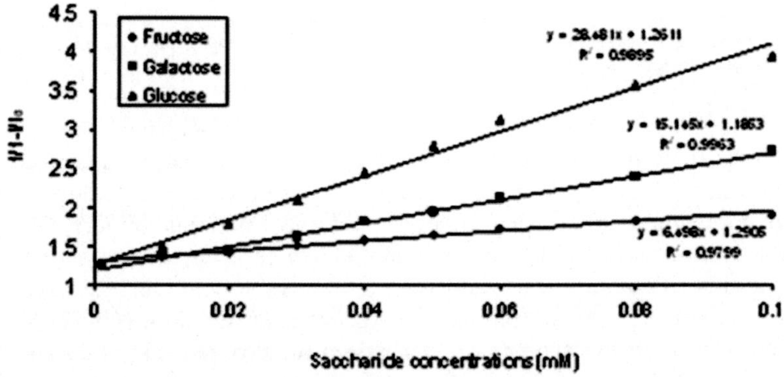

Fig. 7 Prediction of apparent dissociation constants of sensor **9** among different saccharides

synthesized from 3-nitronaphthalic anhydride and 3-aminophenylboronic acid [21]. The incorporation of the boronic acid group in the *meta* position does not lead to significant spectroscopic and photophysical changes in comparison with *ortho* derivative. A relatively smaller dissociation constant was observed for *meta*-derivative (sensor **9**) in the presence of saccharides, which could be explained by the effect of less steric hindrance [22]. Quantum yields for all two probes were within approximately the same order of magnitude. In addition, the apparent dissociation constants for *meta*- (sensor **9**) and *ortho*- (sensor **10**) derivatives are listed in

Table 4 Dissociation constants (K_d) of sensors **9–10** in the presence of monosaccharides

K_d (mmol)			
Probe	D-Fructose	D-Galactose	D-Glucose
9	6.5 ± 0.2	15.2 ± 0.5	28.5 ± 1.2
10	13.2 ± 0.4	22 ± 0.6	35.4 ± 1.5

Table 4. Fluorescence spectra as well as graphical analyses to detail K_d values are also provided in supporting information.

In conclusion, two highly water soluble monoboronic acid sensors that exhibit large fluorescence increases in the presence of monosaccharides show remarkable sensitivity for glucose rather than fructose and galactose. To the best of our knowledge, this is the first highly water soluble monoboronic acid sensor to display the more desirable OFF-ON fluorescence response. By changing position of the boronic acid group from *ortho*- to *meta*-positions of the phenyl ring, there are no significant spectroscopic and photophysical changes in comparison with ortho derivative. Plans are currently underway to extend the absorption wavelength of these dyes to longer wavelength.

References

1. Durrant JR, Haque SA, Palomares E (2006) Photochemical energy conversion: from molecular dyads to solar cells. Chem Commun 37(51):3279–3289
2. Dyakonov V, Sariciftci NS (2003) Organic photovoltaics: concepts and realization, vol 60. Springer, New York
3. Valeur B (2001) Molecular fluorescence: principles and applications. Wiley, New York
4. Klymchenko AS, Demchenko AP (2002) Electrochromic modulation of excited-state intramolecular proton transfer: the new principle in design of fluorescence sensors. J Am Chem Soc 124:12372–12379
5. Klymchenko AS, Ozturk T, Demchenko AP (2002) Synthesis of furanochromones: a new step in improvement of fluorescence properties. Tetrahedron Lett 43:7079–7082
6. Abad S, Kluciar M, Miranda MA, Pischel U (2005) Proton-induced fluorescence switching in novel naphthalimide-dansylamide dyads. J Org Chem 70:10565–10568
7. Badugu R (2005) Fluorescence sensor design for transition metal ions: the role of the PIET interaction efficiency. J Fluoresc 15:71–83
8. Cho DW, Fujitsuka M, Choi KH, Park MJ, Yoon UC, Majima TJ (2006) Photoinduced electron transfer processes in 1,8-naphthalimide-linker-phenothiazine dyads. J Phys Chem B 110:4576–4582
9. Koner AL, Schatz J, Nau WM, Pischel U (2007) Selective sensing of citrate by a supramolecular 1,8-naphthalimide/calix[4]arene assembly via complexation-modulated pK(a) shifts in a ternary complex. J Org Chem 72:3889–3895
10. Miskolczy Z, Nyitrai J, Biczók L, Sebok-Nagy K, Körtvélyesi T (2006) Photophysical properties of novel cationic naphthalimides. J Photochem Photobiol A Chem 182:99–106
11. Takahashi S, Nozaki K, Kozaki M, Suzuki S, Keyaki K, Ichimura A, Matsushita T, Okada K (2008) Photoinduced electron transfer of N-[(3-and 4-diarylamino)phenyl]-1,8-naphthalimide

dyads: orbital-orthogonal approach in a short-linked D-A system. J Phys Chem A 112:2533–2542
12. Frisch MJ et al (2004) Gaussian 03. Gaussian, Wallingford, CT
13. Saha S, Samanta A (2002) Influence of the structure of the amino group and polarity of the medium on the photophysical behavior of 4-amino-1,8-naphthalimide derivatives. J Phys Chem A 106:4763–4771
14. Fromherz P (1995) Monopole dipole model for symmetrical solvatochromism of hemicyanine dyes. J Phys Chem 99:7188–7192
15. Demeter A, Berces T, Biczok L, Wintgens V, Valat P, Kossanyi J (1996) Comprehensive model of the photophysics of N-phenylnaphthalimides: the role of solvent and rotational relaxation. J Phys Chem 100:2001–2011
16. Minta A, Tsien RY (1989) Fluorescent indicators for cystolic sodium. J Biol Chem 264:19449–19457
17. Tsien RY (1989) Fluorescent indicators of ion concentrations. Meth Cell Biol 30:127–156
18. Lakowicz RJ (2006) Principles of fluorescence spectroscopy, 3rd edn. Springer, New York, pp 634–644
19. Szmacinski H, Lakowicz RJ (1999) Potassium and sodium measurements at clinical concentrations using phase-modulation fluorometry. Sens Actuators B Chem 60:8–18
20. Liu HL, Zhang H, Li FA, Xie WJ, Jiang BY (2006) Intramolecular charge transfer dual fluorescent sensors from 4-(dialkylamino)benzanilides with metal binding site within electron acceptor. Tetrahedron 62:10441–10449
21. Malval PJ, Lapouyade R (2001) Derivatization of 4-(dimethylamino)benzamide to dual fluorescent ionophores: divergent spectroscopic effects dependent on N or O amide chelation. Helv Chem Acta 84:2439
22. Crossley R, Goolamali Z, Sammes PG (1994) Synthesis and properties of a potential extracellular fluorescent-probe for potassium. J Chem Soc Perkin Trans 2:1615–1623
23. Kaneda T, Sugihara K, Kamiya H, Misumi S (1981) Synthetic macrocyclic ligands. 4. Lithium ion-characteristic coloration of a crowned dinitrophenylazophenol. Tetrahedron Lett 22:4407
24. de Silva AP, Eilers J, Zlokarnik G (1999) Emerging fluorescence sensing technologies: from photophysical principles to cellular applications. Proc Natl Acad Sci USA 96:8336
25. Shinkai S, Takeuchi M, Ikeda A (2000) Molecular machines useful for the design of chemosensors. In: Osada Y, De Rossi DE (eds) Polymer sensors and actuators. Springer, Berlin, pp 183–206
26. He H, Mortellaro AM, Leiner PJM, Young TS, Fraatz JR, Tusa K (2003) A fluorescent chemosensor for sodium based on photoinduced electron transfer. J Anal Chem 75:549–555
27. Cosnard F, Wintgens V (1998) A new fluoroionophore derived from 4-amino-N-methyl-1,8-naphthalimide. Tetrahedron Lett 39:2751–2754
28. Cao H, Chang V, Hernandez R, Heagy MD (2005) Matrix screening of substituted N-aryl-1,8-naphthalimides reveals new dual fluorescent dyes and unusually bright pyridine derivatives. J Org Chem 70:4929–4934
29. Demchenko PA (2005) The problem of self-calibration of fluorescence signal in microscale sensor systems. Lab Chip 5:1210–1223
30. Cao H, McGill T, Heagy DM (2004) Substituent effects on monoboronic acid sensors for saccharides based on N-phenyl-1,8-naphthalenedicarboximides. J Org Chem 69:2959–2966
31. James TD, Shinkai S (2002) Artificial receptors as chemosensors for carbohydrates. Top Curr Chem 218:159–200
32. Fang H, Kaur G, Wang B (2004) Progress in boronic acid-based fluorescent glucose sensors. J Fluoresc 14:481–489
33. Finney SJ, Zekveld C, Elia A, Evans TW (2003) Glucose control and mortality in critically ill patients. J Am Med Assoc 290:2041–2047
34. Wentholt IM, Vollebregt MA, Hart AA, Hoekstra JB, DcVrics JH (2005) Comparison of a needle-type and a microdialysis continuous glucose monitor in type 1 diabetic patients. Diabetes Care 28:2871–2876

35. Pickup JC, Hussain F, Evans ND, Rolinski OJ, Birch DJS (2005) Fluorescence-based glucose sensors. Biosens Bioelectron 20:2555–2565
36. James TD, Sandanayake KRAS, Shinkai S (1995) Chiral discrimination of monosaccharides using a fluorescent molecular sensor. Nature 374:345–357
37. Wang W, Gao X, Wang B (2002) Boronic acid-based sensors. Curr Org Chem 6:1285–1317
38. Striegler S (2003) Selective carbohydrate recognition by synthetic receptors in aqueous solution. Curr Org Chem 7:81–102
39. Yan J, Fang H, Wang B (2005) Boronolectins and fluorescent boronolectins: an examination of the detailed chemistry issues important for the design. Med Res Rev 25:490–520
40. Heller A (1999) Implanted electrochemical glucose sensors for the management of diabetes. Annu Rev Biomed Eng 1:153–175
41. Yoon JY, Czarnik AW (1992) Fluorescent chemosensors of carbohydrates – a means of chemically communicating the binding of polyols in water based on chelation enhanced quenching. J Am Chem Soc 114:5874–5875
42. Mader HS, Wolfbeis OS (2008) Boronic acid based probes for microdetermination of saccharides and glycosylated biomolecules. Microchim Acta 162:1–34
43. Yamamoto H, Ori A, Ueda K, Dusemund C, Shinkai S (1996) Visual sensing of fluoride ion and saccharides utilizing a coupled redox reaction of ferrocenylboronic acids and dye molecules. Chem Commun 407–408.
44. Shinmori H, Takeuchi M, Shinkai S (1995) Spectroscopic sugar sensing by a stilbene derivative *i*th push-pull ((OH)(2)B) – type substituents. Tetrahedron 51:1893–1902
45. Gabai R, Sallacan N, Chegel V, Bourenko T, Katz E, Willner I (2001) Characterization of the swelling of acrylamidophenylboronic acid-acrylamide hydrogels upon interaction with glucose by faradaic impedance spectroscopy, chronopotentiometry, quartz-crystal microbalance (QCM), and surface plasmon resonance (SPR) experiments. J Phys Chem B 105:8196–8202
46. Shoji E, Freund MS (2002) Potentiometric saccharide detection based on the pKa changes of poly(aniline boronic acid). J Am Chem Soc 124:12486–12493
47. James TD, Linnane P, Shinkai S (1996) Fluorescent saccharide receptors: a sweet solution to the design, assembly and evaluation of boronic acid derived PET sensors. Chem Commun 281–288.
48. DiCesare N, Pinto MR, Schanze KS, Lakowicz JR (2002) Saccharide detection based on the amplified fluorescence quenching of a water-soluble poly(phenylene ethynylene) by a boronic acid functionalized benzyl viologen derivative. Langmuir 18:7785–7787
49. Murakami H, Nagasaki T, Hamachi I, Shinkai S (1993) Sugar sensing utilizing aggregation properties of a boronic acid appended porphyrin. Tetrahedron Lett 34:6273–6276
50. Coskun A, Akkaya EU (2004) Three-point recognition and selective fluorescence sensing of L-DOPA. Org Lett 6:3107–3109

Z-Scan Fluorescence Correlation Spectroscopy: A Powerful Tool for Determination of Lateral Diffusion in Biological Systems

Martin Štefl, Radek Macháň, and Martin Hof

Abstract The characterization of the dynamics of biological membranes is a topic which currently grasps a high level of attention. Biological membranes are extremely important as they are required for both protection and communication of eukaryotic cells. They also play a key role for transportation of nutrients into and out of the cell. Recent studies have proved that biological membranes are not homogeneous but are instead composed of microdomains, which complicate the precise determination of lateral diffusion coefficients. *Z*-scan fluorescence correlation spectroscopy (*Z*-scan FCS), one of the fluorescence fluctuation methods, is a technique which can be employed to determine lateral diffusion coefficients of membrane lipids and also membrane-associated molecules. Moreover, when *Z*-scan FCS is used in combination with Wawrezinieck diffusion law, lipid rafts in heterogeneous membranes can be monitored. This review is focused firstly on the theory of lateral diffusion in biological systems and secondly on FCS, especially *Z*-scan FCS as a very useful approach for determination of lateral diffusion coefficients in planar systems.

1 Introduction

Biological membranes (also referred as phospholipids bilayers) play a crucial role in living organisms. The structural organization of biological membranes, which consists of a hydrophobic core and a hydrophilic surface, creates an extraordinary system and plays a part in the crucial role of maintaining a nonequilibrium state between the extracellular and intracellular environment of the cell. Biological membranes also control the exchange of energy, ions, substrates, metabolites, and

M. Hof (✉)
J. Heyrovský Institute of Physical Chemistry v.v.i., Academy of Sciences
of the Czech Republic, Dolejškova 3, Prague 18223, Czech Republic
e-mail: martin.hof@jh-inst.cas.cz

C.D. Geddes (ed.), *Reviews in Fluorescence 2009*, Reviews in Fluorescence 2009,
DOI 10.1007/978-1-4419-9672-5_12,

also the information between the intracellular and extracellular environment. In order to fulfill those roles, the membrane contains two lipid leaflets, the inner and outer leaflet, in which various proteins, protein complexes, membrane channels, and membrane receptors are incorporated. Furthermore, membranes associate a wide variety of compounds essential for cellular processes. Phospholipids within the membrane are composed of long hydrophobic acyl chains (that fill the center of the membrane) and smaller hydrophilic head groups, which are exposed to the cell cytosol or extracellular matrix. Since biological membranes are very complex systems, the ongoing investigation of the dynamics and interplay between all the components proves to be very difficult; nevertheless, it is very important to achieve a good understanding of all the involved processes.

The basic concept of the fluid mosaic model proposed by Singer and Nicolson [1] assumes randomly distributed components in the membrane, and thus a free diffusion within the plane of the membrane. However, recent studies have revealed that biological membranes are, in fact, not homogeneous but instead contain microdomains (rafts), which differ in protein and lipid composition and structural and dynamical properties [2–5]. This means that the examination of their structure is more complicated than previously thought. For simplification, two main types of membrane model systems have been introduced; supported phospholipid bilayers (SPBs) and giant unilamellar vesicles (GUVs). SPBs are very stable planar systems, prepared usually by the spreading of small unilamellar vesicles (SUVs) on a hydrophilic surface, such as mica [6, 7]. On the other hand, GUVs represent free-standing bilayers of an approximately spherical shape similar in size to eukaryotic cells and are most commonly prepared by electroformation from lipid films [8–11].

One of the most relevant dynamical parameters describing mobility of molecules in biological membranes is their lateral diffusion coefficient D. It can be determined for lipids, proteins, and other membrane-associated molecules. The diffusion of proteins and lipids is affected by their interactions with the cytoskeleton and the presence of domains. As a result, the dependence of lateral diffusion on lipid and protein composition of the membrane is in the focus of many researchers. Experimental techniques used for the investigation of lateral diffusion are usually based on optical microscopy. The four main approaches are fluorescence recovery after photobleaching (FRAP) [12–16], single particle tracking (SPT) [17–20], image correlation spectroscopy (ICS) [21–23] and fluorescence correlation spectroscopy (FCS) [12, 24–27]. In FRAP, a defined area is bleached by a high power laser pulse, and the recovery of fluorescence in the bleached area is monitored. Lateral diffusion and fraction of immobile molecules are calculated from the recovery rate and the intensity of recovered fluorescence, respectively. Samples with high concentration of fluorophores are usually required in this technique. In contrast to FRAP, the SPT method demands extremely diluted samples. The exact positions of the tracer molecule are recorded by video microscopy, and diffusion coefficients are found by the analysis of trajectories of individual particles. One of the most rapidly expanding methods in the last few years is ICS, but is limited in the fact that it usually provides only information about slow dynamics. However, there is a new robust approach in this field (raster scanning image spectroscopy) which covers the whole temporal range of diffusion dynamics in phospholipid membranes [23, 28]. FCS had initially been developed in 1970s [29, 30], and its basic concept is based on recording and analysis of intensity fluctuations caused by fluorophores moving in and out of a small detection volume. FCS like SPT requires a very small concentration of fluorescent molecules (in range of nM) and the parameters of interest are obtained by a mathematical process known as autocorrelation function analysis. An FCS is a powerful tool for lateral diffusion characterization and was one of the most developing methods in this field in the last few years. However, it can be tricky as the determination of lateral diffusion coefficients requires a nontrivial precise external calibration of the detection volume size. To avoid this calibration, several calibration-free methods have been introduced [31–33].

In this review, we focus on one of the calibration-free methods, especially the Z-scan FCS technique and its application for the monitoring of lateral lipid or protein diffusion. Z-scan FCS is based on performing sets of measurements with the sample positioned in different planes along the *z*-axis in respect to the detection volume. Lateral diffusion coefficients are, then, determined from the dependencies of acquired diffusion times on the position of the focus. Moreover, we try to summarize important approaches to Z-scan FCS and show its possible applications in protein and membrane biology.

2 Lateral Diffusion Characteristics and Determination

2.1 *Lateral Diffusion in Membranes: Basic Theory*

According to theory, the free Brownian lateral diffusion in a two-dimensional system is described by the Einstein relation:

$$\langle r^2(t) \rangle = \left\langle (r(t) - r(0))^2 \right\rangle = 4D\,t, \tag{1}$$

where $\langle r^2(t) \rangle$ is the mean square displacement (MSD), D is a constant called the lateral diffusion coefficient, and t is the time. The process described above is driven only by thermal fluctuations around the equilibrium and so we should call it lateral self-diffusion and strictly distinguish it from lateral diffusion driven by concentration gradients [34, 35]. The phenomenological parameter D was in the focus from the beginning and several theories have been developed to relate it to the microscopic properties of diffusing molecules on and within the phospholipid bilayers. In the case of planar lipid membranes, two distinct cases have been described, these cases depend on the size of the diffusing molecules in proportion to the size of the lipids (which are the basic building blocks of the membrane). The diffusion of molecules which are smaller or similar in size to the lipids is usually theoretically treated by free area theory [36–38] based on a two-dimensional random walk. Every step of a molecule of this size is conditioned by the free area available for the molecule to move into and also by a minimal activation energy (E_a) required to perform the step. The total value of the activation energy is influenced by frictional coupling of lipids (one to each other), surrounding aqueous microenvironment and when speaking about SPBs by interaction of lipids with the solid surface (support) [38]. A theoretical model was derived which relates D to E_a, the area per lipid at a given temperature $a(T)$, and also to the minimal cross-sectional area per lipid molecule a_0. Experimental determination of lipid areas $a(T)$ and a_0 allows determination of E_a from temperature dependencies of D. Although the free area theory is rather simple, the good fits of experimental temperature dependencies of D with the theoretical model justify the assumptions involved in derivation of the model [36, 37]. More recent molecular dynamics simulations also proved the validity (at least qualitative) of the predictions of the free area theory [39]. An interesting prediction of the free area theory is that molecules occupying areas similar to lipids or smaller should diffuse with the same D (because a diffusive step is completed by a lipid molecule filling the void left after the tracer molecule). This is naturally an oversimplification, since the collective fashion of membrane lipid movements allows faster filling of smaller voids, but we may at least assume that all molecules similar in occupied area move with the same D [38, 40].

The diffusion of molecules larger than lipids (i.e., peptides, proteins) is usually described as diffusion in viscous continuum [38, 41, 42]. In this theory, the motion of diffusing molecules is driven by random, fluctuating forces provided by unbalanced collisions with the solvent molecules and is resisted by frictional forces inherent in viscous solvents [38]. The diffusion coefficient is defined as:

$$D = \frac{kT}{f}, \tag{2}$$

where k is the Boltzmann constant and f is a frictional coefficient, which for a spherical particle of radius a in a medium of viscosity η equals to $f = 6\pi\eta a$. The resulting relation corresponds to the Stokes–Einstein equation:

$$D = \frac{kT}{6\pi\eta a}. \tag{3}$$

For the lateral diffusion of molecules with larger radii ($a > 10$ Å) (3) has been extended by Saffman and Delbruck, where the thickness of the membrane is taken into account [43, 44].

It has been found by several studies that the description of diffusion in cellular membranes does not fully conform to (1), but to its modification:

$$\langle r^2(t) \rangle = 4D\,t^{\alpha}, \tag{4}$$

where α is called the anomalous exponent and is limited by the interval [0;1]. The diffusion is referred to as anomalous (or sometimes anomalous subdiffusion) to indicate that smaller values of α correspond to slower diffusion. The lateral diffusion coefficient is, in this case, sometimes denoted as Γ [17, 34, 45, 46]. Theoretical studies have shown that the anomalous diffusion can be a result of a broad distribution of jump times, correlations between diffusing particles or multiple diffusion rates. Moreover, anomalous diffusion in cellular membranes has been explained by lipid–protein binding interactions and by hindrance of diffusion by impermeable obstacles (i.e., immobile proteins, lipid microdomains, and the cytoskeleton) [34, 47–50]. Theoretical and experimental analysis of diffusion in heterogeneous systems has revealed how the characteristic length-scale of the measurements ω and its relation to the characteristic size of the obstacles influence values obtained for diffusion coefficients [34, 47, 51–53]. Locally, the tracer molecule may be able to diffuse freely. However, its diffusion is anomalous when probing a larger area ω. For yet larger values of ω, the diffusion satisfies again the relation (1), but with a lower value of D than in the absence of hindrance. With an increasing fraction of the area occupied by the obstacles, ω at which the transition to normal diffusion occurs is shifted to larger values [50, 51]. When the effects of mobile and immobile obstacles are compared, the influence of mobile obstacles is less pronounced [34, 54].

2.2 *Fluorescence Correlation Spectroscopy*

An FCS, as one of the fluorescence fluctuations methods, was introduced in early 1970s by Magde et al. [29, 30] and is based on a statistical analysis of the timescale of fluctuations of fluorescence intensity $I(t)$. Such dependence is described by normalized autocorrelation function $G(\tau)$:

$$G(\tau) = \frac{\langle I(t)\, I(t+\tau)\rangle}{\langle I(t)\rangle^2}, \tag{5}$$

where the angle brackets represent an average over all values of time t. The shape of the autocorrelation function bears information on the timescale of fluorescence intensity fluctuations which originate from a small volume element of the sample and are recorded with a high temporal resolution. The intensity fluctuations mainly result from translational diffusion of molecules in and out of the detection volume and photochemical processes like banned intersystem crossing a triplet state. The detection volume (often called confocal volume) is defined by the optics of the experimental setup and is usually on the order of femtoliters [55–57]. In the case of a two-dimensional sample like a planar lipid bilayer, the confocal volume is actually defined by the intersection of the microscope focus and the plane of the bilayer. In that case, it can be described by diffraction limited two-dimensional Gaussian profile [32, 55, 58]. Fluctuations due to lateral diffusion occur on a hundreds of microsecond to a second timescale while fluctuations due to photochemical processes are much faster. This enables one to separate each of their contributions to the recorded time traces [26, 55, 56, 59]. A typical FCS setup consists of one- or two-photon excitation, where the excitation laser beam comes to the microscope objective through a dichroic mirror and is focused directly on the sample. Since samples used in FCS microscopy are usually dissolved in aqueous solution, water immersion objectives with high numerical aperture are used. Emitted light from the sample is collected by the same objective, passes through the dichroic mirror, emission filter and a pinhole, and finally is focused onto a detector. The pinhole on the emission site is supposed to reject out-of-focus fluorescence light but in the case of membrane lateral diffusion measurements, fully opened pinholes or pinholes with higher diameter can be used, since all the fluorescence intensity is supposed to originate from a single plane.

2.3 *Determination of Lateral Diffusion by Conventional FCS*

To obtain the parameters of interest, the recorded autocorrelation function has to be fitted by a theoretical model function. Interesting parameters are usually the diffusion time of fluorescent molecule τ_D (this gives information about the time the fluorophore stays in the detection volume) and average number of fluorescent molecules in the confocal volume N (particle number), respectively. In the case of two-dimensional Brownian diffusion, the theoretical shape of $G(\tau)$ is given by [29, 58]:

$$G(\tau) = 1 + \frac{1}{N} \frac{1}{1 + \left(\frac{\tau}{\tau_D}\right)}. \tag{6}$$

In some cases, relation (6) can include the geometrical factor γ, which was introduced by Thompson, and this changes the definition of particle number N. In case of the two-dimensional sample or the three-dimensional sample of finite extent in one dimension, the geometrical factor γ is equal 0.5 [58]. Since both definitions of particle number are equivalent in planar systems, in this review we do not use γ factor in further equations. When fluorescence fluctuations which are caused by intersystem crossing to a nonfluorescent triplet state are taken into account, an average fraction of fluorophores in triplet state T and intersystem crossing relaxation time τ_T need to be incorporated into (6) [7, 60, 61], and thus the autocorrelation function has to be fitted by:

$$G(\tau) = 1 + \left[1 - T + T\exp\left(\frac{-\tau}{\tau_T}\right)\right] \frac{1}{N(1-T)} \frac{1}{1 + \left(\frac{\tau}{\tau_D}\right)}. \tag{7}$$

If more than one fluorophores that have different diffusion times contribute to the recorded fluctuations, (7) is changed to:

$$G_M(\tau) = 1 + \frac{\sum_{i=1}^{M} \left(\frac{Q_i}{Q_1}\right)^2 F_i g_i(\tau)}{N\left[\sum_{i=1}^{M} \frac{Q_i}{Q_1} F_i\right]^2}, \quad (8)$$

where F_i is the fraction of fluorophores diffusing with diffusion time τ_{Di} [58, 62], Q_i is the detection efficiency of each fluorescent species and depends on their quantum yields, fluorescence lifetime, and spectral properties, and $g_i(\tau)$is defined as:

$$g_i(\tau) = \frac{1 - T_i + T_i \exp\left(\frac{-\tau}{\tau_{Ti}}\right)}{1 - T_i} \frac{1}{1 + \left(\frac{\tau}{\tau_{Di}}\right)}. \quad (9)$$

Generally, when more components M contribute to the recorded fluctuations, the probability of errors introduced to the fits of autocorrelation functions using (8) increases because of a higher number of free parameters [63, 64].

When using an appropriate theoretical fitting model (6)–(8), the parameters of interest (usually τ_D and N) can be determined from experimentally obtained autocorrelation functions $G(\tau)$ via nonlinear fitting. A plot of reciprocal values of $G(\tau)$ as a function of τ yields a linear form of (6) and the slope and intercept provide a measure of τ_D [58, 65]. The determination of the accuracy of diffusion coefficients is rather complicated because of the highly nonlinear relationship between diffusion coefficients extracted from FCS and measured fluorescence fluctuations [66, 67]. To clear up such problem, several methods have been proposed [67–69]. The signal-to-noise ratio in FCS is the highest when there is on average about one fluorescent molecule in the detection volume [31, 70]. This means that for very low concentrations (in range of tens of nM), when on average only one fluorescent molecule diffuses in the confocal volume, the FCS provides the most relevant information, and thus FCS is sometimes referred to as a single-molecule technique. However, it should be noted that the autocorrelation curve is always calculated from fluctuations caused by a large number of molecules [26, 71, 72]. In the case of two-photon FCS or several other special techniques [53, 71, 73–75], when the size of the confocal volume is reduced, the concentration of fluorescent molecules in detection volume can be higher. For lipid lateral diffusion in the phospholipid bilayers or unilamellar vesicles, the optimal molar ratio of fluorescently labeled tracer to unlabelled native lipid molecules depends on the actual size of the detection area, but for the usual confocal setup varies from 0.5×10^{-5} to 2×10^{-5} [31, 76–78].

The physically most relevant parameter describing lateral diffusion is the lateral diffusion coefficient D and can be extracted directly from τ_D using the following relation:

$$D = \frac{\omega^2}{4\tau_D}, \tag{10}$$

where ω is the radius of the detection volume in the plane of diffusion and is defined as the radial distance from the optical axis, at which the intensity drops by e^{-2} [31, 58, 79]. The surface concentration of fluorescent molecules c_S can be calculated in a similar manner from number of particles in the detection volume N. For a precise determination of diffusion coefficient D, the knowledge of confocal volume radius ω is critical. To obtain this information, a reference FCS measurement of a fluorophore with well-defined diffusion coefficient is applied [25, 80]. As a typical reference, fluorophore rhodamine B or rhodamine 6G were used for many years, but recently rhodamine derivate 5-TAMRA has been found out more suitable for the ω calibration [81]. The calibration measurement is typically performed in a solution, and ω is achieved by fitting the measured autocorrelation function with a model for three-dimensional diffusion assuming a three-dimensional Gaussian shape of the detection volume [25, 59]. Although the radius of the beam waist is calibrated, the real shape of the detection volume may differ from the assumed one (because of beam astigmatism, refractive index mismatch, and other artifacts), and furthermore the radius of the beam waist in the reference solution may vary from that in the sample of interest (because of differences in refractive indices). In such cases, the calibration procedure introduces errors to the determination of D by FCS [26, 31, 55, 56, 82].

2.4 *Inhomogeneities in Planar Membranes*

In FCS analysis, when some impermeable obstacles in the measured system are present, one has to take into account anomalous diffusion, and the anomalous exponent α defined in (4) has to be included in the appropriate fitting model of autocorrelation function. The term τ/τ_D in (6)–(8) has to be replaced by the term $(\tau/\tau_D)^\alpha$ [63, 83–85]. Replacement in (6) leads to the (11):

$$G(\tau) = 1 + \frac{1}{N}\,\frac{1}{1 + \left(\frac{\tau}{\tau_D}\right)^{\alpha}}. \tag{11}$$

In some cases, measured diffusion may be considered as normal even for heterogeneous membranes, particularly when the concentration of inhomogeneities is low and their size is small enough in comparison to the ω value, which is possible to achieve by the given experimental setup. Wawrezinieck et al. and others theoretically investigated such cases and formulated a new analytical model called the FCS diffusion law providing some information on lateral organization of membrane obstacles which have size under the resolution of optical microscopy [83, 86, 87]. This approach takes an advantage from FCS and is based on changing of the detection volume waist radius ω (and thus, the measurement area which is proportional to ω^2) and on analysis of the dependence of the time the tracer molecule takes to diffuse through the detection area (diffusion time τ_D, characteristic for each tracer molecule) on ω^2. For large enough values of ω (for which the diffusion is normal), the dependence is linear [53, 83], and thus can be described by the relation:

$$\tau_D = t_0 + \frac{\omega^2}{4D_{\text{eff}}}, \tag{12}$$

where the intercept t_0 can reach all possible values. If the diffusing molecule is driven by free Brownian diffusion and no inhomogeneities are in the detection area, t_0 equals to 0. Nonzero values of t_0 indicate that the diffusion is hindered. The apparent diffusion coefficient measured for a single value of ω is then different from the effective diffusion coefficient D_{eff}. Two cases of hindered diffusion were investigated, which are most likely encountered in the cellular membranes [83]. The first model consists of isolated lipid microdomains, which diffusing molecules enter with a certain probability (confinement strength S_{conf}) and in which the tracer molecules undergo slower movement. It is believed that the domains are identical disks of radius r and are randomly distributed over the surface. The second system can be considered as a meshwork, which divides the membrane into domains separated by barriers. This situation may be representative of the diffusion of transmembrane proteins in cytoskeletal network. It has been shown that t_0 is positive in the case of isolated microdomains and negative in the case of a meshwork. Moreover, its magnitude yields information on dimensions of the obstacles. The diffusion is normal for $10\omega^2 > \rho^2$ or $\omega^2 > 2\sigma^2$, where ρ and σ are radii of the microdomains or characteristic mesh size, respectively [53, 83, 86]. In Fig. 1, two cases of the dependencies of τ_D on ω^2 are schematically depicted. Presence of fully impermeable obstacles only decreases diffusion coefficient and does not lead to nonzero values of t_0. Several models have been proposed to find the relation between the effective diffusion coefficient D_{eff} in the presence of impermeable obstacles, the fraction of surface area f occupied by the obstacles, and the diffusion coefficient in the absence of obstacles D_0 [35, 83, 88, 89]. According to the effective medium theory, $D_{\text{eff}} = (1 - 2f)\, D_0$ for values of $f \leq 0.2$ [89, 90], but other authors claim that the relation $D_{\text{eff}} = (1 - f)\, D_0$ is more realistic especially for mobile obstacles [88].

2.5 Typical Problems in Lateral Diffusion Coefficients Determination

In FCS measurements, the intensity of the excitation light is supposed to be high enough to reach sufficient molecular brightness (the average number of photons detected per fluorophore per unit of time); but on the other hand, it should be low

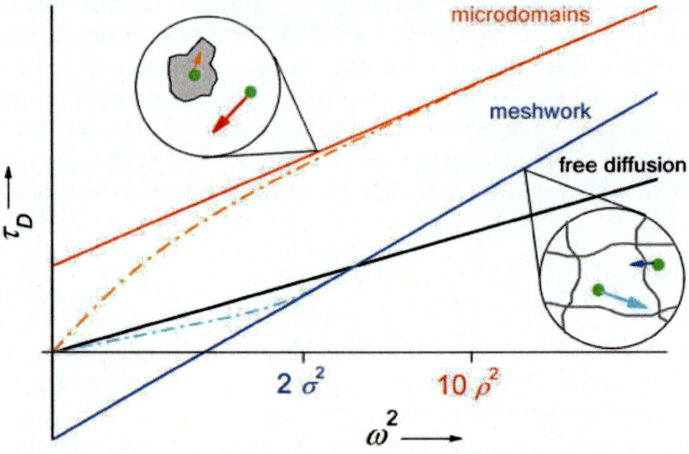

Fig. 1 A schematic depiction of the diffusion law (diffusion time τ_D versus detection area ω^2) for free Brownian diffusion and the two cases of hindered diffusion discussed in [83]: cytoskeleton meshwork and isolated microdomains with dynamic partitioning of tracer molecules. See text for details and meaning of σ and ρ

enough to avoid artifacts caused by photobleaching and optical saturation. The latter occurs when an increased number of molecules within the detection volume are not in the ground state but in an excited or triplet state. This leads to a loss of proportionality between excitation and fluorescence intensities [55, 56, 91]. Moreover, higher powers of the excitation intensity lead to higher values of apparent lateral diffusion coefficients (Fig. 2). Signal-to-noise ratio is strongly dependent on the molecular brightness, the higher the molecular brightness, the higher the signal-to-noise ratio is; a tenfold reduction of excitation power would result in a need for approximately a hundred times longer measurement to reach a comparable statistical accuracy [70, 91, 92]. The maximal excitation intensity at which no photobleaching or saturation artifacts appear is influenced by the photophysics of the fluorophore, ambient conditions and on the average time the molecule spends in the illuminated area (which is dependent on the detection area size and the tracer diffusion rate) [93]. More specifically, for small organic molecules the excitation intensity is supposed to be lower than 30 kW cm^{-2}. Such a value corresponds to the excitation power of approximately 100 μW [55, 91]. Moreover, several authors have found excitation powers below 10 μW optimal for FCS measurements in lipid membranes [76, 78, 94, 95]. In the case of measurement of lateral diffusion coefficients in cellular membranes or in gel-phase domains of model membranes, the photobleaching becomes more important because of slower diffusion of tracer molecule in the detection volume (since the fluorophore resides in the illuminated area for a longer time) [31, 46, 92, 96–98]. Another example of problems with photobleaching and saturation effects is two-photon FCS, where higher intensities

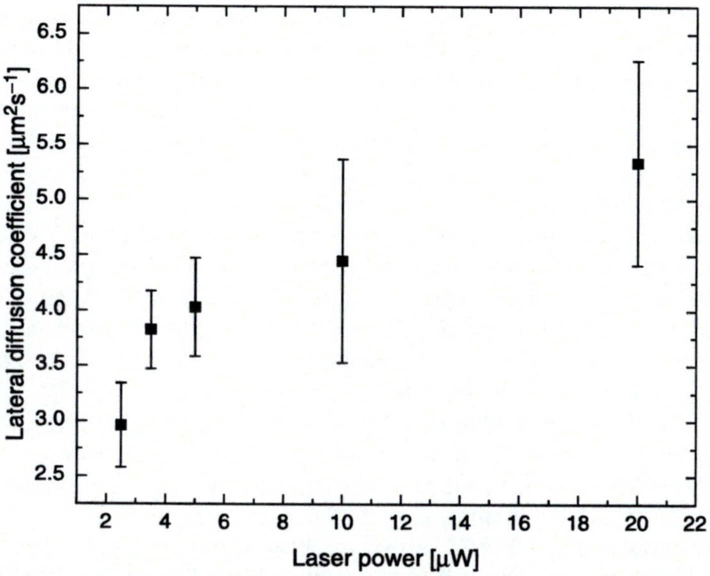

Fig. 2 Changes in lateral diffusion coefficients *D* with different excitation laser intensity. Data collected on the supported phospholipid bilayers (SPBs) containing 1,2-Dioleoyl-sn-Glycero-3-Phosphocholine (DOPC) and 1,2-Dioleoyl-sn-Glycero-3-Phosphoserine (DOPS) in ratio 9:1. For visualization atto488-1,2-Dioleoyl-sn-Glycero-3-Phosphoethanolamine (atto488-DOPE) was used

and pulsed excitation sources are used out of necessity [92, 99] (although it is known to cause less photobleaching outside the focus when compared to standard confocal setup [71, 92, 100, 101]).

To overcome the problems with photobleaching and optical saturation effects, several FCS approaches (scanning FCS (sFCS), ICS) have been developed [33, 96, 102, 103]. When scanning the sample, the time for which the fluorophore is illuminated is reduced, and thus the photobleaching and saturation effects are not introduced in the resulting data. sFCS is based on scanning the excitation laser beam in a line [24, 98, 102, 104] or in a circle [33, 105, 106] across the sample (phospholipid bilayer). FCS data is recorded at each point along the line or circle during the scan. The first instrumentation utilized fixed illumination and a translating sample [102, 107], but nowadays the scanning is usually based on a fixed sample and a mobile excitation light [105, 106, 108]. sFCS is mostly utilized in cases, where the exact localization of the laser beam in a target area is difficult (i.e., membranes). By scanning across the membrane, the laser beam passes through the target area, and if a circular scan is applied, one is sure that the excitation light crosses the membrane twice during each scan [22]. Another possibility to characterize very slow diffusion times in the membranes is ICS, which is considered to be the imaging analog to FCS [21]. The spatial autocorrelation functions are calculated from images of fluorophores acquired on the laser scanning microscope [109].

Several extensions to ICS have been recently developed. Utilization of k-Space ICS (kICS) enables characterization of the dynamics (diffusion and flow) and totally overcomes problems with photobleaching or instable fluorescence signal. Temporal ICS (TICS) correlates an image series in time to provide dynamics, number densities and immobile fraction of fluorophores which are on the timescale of the measurement. On the other hand, raster ICS (RICS) allows measurement and analysis of rapid diffusion, which has not yet been achieved by other ICS methods [28, 110]. Other variants of ICS methods have been developed in the last decade (i.e., ICCS, STICS, PICS) [103, 111, 112], but their detailed description is behind the scope of this review.

In some cases, lateral diffusion of molecules dynamically associating with the membrane is in the focus of interest. In conventional FCS, the absolute axial extension of the detection volume is typically larger than 1 μm; and when the affinity of interacting molecules is low, the fluorescence from unbound molecules contributes to the recorded FCS autocorrelation functions and makes the analysis of such data very complicated. To decrease the axial extension and thus eliminate signal from unattached fluorescent molecules, surface confined FCS techniques [total internal reflection FCS (TIR-FCS), supercritical angle FCS (SA-FCS), and zero mode waveguides] have been developed. In TIR-FCS, an evanescent wave is used for excitation of a small slice above the interface (the evanescent depth is usually lower than ~100 nm) [26, 113, 114]. The disadvantage of this approach is that a significant out-of-focus photobleaching reduces the fluorescence of the fluorophores before they can enter the detection area. Surface confinement achieved by collecting exclusively surface-generated fluorescence emitted into supercritical angles is a basic principle of SA-FCS [115]. Zero mode waveguides are subwavelength optical nanostructures, such as small holes (35–200 nm in diameter) in which membranes can enter. The decay lengths of the evanescent field depend on the diameter of the hole and usually range from 15 to 35 nm [26, 73, 116]. An alternative technique to surface confined FCS methods is Z-scan FCS which is discussed later in this review.

Determination of lateral diffusion coefficients in planar samples using FCS is strongly dependent on the axial position of the sample in the confocal volume. Since the usually used planar systems are very thin (approximately 5 nm in the case of phospholipid bilayers), wrong positioning in the detection volume leads to larger detection area (which increases N and τ_D) and consequently introduces errors to determine lateral diffusion coefficients [26, 55]. It has been shown that positioning the sample by searching for the highest fluorescence intensity does not guarantee reproducibility because the beam waist does not necessarily coincide with the highest fluorescence count rate (see Fig. 3c and read Sect. 3) [31]. Milon et al. have referred about the inaccuracies associated with external calibration and found out that the measured values of D may differ by factor 2 or larger depending on the alignment of sample and focus, and on membrane topography [117]. To avoid such problems, several groups have worked on investigation of calibration-free FCS approaches, such as two-focus FCS, where calibration is not necessary because of the precisely known distance between two foci [32, 118, 119], Z-scan FCS [31] (detection volume is scanned along the z-axis) or various other forms of scanning FCS (mentioned above) [24, 33, 102, 105].

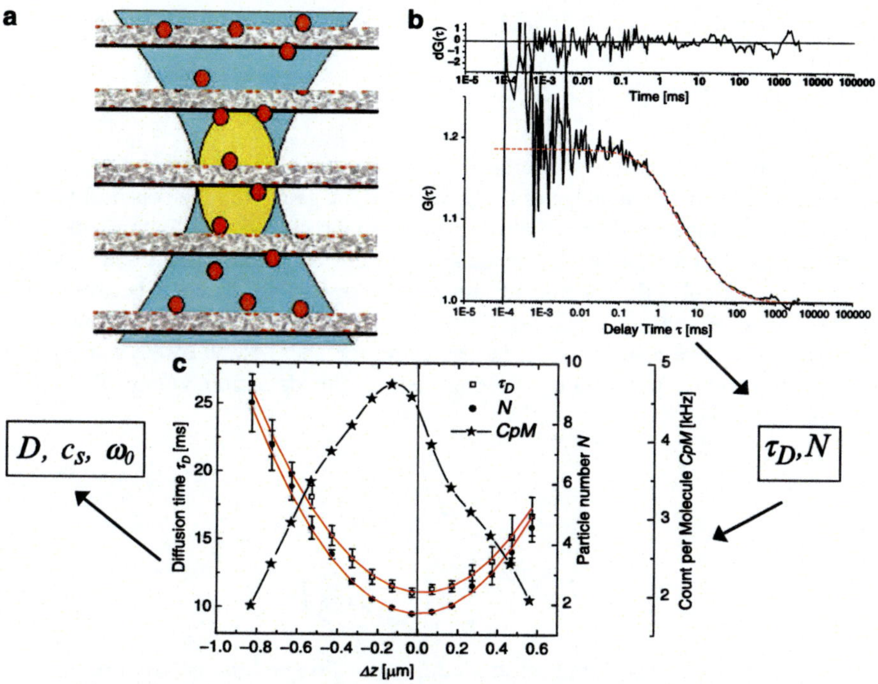

Fig. 3 A schematic illustration of the principle of z-scan FCS; various FCS measurements of the sample are taken along the z-axis in different axial positions (**a**) and fitted by appropriate mathematical model (**b**). Calculated values of τ_D and N are plotted versus relative z-position in the focus (**c**), and thus parameters of interest D, c_s, ω_0 are obtained from the parabolic fit

2.6 Dual-Color FCS and Cross-Correlation Analysis

Study of enzyme–substrate interactions is rather difficult to achieve by conventional FCS because of small differences in molecular mass and hence in diffusion times between enzyme and substrate molecules. For this purpose, dual-color cross-correlation spectroscopy has been developed [111, 120–124], where the enzyme and substrate molecules are labeled with different colors. In instrumentation setup, the sample is excited by two lasers with different wavelength, and emitted light is consequently divided by a dichroic mirror to two detectors. Fluorescence fluctuations obtained from each detector are then cross-correlated and information about interactions of two different species can be extracted from resulting cross-correlation function. Complications with dual-color FCS setup are usually connected with inaccurate overlap of two foci and with finding the best mathematical model for cross-correlation analysis.

3 Z-Scan FCS

3.1 The Principle of Z-Scan FCS

For precise determination of lateral diffusion coefficients and fluorophore surface concentrations, the knowledge of the position of the sample in the confocal volume is critical, and thus the Z-scan approach has been developed [31]. In this technique, the individual autocorrelation functions $G(\tau)$ are acquired in different positions along the optical axis (z-axis) of the microscope. The distance between each measurement along the z-axis (step-size) is typically in the interval from 0.1 to 0.2 μm. When the planar system is parallel to the focal plane of the microscope, diffusion time τ_D and number of particles in the detection volume N can be expressed by the relations:

$$\tau_D(\Delta_Z) = \frac{\omega_0^2}{4D}\left(1 + \frac{\lambda^2 \Delta_Z^2}{\pi^2 \omega_0^4}\right), \tag{13}$$

$$N(\Delta_Z) = N_0 \left(1 + \frac{\lambda^2 \Delta_Z^2}{\pi^2 \omega_0^4}\right), \tag{14}$$

where Δ_Z is the distance between the sample position and the position of the laser beam waist, λ is the excitation wavelength in the sample, ω_0 is the radius of the beam in the focal plane, and $N_0 = \pi {\omega_0}^2 c_s$ is the particle number in the waist of the focus, where the e^{-2} radius of the detection area equals ω_0 [31, 58, 125]. In a typical experiment, the parameters of interest τ_D and N are calculated from the fit of autocorrelation functions (according to (7) or (8) in dependence on the number of fluorophores) obtained for each z-position. Parabolic dependencies of τ_D and N on Δ_Z are subsequently fitted by (13) and (14), respectively, which result in a relevant determination of the lateral diffusion coefficient D, the average concentration of diffusing fluorescence molecules in the illuminated area c_s, and also the radius and z-position of the beam waist ω_0 [31, 58]. Illustration of the principle of D determination on phospholipid bilayer using Z-scan FCS is summarized in Fig. 3.

3.2 Approaches to Z-Scan FCS

Up to now, Z-scan FCS has been introduced in many applications, especially in characterization of biological or model systems like cell membranes, GUVs or SPBs. An interesting approach has been reported by Humpolickova et al. [87] which combines the FCS diffusion law [83] with Z-scan FCS. This technique is based on the combination of (12)–(14) and the resulting relation is:

$$\tau_D = t_0 + \frac{{\omega_0}^2}{4D_{\text{eff}}} \frac{N}{N_0}. \tag{15}$$

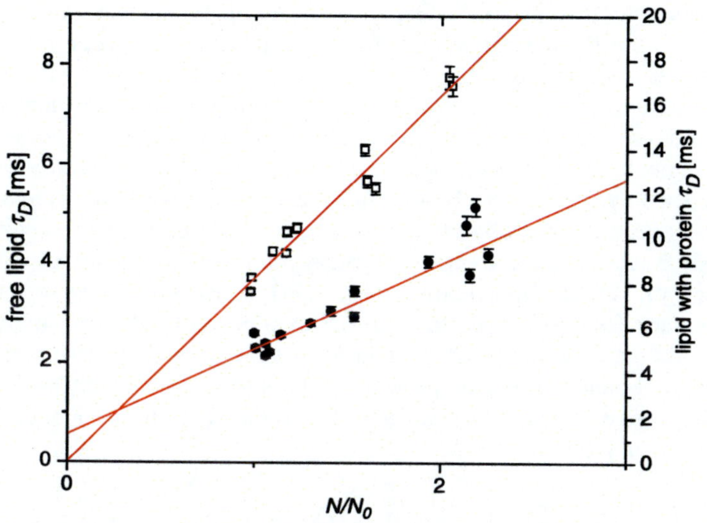

Fig. 4 Plot of diffusion times for the membrane composed of DOPC (90%), DOPS (10%), and atto488-DOPE (10^{-3}%) in dependence on relative particle number N/N_0. Open squares correspond to the free membrane, closed circles to the bilayer with bound protein prothrombin (concentration of prothrombin $c_{prot} = 4$ μM)

This provides a linear dependence of τ_D (Δ_Z) on N (Δ_Z)/N_0 (sometimes called "Humplot") [87, 126, 127]. According to the FCS diffusion law, the position of the intercept t_0 on y-axis gives information about the obstacles present in the membrane (Fig. 1). Note that using the diffusion law theory yields information about the membrane structure at a scale below the diffraction limit (read Sect. 2 for more information). A classic example of "Humplot" is depicted in Fig. 4, where diffusion times for phospholipid bilayers with and without bound protein molecules are compared. When protein is not bound to SPBs, there are no restrictions which affect diffusion, thus the diffusion is considered as free and the intercept t_0 equals 0. In the presence of bound protein, t_0 is positive and according to the theory of FCS diffusion law, interaction of protein molecules with lipid molecules creates slower diffusing microdomains in the membrane.

Shenoy et al. [128] investigated sparsely and densely tethered bilayer lipid membranes (stBLMs and dtBLMs) in which the lipid bilayer is separated from the support by spacer molecule. Using Z-scan FCS, they revealed differences in the lateral diffusion coefficients in proximal ($D = 2.0$ μm^2 s^{-1}) and distal ($D = 7.1$ μm^{2}s^{-1}) leaflets of the membrane when measured separately. On the other hand, lateral diffusion coefficients determined for membranes in which both leaflets contribute to the FCS signal provide single diffusion time and gain values similar to the previously published data. The explanation for this contrast is in different label density in the proximal and distal leaflets and thus a smaller contribution to the decay amplitude of $G(\tau)$ for the proximal leaflet.

When comparing GUVs and SPBs, Przybylo et al. found that the diffusion in GUVs to be more than two times higher than that in SPBs, which suggests coupling of the bilayer with support [76].

Study of dynamics of molecules weakly associated with the membrane is always complicated because of the presence of the unbound fraction of molecules in the sample. In the past, very sophisticated method (TIR/FPPR) has been used to describe such dynamics [13]. Dual-color Z-scan FCS is a simple alternative and has already been successfully applied to the study of lateral diffusion coefficients of weakly bound blood coagulation protein prothrombin associated with SPBs composed of various lipid compositions [95]. Furthermore, using dual-color instrumentation enabled simultaneous observation of lipid and protein lateral diffusion. In the analysis of obtained autocorrelation functions, two-dimensional diffusion of bound molecules as well as three-dimensional diffusion of free molecules has to be taken into account. For such cases, fitting model has been proposed [31, 63, 95]:

$$G(\tau) = 1 + \left[1 - T + T\exp\left(\frac{-\tau}{\tau_T}\right)\right]\frac{1}{(1-T)}$$

$$\left[\frac{A_3}{1+\left(\frac{\tau}{\tau_{D3}}\right)}\frac{1}{\left[1+\frac{\tau}{\tau_{D3}}\left(\frac{\omega_0}{\omega_Z}\right)^2\right]^{\frac{1}{2}}} + \frac{A_2}{1+\left(\frac{\tau}{\tau_{D2}}\right)}\right], \tag{16}$$

where τ_{D2} and τ_{D3} correspond to the diffusion times of molecules diffusing in two and three dimensions, respectively, ω_Z is the characteristic axial dimension of the detection volume and the parameters A_2 and A_3 are amplitudes (degree of contribution) of associated and free fluorescent molecules. τ_{D3} is determined by a measurement in bulk solution before each Z-scan and the amplitudes A_2 (Δ_Z), A_3 (Δ_Z), and the z-position dependent lateral diffusion time τ_{D2} (Δ_Z) are found by fit of the autocorrelation functions obtained during a Z-scan using (16). The dependence of the amplitudes A_2 and A_3 on Δ_Z is nontrivial and reflects the changes in the fraction of membrane-associated fluorescent molecules and also in the ratio of the detection efficiencies of molecules both in aqueous phase and molecules associated with the membrane. For further analysis, the values of the amplitude A_2 and particle number N measured at a well-defined position, i.e., when the membrane is in the focus ($\Delta_Z = 0$) are used and these are plotted versus the protein concentration. Fit with specially corrected Langmuir isotherm directly provides value for dissociation constant K_D [95].

3.3 Challenges in Z-Scan FCS

Because the autocorrelation curves have to be recorded in several positions defined with at least 100 nm accuracy, characterization of very slow diffusion by Z-scan FCS can be affected by temporal instability of the sample position in the confocal volume [26]. Moreover, thermal undulations of free-standing membranes (GUVs) also add to the temporal instability [117]. Small movements of the membrane in the z-axis can then introduce an apparent additional slow diffusion component into the autocorrelation function [26]. Distortions of the detection volume shape cause deviations of $\tau_D(\Delta_Z)$ and $N(\Delta_Z)$ from the assumed parabolic shape (13) and (14) and consequently create another possible source of artifacts in Z-scan FCS.

The position of the focus, especially the axial distance from the microscope objective is different for various excitation wavelengths. For that reason, when performing dual-color Z-scan FCS, the minima of parabolas defined in (13) and (14) are shifted in z-position for each excitation wavelength (Fig. 5) [95]. The shift differs for different excitation wavelength used and represents a problem for dual-color fluorescence cross-correlation spectroscopy in the Z-scan mode. To overcome such problem, the axial position for performing a cross-correlation measurement

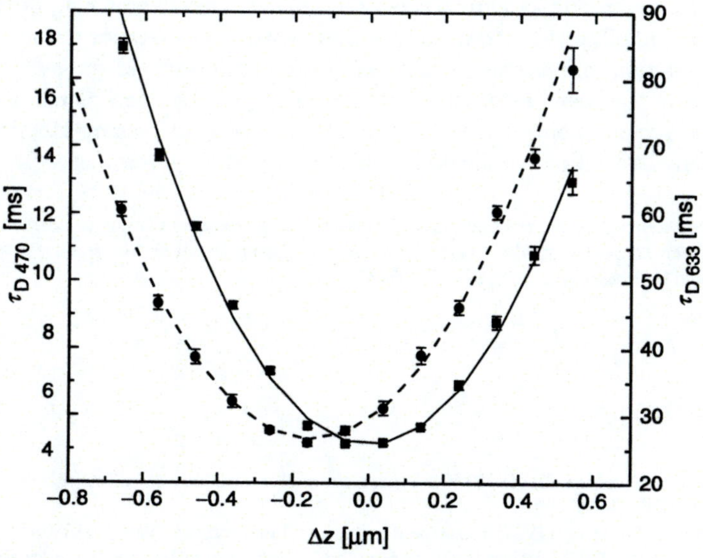

Fig. 5 Dependence of diffusion times on the z-position of the focus for protein labeled by alexa633 (τ_{D633}) and lipids labeled by atto488 (τ_{D488}). The distance between the z-focus positions created by two lasers is about 192 ± 19 nm

has to be selected as a compromise between the minima of the parabolas for the two excitation wavelengths of interest. Other possibility inheres in two photon excitation, which enables excitation of two selected spectrally different dyes with only one infra-red laser line [71].

4 Conclusions

An FCS is a very useful method for investigation of lateral diffusion in planar systems. However, there are a few limitations connected with FCS (the need of external calibration of the detection volume, definition of the exact position of the membrane in the confocal volume and others). When those limitations are not taken into account, inaccurate values of lateral diffusion coefficients are obtained or in the worst case the lateral diffusion determination is almost impossible. *Z*-scan FCS is very comprehensive calibration-free method, which overcomes many problems connected with conventional FCS. Because of the scan along the *z*-axis, exact position of the membrane is monitored and thus lateral diffusion coefficients can be precisely determined. *Z*-scan FCS has been successfully used in investigation of various specific systems where other methods are failing, e.g., diffusion in supported lipid bilayers leaflets, study of membrane microdomains and meshworks of subresolution dimensions, study of lateral diffusion of proteins weakly associated to the membrane, as well as the interplay between membrane lipids and bound protein molecules. Moreover, characterizing of binding kinetics (association and dissociation constants) is achievable by this approach as well.

Acknowledgment The authors thank Mr. Philip Yip for his excellent language corrections. This work has been supported by the Grant Agency of the Czech Republic via grants P208/10/1090 (M. Štefl and R. Macháň) and P208/10/0376 (M. Hof).

References

1. Singer SJ, Nicolson GL (1972) Fluid mosaic model of structure of cell-membranes. Science 175(4023):720–731
2. Simons K, Ikonen E (1997) Functional rafts in cell membranes. Nature 387(6633):569–572
3. Lajoie P et al (2009) Lattices, rafts, and scaffolds: domain regulation of receptor signaling at the plasma membrane. J Cell Biol 185(3):381–385
4. Kusumi A, Koyama-Honda I, Suzuki K (2004) Molecular dynamics and interactions for creation of stimulation-induced stabilized rafts from small unstable steady-state rafts. Traffic 5(4):213–230
5. Lindner R, Naim HY (2009) Domains in biological membranes. Exp Cell Res 315(17): 2871–2878
6. Brian AA, McConnell HM (1984) Allogeneic stimulation of cyto-toxic T-cells by supported planar membranes. Proc Natl Acad Sci USA Biol Sci 81(19):6159–6163

7. Benes M et al (2004) Surface-dependent transitions during self-assembly of phospholipid membranes on mica, silica, and glass. Langmuir 20(23):10129–10137
8. Angelova MI, Dimitrov DS (1986) Liposome electroformation. Faraday Discuss Chem Soc 81:303–312
9. Bagatolli LA, Parasassi T, Gratton E (2000) Giant phospholipid vesicles: comparison among the whole lipid sample characteristics using different preparation methods: a two photon fluorescence microscopy study. Chem Phys Lipids 105(2):135–147
10. Montes LR et al (2007) Giant unilamellar vesicles electroformed from native membranes and organic lipid mixtures under physiological conditions. Biophys J 93(10):3548–3554
11. Pott T, Bouvrais H, Meleard P (2008) Giant unilamellar vesicle formation under physiologically relevant conditions. Chem Phys Lipids 154(2):115–119
12. Adkins EM et al (2007) Membrane mobility and microdomain association of the dopamine transporter studied with fluorescence correlation spectroscopy and fluorescence recovery after photobleaching. Biochemistry 46(37):10484–10497
13. Huang ZP, Pearce KH, Thompson NL (1992) Effect of bovine prothrombin fragment-1 on the translational diffusion of phospholipids in langmuir-blodgett monolayers. Biochim Biophys Acta 1112(2):259–265
14. Ladha S, Mackie AR, Clark DC (1994) Cheek cell-membrane fluidity measured by fluorescence recovery after photobleaching and steady-state fluorescence anisotropy. J Membr Biol 142(2):223–228
15. Gordon GW et al (1995) Analysis of simulated and experimental fluorescence recovery after photobleaching: data for 2 diffusing components. Biophys J 68(3):766–778
16. Sheetz MP (1993) Glycoprotein motility and dynamic domains in fluid plasma-membranes. Annu Rev Biophys Biomol Struct 22:417–431
17. Hong QA, Sheetz MP, Elson EL (1991) Single-particle tracking: analysis of diffusion and flow in 2-dimensional systems. Biophys J 60(4):910–921
18. Saxton MJ (1997) Single-particle tracking: the distribution of diffusion coefficients. Biophys J 72(4):1744–1753
19. Martin DS, Forstner MB, Kas JA (2002) Apparent subdiffusion inherent to single particle tracking. Biophys J 83(4):2109–2117
20. Saxton MJ, Jacobson K (1997) Single-particle tracking: applications to membrane dynamics. Annu Rev Biophys Biomol Struct 26:373–399
21. Kolin DL, Wiseman PW (2007) Advances in image correlation spectroscopy: measuring number densities, aggregation states, and dynamics of fluorescently labeled macromolecules in cells. Cell Biochem Biophys 49(3):141–164
22. Jameson DM, Ross JA, Albanesi JP (2009) Fluorescence fluctuation spectroscopy: ushering in a new age of enlightenment for cellular dynamics. Biophys Rev 1:105–118
23. Brown CM et al (2008) Raster image correlation spectroscopy (RICS) for measuring fast protein dynamics and concentrations with a commercial laser scanning confocal microscope. J Microsc Oxf 229(1):78–91
24. Ries J, Chiantia S, Schwille P (2009) Accurate determination of membrane dynamics with line-scan FCS. Biophys J 96(5):1999–2008
25. Garcia-Saez AJ, Schwille P (2008) Fluorescence correlation spectroscopy for the study of membrane dynamics and protein/lipid interactions. Methods 46(2):116–122
26. Ries J, Schwille P (2008) New concepts for fluorescence correlation spectroscopy on membranes. Phys Chem Chem Phys 10(24):3487–3497
27. Ohsugi Y et al (2006) Lateral mobility of membrane-binding proteins in living cells measured by total internal reflection fluorescence correlation spectroscopy. Biophys J 91 (9):3456–3464
28. Digman MA et al (2005) Measuring fast dynamics in solutions and cells with a laser scanning microscope. Biophys J 89(2):1317–1327
29. Elson E, Magde D (1974) Fluorescence correlation spectroscopy. I. Conceptual basis and theory. Biopolymers 13:1–27

30. Magde D, Elson EL, Webb WW (1974) Fluorescence correlation spectroscopy. 2. Experimental realization. Biopolymers 13:29–61
31. Benda A et al (2003) How to determine diffusion coefficients in planar phospholipid systems by confocal fluorescence correlation spectroscopy. Langmuir 19(10):4120–4126
32. Dertinger T et al (2007) Two-focus fluorescence correlation spectroscopy: a new tool for accurate and absolute diffusion measurements. Chemphyschem 8(3):433–443
33. Petrasek Z, Schwille P (2008) Precise measurement of diffusion coefficients using scanning fluorescence correlation spectroscopy. Biophys J 94(4):1437–1448
34. Ratto TV, Longo ML (2003) Anomalous subdiffusion in heterogeneous lipid bilayers. Langmuir 19(5):1788–1793
35. Saxton MJ (1989) Lateral diffusion in an archipelago: distance dependence of the diffusion-coefficient. Biophys J 56(3):615–622
36. Almeida PFF, Vaz WLC, Thompson TE (1992) Lateral diffusion in the liquid-phases of dimyristoylphosphatidylcholine cholesterol lipid bilayers: a free-volume analysis. Biochemistry 31(29):6739–6747
37. Vaz WLC, Clegg RM, Hallmann D (1985) Translational diffusion of lipids in liquid-crystalline phase phosphatidylcholine multibilayers: a comparison of experiment with theory. Biochemistry 24(3):781–786
38. Vaz WLC, Goodsaid-Zalduondo F, Jacobson K (1984) Lateral diffusion of lipids and proteins in bilayer-membranes. FEBS Lett 174(2):199–207
39. Falck E et al (2004) Lessons of slicing membranes: interplay of packing, free area, and lateral diffusion in phospholipid/cholesterol bilayers. Biophys J 87(2):1076–1091
40. Liu CH, Paprica A, Petersen NO (1997) Effects of size of macrocyclic polyamides on their rate of diffusion in model membranes. Biophys J 73(5):2580–2587
41. Petrov EP, Schwille P (2008) Translational diffusion in lipid membranes beyond the Saffman–Delbruck approximation. Biophys J 94(5):L41–L43
42. Hughes BD, Pailthorpe BA, White LR (1981) The translational and rotational drag on a cylinder moving in a membrane. J Fluid Mech 110:349–372
43. Saffman PG, Delbruck M (1975) Brownian-motion in biological-membranes. Proc Natl Acad Sci USA 72(8):3111–3113
44. Saffman PG (1976) Brownian-motion in thin sheets of viscous-fluid. J Fluid Mech 73:593–602
45. Schutz GJ, Schindler H, Schmidt T (1997) Single-molecule microscopy on model membranes reveals anomalous diffusion. Biophys J 73(2):1073–1080
46. Schwille P, Korlach J, Webb WW (1999) Fluorescence correlation spectroscopy with single-molecule sensitivity on cell and model membranes. Cytometry 36(3):176–182
47. Saxton MJ (1996) Anomalous diffusion due to binding: a Monte Carlo study. Biophys J 70(3):250–1262
48. Saxton MJ (2007) A biological interpretation of transient anomalous subdiffusion. I. Qualitative model. Biophys J 92(4):1178–1191
49. Saxton MJ (2001) Anomalous subdiffusion in fluorescence photobleaching recovery: a Monte Carlo study. Biophys J 81(4):2226–2240
50. Saxton MJ (1994) Anomalous diffusion due to obstacles: a Monte-Carlo study. Biophys J 66(2):394–401
51. Deverall MA et al (2005) Membrane lateral mobility obstructed by polymer-tethered lipids studied at the single molecule level. Biophys J 88(3):1875–1886
52. Saxton MJ (1995) Single-particle tracking: effects of corrals. Biophys J 69(2):389–398
53. Wenger J et al (2007) Diffusion analysis within single nanometric apertures reveals the ultrafine cell membrane organization. Biophys J 92(3):913–919
54. Saxton MJ (1990) Lateral diffusion in a mixture of mobile and immobile particles: a Monte-Carlo study. Biophys J 58(5):1303–1306
55. Enderlein J et al (2005) Performance of fluorescence correlation spectroscopy for measuring diffusion and concentration. Chemphyschem 6(11):2324–2336

56. Enderlein J et al (2004) Art and artefacts of fluorescence correlation spectroscopy. Curr Pharm Biotechnol 5(2):155–161
57. Rigler R (1995) Fluorescence correlations, single-molecule detection and large number screening – applications in biotechnology. J Biotechnol 41(2–3):177–186
58. Thompson NL (1991) Fluorescence correlation spectroscopy. In: Lakowicz JR (ed) Topics in fluorescence spectroscopy. Plenum, New York, pp 337–378
59. Hess ST et al (2002) Biological and chemical applications of fluorescence correlation spectroscopy: a review. Biochemistry 41(3):697–705
60. Schwille P, Oehlenschlager F, Walter NG (1996) Quantitative hybridization kinetics of DNA probes to RNA in solution followed by diffusional fluorescence correlation analysis. Biochemistry 35(31):10182–10193
61. Widengren J, Mets U, Rigler R (1995) Fluorescence correlation spectroscopy of triplet-states in solution: a theoretical and experimental-study. J Phys Chem 99(36):13368–13379
62. Yu L et al (2005) Investigation of a novel artificial antimicrobial peptide by fluorescence correlation spectroscopy: an amphipathic cationic pattern is sufficient for selective binding to bacterial type membranes and antimicrobial activity. Biochim Biophys Acta Biomembr 1716 (1):29–39
63. Donsmark J, Rischel C (2007) Fluorescence correlation spectroscopy at the oil-water interface: hard disk diffusion behavior in dilute beta-lactoglobulin layers precedes monolayer formation. Langmuir 23(12):6614–6623
64. Provencher SW (1982) A constrained regularization method for inverting data represented by linear algebraic or integral-equations. Comput Phys Commun 27(3):213–227
65. Zhang LF, Granick S (2007) Interleaflet diffusion coupling when polymer adsorbs onto one sole leaflet of a supported phospholipid bilayer. Macromolecules 40(5):1366–1368
66. Enderlein J et al (2005) Statistical analysis of diffusion coefficient determination by fluorescence correlation spectroscopy. J Fluoresc 15(3):415–422
67. Koppel DE (1974) Statistical accuracy in fluorescence correlation spectroscopy. Phys Rev A 10(6):1938–1945
68. Wohland T, Rigler R, Vogel H (2001) The standard deviation in fluorescence correlation spectroscopy. Biophys J 80(6):2987–2999
69. Qian H (1990) On the statistics of fluorescence correlation spectroscopy. Biophys Chem 38 (1–2):49–57
70. Kask P, Gunther R, Axhausen P (1997) Statistical accuracy in fluorescence fluctuation experiments. Eur Biophys J Biophys Lett 25(3):163–169
71. Schwille P, Heinze KG (2001) Two-photon fluorescence cross-correlation spectroscopy. Chemphyschem 2(5):269–272
72. Guo L et al (2008) Molecular diffusion measurement in lipid bilayers over wide concentration ranges: a comparative study. Chemphyschem 9(5):721–728
73. Samiee KT et al (2006) Zero mode waveguides for single-molecule spectroscopy on lipid membranes. Biophys J 90(9):3288–3299
74. Blom H, Kastrup L, Eggeling C (2006) Fluorescence fluctuation spectroscopy in reduced detection volumes. Curr Pharm Biotechnol 7(1):51–66
75. Ringemann C et al (2009) Exploring single-molecule dynamics with fluorescence nanoscopy. New J Phys 11:29
76. Przybylo M et al (2006) Lipid diffusion in giant unilamellar vesicles is more than 2 times faster than in supported phospholipid bilayers under identical conditions. Langmuir 22(22): 9096–9099
77. Benda A et al (2006) Fluorescence lifetime correlation spectroscopy combined with lifetime tuning: new perspectives in supported phospholipid bilayer research. Langmuir 22(23): 9580–9585
78. Renner L et al (2008) Supported lipid bilayers on spacious and pH-responsive polymer cushions with varied hydrophilicity. J Phys Chem B 112(20):6373–6378

79. Chiantia S, Ries J, Schwille P (2009) Fluorescence correlation spectroscopy in membrane structure elucidation. Biochim Biophys Acta Biomembr 1788(1):225–233
80. Benes M et al (2002) Muscovite (mica) allows the characterisation of supported bilayers by ellipsometry and confocal fluorescence correlation spectroscopy. Biol Chem 383(2):337–341
81. Sachl R et al (2009) A comparative study on ganglioside micelles using electronic energy transfer, fluorescence correlation spectroscopy and light scattering techniques. Phys Chem Chem Phys 11(21):4335–4343
82. Hess ST, Webb WW (2002) Focal volume optics and experimental artifacts in confocal fluorescence correlation spectroscopy. Biophys J 83(4):2300–2317
83. Wawrezinieck L et al (2005) Fluorescence correlation spectroscopy diffusion laws to probe the submicron cell membrane organization. Biophys J 89(6):4029–4042
84. Vats K, Kyoung M, Sheets ED (2008) Characterizing the chemical complexity of patterned biomimetic membranes. Biochim Biophys Acta Biomembr 1778(11):2461–2468
85. Sisan DR et al (2006) Spatially resolved fluorescence correlation spectroscopy using a spinning disk confocal microscope. Biophys J 91(11):4241–4252
86. Destainville N (2008) Theory of fluorescence correlation spectroscopy at variable observation area for two-dimensional diffusion on a meshgrid. Soft Matter 4(6):1288–1301
87. Humpolickova J et al (2006) Probing diffusion laws within cellular membranes by Z-scan fluorescence correlation spectroscopy. Biophys J 91(3):L23–L25
88. Almeida PFF, Vaz WLC, Thompson TE (1992) Lateral diffusion and percolation in 2-phase, 2-component lipid bilayers: topology of the solid-phase domains in-plane and across the lipid bilayer. Biochemistry 31(31):7198–7210
89. Forstner MB et al (2006) Lipid lateral mobility and membrane phase structure modulation by protein binding. J Am Chem Soc 128(47):15221–15227
90. Xia W, Thorpe MF (1988) Percolation properties of random ellipses. Phys Rev A 38 (5):2650–2656
91. Gregor I, Patra D, Enderlein J (2005) Optical saturation in fluorescence correlation spectroscopy under continuous-wave and pulsed excitation. Chemphyschem 6(1):164–170
92. Petrášek Z, Schwille P (2008) Photobleaching in two-photon scanning fluorescence correlation spectroscopy. Chemphyschem 9(1):147–158
93. Widengren J, Rigler R (1996) Mechanism of photobleaching investigated by fluorescence correlation spectroscopy. Bioimaging 4:149–157
94. Chiantia S et al (2006) Effects of ceramide on liquid-ordered domains investigated by simultaneous AFM and FCS. Biophys J 90(12):4500–4508
95. Štefl M, Kulakowska A, Hof M (2009) Simultaneous characterization of lateral lipid and prothrombin diffusion coefficients by Z-scan fluorescence correlation spectroscopy. Biophys J 97(3):L1–L3
96. Satsoura D et al (2007) Circumvention of fluorophore photobleaching in fluorescence fluctuation experiments: a beam scanning approach. Chemphyschem 8(6):834–848
97. Petersen NO (1984) Diffusion and aggregation in biological-membranes. Can J Biochem Cell Biol 62(11):1158–1166
98. Ries J, Schwille P (2006) Studying slow membrane dynamics with continuous wave scanning fluorescence correlation spectroscopy. Biophys J 91(5):1915–1924
99. Dittrich PS, Schwille P (2001) Photobleaching and stabilization of fluorophores used for single-molecule analysis with one- and two-photon excitation. Appl Phys B Lasers Opt 73 (8):829–837
100. Zipfel WR, Williams RM, Webb WW (2003) Nonlinear magic: multiphoton microscopy in the biosciences. Nat Biotechnol 21(11):1368–1376
101. Mutze J, Petrasek Z, Schwille P (2007) Independence of maximum single molecule fluorescence count rate on the temporal and spectral laser pulse width in two-photon FCS. J Fluoresc 17(6):805–810
102. Petersen NO (1986) Scanning fluorescence correlation spectroscopy. 1. Theory and simulation of aggregation measurements. Biophys J 49(4):809–815

103. Hebert B, Costantino S, Wiseman PW (2005) Spatiotemporal image correlation spectroscopy (STICS) theory, verification, and application to protein velocity mapping in living CHO cells. Biophys J 88(5):3601–3614
104. Petersen NO, Johnson DC, Schlesinger MJ (1986) Scanning fluorescence correlation spectroscopy. 2. Application to virus glycoprotein aggregation. Biophys J 49(4):817–820
105. Ruan QQ et al (2004) Spatial-temporal studies of membrane dynamics: scanning fluorescence correlation spectroscopy (SFCS). Biophys J 87(2):1260–1267
106. Skinner JP, Chen Y, Muller JD (2005) Position-sensitive scanning fluorescence correlation spectroscopy. Biophys J 89(2):1288–1301
107. Weissman M, Schindler H, Feher G (1976) Determination of molecular-weights by fluctuation spectroscopy: application to DNA. Proc Natl Acad Sci USA 73(8):2776–2780
108. Berland KM et al (1996) Scanning two-photon fluctuation correlation spectroscopy: particle counting measurements for detection of molecular aggregation. Biophys J 71(1):410–420
109. Petersen NO et al (1993) Quantitation of membrane-receptor distributions by image correlation spectroscopy: concept and application. Biophys J 65(3):1135–1146
110. Digman MA et al (2005) Fluctuation correlation spectroscopy with a laser-scanning microscope: exploiting the hidden time structure. Biophys J 88(5):L33–L36
111. Wiseman PW, Squier JA, Wilso KR (2000) Dynamic image correlation spectroscopy (ICS) and two-color image cross-correlation spectroscopy (ICCS): concepts and application. In: Conchello JA, Cogswell CJ, Wilson T (eds) Three-dimensional and multidimensional microscopy: image acquisition processing, vol VII. SPIE, Bellingham, pp 14–20
112. Semrau S, Schmidt T (2007) Particle image correlation spectroscopy (PICS): retrieving nanometer-scale correlations from high-density single-molecule position data. Biophys J 92(2):613–621
113. Thompson NL, Steele BL (2007) Total internal reflection with fluorescence correlation spectroscopy. Nat Protoc 2(4):878–890
114. Ries J, Petrov EP, Schwille P (2008) Total internal reflection fluorescence correlation spectroscopy: effects of lateral diffusion and surface-generated fluorescence. Biophys J 95(1): 390–399
115. Ries J et al (2008) Supercritical angle fluorescence correlation spectroscopy. Biophys J 94(1): 221–229
116. Samiee KT et al (2005) Lambda-repressor oligomerization kinetics at high concentrations using fluorescence correlation spectroscopy in zero-mode waveguides. Biophys J 88(3): 2145–2153
117. Milon S et al (2003) Factors influencing fluorescence correlation spectroscopy measurements on membranes: simulations and experiments. Chem Phys 288(2–3):171–186
118. Chiantia S et al (2006) Combined AFM and two-focus SFCS study of raft-exhibiting model membranes. Chemphyschem 7(11):2409–2418
119. Digman MA, Gratton E (2009) Imaging barriers to diffusion by pair correlation functions. Biophys J 97(2):665–673
120. Berland KM (2001) Dual-color two-photon fluorescence correlation spectroscopy. In: Periasamy A, So PTC (eds) Multiphoton microscopy in the biomedical sciences. SPIE, Bellingham, pp 268–278
121. Swift K, Anderson SN, Matayoshi E (2001) Dual laser fluorescence correlation spectroscopy as a biophysical probe of binding interactions – evaluation of new red-fluorescent dyes. In: Lakowicz JR, Thompson RB (eds) Advances in fluorescence sensing technology, vol V. SPIE, Bellingham, pp 47–58
122. Bohmer M et al (2002) Time-resolved fluorescence correlation spectroscopy. Chem Phys Lett 353(5–6):439–445
123. Medina MA, Schwille P (2002) Fluorescence correlation spectroscopy for the detection and study of single molecules in biology. Bioessays 24(8):758–764
124. Rigler R et al (1998) Fluorescence cross-correlation: a new concept for polymerase chain reaction. J Biotechnol 63(2):97–109

125. Sorscher SM, Klein MP (1980) Profile of a focused collimated laser-beam near the focal minimum characterized by fluorescence correlation spectroscopy. Rev Sci Instrum 51(1):98–102
126. Gielen E et al (2009) On the use of Z-scan fluorescence correlation experiments on giant unilamellar vesicles. Chem Phys Lett 469(1–3):110–114
127. Miszta A et al (2008) Combination of ellipsometry, laser scanning microscopy and Z-scan fluorescence correlation spectroscopy elucidating interaction of cryptdin-4 with supported phospholipid bilayers. J Pept Sci 14(4):503–509
128. Shenoy S et al (2010) In-plane homogeneity and lipid dynamics in tethered bilayer lipid membranes (tBLMs). Soft Matter 6:1263–1274

Total Internal Reflection with Fluorescence Correlation Spectroscopy*

Nancy L. Thompson, Punya Navaratnarajah, and Xiang Wang

Abstract The method in which total internal reflection illumination is combined with fluorescence correlation spectroscopy (TIR-FCS) is reviewed. Included are descriptions of the conceptual basis and experimental design, as well as a variety of applications and the theoretical models associated with these measurements. Emerging, more complex versions of TIR-FCS, related techniques, and possible future directions for TIR-FCS are also outlined.

1 Introduction

Total internal reflection fluorescence microscopy (TIRFM) is a mature and widely applied method. In this technique, only fluorescent molecules very near or bound to an optically transparent (and usually planar) substrate are excited and, therefore, observed. The preferential discrimination for fluorescent molecules that are adjacent to the surface rather than far from the surface in the bulk is achieved by using a surface-associated evanescent wave as the excitation source for fluorescence. The evanescent wave is generated by internally reflecting a light source at the planar interface between the surface and the adjacent bulk medium (usually fused silica or glass and air or an aqueous solution, respectively). The surface-selectivity of TIRFM has been used in a variety of contexts, most of which are now well established. Some of the more well-developed TIRFM methods include the use of evanescent illumination to measure surface densities of bound, fluorescent

*Submitted to *Reviews in Fluorescence*, April 2010.

N.L. Thompson (✉)
Department of Chemistry, University of North Carolina at Chapel Hill,
Chapel Hill, NC 27599-3290, USA
e-mail: nlt@unc.edu

C.D. Geddes (ed.), *Reviews in Fluorescence 2009*, Reviews in Fluorescence 2009,
DOI 10.1007/978-1-4419-9672-5_13,

molecules for the purpose of measuring surface binding isotherms; to monitor fluorescence recovery after photobleaching for the purpose of examining the kinetics of surface association and dissociation, including surface rebinding events; and to examine the sensitivity of the steady-state fluorescence intensity to the evanescent wave polarization for the purpose of measuring order parameters for molecules in thin films which reside on the surface such as supported phospholipid bilayers. TIRFM methods are particularly useful for examining events that occur on or near the basal membranes of live biological cells that are adhered to the surface at which internal reflection occurs. TIRFM has also been used in a variety of other contexts, including biosensors and polymer chemistry. General and comprehensive descriptions of TIRFM have been provided in numerous review papers (most recently, [6, 31, 50, 63, 107, 110, 124]).

Fluorescence correlation spectroscopy (FCS) is also a mature and widely applied method. In this technique, most usually, the fluorescence arising from a small volume is monitored with time. If the observed volume is small enough to contain only a low number of fluorescent molecules, the measured fluorescence fluctuates significantly with time as the fluorescence collected from individual molecules stochastically changes (e.g., through motion into and out of the observed volume, or through changes in the magnitude of the emitted fluorescence arising from intramolecular spectroscopy or intermolecular binding). Autocorrelating the temporal fluorescence fluctuations provides information about the process or processes giving rise to the fluctuations. In the most common type of FCS, the fluorescence fluctuations arise from diffusion through the observed volume. In this case, the autocorrelation function decreases with time to zero. The magnitude of the autocorrelation function is inversely proportional to the average number of fluorescent molecules in the observed volume, and the characteristic decay time depends on the size of the observed volume and the diffusion coefficient of the fluorescent molecules. Many other types of FCS have also been developed. These methods include, but are not limited to, two-color fluorescence fluctuation cross-correlation for quantifying bimolecular binding, photon counting histograms for characterizing molecular aggregation, and spatial fluorescence fluctuation analysis for detecting molecular clustering. FCS is most commonly carried out with confocal optics, rather than evanescent illumination (as described herein). General and comprehensive descriptions of FCS have been provided in numerous review papers (most recently, [10, 16, 17, 30, 31, 40, 60, 130]).

In this paper, the technique in which evanescent illumination is combined with fluorescence correlation spectroscopy (TIR-FCS) is reviewed. Included are descriptions of the conceptual basis and experimental design of conventional TIR-FCS, as well as a variety of applications and the theoretical models associated with these measurements. New, emerging, more complex versions of TIR-FCS, related techniques, and possible future directions for TIR-FCS are also outlined. Other papers reviewing TIR-FCS have also been presented recently [2, 125–127].

2 Conceptual Basis and Experimental Design

Total internal reflection fluorescence microscopy. When a plane wave traveling in a medium with refractive index (n_1) impinges on an interface with a lower refractive index (n_2) medium, at an angle greater than the critical angle, α_c, where:

$$\alpha_c = \sin^{-1}\left(\frac{n_2}{n_1}\right), \tag{1}$$

the plane wave is completely reflected back into the first medium in a process called total internal reflection. During TIR, a component of the incident light propagates parallel to the TIR interface and penetrates into the lower refractive index medium, where the intensity, $I(z)$, decays (theoretically) exponentially with increasing distance, z, from the interface:

$$I(z) = I_0 e^{-z/d}. \tag{2}$$

The depth, d, of this exponentially decaying evanescent wave is determined by the vacuum wavelength of the incident light, λ_0, the angle of incidence, α, and refractive indices, n_1 and n_2:

$$d = \frac{\lambda_0}{4\pi\sqrt{n_1^2 \sin^2\alpha - n_2^2}}. \tag{3}$$

The depth of the evanescent field is typically on the order of 100 nm. TIR illumination thereby allows one to selectively excite fluorophores close to surfaces. The depth, d, is the distance from the interface at which the intensity of the evanescent wave is 37% [corresponding to $1/e$ in (2)] of its maximum intensity at the surface. While (3) yields the evanescent field depth for an ideal system, the actual penetration profile and/or depth of the excitation light may deviate from the theoretical predictions due to scattering and other aberrations. Hence, several approaches have been used to experimentally determine the evanescent intensity profile and depth. This type of measurement has been carried out only on systems in which the evanescent field is generated by through-objective total internal reflection (see below).

The most direct approach is to measure the fluorescence intensity of a small object as a function of its distance from the TIR interface. In one such experiment, an atomic force microscope tip was made to be fluorescent by attaching a quantum dot or fluorescent bead. The tip was then imaged as it was moved along the vertical axis of the evanescent field with nanometer precision [105]. The resulting intensity profile agreed well with a single exponential decay, and the measured depth changed with the incidence angle in a manner qualitatively consistent with (3), although quantitative comparison was not possible because of uncertainties in the incidence angle. Oreopoulos and Yip [87] also used scanning probe microscopy (SPM) in combination with TIRFM to probe the evanescent field. They attached a 20-nm fluorescent bead to the tip of an SPM cantilever, lowered the tip in incremental steps toward a surface containing an evanescent field generated by through-objective TIR illumination, and recorded the fluorescence intensity as a function of the distance from the interface. In this later work, two new effects were observed. First, the shape of the intensity as a function of z was best described by (2) summed to a second exponential (with much lower initial magnitude). The second exponential was attributed to light scattering. Second, the measured (primary) d-value was significantly higher than the value predicted by (3). This effect was most likely a result of uncertainty in the incidence angle. Also, neither of these studies accounted for deviations of the angular emission profile of fluorescent molecules near dielectric interfaces from the emission pattern in bulk solution [43]. This displacement method of calibrating the evanescent wave profile has also recently been used in an additional study directed toward understanding the assembly of clathrin-coated pits [101].

Mattheyses and Axelrod [77] report a different method for examining the evanescent profile. In this work, large, low refractive index beads labeled on their surfaces with a fluorescent dye and suspended in an index-matched liquid were allowed to adhere gently (by gravity) to a surface illuminated by through-objective internal reflection. Because the beads were much larger than the evanescent wave depth, the fluorescence as a function of the distance from the center of the bead, given the spherical bead geometry, reported the fluorescence at different z-values. Plots of the fluorescence intensity versus z revealed that the axial intensity profile of the evanescent field was best described by two exponentials. The exponential that accounted for 90% of the decay had the theoretically predicted depth, while a more slowly decaying component was attributed to scattering. This work does incorporate careful calibration of the incidence angle and accounts for the unusual angular profile of emission for fluorophores next to dielectric surfaces. A similar approach using a curved lens coated with small fluorescent beads has also been used to calibrate the evanescent intensity profile [115].

Gell et al. [28] used fluorescently labeled microtubules immobilized at an angle on microscope slides to characterize the axial intensity profile of the evanescent field. A series of TIRFM images obtained at different incident angles were used in combination with the tilt angles of the microtubules to determine the depth of the evanescent field. They found that the evanescent intensity was well described by a single exponential as in (2), and the evanescent wave depth was close to the theoretically predicted value.

Harlepp et al. [36] employed TIR-FCS to measure the depth of the evanescent intensity. As will be described below, TIR-FCS allows one to monitor the pattern of fluorescence fluctuations that arises from the diffusion of fluorescent molecules in the evanescent field. This pattern, obtained in the form of an autocorrelation curve, can be fit to a theoretical fluorescence fluctuation autocorrelation function describing diffusion in the evanescent field to obtain diffusion coefficients. When the diffusion coefficient of the fluorescent species is known, the fit can yield the depth of the evanescent wave. The depths measured by Harlepp et al. [36] using this method differed from those predicted by (3) by only 0.7%. However, this approach does not account for deviation from ideal diffusion for molecules close to surfaces (see below). In addition, unlike other approaches discussed above, TIR-FCS does not directly provide information about the axial intensity profile of the evanescent field.

The polarization of the electric field in an evanescent wave differs from the polarization observed in freely propagating plane waves. Incident light that is *s*-polarized (along the *y*-axis), perpendicular to the incidence plane, yields evanescent fields polarized along the *y*-axis. When incident light is *p*-polarized, polarized in the incidence plane (here denoted as the *x*–*z* plane), the evanescent field consists of longitudinal (*x*) and transverse (*z*) components. This polarization contrasts with that of freely propagating light, which only has transverse components. The unique polarization properties of evanescent fields obtained from *p*-polarized incident light can be used to characterize orientation changes in molecules. While the above-mentioned concept is a description of the polarization of single plane waves, accurate intensity and polarization profiles of evanescent fields generated by different optical apparatuses are described extensively in the literature (e.g., [1, 2, 12, 13, 29, 43, 46, 57, 77]).

The polarization of the incident light also affects the intensity profiles of evanescent fields. Linearly polarized light gives a focus that is split into two lobes in the direction of polarization. Depending on the angle of incidence of the polarized incident light, the evanescent field undergoes a phase change that serves to increase the spread of the evanescent field in the lower refractive index medium, and thereby increase the size of the observation volume. Radially polarized light can be focused to a much tighter spot, although it too has side lobes [22]. Ivanov et al. [49] showed that the side lobes resulting from the radially polarized light can be suppressed by using two-photon excitation to achieve observation volumes on the order of 5 attoliters (10^{-18} L). As will be discussed below, reduced observation volumes are particularly desirable in FCS experiments, as they serve to reduce the average number of fluorescent molecules in the observation volume and increase the magnitude of fluorescence fluctuations.

TIR illumination can be achieved via through-prism and through-objective setups (Fig. 1). Through-prism TIRFM enables one to have ready access to and complete control over the excitation pathway. High refractive index substrates might be easily introduced to achieve greater axial confinement of the excitation field [112]. Through-prism TIRFM also allows for more easy construction of evanescent interference patterns [47]. As the incident light never enters the microscope, there is usually a lower background associated with this setup. Through-objective TIRFM offers several advantages of its own; chief among them are greater image quality and emission collection efficiency. The latter phenomenon is due to the fact that the interaction between the planar dielectric interface and nearby fluorophores results in more fluorescence being emitted through the higher refractive index medium. It has also been reported that there is somewhat greater axial confinement of the evanescent field in through-objective TIRFM [116]. In addition, this apparatus allows for greater sample-top accessibility.

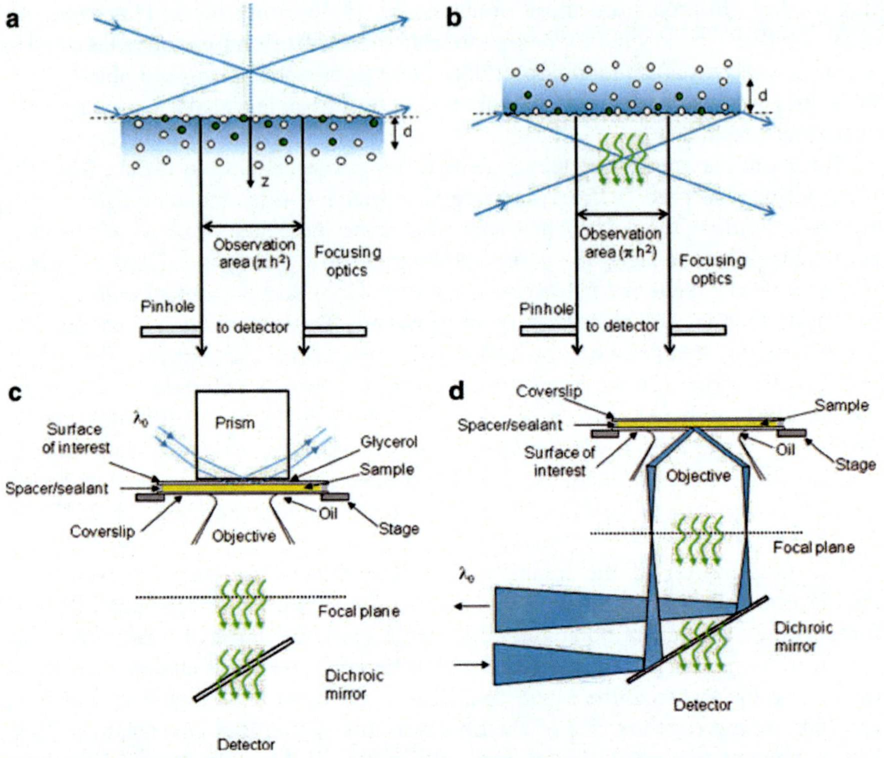

Fig. 1 Through-prism and through-objective TIR-FCS. **(a, c)** In through-prism TIR-FCS, a laser beam is totally internally reflected at a sample plane through a prism that is mounted on the stage of an inverted microscope. **(b, d)** In through-objective TIR-FCS, a laser beam is directed through a microscope port, reflected toward the sample plane with a dichroic mirror, and is focused through the periphery of a high numerical aperture objective so that the beam internally reflects at the sample plane. In all cases, the evanescently excited fluorescence is collected through the high-aperture objective and passed through a spectral filter to remove scattered evanescent light. The light is passed through a small aperture or collected with a fiber optic, placed at a back image plane of the microscope, before reaching a single-photon counting detector. The detector signal is processed using either a computer with a correlator card and associated software or similar in-house constructed signal processing equipment. Reproduced with permission from J Struct Biol 168:95–106 (2009)

Fluorescence correlation spectroscopy. FCS is used to monitor the temporal pattern of spontaneous fluorescence fluctuations from the time-averaged fluorescence in a given system and, thereby, to gain insight into the dynamics of the fluorescent species being studied. This temporal pattern of fluorescence fluctuations is obtained by autocorrelating the fluctuations in the fluorescence signal:

$$G(\tau) = \frac{\langle \delta F(t)\delta F(t+\tau)\rangle}{\langle F\rangle^2}, \quad (4)$$

where $\langle F \rangle$ is the time-averaged fluorescence, $\delta F(t)$ is the difference between the instantaneous fluorescence, $F(t)$, and $\langle F \rangle$, and τ is a given correlation lag-time. The autocorrelation function, $G(\tau)$, is independent of the time, t, when the system of interest is at equilibrium. $G(\tau)$ decays to zero as τ approaches infinity. The magnitude of this function is inversely related to the average number of fluorescent molecules in the observation volume. The rate of decay and shape of $G(\tau)$ contain information about the processes that affect the pattern of fluorescence fluctuations, including photophysical dynamics (e.g., triplet state kinetics), solution diffusion, association and dissociation with surface-binding sites, and/or enzyme kinetics.

As FCS monitors the deviation of the instantaneous fluorescence from the time-averaged fluorescence, recording the fluorescence emanating from a large number of fluorescent molecules will only yield ensemble-averaged fluorescence measurements and fluorescence fluctuation autocorrelation functions that are small in magnitude. To obtain autocorrelation functions with magnitudes large enough to be accurately measured, the average number of fluorescent molecules in the observed volume has to be small. Most FCS measurements are carried out with at most 100 fluorescent molecules in the observed volume. Several strategies have been proposed to improve signal-to-noise ratios [18, 56, 58, 61, 81, 92, 100, 134]. Increasing the number of photons collected per molecule per sample time increases signal-to-noise ratios, as does increasing the total time over which fluctuations are correlated. High aperture objectives achieve high fluorescence collection efficiency.

As the concentration of fluorescent molecules required for successful FCS is often below biologically relevant levels, there is often a need to circumvent this difficulty. One way to reduce the average number of fluorescent molecules in the observation volume, other than by simply decreasing the concentration, is to reduce the size of the volume. Another method is to mix unlabeled molecules with fluorescently labeled ones (see below). Photoactivatable fluorophores also present a means to simultaneously have higher concentrations of the molecule of interest and low concentrations of fluorescent molecules [23]. In this approach, the proportion of molecules labeled with a photoactivatable fluorophore that fluoresce can be varied by controlling the intensity of the light used to switch on and off the fluorescence of these molecules.

Total internal reflection with fluorescence correlation spectroscopy. In conventional TIR-FCS, FCS is coupled to TIRFM and used to study the dynamics of fluorescent molecules at surfaces (Fig. 1). Incident light that is totally internally reflected at a dielectric interface excites fluorophores in an observation volume defined, in part, by the evanescent field. However, axial confinement achieved by TIR illuminations does not sufficiently reduce the observation volume. Lateral (*x*–*y* sample plane) confinement that reduces the observed area to about 1 μm^2 is achieved by using a circular pinhole [69, 114] or fiber-optic cable [37, 38] placed at a back image plane of the microscope. The fluorescence fluctuates as fluorescent species move in and out of the detection volume, undergo photophysics and/or reversibly associate with the surface. The fluorescence fluctuations are autocorrelated to obtain information about the dynamics of the fluorescent species (Fig. 2).

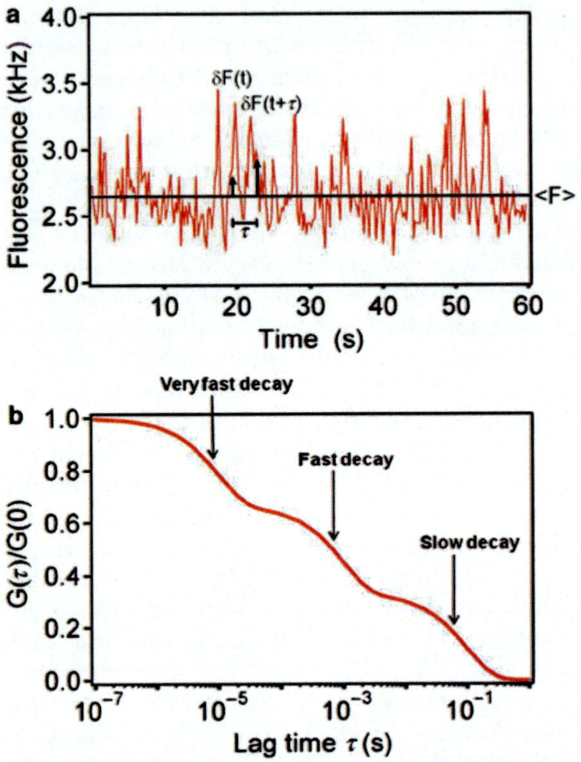

Fig. 2 Fluorescence fluctuation autocorrelation functions. (**a**) The fluorescence measured from the observation volume fluctuates with time as individual fluorescent molecules diffuse through the evanescent field, associate with and dissociate from surface binding sites, and/or undergo transitions between states with different detected fluorescence intensities. (**b**) The temporal fluorescence fluctuations are autocorrelated [see (1)]. This plot shows an idealized case in which three components with different characteristic rates are present in $G(\tau)$, e.g., a very fast decay arising from photophysics, a fast decay arising from free diffusion of the fluorescent molecules through the depth of the evanescent field, and a slow decay resulting from reversible association of the fluorescent molecules with surface sites

Evanescent illumination in TIR-FCS serves many purposes: surface selectivity; reduction in observation volume via axial confinement and lower background signal [48].

Although evanescent illumination reduces background, the sensitivity of TIR-FCS still requires that the measured $G(\tau)$ be corrected for background contributions. Uncorrelated background light will affect the magnitude of the autocorrelation function [121]. Methods of correcting for uncorrelated background light are available [123] and can be extended to correlated backgrounds. The primary sources of background light in TIR-FCS and TIRFM, in general, are scattered evanescent light and substrate luminescence. The use of fused silica (when possible) in the place of glass helps reduce the latter. In through-objective

TIRFM, the excitation light passing through the microscope interior also contributes to background.

Once autocorrelation functions with large magnitudes and high signal-to-noise ratios are obtained, they are usually fit to theoretical autocorrelation functions derived for a given system, to obtain values of the parameters of interest. These theoretical models take into account the optical arrangement of the excitation and emission pathways, as well as the processes giving rise to the observed fluorescence fluctuations. Many different theoretical models have been described in the literature: triplet state kinetics [133], diffusion in the evanescent field [11, 45, 54, 62, 91, 113], binding kinetics [70, 119], enzyme kinetics [39], cross-correlation TIR-FCS [67], and other situations [37, 38, 94, 120].

3 Applications

Motion of fluorescent molecules near surfaces. To date, TIR-FCS has been most widely applied to the examination of fluorescence fluctuation autocorrelation functions arising from fluorescent molecules moving close to a surface at which an evanescent wave has been created. As described above, these $G(\tau)$ contain information about the molecular mobility in solution close to the surface as well as the depth of the evanescent intensity. They also contain information about the average number of fluorescent molecules in the observed volume and fluorophore photophysics. The form of the autocorrelation function most often used for these measurements is [37, 38, 113]:

$$G(\tau) = [\frac{1}{2N_A}][1 + \frac{p}{1-p}\exp(-R_t\tau)][(1 - 2R_z\tau)\exp(R_z\tau)erfc(\sqrt{R_z\tau}) + 2\sqrt{\frac{R_z\tau}{\pi}}][\frac{1}{1+R_{x,y}\tau}] \tag{5}$$

In (5), N_A is the average number of molecules in the observed volume close to the surface; p is the probability that (for a given excitation intensity) the fluorophore is in the triplet state; R_t is the triplet state relaxation rate; $R_z = D/d^2$ is the rate associated with diffusion through the evanescent wave in the direction perpendicular to the surface; D is the fluorophore diffusion coefficient in solution; the evanescent wave depth d is defined and discussed above; $R_{x,y} = 4D/s^2$ is the rate associated with diffusion in a direction parallel to the surface; and s is the characteristic length of the observed area. Although the third factor in (5) contains an exponential that approaches infinity with large τ, this factor is multiplied by a complementary error function that approaches zero with large τ, and the product of these two functions approaches zero with large τ as the second function has a stronger dependence on τ. At $\tau = 0$, $G(0) = [1/(2N_A)][1/(1 - p)]$. As $\tau \to \infty$, each of the two last factors in (5) approach zero.

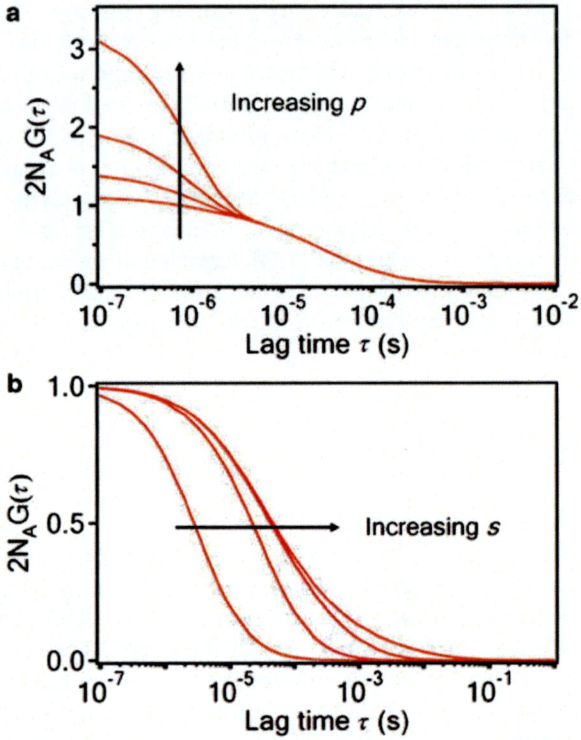

Fig. 3 TIR-FCS autocorrelation function for diffusion through the evanescent wave. Shown are values of $G(\tau)$ as theoretically predicted by (5). In all plots, $D = 7 \times 10^{-6}\ \text{cm}^2\ \text{s}^{-1}$, $d = 0.1\ \mu\text{m}$, and $R_z = D/d^2 = 7 \times 10^4\ \text{s}^{-1}$. Panel (**a**) shows the product of the second and third factors in (5) with $R_t = 10^6\ \text{s}^{-1}$ and p equal to 0.1, 0.3, 0.5, or 0.7. Panel (**b**) shows the product of the third and fourth factors in (5) with s equal to 0.1, 0.5, 2, and 10 μm, respectively. For these values of s, $R_{x,y}$ ranges from $2.8 \times 10^5\ \text{s}^{-1}$ to $28\ \text{s}^{-1}$

Figure 3 illustrates $2N_AG(\tau)$ calculated from (5). In Fig. 3a, only the factors describing photophysics and diffusion through the evanescent wave in a direction perpendicular to the surface are included. As shown, if the excitation intensity is low enough or if the measured $G(\tau)$ can be cropped to exclude fast time lags τ, then $G(\tau)$ is equivalent to the term describing diffusion perpendicular to the surface and the photophysical term can be excluded. This condition can be experimentally confirmed by insuring that $G(\tau)$, for the time lags τ examined, does not change with the excitation intensity. However, lowering this intensity enough so that this condition is satisfied will compromise the signal-to-noise ratio of $G(\tau)$, as this ratio depends on the count rate per molecule. In Fig. 3b, only the factors describing diffusion through the evanescent wave in directions parallel and perpendicular to the surface are plotted. As shown, if the characteristic length of the observation area, s, is large enough, diffusion through the evanescent wave in a direction parallel to the surface can be neglected. However, increasing s to the extent that the decay of $G(\tau)$ is essentially described by one-dimensional diffusion through the evanescent wave in a direction parallel to the surface may compromise the magnitude of $G(\tau)$ and, therefore, its signal-to-noise ratio, as increasing the size of the observed volume increases the average number of fluorescent molecules contained in this volume.

The form of $G(\tau)$ shown in (5) and illustrated in Fig. 3 contains a number of assumptions which may not be adequate for some measurements. (1) The factor describing fluorophore photophysics addresses only triplet state involvement and not other photophysical phenomena such as, for example, blinking arising from redox reactions [52] or from *cis–trans* isomerizations [73]. Also, this factor does not account for changes in fluorescence emission and/or collection efficiencies arising from energy transfer, fluorophore orientational motions (e.g., [122]), binding to other molecules (e.g., [74]), or changes of the spectroscopic properties of the fluorophore with respect to distance from the interface [43]. (2) In some cases, the factor describing photophysics will depend on the precise excitation intensity. As shown previously, the intensity of the evanescent wave at the interface, $I(0)$ (2), depends on a number of optical parameters and is not equal to the intensity of the beam which is internally reflected [5]. In most apparatuses, it is difficult to precisely measure $I(0)$, but sufficiently accurate calculations may be possible by considering the factors described in Axelrod et al. [5] along with knowledge of the details of the optical apparatus. When intermediate layers such as supported phospholipid membranes or polymer films are present, this issue becomes more difficult to address. (3) The form of the fourth factor in (5) arises from assuming that the excitation and detection spatial profile is Gaussian in shape [74]. For most experimental apparatuses (Fig. 1), this assumption may not be entirely accurate. (4) Equation (5) was derived assuming that the evanescent wave decays with distance from the interface in a strictly single exponential fashion (2). As described above, some measurements have indicated that this assumption may not be correct. (5) The form of (5) also assumes that the concentration of fluorophores in solution close to the interface is constant and does not depend on the distance from the interface. This assumption could be seriously violated under certain conditions such as, for example, strong electrostatic effects between the molecules of interest and the surface. (6) Finally, (5) was derived with the assumption that there is a single diffusion coefficient, D, close to the surface that does not depend on the distance of the fluorescent molecule from the interface. At present, this assumption is known to be incorrect and has been taken into account only in some situations (see below).

Nonetheless, with all of these caveats, the simplest use of TIR-FCS to date has been to autocorrelate fluorescence fluctuations arising from small fluorophores diffusing through the depth of the evanescent wave, in cases for which the fluorophores do not bind to the surface, to demonstrate that the method is at least approximately working properly (e.g., [2, 36–38, 67, 85, 86, 98]). In addition, TIR-FCS has been used to extensively characterize molecular transport in sol–gel films [78–80].

The concentrations and diffusion coefficients of fluorescently labeled IgG close to supported phospholipid membranes have been examined in detail by using TIR-FCS (Fig. 4) [114]. These measurements were carried out for membranes with

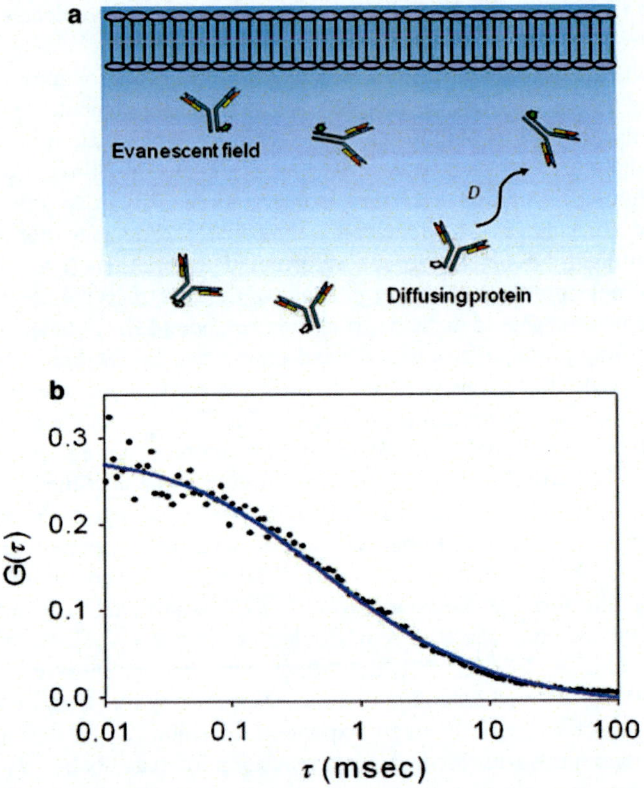

Fig. 4 Diffusion of IgG close to model membrane surfaces. (**a**) Schematic of antibody diffusion close to a supported planar membrane. Fluorophores within the evanescent wave are excited and fluoresce. (**b**) Representative TIR-FCS data and the best fit to an appropriate theoretical expression [see (5)]. The data are for a mixture of 10 nM Alexa448-labeled IgG, 1 μM unlabeled IgG and 10 mg mL^{-1} ovalbumin in phosphate-buffered saline close to surfaces coated with supported phospholipid bilayers. Part (**b**) is reproduced with permission from Biophys J 85:3294–3302 (2003)

different lipid compositions and surface charges and for solutions of varying pH and ionic strength values. The results showed no statistically significant change in the local IgG concentration or diffusion coefficient for membranes with different charges or for solutions with different pH values spanning the IgG isoelectric point. However, a significant decrease in the local mobility was detected with increasing ionic strength.

The result that the local mobility of IgG close to membrane surfaces is reduced in solutions with higher ionic strengths suggested that strong hydrodynamic interactions occur between proteins and membrane surfaces and that these interactions are amplified by increasing ionic strength. In fact, previous theoretical and experimental work has provided approximate analytical solutions for the predicted diffusion coefficients of spheres next to planar walls as a function of

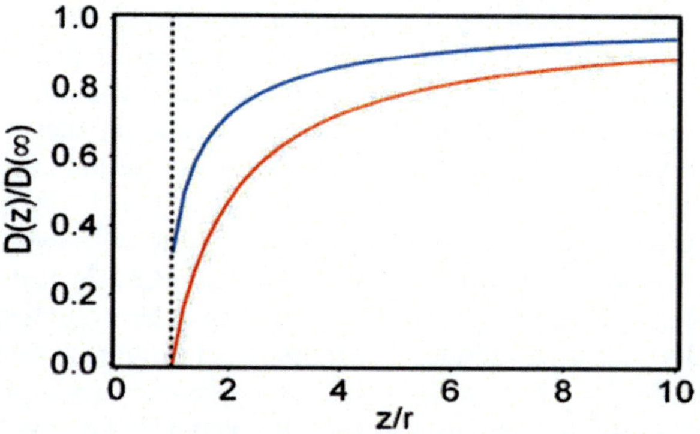

Fig. 5 Sphere diffusion close to planar surfaces. This plot show the theoretically predicted diffusion coefficients of spheres close to walls (6) as a function of the ratio of the distance from the wall (z) and the sphere radius (r), and relative to the bulk diffusion coefficient. For motion in a direction perpendicular to the wall (*red*) the ratio (and diffusion coefficient) approach zero when the sphere is at the wall. For motion in a direction parallel to the wall (*blue*), the ratio approaches approximately 0.32

the distance between the sphere and the wall, as well as the sphere radius [62, 72, 91]. These approximations are illustrated in Fig. 5 and are as follows:

$$\frac{D_{\text{per}}(z)}{D(\infty)} = \frac{2\left(3 - 5\left(\frac{r}{z}\right) + 2\left(\frac{r}{z}\right)^2\right)}{6 - 3\left(\frac{r}{z}\right) - \left(\frac{r}{z}\right)^2}$$

$$\frac{D_{\text{par}}(z)}{D(\infty)} = 1 - \frac{9}{16}\left(\frac{r}{z}\right) + \frac{1}{8}\left(\frac{r}{z}\right)^3 - \frac{45}{256}\left(\frac{r}{z}\right)^4 - \frac{1}{16}\left(\frac{r}{z}\right)^5 \ldots \qquad (6)$$

where D_{per} describes diffusion in a direction perpendicular to the wall, D_{par} describes diffusion in a direction parallel to the wall, r is the sphere radius, and $z \geq r$ is the distance between the center of the sphere and the wall.

To test the hypothesis that hydrodynamic effects are the primary factor governing protein diffusion close to model membrane surfaces, TIR-FCS was used to examine the mobilities of fluorescently labeled antibodies, antibody fragments and antibody complexes with a range of hydrodynamic radii, close to supported planar membranes [91]. It was found that planar lipid bilayers reduced the diffusion coefficients of these molecules in a size-dependent manner. Specifically, diffusion coefficients close to membranes were seen to decrease with hydrodynamic radii above what would be predicted in solution and in agreement with $D_{\text{per}}(z)$ as defined in (6). This work required data analysis more complex than simply fitting to the appropriate factor in (5) because this factor does not account for distance-dependent diffusion coefficients.

Motion of fluorescent particles near surfaces. TIR-FCS has also been used to examine the mobilities of fluorescent particles, rather than molecules close to membrane surfaces. Specifically, the diffusive behavior of phospholipid vesicles very close to supported membrane surfaces was examined for a wide range of conditions including different ionic strengths, different solution pH values, and different lipid compositions [62]. The results described in this work provide a comprehensive view of the factors that determine vesicle motions very close to membrane surfaces.

The mobility of intracellular vesicles very close to the basal membranes of live, adherent cells has also been examined by using TIR-FCS. First, the dynamics of fluorescently labeled secretory granules in the cytoplasms of adherent bovine chromaffin cells near their plasma membranes were probed in detail, before and during exocytosis, by using various TIRFM approaches [54]. As part of this comprehensive work, the granule velocities, as measured with evanescent excitation and sequential imaging, were autocorrelated to determine if granule motion was Brownian in nature or if the autocorrelation data showed evidence of possible granule caging or tethering. The data, quite interestingly, were indicative of the latter situation in some cases. Second, the mobilities of fluorescently labeled synaptic vesicles near ribbon synapses in retinal bipolar cells were characterized by using TIR-FCS [45].

Kinetics of fluorescent molecules that reversibly associate with evanescently illuminated surfaces. In addition to its ability to examine the properties of the mobilities of fluorescent molecules in solution but close to surfaces, TIR-FCS can also be employed to characterize the kinetic properties of the reversible association of fluorescent molecules with surfaces. The notion underlying this use of TIR-FCS is that $G(\tau)$ will contain information related to the rates of surface association and/or dissociation because if a fluorescent molecule transiently binds to the surface in addition to transporting through the evanescent wave, the average duration of a fluorescence fluctuation will depend on the average surface residency time in addition to the residency time within the evanescent wave while moving in solution close to the surface.

Worth noting is that it is possible to characterize mobility close to surfaces in the absence of surface interactions, but it is not possible to characterize the kinetics of surface interactions in the absence of mobility in solution. If the rate associated with mobility through the evanescent wave (D/d^2; see above) is very much faster than the rates(s) associated with surface association/dissociation, then, to obtain information about the kinetic properties of surface association/dissociation, the measured $G(\tau)$ might be cropped to include only time lags τ much greater than d^2/D. If the rate associated with mobility through the evanescent wave is comparable to or faster than the rates(s) associated with surface association/dissociation, then to obtain information about the kinetic properties of surface association/dissociation, the measured $G(\tau)$ must be analyzed with theoretical models that account for both processes (see below).

Kinetics of fluorescent molecules reversibly but nonspecifically associating with surfaces. In the first of these types of systems to which TIR-FCS was applied, the reversible association with the surface was "nonspecific" in the sense that the fluorescent molecules in solution interacted with diverse sites on the surface having heterogeneous densities and kinetic properties. This class of TIR-FCS applications includes one of the first experimental demonstrations of TIR-FCS in which the kinetics associated with the nonspecific, reversible adsorption of fluorescently labeled IgG and insulin to serum albumin-coated fused silica slides were examined [121]. The measured half-times for the decay of $G(\tau)$ were on the order of 30–40 ms.

Accompanying this work was a derivation of the expected form of $G(\tau)$ [119], which is:

$$G(\tau) = \frac{\beta}{N_C} \exp[-(k_a A + k_d)\tau)] \tag{7}$$

where β is the average fraction of surface binding sites that are not occupied; N_C is the average number of fluorescent molecules on the surface within the observed area; k_a and k_d are the apparent association and dissociation kinetic rates, respectively, for surface binding; and A is the concentration of fluorescent molecules in solution. The factor of β arises because there is a finite number of binding sites in the observed area, and therefore, binomial rather than Poisson statistics are applicable. Equation (7) is a simplified form of a more general function and is based on the assumptions that (1) the fluorescent molecules do not have a high propensity for rebinding to the surface after dissociation; (2) the surface binding sites are not laterally mobile; and (3) contributions to $G(\tau)$ arising from diffusion through the evanescent wave and photophysics are negligible. In theory, according to (7), if these assumptions are met by given experimental conditions and if the concentration A can be increased enough so that $k_a A$ is comparable to k_d but β/N_C does not become too small, one would predict from this early equation that both kinetic rates can be obtained from $G(\tau)$ measured for different solution concentrations A. Because TIR-FCS is a relaxation method, more kinetic rate constants are found in the signal relative to perturbation methods such as fluorescence photobleaching recovery with evanescent illumination [119]. The conditions required for the use of (7) were not entirely met in the accompanying experimental work [121].

This class of TIR-FCS applications including nonspecific surface adsorption and desorption also includes two later studies designed to further understand surface interactions in reversed-phase chromatography. First, TIR-FCS autocorrelation functions were measured for rhodamine 6G in water/methanol solutions at bare silica surfaces and at silica surfaces derivatized with C-18 alkyl chains [34]. The measured values of $G(0)$ were used to obtain average fluorophore surface densities and the measured surface densities were used to determine average equilibrium constants for the reversible adsorption of rhodamine 6G to the chromatographic support mimics for the different solution conditions. A second study extended the work to measure dye adsorption and desorption kinetic rates [35].

Again with the primary motivation associated with understanding chromatography, TIR-FCS has recently been used to study the behavior of proteins near modified surfaces in the presence of a surfactant [111]. The proteins examined were fluorescein-labeled bovine serum albumin (BSA) and fluorescein-labeled *Thermomyces lanuginosus* lipase; the solid support was C_{18}-coated glass; and the nonionic/anionic detergent was a mixture of hexaethylene glycol dodecyl ether and linear alkylbenzene sulfonate. The measured $G(\tau)$ were analyzed in terms of previously developed theoretical models for surface interactions, free diffusion within the evanescent field, and photophysical kinetics. Overall, the results implied that the average surface residency time of the proteins decreased as the surfactant concentration was increased.

Kinetics of fluorescent molecules reversibly and specifically associating with surfaces. Although the kinetic properties of reversible, nonspecific association of fluorescent molecules with diverse surface sites is of interest in a variety of contexts, for many situations, one is more interested in understanding the kinetic properties of reversible, specific association of fluorescent ligands with well-defined surface sites having a well-defined surface density. Conceptually, the most simple method of designing a sample for such studies by TIR-FCS is by irreversibly depositing specific receptor molecules on the surface to be illuminated and then allowing fluorescently labeled ligands to freely diffuse, bind to, and dissociate from the surface-associated receptors (Fig. 6). A number of methods can be used to attach receptors to surfaces, including direct covalent conjugation (e.g., on fused silica, by using silanizing reagents), biotinylating the receptors in a region far from the ligand binding site and attaching them to surface-deposited avidin, and incorporating receptors into substrate-supported planar membranes. This last approach has been used to examine the kinetics of fluorescently labeled IgG specifically and reversibly associating with the mouse Fc receptor FcγRII, which was purified and reconstituted into supported membranes [69].

The experimental parameters required for successful implementation of this type of measurement are not completely straightforward. First, to ensure that a high enough fraction of the evanescently excited fluorescence arises from surface-bound fluorescent ligands, as opposed to those merely close to the surface, a high enough receptor density must be used. Second, large enough ligand concentrations must be used to avoid working far below the midpoint of the binding isotherm, where rare, tight, nonspecific binding sites might dominate the surface-bound species. However, these constraints can be contradictory to the general mandate of FCS where a small number of observed, fluorescent molecules is required so that the magnitude of the fluorescence fluctuations relative to the mean fluorescence value is large enough to be accurately measured. This difficulty was addressed in the work cited above [69] by placing in solution a mixture of nonfluorescent IgG at a concentration approximately equivalent to the equilibrium dissociation constant for surface binding along with a trace amount of fluorescently labeled IgG.

The theoretically predicted autocorrelation function for the situation in which fluorescent and nonfluorescent molecules compete for surface binding sites is described in detail in a published theoretical work [70] where a general expression for the form of $G(\tau)$ is given. This general expression is rather complex, consisting of six terms. Two terms are related to diffusion through the evanescent wave [see (5)].

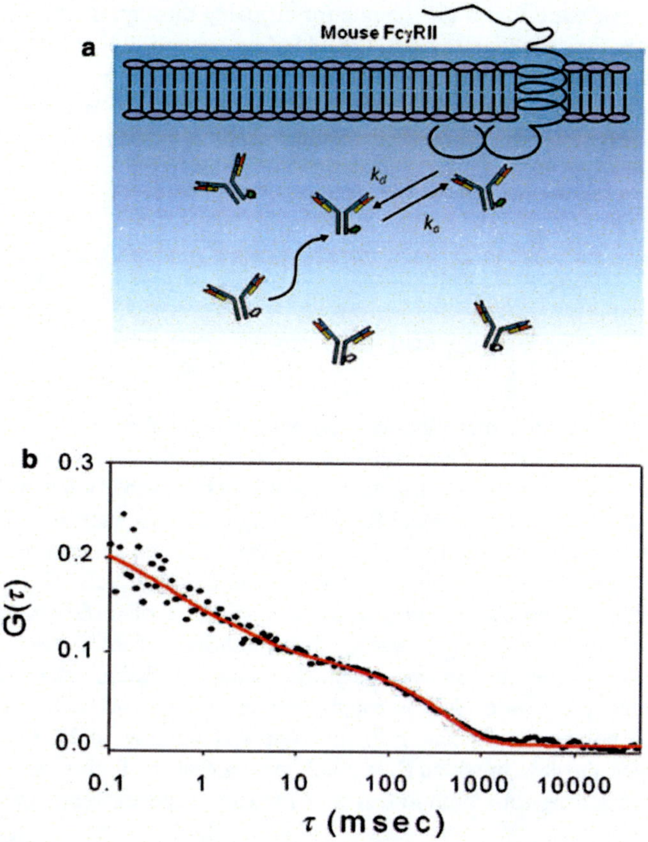

Fig. 6 Reversible association of IgG with receptors in planar membranes. (**a**) Schematic of antibodies diffusing close to and reversibly associating with a supported planar membrane containing purified and reconstituted Fc receptors. The parameters k_a and k_d are the association and dissociation rate constants, respectively. (**b**) Representative TIR-FCS data and the best fit to an appropriate theoretical expression [see (8)]. The data are for a mixture of 10 nM Alexa448-labeled IgG, 1 μM unlabeled IgG and 10 mg mL^{-1} ovalbumin in phosphate-buffered saline close to surfaces coated with supported phospholipid bilayers containing approximately 800 receptors per μm^2. Part (**b**) is reproduced with permission from Biophys J 85:3294–3302 (2003)

The other four terms take the form of the product of a positively going exponential and a negatively going complementary error function (see above). The rates associated with these terms are the roots of a quartic equation which contain the association and dissociation rate constants for the fluorescent and the nonfluorescent molecules, the concentrations of the fluorescent and the nonfluorescent molecules, and factors related to the propensity for rebinding to the surface after dissociation, for both molecule types.

It is desirable to have a more simple form for the autocorrelation function for curve-fitting, when conditions exist such that the general expression can be reduced. It was found that if (1) the rate for diffusing through the evanescent wave is much faster than the kinetic rates, (2) rebinding is negligible, (3) the association and dissociation kinetic rates for the fluorescent and nonfluorescent molecules are equivalent (e.g., labeled and nonlabeled IgG), and (4) there is a large excess of nonfluorescent molecules compared to fluorescent ones, then $G(\tau)$ assumes the following form [70]:

$$G(\tau) = \frac{N_A}{2(N_C + N_A)^2}\left\{(1 - 2R_z\tau)\exp(R_z\tau)\mathrm{erfc}(\sqrt{R_z\tau}) + 2\sqrt{\frac{R_z\tau}{\pi}}\right\} + \frac{\beta N_C}{(N_C + N_A)^2}\exp(-k_d\tau) \tag{8}$$

The parameters in this expression have all been defined above.

Figure 7 compares expected results predicted from (8) when $G(\tau)$ are measured for surfaces containing or not containing receptors, but the solution concentration of fluorescent molecules is equivalent for both sample types. As shown, in comparing these two measured autocorrelation functions, there are two general features which one should expect to observe. First, somewhat counter-intuitively, $G(0)$ is predicted to be higher in the absence, as compared to the presence, of surface receptors. This result can be understood in that samples containing receptors will have more fluorescent molecules on the average in the observed volume than samples not containing receptors, and $G(0)$ is roughly inversely proportional to the average number of fluorescent molecules in the observed volume in FCS. Second, (8) applies only to the case in which R_z is much greater than k_d. In this case, one should expect to see a long-time component in samples containing receptors that is not

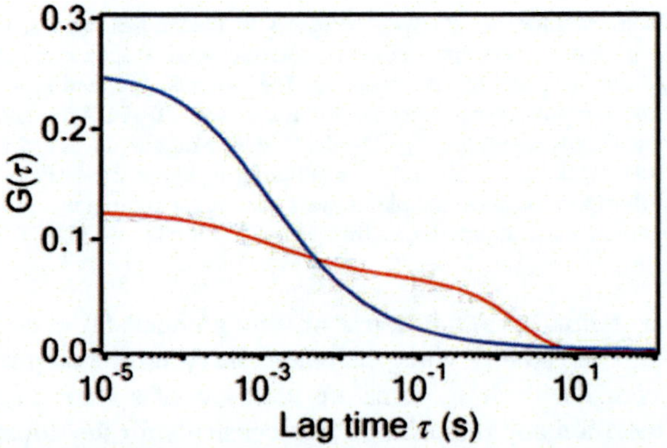

Fig. 7 Comparison of theoretically predicted autocorrelation functions for surfaces containing and not containing specific surface binding sites. $G(\tau)$ were calculated and plotted by using (8). The *blue line* shows the values of (8) with $N_A = 2$, $R_z = 2{,}000\ \mathrm{s}^{-1}$, and $N_C = 0$ (surfaces without binding sites). The *red line* shows the values of (8) with $N_A = 2$, $R_z = 2{,}000\ \mathrm{s}^{-1}$, $N_C = 2$, $\beta = 0.5$, and $k_d = 0.5\ \mathrm{s}^{-1}$ (surfaces with binding sites)

present in samples not containing receptors. Both of these qualitative features are visible in Fig. 7 and were observed for the data described in Lieto et al. [69]. Curve-fitting the data to (8) gave a value of k_d (on the order of 1 s^{-1}) consistent with that measured by other methods for the weak and reversible association of mouse IgG with its receptor FcγRII.

Enzyme kinetics. TIR-FCS has been used to study the catalytic cycle of horseradish peroxidase (HRP) [39]. In these measurements, the reduction of a fluorogenic agent catalyzed by immobilized HRP was followed. Single enzymes were located by using a CCD camera, and time-dependent fluorescence intensity data were obtained using a single-photon counting silicon avalanche photodiode. The autocorrelated fluorescence data were fit to an appropriate theoretical expression for $G(\tau)$ to obtain the rate constant for the formation of the enzyme-product complex, the only fluorescent state in the enzyme's catalytic cycle. This rate constant was distributed over several orders of magnitude, suggesting that conformational heterogeneity of individual HRP molecules contributed to the range of catalytic rates obtained from ensemble measurements.

Triplet state kinetics. The presence of triplet states in fluorescent molecules can lead to a "blinking" in fluorescence measurements as well as emission saturation. Triplet state population is also closely connected to photodegradation. These effects become more prominent at higher excitation intensities. An intriguing topic in FCS has been to explore the impact of triplet state population on fluorescence fluctuation autocorrelation functions. Triplet states of fluorescent molecules were successfully examined by conventional solution FCS a number of years ago [133]. However, there have been few further works in this area since that time even though this topic is of high importance because triplet state involvement can significantly change the nature of $G(\tau)$ in FCS.

Recently, Blom et al. [9] thoroughly investigated the triplet states of small fluorescent molecules next to dielectric interfaces (glass–water) by using TIR-FCS. In this work, the population probabilities and photophysical rates of the triplet states of several dyes, including fluorescein, ATTO 488, and a number of rhodamine variants, were examined. These measurements, for most of the investigated molecules, gave slightly higher (but similar) intersystem crossing rates and triplet state relaxation rates as compared to the rates measured in bulk solution by using confocal FCS. It was concluded that the higher rates were most probably explained by uncertainties in the magnitude and axial profile of the evanescent intensity, as well as the anisotropic angular distribution of fluorescence emission next to the interface. An exception was rhodamine 6G, which displayed a significantly higher triplet relaxation rate as well as a significantly different intersystem crossing rate and triplet state population probability. Additional measurements suggested that the changes were a result of direct dye–surface interactions, dye–solvent interactions, and/or dye aggregation.

Most recently, a TIR-FCS study of the electrostatic interactions of fluorescent molecules with nearby glass surfaces revealed more details on the impact of the surface on triplet state properties [8]. Dyes with different charges (including those that were cationic, anionic, zwitterionic, and neutral) in solutions with various ionic strengths were investigated. Emphasis was placed on defining changes in triplet state population probabilities and photophysical rate constants, with attention also placed on the number of fluorescent molecules in the detection volume, the axial diffusion time, the average fluorescence intensity and the count-rate-per-molecule. This study provides a new approach and detailed guide for characterizing fluorescent molecules with different charges diffusing near surfaces, which could prove to be applicable to more biologically relevant systems such as binding of charged ligands to membrane surfaces or transport of ions through membrane channels.

Live cells. TIR-FCS has been used to study molecular dynamics at the basal membranes of intact cells plated on glass surfaces. A version of this technique was used in early studies to study the behavior of intracellular vesicles at the plasma membranes of live, adherent cells (see above [45, 54]). In another live cell application, TIR-FCS was used to characterize the mobility of farnesylated enhanced green fluorescent protein (F-EGFP) at the plasma membranes of COS7 and HeLa cells [86]. The basal membranes of cells adhered to glass were imaged using through-objective TIRFM. The autocorrelation curve of the resulting fluorescence trace was fit by a functional form containing terms for both two- and three-dimensional diffusion. The three-dimensional diffusion component, which was attributed to the movement of free F-EGFP in the cytosol, was on the order of 10^{-7} cm^2 s^{-1}. The two-dimensional diffusion component, with a value on the order of 10^{-9} cm^2 s^{-1}, was thought to correspond to the slower movement of F-EGFP in the plane of the plasma membrane. Similar two- and three-dimensional diffusion components have been obtained in TIR-FCS experiments that looked at the movement of EGFP-labeled protein kinase C in or near the membranes of adherent HeLa cells [85]. Worth noting is that cellular dynamics at the basal membrane are perhaps better investigated by using TIR-FCS with a fast imaging camera. A variety of recent works along these lines are described below.

4 More Complex Versions of TIR-FCS

Cross-correlation TIR-FCS. Fluorescence cross-correlation spectroscopy (FCCS) is an FCS variant in which fluorescence fluctuations observed in distinct detector channels are cross-correlated [7, 24]. In this method, the fluorescence fluctuation correlation function is calculated as:

$$G(\tau) = \frac{<\delta F_1(t)\delta F_2(t+\tau)>}{<F_1><F_2>} \tag{9}$$

where $F_1(t)$ and $F_2(t)$ are the two independently measured signals, $\delta F_1(t) = F_1(t) - \langle F_1\rangle$ and $\delta F_2(t) = F_2(t) - \langle F_2\rangle$.

The combination of FCCS with evanescent illumination, TIR-FCCS, was first demonstrated in a two-color study that looked at the diffusion of Alexa 488- and Cy5-labeled DNA diffusing in solution [67]. Recent advancements in TIR-FCCS have sought to cross-correlate fluorescence fluctuations in spatially and, sometimes, temporally resolved observation volumes. Sankaran et al. [104] cross-correlated signals from different areas of arbitrary shape and size on a camera (EMCCD). A theoretical framework in which this type of TIR-FCCS can be used to study diffusion and transport at or near cell membranes is also presented in this work. In another application of the technique, TIR-FCCS was used to study the flow of aqueous electrolyte solutions, containing fluorescent tracers, close to a hydrophilic surface [137]. In these measurements, the fluorescence signals from two observation volumes, defined by two confocal pinholes, were independently measured and then cross-correlated. The pinholes were mounted on translation stages and used to create observation volumes that were laterally shifted in the direction of flow by a given distance from each other. During flow, tracer particles passed through the two observation volumes, producing time-resolved fluorescence intensities that were independently recorded. The measured cross-correlation curves had a local maximum at the average time it took for a tracer particle to move from one observation volume to the next. Flow velocities were obtained by fitting experimentally measured cross-correlation curves with a theoretical expression for $G(\tau)$. In addition, by changing the depth of the evanescent field, the flow velocities averaged over different distances from the interface were obtained.

Imaging TIR-FCS. A new and extremely promising extension of conventional TIR-FCS is to image the fluorescence fluctuations as a function of space and time by using a fast camera (e.g., an EMCCD). In the most simple case, one can autocorrelate the temporal fluorescence fluctuations from single pixels to obtain a very large number of $G(\tau)$. If the sample is heterogeneous with respect to space, then in principle one can obtain autocorrelation functions for all of the spatially heterogeneous conditions from a single image sequence. This multiplexing advantage means that the spatial distributions of system dynamics can be much more rapidly determined as compared to single-point TIR-FCS.

Combining TIR-FCS with an EMCCD was first demonstrated by Kannan et al. [55] and called imaging TIR-FCS or ITIR-FCS. In this work, feasibility was demonstrated by measuring the apparent diffusion coefficients of fluorescently labeled phospholipids in supported membranes and of epidermal growth factor receptors tagged with fluorescent proteins and present in the membranes of CHO cells. This work was novel in that it introduced the use of fast cameras in TIR-FCS and the described measurements are very useful because they explore technical aspects related to this new method.

In a subsequent work [33], ITIR-FCS was further tested by measuring the diffusion coefficients of labeled lipids in supported lipid bilayers. The results were compared with measurements made by other available methods including confocal FCS, *z*-scan confocal FCS, fluorescence recovery after photobleaching, and single particle tracking. Similar, although not identical, results were obtained with the different methods. This work includes an extensive and very useful discussion in which the different methods are compared both in terms of advantages and disadvantages as well as possible artifacts.

Another demonstration of combining TIR-FCS with an EMCCD camera was a study in which the dynamics of the protein paxillin in focal adhesions of migratory cells were explored [19]. Specifically, the spatial and temporal dependence of the dynamics of EGFP-labeled paxillin at the basal membranes of motile CHO cells were determined. This work is unusual because it not only introduces a new method but also provides new information about a specific biological process. As the images were obtained at an acquisition rate of 90 frames s^{-1}, the diffusion of paxillin in the cytosol was too fast to be detected. The imaging conditions thereby ensured that the observed fluorescence fluctuations pertained to the behavior of EGFP-paxillin at focal adhesions. To characterize the spatial dependence of paxillin dynamics, fluorescence fluctuations in each pixel were independently autocorrelated and the resulting curves were fit to functional forms that described binding kinetics. Both scanning FCS [20, 93] and ITIR-FCS showed that the fastest exchange rates occurred at the leading edge of the cell where paxillin is incorporated during adhesion assembly. Slower exchange rates were observed at the trailing edge where paxillin is released during adhesion disassembly. Extrapolated $G(0)$ and intensity values indicated that paxillin monomers are incorporated during adhesion assembly, whereas clusters of proteins containing many paxillin molecules dissociate during disassembly.

TIR with STICS. Evanescent illumination has also been combined with spatiotemporal image correlation spectroscopy (STICS; [41]). Although technically a version of ITIR-FCS, this method has to date usually been employed using CCD cameras with slower time resolution than EMCCD cameras and the evanescent wave itself has not been an integral component of the measurements but used primarily to eliminate background from cell interiors. Two recent examples include studies of the dynamics of GFP-labeled integrins in CHO cell membranes [59] and of different flow pathways of labeled actin and myosin during cytokinesis [138].

Imaging TIR-FCS with cross-correlation. Imaging TIR-FCS has very recently been extended to cross-correlation in a method the authors call ITIR-FCCS [104]. In this technique, the fluorescence fluctuations from different areas in the image are cross-correlated. The method was demonstrated by a thorough series of measurements on supported planar membranes in which lipid diffusion naturally occurs; membrane flow was simulated by translating the microscope stage. Applicability to live cells was shown by examining the transport properties of labeled sphingolipid binding domains from the amyloid peptide Aβ bound to neuroblastoma cells. A key capability of this method is its ability to distinguish random diffusion from directed motion, and to characterize the properties of both manners of mobility when they occur in concert. The way in which this distinction is accomplished is more powerful than simply fitting to models that account for both properties; in addition, differences between "forward" and "backward" cross-correlation functions are analyzed.

Multipoint TIR-FCS. A method related to the one now called ITIR-FCS is one in which spatially resolved dynamics are measured not by using a fast imaging camera but by using multiple single detectors. This technique is called multipoint TIR-FCS (M-TIR-FCS) and is described by Ohsugi and Kinjo [84]. As in ITIR-FCS, M-TIR-FCS allows for simultaneous correlation and analysis of fluorescence fluctuations from multiple detection volumes in a single evanescently illuminated image, and can be used to map spatially heterogeneous behavior of fluorescent molecules at surfaces. In the work described by Ohsugi and Kinjo [84], seven detection volumes were defined by an evanescent field and seven optical fibers placed at an image plane. The optical fibers were in turn coupled to seven PMTs, and a specially constructed multichannel correlator was employed to generate autocorrelation curves from the fluorescence emanating from the seven detection volumes. As a proof of principle, M-TIR-FCS was used to characterize the dynamics of farnesylated monomeric EGFP (F-mEGFP) at the basal membranes of COS7 cells. The two- and three-dimensional diffusion components of F-mEGFP did not vary much between the seven detection volumes, as expected for a protein that has no biological role in COS7 cells. M-TIR-FCS improves on the millisecond time resolution of EMCCD cameras by achieving temporal resolution on the order of microseconds but has the disadvantage of providing fewer simultaneously observed areas.

5 Related Methods

Overview. As FCS is concerned with the manner in which the fluorescence fluctuates from its time-averaged value, the average number of fluorescent molecules in the observation volume has to be small. In vitro, this requirement can often be met by working with low concentrations of the fluorescent reporter. In in situ experiments, where concentrations are fixed, the size of the observation volume has often to be reduced to ensure that the average number of fluorescent molecules is small. In TIR-FCS, when evanescent illumination is used in conjunction with a pinhole, an observation volume on the order of a femtoliter (10^{-15} L) is achieved. Many other illumination and detection schemes that further reduce this volume have been recently combined with FCS.

Focused TIRFM. Recent work has demonstrated that focused TIRFM (fTIRFM) can be used to reduce observation volumes. In one approach, a parabolic mirror objective (PMO) was used to create focused evanescent illumination in the sample plane [97]. Subsequent measurements demonstrated that fTIRFM, when combined with confocal detection, yielded observation volumes less than 5 atto-L [98]. A similar method used only a conventional high aperture objective, associated optics, and a confocal pinhole to generate detection volumes as small as 3 atto-L [13].

Supercritical angle fluorescence. Supercritical angle fluorescence (SAF) detection is another technique that can be used to reduce observation volumes. When a fluorescent molecule is close to a planar dielectric interface, the near-field of the emitted radiation couples to the interface and a significant fraction of the fluorescence propagates as far-field radiation into the high refractive index side of the interface, at angles greater than the critical angle. Therefore, by restricting fluorescence detection to supercritical angles, one can achieve axial confinement and reduce detection volumes. In addition to limiting the lateral extent of the observation volume in objective-type fTIRFM, PMOs also enhance axial confinement because they can be made to be selective for fluorescence emitted at supercritical angles [97, 98].

TIR-FCS with PMO has been employed to detect single nucleotide polymorphisms (SNPs) between short, single-stranded DNA oligomers immobilized on microscope slides and fluorescently tagged ones in solution [99]. As hybridization between short DNA strands is weak, high DNA solution concentrations were required to generate detectable amounts of SNPs. However, as described above, high concentrations of fluorescent molecules in solution, especially for weak surface binding, yield untenably high fluorescence background signals. The use of a PMO and subsequent SAF detection restricted fluorescence detection to that generated very close to the surface, where fluorescent, SNP-related oligomers were present.

A related but slightly different and novel objective that can be used to collect fluorescence at supercritical angles has recently been described [95]. This objective consists of a parabolic reflector and an embedded aspheric lens. The parabolic reflector serves to invert the fluorescence exiting the objective such that the lowest angle collected lies at the outer edges of the collimated beam. This arrangement facilitates the use of an iris aperture to alter the lower limit of collected angles, thereby allowing one to easily alter the axial confinement and detection volume. This work also provides a model with which to fit supercritical angle-FCS data.

Surface plasmon coupled emission. The use of metal films in fluorescence microscopy presents yet more techniques that take advantage of the near-field to achieve axial confinement and reduce observation volumes. Similar to through-glass TIRFM, when plane polarized light is totally internally reflected at a glass–metal–water interface, incident light at the surface plasmon resonance (SPR) angle induces resonant excitation of electron oscillations or surface plasmons, which penetrate into the lower refractive index medium (water) as a nonpropagating, evanescent field. Excited fluorophores emit both propagating transverse waves in the far-field and nonpropagating longitudinal waves in the near-field. In TIRFM, one can detect some combination of the far- and near-field. In SPR-modified through-objective TIRFM, however, the far-field is reflected by the metal film, and only the near-field is detected. As before, the near-field of excited fluorophores close to the glass–metal–water dielectric interface interacts with the metal and is converted to a propagating far-field that penetrates the glass at large angles. This process is called surface plasmon coupled emission (SPCE), and it has been coupled to FCS [11, 15]. In addition to achieving axial confinement via the SPCE phenomenon, the metal film further reduces the depth of the observation volume by quenching fluorophores that are very close to it. SPCE-FCS was first demonstrated in a study looking at the diffusion of rhodamine-labeled microspheres close to glass coated with either a gold or silver film [11]. The gold (48 nm) and silver (52 nm) films yielded detection depths of 35 and 50 nm, with characteristic quenching distances of 10 and 15 nm, respectively.

Near-field scanning optical microscopy. In near-field scanning optical microscopy (NSOM), the diffraction-limit is overcome and observation volumes in the atto-L to zepto-L (10^{-18} to 10^{-21} L) range are attained by using the near-field of incident light to illuminate samples. This type of illumination is achieved by using NSOM probes with subwavelength apertures. When the aperture radius is less than about a third of the wavelength, the light that exits the probe is primarily in the near-field and is characterized by a nonpropagating, evanescent field that decays approximately exponentially with increasing distance from the tip of the probe. To work in the near-field, NSOM probes have to be placed at subwavelength distances from the sample. Piezoelectric stages are used to maintain these distances and obtain raster-scanned NSOM images of the sample.

NSOM has been combined with FCS to study three-dimensional diffusion in solution [68] and two-dimensional diffusion in supported planar membranes [129]. NSOM-FCS has also been used to study the kinetics of protein transport across the nuclear envelope (NE), through the nuclear pore complex (NPC) [44]. NSOM was used to image single NPC in unsupported native NE.

Optical nano-antennas. Optical antennas focus light to subwavelength dimensions. Metallic nanoparticles can serve as optical antennas when the far-field excites surface plasmons and gives rise to near-field radiation. Estrada et al. [25] used spherical gold nanoparticles attached to a coverslip to bring about a drastic reduction in the observation volume for FCS. This paper also derives an expression for the theoretical autocorrelation function that describes diffusion close to spherical nanoparticles.

Zero-mode waveguides. Zero-mode waveguides (ZMW) are subwavelength nanostructures milled into thin metal films. These devices yield observation volumes in the zepto-L range by means of extreme axial confinement of the incident light [66]. ZMW disallow propagating modes above a cut-off wavelength that is determined by the size and shape of the guide. The resulting nonpropagating, evanescent field is but tens of nanometers deep. ZMW are amenable for use in single molecule studies, particularly when concentrations are high. From their inception, ZMW have been used in combination with FCS [66, 71, 102, 103]. ZMW-FCS has been used to study oligomerization of a bacteriophage protein in a 1 μM solution [102] and to study ligand–receptor interactions [103]. In depositing planar membranes on nanostructures for the purposes of the second study, it was found that liquid-disordered phase membranes invaginate into the guides, whereas gel phase membranes do not. ZMW-FCS was then used to measure the equilibrium constant of a fluorescently labeled tetanus toxin C fragment interacting with gangliosides embedded in liquid disordered phase membranes. These experiments were conducted at tetanus toxin fragment concentrations of 500 nM.

ZMW-FCS has also been extended to measure diffusion coefficients of fluorescent microspheres undergoing rapid flow in a continuous flow microfluidic mixer patterned with ZMW [71]. The walls of these waveguides protect resident microspheres from the advection of the flow above, allowing FCS to be used to measure diffusion coefficients.

ZMW can come in a variety of different geometries, which can confer unique properties to the nanostructures. Unlike circular guides, rectangular ZMW have cut-off wavelengths that differ with each axis. This property allows one to change the penetration depth of the evanescent field by changing the polarization of the incident light relative to the axes [132]. ZMW-FCS was used to monitor the change in observation volume as the polarization was altered.

In addition to reducing observation volumes, ZMW, as well as other metallic devices like silver island films (SIF), enhance fluorescence emission and reduce lifetimes [82, 96, 132]. Metallic structures accomplish these feats by increasing the rates of radiative decay of adjacent fluorophores. Shorter fluorescence lifetimes reduce the likelihood that fluorophores will react with dissolved oxygen while in an excited state, leading to fluorophore photobleaching. The likelihood of photobleaching is also reduced because enhanced fluorescence emission allows for the use of lower excitation powers.

For a recent review of the applications of various nanometer-sized apertures in imaging, the reader is referred to an article by Wenger et al. [131].

Evanescent illumination with light scattering. Thus far, we have discussed related methods that combine FCS with various types of illumination and detection schemes that primarily serve to reduce the observation volume. Yet many more methods combine TIR illumination with data processing and analysis techniques, other than FCS, to characterize surfaces (reviewed in [6]). One such technique that is particularly related to TIR-FCS is evanescent wave light scattering (EWLS). In EWLS, incident light is totally internally reflected at the interface between the sample and a second medium. The resulting evanescent field that penetrates into the lower refractive index medium is scattered. Scattered light can be detected through either of the two media. In evanescent wave dynamic light scattering (EWDLS), the intensity of the scattered light is correlated much in the same way that fluorescence fluctuations are correlated in FCS. The autocorrelation curve obtained from the scattering intensity trace is fit to theoretical functional forms that describe the particular system. Similar to TIR-FCS, EWDLS can be used to measure diffusion coefficients of solute particles close to surfaces. An extensive review of EWLS in relation to its application to the study of colloidal dispersions is provided by Sigel [109]. An example of a recent work is one in which EWLS was used to study the air–water interface in the presence and absence of particles that served as bulk and surface scattering components [117]. In this experiment, scattered evanescent light contained information about capillary wave-driven dynamics at the air–water interface, whereas in standard light scattering experiments, the scattered light would have contained information about the Brownian motion of the scattering components in the water. The primary advantage of EWLS as compared to TIR-FCS is that light scattering does not require the introduction of an extrinsic label; however, this feature is also a disadvantage because without fluorescent labels, specificity is lost. Also worth noting is the fact that molecules and particles can scatter evanescent light is a matter that needs attention in the design and interpretation of TIR-FCS measurements because the scattered light can excite fluorescence from the bulk (see above).

6 Future Directions

TIR-FCS, like most if not all FCS measurements, is particularly applicable to studies that require low amounts of material because of the intrinsic mandate that the fluctuations in the average number of observed fluorescent molecules be large relative to the average fluorescence. This method is especially amenable to studies requiring low amounts of material because the technique is surface-selective and, therefore, involves consideration of the two-dimensional density of fluorescent molecules on or near a surface as compared to techniques working in the bulk, which involve consideration of the three-dimensional concentration of fluorescent molecules. Given the considerable current difficulty of obtaining preparations of biological macromolecules in high amounts with high purity and high stability, it appears on these bases alone that TIR-FCS will have a very rich future.

A positive feature of conventional TIR-FCS is that the instrumentation is extremely simple. As illustrated in Fig. 1 and described in detail in a previous methods paper [126], the apparatus for conventional TIR-FCS requires only an excitation source (which can be a low-cost diode laser in that the power requirements are low), a very basic fluorescence microscope, a method for generating either through-objective or through-prism internal reflection, a method for restricting detection to a small area on the surface (e.g., either with an aperture or fiber optic placed at a back image plane), a sensitive photon detector (e.g., a silicon avalanche photodiode or a photomultiplier), an off-the-shelf correlator, and various optical components required for light direction and vibration isolation. Not to be neglected, however, is the requirement for a high numerical aperture objective. Such an objective is required, first, because the fluorescence collection efficiency must be very high to generate an autocorrelation function with a high enough signal-to-noise ratio. A high numerical aperture objective is also required for through-objective evanescent illumination, although not required for through-prism evanescent illumination, which requires only a simple fused silica prism. Compared to other emerging methods in fluorescence microscopy, which can require multiple excitation sources, motorized microscopes synchronized with other moving components, sophistocated optical focusing components, and/or complex biochemical constructions, the components required for basic TIR-FCS measurements are negligible. Thus, the technique should be accessible to a large number of laboratories.

There are, of course, caveats. First, FCS, in general, is not always simple to carry out. When one considers the requirement for low average numbers of fluorescent molecules in the observed volume along with other requirements for a given application, often it is found that only a small number of experimental conditions can produce data of quality high enough for adequate analysis and reliable results. An example of this dilemma was found in the measurements described above in which the kinetics of the reversible association of fluorescently labeled IgG with an IgG receptor were investigated [69], and the solution produced only the kinetic dissociation rate and not association rate constant for surface binding. Other more subtle difficulties can arise, as described in the main text, because of a variety of factors, including, e.g., the effects of fluorescence fluctuations arising from the intrinsic photophysical spectroscopy of fluorescent probes (e.g., triplet state population or isomerization between states with different fluorescence quantum efficiencies), diffusion of the observed fluorescent molecules through the evanescent field, and photobleaching arising from prolonged exposure to excitation light. Inadequate sampling of surface sites can also occur in that, at low solution concentrations of fluorescent molecules, unusally tight binding might be observed because rare, nonspecific, and unblockable surface sites exist. Complications can also arise if fluorescent protein preparations contain small amounts of dimerized or oligomerized species that bind tightly to the surface of interest because of their multivalency, or if the preparations contain small fluorescence contaminants. Nonetheless, TIR-FCS has the possibility of a bright future when used in an appropriate manner.

One direction worth pursuing is to use TIR-FCS to examine surface binding kinetics when the mechanism is more complex than a simple bimolecular reaction between fluorescent molecules in solution and surface binding sites. One particular case of interest is the one in which a fluorescent species in solution competes with a different, nonfluorescent species in solution for the same surface binding sites. In this case, because of cross talk, the autocorrelation function contains information about the kinetic association and dissociation rate constants of the nonfluorescent ligands. Thus, this approach might be developed into a kinetic screen for nonfluorescent ligands. Another possible application is to characterize the role of solvent flow in biochemical mechanisms thought to be influenced by alignment of reactive species in solution in response to flow. TIR-FCS should also be particularly amenable to elucidating interaction mechanisms involving oligomerization, as the magnitude of the fluorescence fluctuation autocorrelation function is particularly sensitive to clustering of the fluorescent molecules being examined.

A significant, current limitation to understanding biochemical molecular mechanisms which involve multiple components is that characterization of these mechanisms using conventional methods often requires not only the generation and purification of large amounts of material but also unmanageable time expenses. This bottleneck might potentially be overcome by using a combination of nanofabricated microfluidic devices, fast EMCCD cameras, global data analysis, and TIR-FCS. The notion is to create surfaces containing gradients in the density of an immobilized species in one lateral direction and gradients in the concentration of a species in solution in the other lateral direction by using microfludic devices.

Microfluidic chambers have previously been used both with FCS (e.g., [21, 26, 27]) and TIRFM [51, 53, 135, 136]. In this imagined configuration, fluorescence fluctuation correlation functions calculated in a pixel-dependent manner could at least in theory provide a full spectrum of the surface association and dissociation kinetic rates as a function of the density of the immobilized species as well as the concentration of the solubilized species. Not too many repetitions would be required to examine the full space defined by all combinations of all concentrations of all reactants in the mechanism.

It seems likely that one of the areas in which TIR-FCS will be applied most widely in the future is the investigation of events at the membranes of live cells adhered to evanescently excited surfaces. In these types of systems, only fluorescent molecules close to or in the basal membrane are illuminated and fluoresce [4]. Thus, if intracellular fluorescent molecules are reversibly associating with the cytoplasmic face of the plasma membrane, the kinetics of this process can be characterized by TIR-FCS [85, 86]. Similar measurements have been carried out either by combining evanescent illumination with fluorescence recovery after photobleaching [118] or by imaging single fluorescent molecules as they bind to and dissociate from the cytoplasmic membrane inner leaflet [76, 128]. When combined with an imaging format, this type of measurement produces a kinetic map of the basal cell membrane. This map could then be interpreted in terms of the spatial distributions of other membrane components. Second, for adherent cells that have very flat extensions, the evanescent wave might penetrate through the entire cell and illuminate the exterior face of the apical plasma membrane. This arrangement would allow use of TIR-FCS to examine the behavior of fluorescent molecules in solution which reversibly interact with sites on the cell exterior. Finally, because evanescent illumination provides much higher resolution than epi-illumination in that fluorescence that might arise from the cell interior does not occur, TIR-FCS might provide in many cases a superior method for examining the dynamics of fluorescent molecules moving laterally in the basal membrane.

In most TIR-FCS studies to date, only the first-order fluorescence fluctuation autocorrelation function (1) has been examined although some measurements have employed the first-order fluorescence fluctuation cross-correlation function (9). However, there are a number of alternative and more sophisticated methods for analyzing fluorescence fluctuations, many of which have been used in solution-based FCS, which may prove to be advantageous for TIR-FCS studies. One method for analyzing fluorescence fluctuations which is particularly sensitive to molecular clustering is to calculate high-order fluorescence fluctuation autocorrelation functions, defined as:

$$G(\tau) = \frac{<\delta F^m(t)\delta F^n(t+\tau)>}{<F>^{m+n}} \tag{10}$$

where m and/or n are integers greater than one. This method has been used previously to characterize polydisperse solutions [88–90, 108] but has not yet been extensively explored with respect to evanescent excitation. One can also imagine exploring mutliple time lags, e.g., by calculating functions such as:

$$G(\tau_1, \tau_2) = \frac{<\delta F(t)\delta F(t+\tau_1)\delta F(t+\tau_2)>}{<F>^3} \tag{11}$$

Functions like those shown in (10) and (11) might also be generalized to dual or multiple signals arising from spectrally or spatially resolved sample volumes. The conditions for which these more sophistocated analysis methods will prove to be useful remain to be determined.

A variety of experimental TIR-FCS variants, as described above, hold considerable promise. These include the use of fast EMCCD cameras, thin metal films, or other types of metal constructions [3, 10, 11, 14, 64, 65, 75], high refractive index substrates to create very thin evanescent waves [11, 82, 112], two-photon excitation, either to reduce background from scattered light or for excitation of two different fluorophores with a single light source [32, 42, 46, 49, 83, 106], and the unusual evanescent field polarization [122]. Related methods such as supercritical angle fluorescence [13, 95, 97, 99], the use of novel optical collection schemes [13, 95], surface plasmon coupled emission [11, 15], near-field scanning optical microscopy [44, 68, 129], optical nano-antennas [25], and zero-mode waveguides and related metallic devices [66, 71, 102, 103] also hold particular promise when combined with FCS.

Acknowledgments This work was supported by NSF grant MCB-0641087 and NIH grant GM-041402.

References

1. Agudin JL, Platzeck AM (1978) Fermat's principle and evanescent waves. J Opt 9:101–106
2. Anhut T, Hassler K, Lasser T, Konig K, Rigler R (2005) Fluorescence correlation spectroscopy on dielectric surfaces in total internal reflection geometries. Proc SPIE 5699:159–166
3. Aslan K, Gryczynski I, Malicka J, Matveeva E, Lakowicz JR, Geddes CD (2005) Metal-enhanced fluorescence: an emerging tool in biotechnology. Curr Opin Biotechnol 16:55–62
4. Axelrod D (1981) Cell-substrate contacts illuminated by total internal reflection fluorescence. J Cell Biol 89:141–145
5. Axelrod D, Burghardt TP, Thompson NL (1984) Total internal reflection fluorescence. Annu Rev Biophys Bioeng 13:247–268
6. Axelrod D (2008) Total internal reflection fluorescence microscopy. Meth Cell Biol 89:169–221
7. Bacia K, Schwille P (2007) Practical guidelines for dual-color fluorescence cross-correlation spectroscopy. Nat Protoc 2:2842–2856
8. Blom H, Chmyrov A, Hassler K, Davis LM, Widengren J (2010) Electrostatic interactions of fluorescent molecules with dielectric interfaces studied by total internal reflection fluorescence correlation spectroscopy. Int J Mol Sci 11:386–406

9. Blom H, Chmyrov A, Hassler K, Davis LM, Widengren J (2009) Triplet-state investigations of fluorescent dyes at dielectric interfaces using total internal reflection fluorescence correlation spectroscopy. J Phys Chem A 113:5554–5566
10. Blom H, Kastrup L, Eggeling C (2006) Fluorescence fluctuation spectroscopy in reduced detection volumes. Curr Pharm Biotechnol 7:51–66
11. Borejdo J, Calander N, Grycznski Z, Grycznski I (2006) Fluorescence correlation spectroscopy in surface plasmon coupled emission microscope. Opt Express 14:7878–7888
12. Burghardt TP, Thompson NL (1984) Evanescent intensity of a focused Gaussian light beam undergoing total internal reflection in a prism. Opt Eng 23:62–67
13. Burghardt TP, Ajtai K, Borejdo J (2006) In situ single-molecule imaging with attoliter detection using objective total internal reflection confocal microscopy. Biochemistry 45:4058–4068
14. Burghardt TP, Charlesworth JE, Halstead MF, Tarara JE, Ajtai K (2006) In situ fluorescent protein imaging with metal film-enhanced total internal reflection microscopy. Biophys J 90:4662–4671
15. Calander N, Muthu P, Gryczynski Z, Gryczynski I, Borejdo J (2008) Fluorescence correlation spectroscopy in a reverse Kretchmann surface plasmon assisted microscope. Opt Express 16(17):13381–13390
16. Chen H, Farkas ER, Webb WW (2008) In vivo applications of fluorescence correlation spectroscopy. Meth Cell Biol 89:3–35
17. Chiantia S, Reis J, Schwille P (2009) Fluorescence correlation spectroscopy in membrane structure elucidation. Biochim Biophys Acta 1788:225–233
18. Costantino S, Comeau JWD, Kolin DL, Wiseman PW (2005) Accuracy and dynamic range of spatial image correlation and cross-correlation spectroscopy. Biophys J 89:1251–1260
19. Digman MA, Brown CM, Horwitz AR, Mantulin WW, Gratton E (2008) Paxillin dynamics measured during adhesion assembly and disassembly by correlation spectroscopy. Biophys J 94:2819–2831
20. Digman MA, Brown CM, Sengupta P, Wiseman PW, Horwitz AR, Gratton E (2005) Measuring fast dynamics in solutions and cells with a laser scanning microscope. Biophys J 89:1317–1327
21. Dittrich PS, Schwille P (2002) Spatial two-photon fluorescence cross-correlation spectroscopy for controlling molecular transport in microfluidic structures. Anal Chem 74:4472–4479
22. Dorn R, Quabis S, Leuchs G (2003) Sharper focus for a radially polarized light beam. Phys Rev Lett 91(23):233901-1–233901-4
23. Eggeling C, Hilbert M, Bock H, Ringemann M, Hoffmann M, Stiel AC, Andresen M, Jakobs S, Egner A, Schonle A, Hell SW (2007) Reversible photoswitching enables single-molecule fluorescence fluctuation spectroscopy at high molecular concentration. Microsc Res Tech 70:1003–1009
24. Eigen M, Rigler R (1994) Sorting single molecules: applications to diagnostics and evolutionary biotechnology. Proc Natl Acad Sci USA 91:5740–5747
25. Estrada LC, Aramednia PF, Martinez OE (2008) 1000 times volume reduction for fluorescence correlation spectroscopy using nano-antennas. Opt Express 16(25):20597–20602
26. Foquet M, Korlach J, Zipfel WR, Webb WW, Craighead HG (2002) DNA fragment sizing by single molecule detection in submicrometer-sized closed fluidic channels. Anal Chem 74:1415–1422
27. Foquet M, Korlach J, Zipfel WR, Webb WW, Craighead HG (2004) Focal volume confinement by submicrometer-sized fluidic channels. Anal Chem 76:1618–1626
28. Gell C, Berndt M, Enderlein J, Diez S (2009) TIRF microscopy evanescent field calibration using titled fluorescent microtubules. J Microsc 234:38–46
29. Girard C, Joachim C, Gauthier S (2000) The physics of the near-field. Rep Prog Phys 63:893–938

30. Gosch M, Rigler R (2005) Fluorescence correlation spectroscopy of molecular motions and kinetics. Adv Drug Deliv Rev 57:169–190
31. Groves JT, Parthasarathy R, Forstner MB (2008) Fluorescence imaging of membrane dynamics. Annu Rev Biomed Eng 10:311–338
32. Gryczynski I, Gryczynski Z, Lakowicz JR (1997) Two-photon excitation by the evanescent wave from total internal reflection. Anal Biochem 247:69–76
33. Guo L, Har JY, Sankaran J, Hong YM, Kannan B, Wohland T (2008) Molecular diffusion measurement in lipid bilayers over wide concentration ranges: a comparative study. Chemphyschem 9:721–728
34. Hansen RL, Harris JM (1998) Total internal reflection fluorescence correlation spectroscopy for counting molecules at solid/liquid interfaces. Anal Chem 70:2565–2575
35. Hansen RL, Harris JM (1998) Measuring reversible adsorption kinetics of small molecules at solid/liquid interfaces by total internal reflection fluorescence correlation spectroscopy. Anal Chem 70:4247–4256
36. Harlepp S, Robert J, Darnton NC, Chatenay D (2004) Subnanometric measurements of evanescent wave penetration depth using total internal reflection microscopy combined with fluorescent correlation spectroscopy. Appl Phys Lett 85:3917–3919
37. Hassler K, Anhut T, Rigler R, Gösch M, Lasser T (2005) High count rates with total internal reflection fluorescence correlation spectroscopy. Biophys J 88:L01–L03
38. Hassler K, Leutenegger M, Rigler P, Rao R, Rigler R, Gösch M, Lasser T (2005) Total internal reflection fluorescence correlation spectroscopy (TIR-FCS) with low background and high count-rate per molecule. Opt Express 13:7415–7423
39. Hassler K, Rigler P, Blom H, Rigler R, Widengren J, Lasser T (2007) Dynamic disorder in horseradish peroxidase observed with total internal reflection fluorescence correlation spectroscopy. Opt Express 15:5366–5375
40. Haustein E, Schwille P (2007) Fluorescence correlation spectroscopy: novel variations of an established technique. Annu Rev Biophys Biomol Struct 36:151–169
41. Hebert B, Costantino S, Wiseman PW (2005) Spatiotemporal image correlation spectroscopy (STICS): theory, verification, and application to protein velocity mapping in living CHO cells. Biophys J 88:3601–3614
42. Heinze KG, Koltermann A, Schwille P (2000) Simultaneous two-photon excitation of distinct labels for dual-color fluorescence cross-correlation analysis. Proc Natl Acad Sci USA 97:10377–10382
43. Hellen EH, Axelrod D (1987) Fluorescence emission at dielectric and metal-film interfaces. J Opt Soc Am B 4:337–350
44. Herrmann M, Neuberth N, Wissler J, Perez J, Gradl D, Naber A (2009) Near-field optical study of protein transport kinetics at a single nuclear pore. Nano Lett 9(9):3330–3336
45. Holt M, Cooke A, Neef A, Lagnado L (2004) High mobility of vesicles supports continuous exocytosis at a ribbon synapse. Curr Biol 14:173–183
46. Huang Z, Thompson NL (1993) Theory for two-photon excitation in pattern photobleaching with evanescent illumination. Biophys Chem 47:241–249
47. Huang Z, Pearce KH, Thompson NL (1994) Translational diffusion of bovine prothrombin fragment 1 weakly bound to supported planar membranes: measurement by total internal reflection with fluorescence pattern photobleaching recovery. Biophys J 67:1754–1766
48. Huang Z, Thompson NL (1996) Imaging fluorescence correlation spectroscopy: nonuniform IgE distributions on planar membranes. Biophys J 70:2001–2007
49. Ivanov D, Shcheslavskiy V, Marki I, Leutenegger M, Lasser T (2009) High volume confinement in two-photon total-internal-reflection fluorescence correlation spectroscopy. Appl Phys Lett 94:083902-1–083902-3
50. Jaiswal JK, Simon SM (2007) Imaging single events at the cell membrane. Nat Chem Biol 3:92–98
51. Jakeway SC, de Mello AJ (2001) Chip-based refractive index detection using a single point evanescent wave probe guide. Analyst 126:1505–1510

52. Jeong S, Achermann M, Nanda J, Ivanov S, Klimov V, Hollingsworth JA (2005) Effect of the thiol-thiolate equilibrium on the photophysical properties of aqueous CdSe/ZnS nanocrystal quantum dots. J Am Chem Soc 127:10126–10127
53. Jin S, Huang P, Park J, Yoo JY, Breuer KS (2004) Near-surface velocimetry using evanescent wave illumination. Exp Fluids 37:825–833
54. Johns LM, Levitan ES, Shelden EA, Holz RW, Axelrod D (2001) Restriction of secretory granule motion near the plasma membrane of chromaffin cells. J Cell Biol 153:177–190
55. Kannan B, Guo L, Sudhaharan T, Ahmed S, Maruyama I, Wohland T (2007) Spatially resolved total internal reflection fluorescence correlation microscopy using an electron multiplying charge-coupled device camera. Anal Chem 79:4463–4470
56. Kask P, Gunther R, Axhausen P (1997) Statistical accuracy in fluorescence fluctuation experiments. Eur Biophys J 25:163–169
57. Knoll W (1998) Interfaces and thin films as seen by bound electromagnetic waves. Annu Rev Phys Chem 49:569–638
58. Kolin DL, Costantino S, Wiseman PW (2006) Sampling effects, noise and photobleaching in temporal image correlation spectroscopy. Biophys J 90:628–639
59. Kolin DL, Ronis D, Wiseman PW (2006) k-Space image correlation spectroscopy: a method for accurate transport measurements independent of fluorophore photophysics. Biophys J 91:3061–3075
60. Kolin DL, Wiseman PW (2007) Advances in image correlation spectroscopy: measuring number densities, aggregation states, and dynamics of fluorescently labeled macromolecules in cells. Cell Biochem Biophys 49:141–164
61. Koppel DE (1974) Statistical accuracy in fluorescence correlation spectroscopy. Phys Rev A 10:1938–1945
62. Kyoung M, Sheets ED (2008) Vesicle diffusion close to a membrane: intermembrane interactions measured with fluorescence correlation spectroscopy. Biophys J 95:5789–5797
63. Lagerholm BC, Weinreb GE, Jacobson K, Thompson NL (2005) Detecting microdomains in intact cell membranes. Annu Rev Phys Chem 56:309–336
64. Lakowicz JR (2001) Radiative decay engineering: biophysical and biomedical applications. Anal Biochem 298:1–24
65. Lakowicz JR, Geddes CD, Gryczynski I, Malicka J, Gryczynski Z, Aslan K, Lukomska J, Matveeva E, Zhang JA, Badugu R, Huang J (2004) Advances in surface-enhanced fluorescence. J Fluoresc 14:425–441
66. Levene MJ, Korlach J, Turner SW, Foquet M, Craighead HG, Webb WW (2003) Zero-mode waveguides for single-molecule analysis at high concentrations. Science 299:682–686
67. Leutenegger M, Blom H, Widengren J, Eggeling C, Gösch M, Leitgeb RA, Lasser T (2006) Dual-color total internal reflection fluorescence cross-correlation spectroscopy. J Biomed Opt 11:1–3
68. Lewis A, Kuttner YY, Dekhter R, Polhana M (2007) Fluorescence correlation spectroscopy at 100 nM concentrations using near-field scanning optical microscopic (NSOM) geometries and highly diffracting force sensing fiber probes. Isr J Chem 47:171–176
69. Lieto AM, Cush RC, Thompson NL (2003) Ligand–receptor kinetics measured by total internal reflection with fluorescence correlation spectroscopy. Biophys J 85:3294–3302
70. Lieto AM, Thompson NL (2004) Total internal reflection with fluorescence correlation spectroscopy: nonfluorescent competitors. Biophys J 87:1268–1278
71. Liao D, Galajda P, Riehn R, Ilic R, Puchalla JL, Yu HG, Craighead HG, Austin RH (2008) Single molecule correlation spectroscopy in continuous flow mixers with zero-mode waveguides. Opt Express 16(14):10077–10090
72. Lin BH, Yu J, Rice SA (2000) Direct measurements of a constrained Brownian motion of an isolated sphere between two walls. Phys Rev E 62(2):3909–3919
73. Liu Y, Kim H-R, Heikal AA (2006) Structural basis of fluorescence fluctuation dynamics of green fluorescent proteins in acidic environments. J Phys Chem B 110(47):24138–24146

74. Magde D, Elson EL, Webb WW (1974) Fluorescence correlation spectroscopy. 2. Experimental realization. Biopolymers 13:29–61
75. Maliwal BP, Malicka J, Gryczynski I, Gryczynski Z, Lakowicz JR (2003) Fluorescence properties of labeled proteins near silver colloid surfaces. Biopolymers 70:585–594
76. Mashanov GI, Tacon D, Peckham M, Molloy JE (2004) The spatial and temporal dynamics of pleckstrin homology domain binding at the plasma membrane measured by imaging single molecules in live mouse myoblasts. J Biol Chem 279:15274–15280
77. Mattheyses AL, Axelrod D (2006) Direct measurement of the evanescent field profile produced by objective-based total internal reflection. J Biomed Opt 11:014006
78. McCain KS, Harris JM (2003) Total internal reflection fluorescence-correlation spectroscopy study of molecular transport in thin sol–gel films. Anal Chem 75:3616–3624
79. McCain KS, Schluesche P, Harris JM (2004) Modifying the adsorption behavior of polyamidoamine dendrimers at silica surfaces investigated by total internal reflection fluorescence correlation spectroscopy. Anal Chem 76:930–938
80. McCain KS, Schluesche P, Harris JM (2004) Poly(amidoamine) dendrimers as nanoscale diffusion probes in sol-gel films investigated by total internal reflection fluorescence spectroscopy. Anal Chem 76:939–946
81. Meseth U, Wohland T, Rigler R, Vogel H (1999) Resolution of fluorescence correlation measurements. Biophys J 76:1619–1631
82. Muthu P, Gryczynski I, Gryczynski Z, Talent J, Akopova I, Jain K, Borejdo J (2007) Decreasing photobleaching by silver island films: application to muscle. Anal Biochem 366:228–236
83. Oheim M, Schapper F (2005) Non-linear evanescent-field imaging. J Phys D 38:R185–R197
84. Ohsugi Y, Kinjo M (2009) Multipoint fluorescence correlation spectroscopy with total internal reflection fluorescence microscope. J Biomed Opt 14(1):014030-1–014030-4
85. Ohsugi Y, Kinjo M (2006) Analysis of membrane-binding protein mobility in living cells using total internal reflection fluorescence correlation spectroscopy. Biophys Rev Lett 1: 293–299
86. Ohsugi Y, Saito K, Tamura M, Kinjo M (2006) Lateral mobility of membrane-binding proteins in living cells measured by total internal reflection fluorescence correlation spectroscopy. Biophys J 91:3456–3464
87. Oreopoulos J, Yip CM (2008) Combined scanning probe and total internal reflection fluorescence microscopy. Methods 46:2–10
88. Palmer AG, Thompson NL (1987) Molecular aggregation characterized by high order autocorrelation in fluorescence correlation spectroscopy. Biophys J 52:257–270
89. Palmer AG, Thompson NL (1989) Fluorescence correlation spectroscopy for detecting submicroscopic clusters of fluorescent molecules in membranes. Chem Phys Lett 50:253–270
90. Palmer AG, Thompson NL (1989) High-order fluorescence fluctuation analysis of model protein clusters. Proc Natl Acad Sci USA 86:6148–6152
91. Pero JK, Hass EM, Thompson NL (2006) Size dependence of protein diffusion very close to membrane surfaces: measurement by total internal reflection with fluorescence correlation spectroscopy. J Phys Chem B 110:10910–10918
92. Qian H (1990) On the statistics of fluorescence correlation spectroscopy. Biophys Chem 38:49–57
93. Ries J, Chiantia S, Schwille P (2009) Accurate determination of membrane dynamics with line-scan FCS. Biophys J 96:1999–2008
94. Ries J, Petrov EP, Schwille P (2008) Total internal reflection fluorescence correlation spectroscopy: effects of lateral diffusion and surface-generated fluorescence. Biophys J 95:390–399
95. Ries J, Ruckstuhl T, Verdes D, Schwille P (2008) Supercritical angle fluorescence correlation spectroscopy. Biophys J 94:221–229
96. Rigneault H, Capulade J, Dintinger J, Wenger J, Bonod N, Popov E (2005) Enhancement of single-molecular fluorescence detection in subwavelength apertures. Phys Rev Lett 95:117401-1–117401-4

97. Ruckstuhl T, Seeger S (2003) Confocal total-internal-reflection fluorescence microscopy with a high-aperture parabolic mirror lens. Appl Opt 42:3277–3283
98. Ruckstuhl T, Seeger S (2004) Attoliter detection volumes by confocal total-internal-reflection fluorescence microscopy. Opt Lett 29:569–571
99. Ruckstuhl T, Krieg A (2005) Microscope objective for large-angle fluorescence used for rapid detection of single nucleotide polymorphisms in DNA hybridization. Anal Chem 77:2656–2661
100. Saffarian S, Elson EL (2003) Statistical analysis of fluorescence correlation spectroscopy: the standard deviation and bias. Biophys J 84:2030–2042
101. Saffarian S, Kirchhausen T (2008) Differential evanescence nanometry: live-cell fluorescence measurements with 10-nm axial resolution on the plasma membrane. Biophys J 94:2333–2342
102. Samiee KT, Foquet M, Guo L, Cox EC, Craighead HG (2005) λ-Repressor oligomerization kinetics at high concentrations using fluorescence correlation spectroscopy in zero-mode waveguides. Biophys J 88:2145–2153
103. Samiee KT, Moran-Mirabal JM, Cheung YK, Craighead HG (2006) Zero mode waveguides for single-molecule spectroscopy on lipid membranes. Biophys J 90:3288–3299
104. Sankaran J, Manna M, Guo L, Kraut R, Wohland T (2009) Diffusion, transport, and cell membrane organization investigated by imaging fluorescence cross-correlation spectroscopy. Biophys J 97:2630–2639
105. Sarkar A, Robertson RB, Fernandez JM (2004) Simultaneous atomic force microscope and fluorescence measurements of protein unfolding using a calibrated evanescent wave. Proc Natl Acad Sci USA 101:12882–12886
106. Schapper F, Goncalves JT, Oheim M (2003) Fluorescence imaging with two-photon evanescent wave excitation. Eur Biophys J 32:635–643
107. Schneckenburger H (2005) Total internal reflection fluorescence microscopy: technical innovations and novel applications. Curr Opin Biotechnol 16:13–18
108. Sergeev M, Costantino S, Wiseman PW (2006) Measurement of monomer-oligomer distributions via fluorescence moment image analysis. Biophys J 91:3884–3896
109. Sigel R (2009) Light scattering near and from interfaces using evanescent wave and ellipsometric light scattering. Curr Opin Colloid Interface Sci 14:426–437
110. Simon SM (2009) Partial internal reflections on total internal reflection fluorescent microscopy. Trends Cell Biol 19:661–668
111. Sonesson AW, Blom H, Hassler K, Elofsson UM, Callisen TH, Widengren J, Brismar H (2008) Protein–surfactant interactions at hydrophobic interfaces studied with total internal reflection fluorescence correlation spectroscopy (TIR-FCS). J Colloid Interface Sci 317:449–457
112. Starr TE, Thompson NL (2000) Formation and characterization of planar phospholipid bilayers supported on TiO2 and SrTiO3 single crystals. Langmuir 16:10301–10308
113. Starr TE, Thompson NL (2001) Total internal reflection with fluorescence correlation spectroscopy: combined surface reaction and solution diffusion. Biophys J 80:1575–1584
114. Starr TE, Thompson NL (2002) Local diffusion and concentration of IgG near planar membranes: measurement by total internal reflection with fluorescence correlation spectroscopy. J Phys Chem B 106:2365–2371
115. Steyer JA, Almers W (1999) Tracking single secretory granules in live chromaffin cells by evanescent-field fluorescence microscopy. Biophys J 76:2262–2271
116. Steyer JA, Almers W (2001) A real-time view of life within 100 nm of the plasma membrane. Nat Rev Mol Cell Biol 2:268–275
117. Stocco A, Tauer K, Pispas S, Sigel R (2009) Dynamics at the air-water interface revealed by evanescent wave light scattering. Eur Phys J E 29:95–105
118. Sund SE, Axelrod D (2000) Actin dynamics at the living cell submembrane imaged by total internal reflection fluorescence photobleaching. Biophys J 79:1655–1669

119. Thompson NL, Burghardt TP, Axelrod D (1981) Measuring surface dynamics of biomolecules by total internal reflection fluorescence with photobleaching recovery or correlation spectroscopy. Biophys J 33:435–454
120. Thompson NL (1982) Surface binding rates of nonfluorescent molecules may be obtained by total internal reflection with fluorescence correlation spectroscopy. Biophys J 38:327–329
121. Thompson NL, Axelrod D (1983) Immunoglobulin surface-binding kinetics studied by total internal reflection with fluorescence correlation spectroscopy. Biophys J 43:103–114
122. Thompson NL, McConnell HM, Burghardt TP (1984) Order in supported phospholipid monolayers detected by the dichroism of fluorescence excited by polarized evanescent illumination. Biophys J 46:739–747
123. Thompson NL (1991) Fluorescence correlation spectroscopy. In: Lakowicz JR (ed) Topics in fluorescence spectroscopy, vol 3. Plenum, New York, pp 337–378
124. Thompson NL, Pero JK (2005) Fluorescence spectroscopy in biology: advanced methods and their applications to membranes, proteins, DNA and cells. Springer, Berlin, pp 79–103
125. Thompson NL, Pero JK (2006) Total internal reflection-fluorescence correlation spectroscopy. In: Geddes CD, Lakowicz JR (eds) Reviews in fluorescence, vol 3. Springer, New York, pp 215–237
126. Thompson NL, Steele BL (2007) Total internal reflection with fluorescence correlation spectroscopy. Nat Protocol 2:878–890
127. Thompson NL, Wang X, Navaratnarajah P (2008) Total internal reflection with fluorescence correlation spectroscopy: applications to substrate-supported planar membranes. J Struct Biol 168:95–106
128. Ueda M, Sako Y, Tanaka T, Devreotes P, Yanagida T (2001) Single molecule analysis of chemotactic signaling in *Dictyostelium* cells. Science 294:864–867
129. Vobornik D, Banks DS, Lu Z, Fradin C, Taylor R, Johnston LJ (2008) Fluorescence correlation spectroscopy with sub-diffraction-limited resolution using near-field optical probes. Appl Phys Lett 93:163904-1–163904-3
130. Vukojevic V, Pramanik A, Yakovleva T, Rigler R, Terenius L, Bakalkin G (2005) Study of molecular events in cells by fluorescence correlation spectroscopy. Cell Mol Life Sci 62:535–550
131. Wenger J, Gerard D, Lenne P-F, Rigneault H, Bonod N, Popov E, Marguet D, Nelep C, Ebbesen TW (2009) Biophotonics applications of nanometric apertures. Int J Mater Prod Technol 34(4):488–506
132. Wenger J, Lenne P-F, Popov E, Rigneault H (2005) Single molecule fluorescence in rectangular nano-apertures. Opt Express 2005:7035–7044
133. Widengren J, Mets U, Rigler R (1995) Fluorescence correlation spectroscopy of triplet states in solution: a theoretical and experimental study. J Phys Chem 99:13368–13379
134. Wohland T, Rigler R, Vogel H (2001) The standard deviation in fluorescence correlation spectroscopy. Biophys J 80:2987–2999
135. Yang TL, Baryshnikova OK, Mao HB, Holden MA, Cremer PS (2003) Investigations of bivalent antibody binding on fluid-supported phospholipid membranes: the effect of hapten density. J Am Chem Soc 125:4779–4784
136. Yang TL, Jung SY, Mao HB, Cremer PS (2001) Fabrication of phospholipid bilayer coated microchannels for on-chip immunoassays. Anal Chem 73:165–169
137. Yordanov S, Best A, Butt H-J, Koynov K (2009) Direct studies of liquid flows near solid surfaces by total internal reflection fluorescence cross-correlation spectroscopy. Opt Express 17(23):21149–21158
138. Zhou M, Wang YL (2008) Distinct pathways for the early recruitment of myosin II and actin to the cytokinetic furrow. Mol Biol Cell 19:318–326

Index

C.D. Geddes (ed.), *Reviews in Fluorescence 2009*, Reviews in Fluorescence 2009, DOI 10.1007/978-1-4419-9672-5,